Practical Pharmaceutical Chemistry

Practical Pharmaceutical Chemistry

Third Edition, in Two Parts

by

A. H. BECKETT
Ph.D., D.Sc., F.P.S., F.R.I.C.
Professor of Pharmaceutical Chemistry
Chelsea College of Science and Technology
University of London

and

J. B. STENLAKE
Ph.D., D.Sc., F.P.S., F.R.I.C., F.R.S.E.
Professor of Pharmacy and Pharmaceutical Chemistry
University of Strathclyde, Glasgow

Part Two

THE ATHLONE PRESS *of the University of London* 1976

Published by
THE ATHLONE PRESS
UNIVERSITY OF LONDON
4 Gower Street London WC 1

Distributed by Tiptree Book Services Ltd
Tiptree Essex

U.S.A. and Canada
Humanitie Press Inc
New Jersey

ISBN 0 485 11159 4

Set in 10/11 pt Monophoto Times
and printed by photolithography
in Great Britain by J. W. Arrowsmith Ltd, Bristol

Preface to the Second Edition

The rapid development of chemical, physico-chemical and instrumentation techniques since the publication of the first edition, and their widespread application in pharmaceutical analysis, together with the introduction of many new drugs, has necessitated a complete revision of the book. The selection of new material for inclusion has not been easy, and in order to contain the book within a reasonable compass, and in an attempt to meet the differing needs of students at elementary and advanced levels, the 2nd Edition of Practical Pharmaceutical Chemistry has been produced in two parts.

Part 1, which is concerned solely with general analytical methods, provides an introduction to Pharmaceutical Analysis in a completely revised chapter on the general principles of quality control. Part 1 is also intended to provide the basis for more elementary instructional laboratory classes in pharmaceutical analysis. The selection of examples for inclusion has been based on the current editions of the British Pharmacopoeia, and its Addenda, and whilst coverage is, intentionally, not comprehensive, a wide selection of examples has been achieved.

As in the first edition, a number of accepted pharmaceutical conventions and certain pharmacopoeial practices have been adopted to avoid confusing the student should he wish to refer to the Pharmacopoeia or other texts. In particular, substances which are the subject of official monographs in the Pharmacopoeia are indicated by the use of initial capital letters, e.g. Sodium Chloride. For similar reasons, we have continued to express volumes as litres and millilitres, and to ignore the recent I.S.O. ruling on dm^3 as the unit of volume. The widespread use of demineralised water as an alternative to distilled water, has also been recognised and the text amended appropriately by the adoption of the term Water (spelt with an initial capital letter) to indicate either, in distinction to tap water.

Part 2 is intended to cater for the needs of the more advanced students. It covers the physical and instructional methods of analysis included in the first edition, extensively revised and expanded, and in addition includes a separate treatment of physical criteria of pharmaceutical chemicals, particle size analysis, chromatography, infrared spectrometry, nuclear magnetic resonance spectrometry, mass spectrometry and radiochemistry. We are greatly indebted to our colleagues for these expert contributions, which are separately acknowledged at the head of each chapter. The extent to which the treatment of these subjects has been linked to examples of the British Pharmacopoeia has been deliberately limited, as it is hoped there may well be other users of Part 2 for whom this aspect is unimportant. Differences in the emphasis on pharmacopoeial requirements also reflect the extent of progress in the application of the various techniques to pharmaceutical analysis at the time of publication. It is felt, however, that the potential use for such techniques as nuclear magnetic resonance spectroscopy, and mass spectrometry, warrants the inclusion of chapters at this stage.

A. H. B.
J. B. S.

Preface to the Third Edition

Part 1

Since the publication of the second edition in 1968, there have been a number of important developments affecting the practice of pharmaceutical analysis in the United Kingdom. These have included the passing into Law of the Medicines Act, 1968, the entry of the United Kingdom into the European Common Market, the publication of the European Pharmacopoeia in a series of volumes and supplements commencing in 1969, and the publication of the British Pharmacopoeia.

The third edition endeavours to reflect the repercussion of these developments. Accordingly, the requirements for drug and product assessment imposed by the Medicine Act, 1968, and the implications for the analyst and compiler of product applications form the subject of an entirely new Chapter. For similar reasons, we have also greatly extended our treatment of general analytical procedures applicable to pharmaceutical dosage forms in another entirely new Chapter. This covers all the important product types including pressurised aerosols, tablets, capsules, injections, ointments, creams, lozenges, mixtures, solutions, suspensions, suppositories, and relevant analytical developments relating to particle size control, dissolution testing of bioavailability and the examination of sustained release preparations.

In other respects, the character of the book has been largely retained, though with a number of deletions, minor amendments, and additions which reflect trends and developments in pharmaceutical analysis over the last few years.

Part 2

Advances in the application of physico-analytical techniques to pharmaceutical analysis since the publication of the second edition in 1970 have been less rapid than those in other general aspects of pharmaceutical analysis covered in Part 1. Nevertheless, we have taken the opportunity to revise the text of Part 2 to include such important pharmaceutical analytical developments as the application of circular dichroism spectrometry, differential thermal analysis, differential scanning calorimetry, coulometry and high pressure liquid chromatography. We have also considerably extended the treatment of flame photometry and atomic absorption spectroscopy, because of their importance in the precise measurement of metallic ions and organometallics in haemodialysis solutions, contact lens fluids and various other pharmaceutical products. The combination of gas chromatography with mass spectrometry has, similarly, been given extended treatment in recognition of its established use in drug metabolism studies and toxicological analysis.

Other additions to this edition include an extended discussion of NMR double resonance techniques, and brief descriptions of Fourier transform NMR, carbon-13 NMR, lanthanide shift reagents in NMR, mass fragmentometry and chemical ionisation mass spectrometry. The chapter on radiochemical techniques has also been briefly extended to include a description of radionuclide generators used in the preparation of short-lived radionuclide injection solutions, and the necessary analytical control procedures associated with their use. The principles of the radio-immune assay procedure and its application are also described.

As in the revision of Part 1, we have continued to draw attention to applications of pharmaceutical analytical methods in pharmacokinetic and drug metabolism studies. Accordingly, we have included a number of examples showing the applications of physico-chemical techniques in the separation from and measurement of drug substances and their metabolism in biological fluids and tissues. These include application of ultraviolet absorption spectroscopy, spectrofluorimetry, nuclear magnetic resonance spectroscopy, thin-layer, gas-liquid and high pressure liquid chromatography, and radio-chemical techniques, among others.

A. H. B.
J. B. S.

Part 2

Physical Methods of Analysis

Revised by

A. H. BECKETT
Ph.D., D.Sc., F.P.S., F.R.I.C.

and

J. B. STENLAKE
Ph.D., D.Sc., F.P.S., F.R.I.C., F.R.S.E.

with contributions by

A. T. FLORENCE
B.Sc., Ph.D., A.R.C.S.T., A.R.I.C., M.P.S.

N. D. HARRIS
B.Pharm., Ph.D., D.I.C., F.P.S.

G. O. JOLLIFFE
B.Pharm., Ph.D., A.R.I.C.

R. T. PARFITT
B.Pharm., Ph.D., M.P.S., A.R.I.C.

W. D. WILLIAMS
B.Pharm., Ph.D., F.P.S., A.R.I.C.

G. C. WOOD
B.Sc., Ph.D.

Acknowledgements

Permission to reproduce the following Tables and Figures is gratefully acknowledged:

Tables
14, Stanton-Redcroft, London; 51, *Analytical Chemistry*; 54 and 55, calculated by Dr C. L. Hart on the London University Atlas computer.

Figures
38, Stanton-Redcroft Ltd; 41, 44, 46, 178, 219 and 229, Perkin-Elmer Ltd, Beaconsfield, Bucks.; 147 Pye Unicam Ltd; 102, 164 and 165, Corning-EEL, Halstead, Essex; 168, American Instrument Co. Inc., 8030 Georgia Avenue, Silver Spring, Maryland, U.S.A.; 169, Dr C. A. Parker, Admiralty Materials Laboratory, Holton Heath, Poole, Dorset; 172, Hilger and Watts Ltd; 179(A) and 179(B), N. B. Colthup, American Cyanamide Co., Stamford Research Laboratories, 1937 West Main Street, Stamford, Conn., U.S.A.; 207 and 244, Varian Associates; 208, *Angew. Chemie*, International Edition, and Professor J. D. Roberts; 209 and 210, *Chemical and Engineering News*; 211, 222b/222c, from *Introduction to Practical High Resolution Nuclear Magnetic Resonance*, by D. Chapman and P. D. Magnus, Academic Press; 226, 232 and 235, *Journal of Chemical Education*; 230, *Chemical and Engineering News* and Dr F. A. Bovey; 231, adapted from Chan and Hill, *Tetrahedron*, 1965, **21**, 2017, Pergamon Press; 233 and 234, *Journal of Pharmacy and Pharmacology*; 236, *Journal of Pharmaceutical Sciences*; 237, 247 and 255, Wiley-Interscience; 238, adapted from K. Biemann, *Mass Spectrometry*, McGraw-Hill Publishing Co. Ltd; 239, John Wiley & Sons and Drs R. M. Silverstein and G. C. Bassler; 240, A.E.I. Ltd and Heyden and Son Ltd; 241 and 243, Finnegan Corporation; 245, *Journal of the American Chemical Society*; 248, Heyden and Son Ltd; 249, The Chemical Society and Dr R. I. Reed; 250, McGraw-Hill Publishing Co. Ltd and Professor K. Biemann; 256, *Analytical Chemistry*; 258(A) and 262, Labgear Ltd, Cambridge; 258(B), Panax Equipment Ltd, Redhill, Surrey; 261, The Director, National Physical Laboratory; 266, Nuclear Enterprises Ltd; 268, The Director, Radiological Protection Service, Sutton, Surrey; 269, R. A. Stephen & Co. Ltd, Mitcham, Surrey; 270, Eberline Instrument Co. Ltd; 306, J. R. French and The Radiochemical Centre, Amersham.

Contents

1 General Physical Methods   1

Density, 1; solubility, 6; molecular weight, 8; refractometry, 14; optical activity, 20; viscosity, 29; surface tension, 35; metrication in scientific literature 40

2 Analysis of Drugs in the Solid State   41

Particle size analysis, 41; methods of particle size analysis, 48; thermal methods of analysis, 63

3 Chromatography   75

Column chromatography, 75; paper chromatography, 86; thin-layer chromatography, 96; gas chromatography, 109; high pressure liquid chromatography, 131

4 Measurement of E.M.F. and pH   150

Introduction, 150; indicator electrodes, 152; reference electrodes, 156; measurement of EMF and pH, 158; potentiometric titrations, 162; redox titrations, 169; precipitation reactions, 170; specific ion electrodes, 170

5 Conductimetric Titrations   172

Experimental methods, 175; high frequency titrations, 178

6 Polarography and the Elements of Coulometry   183

Introduction, 183; theoretical considerations, 185; organic polarography, 192; the dropping mercury electrode, 193; basic principles of polarographic instrumentation, 196; polarographic methods of analysis, 204; experiments, 205; amperometric titrations, 211; experiments, 216; coulometric analysis, 217; coulometry experiments, 218

7 The Basis of Spectrophotometry   221

· 8 Analytical Applications of Absorption Spectra   235

Introduction, 235; absorptiometric assays of the British Pharmacopoeia, 238; spectrophotometry, 246; experiments, 263

9 Flame Photometry and Atomic Absorption Spectrophotometry   297

Flame photometry, 297; atomic absorption spectrophotometry, 301

10 Spectrofluorimetry   311

Experiments, 321

11   Infrared Spectrophotometry   331

Introduction, 331; instrumentation, 332; qualitative uses, 333; interpretation of infrared spectra, 336; quantitative analysis, 340; experiments, 342; routine maintenance, 358

12   Nuclear Magnetic Resonance Spectroscopy   361

Introduction, 361; practical considerations, 367; chemical shift, 371; spin-spin coupling, 383; applications, 397; investigation of dynamic properties of molecules, 398; quantitative analysis, 406

13   Mass Spectrometry   417

Practical considerations, 422; combined gas chromatography–mass spectrophotometry, 424; applications, 429

14   Radiochemical Techniques   447

Counting equipment, 447; body scanning, 460; source preparation, 462; practical experiments, 469

Index   535

1 General Physical Methods

G. C. WOOD

Density

Density, ρ, is the mass of a unit volume of a material. The millilitre is usually chosen to express volume, this being the volume of 1 g water at 3.98°C, the temperature of maximum density of water. Thus the density of water at 3.98°C is 1.0000 g/ml (for most purposes the difference between densities expressed in g/ml and in g/cm^3 can be neglected). Density depends on temperature which is therefore specified by a subscript (ρ_t), t being in degrees centigrade. The density of water at various temperatures is given in Table 1.

Table 1. Densities of Water (g/ml) at Various Temperatures

t°C	0	3.98	10	15	20	25
ρ_t	0.99987	1.0000	0.99973	0.99913	0.99823	0.99707

t°C	30	40	50	60	70	80
ρ_t	0.99569	0.99224	0.98807	0.98324	0.97781	0.97183

Specific Gravity

Specific gravity or relative density, $d_{t_1}^{t_2}$, is the ratio of the mass of a certain volume of the material at a particular temperature (t_2) to the mass of an equal volume of water at the same or some other specified temperature (t_1).

Thus, $d_{t_1}^{t_2} = \rho_{t_2}/\rho_{t_1,\mathrm{H_2O}}$. It follows that d_4^t is numerically equal to ρ_t (the difference between the densities of water at 4°C and 3.98°C being neglected) though, unlike density, it is dimensionless.

Molar Volumes

The molar volume of a compound, a quantity often used in calculations, is expressed as:

$$\text{molar volume} = \text{molecular weight/density}$$

Physically the molar volume is a measure of molecular volume plus any free space between the molecules. Attempts have been made to use molar volumes as an additive property of the number and types of atoms and groupings in a molecule, but the additivity is only approximate.

Partial Molar Volumes

When solutions are formed from pure components, unless the solution is ideal (obeying Raoult's Law, and giving no heat or volume changes on mixing the components), the final volume is not simply the sum of the constituent volumes. Volume changes nearly always occur on mixing. It is rare to find that:

$$V = n_1 V_1 + n_2 V_2 \tag{1}$$

where V is the total volume of solution, and V_1 and V_2 are the molar volumes of components 1 and 2 respectively. For non-ideal solutions, which are those normally encountered, eq. (1) becomes:

$$V = n_1 \bar{V}_1 + n_2 \bar{V}_2 \tag{2}$$

where $\bar{V}_1$ and $\bar{V}_2$ are the partial molar volumes of components 1 and 2 and n_1 and n_2 are the numbers of moles of components 1 and 2. A partial molar volume is the change in volume when one mole of a particular component is added to an infinitely large volume of solution at constant temperature and pressure. For component 1:

$$\bar{V}_1 = \left(\frac{\delta V}{\delta n_1} \right)_{T,P,n_2} \tag{3}$$

A simple means of evaluating the partial molar volume is to plot V, the volume of solution containing 1000 g solvent, against the molality, m, of the solute. The slope of (or tangent to) the line at a particular value of m, gives $\bar{V}_2$ at this concentration.

Interconversion for Concentrations

Solution concentrations are expressed in a number of different ways, and it is often useful to be able to convert from one system to another. Consider a solution of total volume, V, and density, ρ, in which:

$W_1 =$ weight of solvent of molecular weight M_1
$W_2 =$ weight of solute of molecular weight M_2
Total weight of solution, $\quad W = W_1 + W_2$
Number of moles of solvent, $n_1 = W_1/M_1$
Number of moles of solute, $\quad n_2 = W_2/M_2$

Molarity (C):

$C =$ number of moles of solute per litre of solution

$$C = 1000 n_2/V = \frac{1000 n_2 \rho}{W_1 + W_2} = \frac{1000 \rho W_2}{M_2(W_1 + W_2)} \tag{4}$$

Molality (m):

$m =$ number of moles solute per 1000 g solvent

$$m = \frac{1000 n_2}{W_1} = \frac{1000 W_2}{M_2 W_1} \tag{5}$$

Interconversion of molarity and molality:

$$C = \frac{1000 m \rho}{1000 + M_2 m} \tag{6}$$

Interconversion of percentages:

$$(\% \text{ w/w}) = (\% \text{ w/v})/\rho \tag{7}$$

Mole fraction:

$$X = \frac{\text{number of moles of component}}{\text{total number of moles present}}$$

$$\text{Mole fraction of solvent} = X_1 = n_1/(n_1 + n_2) \tag{8a}$$

$$\text{Mole fraction of solute} = X_2 = n_2/(n_2 + n_2) \tag{8b}$$

Interconversion of molarity and mole fraction:

$$C = \frac{1000 X_2 \rho}{(X_1 M_1 + X_2 M_2)} \tag{9}$$

Experiment 1. *Determination of the specific gravity and the density of a liquid*
A pycnometer is used for the accurate determination of liquid densities. A convenient form is shown in Fig. 1, although many different types are available.

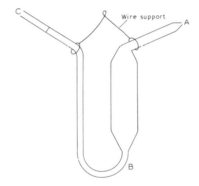

Fig. 1. Pycnometer for determining liquid densities

For example the vessels for volatile liquids are fitted with ground glass caps to prevent evaporation. For rapid, low precision determination of liquid densities hydrometers are convenient. As the temperature coefficient of density is about 0.1% per degree, the temperature must be carefully controlled during measurements. To obtain an accuracy of about 0.0001 g/ml in ρ, temperature should be controlled to $\pm 0.1°$.

Method. Clean the pycnometer with chromic acid, and wash it with water. Rinse it with ethanol and then acetone, and dry by drawing air through it by means of a filter pump. Determine the weight (W_1) of the pycnometer plus the support by which it is attached to the balance arm. Fill the pycnometer by slowly drawing water in through A; hold it inverted until the liquid level reaches B, then turn it upright and allow the water to reach mark C. Suspend it in a thermostat bath at 25°, so that the ends A and C are clear of the water in the thermostat. After temperature equilibrium has been attained (20 min), adjust the volume of water in the pycnometer so that it is completely full from

SOLUBILITY

In the British Pharmacopoeia, a statement of solubility as weight of solute in a specified volume of solvent at a particular temperature only applies at this temperature (see 'Effect of Temperature' below). If the temperature is not stated, the solubilities stated apply at room temperature. Solubilities are also expressed as 'one part of solute in x parts solvent', meaning either 1 g solid solute or 1 ml liquid solute in x ml solvent.

The usual expression is in terms of weight of solute per volume or per weight of solvent or solution; usually as g solute/100 g solvent, although the normal means of expressing solution concentration such as molar, molal, are also used. Only solutions of liquids in liquids, and solids in liquids will be dealt with here.

Effect of Temperature

Whether solubility increases or decreases with increasing temperature will depend on the nature of the solute and solvent, but in general solubility increases with temperature rise. The rate of solution is also increased by heating. Studies of the variation of solubility with temperature can be used to evaluate the heat of solution, ΔH, from *either*

$$\frac{d \ln S}{dT} = \frac{\Delta H}{RT^2} \tag{12a}$$

or from the integral form (12b), derived on the assumption that ΔH is independent of temperature

$$2.303[\log S_2 - \log S_1] = \frac{\Delta H(T_2 - T_1)}{RT_2 T_1} \tag{12b}$$

where S_1 and S_2 are the solubilities at two temperatures T_1 and T_2. The value of ΔH obtained can give information on the mechanism of solution.

Effect of Particle Size

While small particles of a solid solute may often dissolve more rapidly than large ones, particle size has little effect on solubility as such in the size range normally encountered. Only in the colloidal particle size range does solubility increase significantly with decrease of particle size.

Solubility and Purification Procedures

Recrystallisation is used for purifying a vast range of organic compounds; the basis of the technique is that for a mixture of solutes, the required one will crystallise from a particular solvent, while the others remain (largely!) in solution. Recrystallisation is only really effective when most of the major contaminants have already been removed from the solute; as large amounts of impurities may either crystallise out with the desired material, or may inhibit its crystallisation.

The decreasing solubility of polymers with increasing molecular weight is used as a means of fractionation. The polymer is dissolved in a solvent, some precipitating liquid added, and the fraction of polymer precipitated is centrifuged off. This fraction has the highest molecular weight. Addition of a little more precipitating liquid gives a fraction of lower molecular weight, and a range of fractions can be obtained by repeating this process. In protein chemistry much use is made of precipitation for purification, generally using salts as precipitants, and arranging experimental conditions which will leave the impurities in solution as far as possible.

Solubility of Liquids in Liquids

Information on the mutual solubility of immiscible liquids can be obtained by shaking them together at a controlled temperature, allowing separation to occur, and analysing samples from each layer for each component. For liquids which become miscible at accessible temperatures, it is often convenient to determine the mutual solubility curve. Mixtures of the two liquids of known composition are prepared, and heated until a homogeneous phase appears (a saturated solution of one liquid in the other). The highest temperature at which two phases disappear, measured as a function of composition, is the critical solution temperature; above it the two liquids are miscible in all proportions, no matter what the composition.

Experiment 4. *Determination of the mutual solubility curve of phenol and water*

Method. Use test tubes of 20 ml capacity, fitted with ground glass stoppers. Weigh accurately about 1, 2, 3, 4, 5, 6, 7, 8, and 9 g phenol into the tubes. Add sufficient water to each tube to make the total contents weigh about 10 g, and weigh accurately. Fit up a 2 l beaker with a hand stirrer, and a thermometer (0–100°, graduated to 0.1°). Attach a piece of copper wire to the tube containing the most dilute solution of phenol in water, and hang it on the rim of the beaker. Place sufficient water in the beaker to bring the level to the bottom of the ground glass stopper of the tube. Heat the water in the beaker, shaking the tube frequently, and raising the temperature slowly when the turbidity of the phenol-water mixture shows signs of disappearing. Determine the temperature at which the turbidity disappears on shaking. Return the tube to the beaker, allow the temperature to fall, and note the temperature at which the turbidity just reappears. Repeat the heating and cooling, and take the mean of all four temperature readings. Determine the temperature at which miscibility occurs for the other solutions (cooling may be required for the most concentrated solutions).

Treatment of results: Plot the temperatures at which a single phase occurs as ordinates, against percentage (w/w) phenol in water as abscissae. Determine the critical solution temperature from this graph.

Experiment 5. *Determination of solubility of adipic acid in water*

A simple quick method for determining the solubility of a solid in a liquid is to prepare a set of test tubes each containing a weighed amount of solid, add varying quantities of solvent, shake thoroughly, and note the concentration at which all the solid dissolves.

For accurate solubility determinations it is necessary to ensure that the solvent is completely saturated with solute at a particular temperature; withdraw a sample of saturated solution without disturbing the equilibrium, and analyse it to determine concentration. The solid and solvent are mixed by stirring or shaking in a thermostat, samples withdrawn at intervals, and assayed. Equilibrium is reached when there is no further uptake of solute by solvent. This method can also be applied to the determination of liquid-liquid solubilities.

Method. Place adipic acid (about 4 g, roughly weighed) in each of two flasks (100 ml) fitted with B24 size ground glass stoppers, and add water (about 60 ml) to each flask. Immerse one flask at a level just above the bottom of its stopper in a thermostat bath set at 20°, warm the second flask to 50° for 10 min shaking it occasionally, and then place it similarly in the thermostat bath. This procedure is adopted so that equilibrium is approached from both over- and under-saturation. Shake the flasks every 20 min for 2 hr. Allow the flasks to remain undisturbed for 10 min. Fit a one inch length of polythene tubing to the tip of a 10 ml pipette, and pack the tubing with cotton wool to act as a filter. Cautiously withdraw a sample of solution from the flask, examining the contents of the pipette against a bright light to make sure the filtration was efficient. Remove the filter, adjust the volume of the solution in the pipette to 10 ml, run the solution out into a tared flask. Weigh the sample.

Titrate the sample with standard *0.2N* NaOH using phenolphthalein as indicator. Continue to shake the flasks during a further hour, withdraw a sample from each, and re-assay. Equilibrium should have been reached after two hours, as shown by agreement between the four sets of results.

If the heat of solution is to be determined, repeat the experiment at 30°. The pipette should be preheated to the temperature of the experiment.

Treatment of results. 1. Calculate the solubility of adipic acid in g solute/100 g water, and in moles/l solution (molecular weight of adipic acid = 146). Present the results to show that equilibrium has been reached.

2. Calculate ΔH from eq. (12b) using the solubility in moles/l.

MOLECULAR WEIGHT

Excluding very large molecules, the determination of the molecular weight of compounds is based on the use of colligative properties, generally freezing point depression or boiling point elevation. Osmotic pressure can be used in the 5000–300 000 molecular weight range.

The presence of a non-volatile solute in a solution lowers the vapour pressure from p_1^0 (that of the pure solvent) to p_1 (that of the solution). For ideal solutions, Raoult's law gives

$$p_1 = X_1 p_1^0 = (1 - X_2)p_1^0 \tag{13}$$

where X_1 is the mole fraction of solvent, and X_2 that the solute. In dilute solution the term W_2/M_2 in the denominator of X_2 can be neglected:

$$X_2 = \frac{W_2/M_2}{W_1/M_1 + W_2/M_2} \simeq \frac{W_2 M_1}{M_2 W_1} \tag{14}$$

giving for the relative lowering of the vapour pressure

$$\frac{p_1^0 - p_1}{p_1^0} = \frac{W_2 M_1}{M_2 W_1} \tag{15}$$

Equation (15) can be used to determine M_2 from measurements of vapour pressure, assuming that no dissociation or association of solute occurs.

Using Fig. 2 consider the boiling point elevation and freezing point depression. AB represents the sublimation curve of pure solid solvent, and BC the vapour pressure curve of liquid solvent, reaching the boiling point at C. When solute is present, the vapour pressure of the solution is lower than that of the solvent (DE), giving a freezing point at T_f, a depression of the freezing point $\Delta T_f = T_0 - T_f$; the boiling point of the solution is raised to T_b, giving elevation of boiling point $\Delta T_b = T_b - T_{bo}$.

At the freezing point of a solution, there is an equilibrium between solid solvent and solution, such that the chemical potential of the separated (frozen) solvent and of solvent in the solution are equal. By considering the variation of chemical potential with temperature the following relationship is obtained

$$\Delta T_f = \frac{RT_0^2 X_2}{\Delta H_{fus}} \tag{16}$$

where ΔH_{fus} is the latent heat of fusion per mole of solvent. In dilute solution eq. (14) is used for the mole fraction of solute, and taking molality from eq. (5):

$$X_2 = M_1/m1000$$

hence:

$$\Delta T_f = \frac{RT_0^2 M_1 m}{\Delta H_{fus}1000}$$

and

$$\Delta T_f = K_f m \tag{17}$$

in which all the constants for a particular solvent are combined in a new constant K_f.

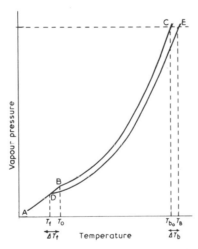

Fig. 2. Vapour pressure–temperature curves for pure solvent and solution

By an analogous treatment, considering the equilibrium between solvent vapour and liquid solvent in solution at the boiling point, the chemical potential of solvent in the vapour phase must be the same as that in the liquid phase,

and

$$\Delta T_b = \frac{RT_b^2 X_2}{\Delta H_{vap}} \tag{18}$$

where ΔH_{vap} is the molar latent heat of vaporisation. Simplifying as before for dilute solution:

$$\Delta T_b = \frac{RT_b^2 M_1 m}{\Delta H_{vap} 1000}$$

and

$$\Delta T_b = K_b m \tag{19}$$

Examples of molal cryoscopic and ebullioscopic constants are given for commonly used solvents in Table 2. The table shows that for a certain solute con-

Table 2

Solvent	T_f	T_b	K_f	K_b
Acetic acid	16.7	118	3.9	2.93
Benzene	5.4	80	5.12	2.53
Chloroform		61		3.63
Ethanol		79		1.22
Water	0	100	1.86	0.51

centration, the depression of the freezing point will be greater than the elevation of the boiling point; freezing points are also easier to determine than boiling points.

The osmotic pressure can be considered in relation to the other colligative properties using Fig. 3. Solvent and solution are placed in the two arms of the vessel shown. As $p_1^0 > p_1$ solvent distils through the vapour space, which in this case is acting like a semi-permeable membrane. To prevent distillation, the pressure on the solution must be increased; this raises the chemical potential of the solvent in solution, so that at equilibrium there is no further distillation.

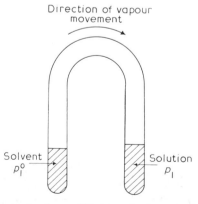

Fig. 3. Apparatus for equilibrium of solvent and solution

From a thermodynamic consideration of the effect of pressure

$$\Pi \overline{V}_1 = RT \ln (p_1^0/p_1) \tag{20}$$

where Π = osmotic pressure and $\overline{V}_1$ = partial specific volume of solvent. If it is assumed that Raoult's law holds, and that solutions are dilute, this equation can be simplified:

$$\Pi = W_2 RT/M_2 \tag{21}$$

This enables the molecular weight of the solute to be calculated from measurements of osmotic pressure, W_2 being the concentration of solute in g/l.

Solute Mixtures. If the solute under test is impure an average molecular weight will be obtained. The average is due to materials of different molecular weight depressing the freezing point (or elevating the boiling point) by different amounts. Colligative methods depend on the number of molecules present in the solution, and the methods give a number average molecular weight, M_N,

$$M_N = \frac{\Sigma n_i M_i}{\Sigma n_i}$$

where n_i = number of molecules of molecular weight M_i, and so on. Hence contamination of the material under test with low molecular weight impurities can give erroneous values for the molecular weight.

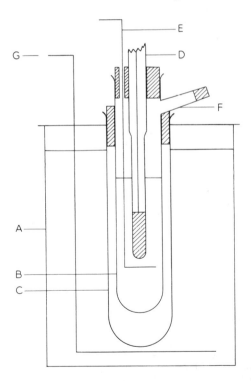

Fig. 4. Beckmann freezing point depression apparatus

Experiment 6. *Determination of molecular weight by freezing point depression*

The conventional apparatus of Beckmann (Fig. 4) consists of an outer bath, A, holding the freezing mixture, which cools the solution in the tube, B, which is surrounded by an air jacket, C. A Beckmann thermometer, D, and stirrer, E, dip into the solution. Samples may be introduced through the side arm, F. The freezing bath is covered by a lid, and kept mixed by a stirrer, G.

When using this apparatus, some supercooling of the sample occurs, as separation of solid solvent does not take place until the temperature is a little below the freezing point. The latent heat of crystallisation then raises the temperature. This effect may be minimised if the temperature of the freezing mixture is too low, and a temperature lower than the true freezing point may be recorded. The freezing mixture should be kept 3–4° below the freezing point of the solution. The degree of supercooling should not be allowed to exceed 0.3–0.5°, otherwise a great deal of solvent may crystallise out, and seriously affect the concentration of the solution.

The Beckmann thermometer. This is a sensitive thermometer, graduated in 0.01°, and capable of being read to 0.002°. The scale length of the thermometer is 5–6°, and provision is made for adjusting the amount of mercury in the bulb, so that the thermometer can be used over different temperature ranges. Place the bulb in a beaker of water whose temperature has been adjusted to the freezing point of the solvent to be used. The mercury should stand near the top of the scale. If it stands above, place the thermometer in water warm enough to cause the mercury to rise up and form a drop at the end of the capillary (A) (Fig. 5). Shake off the drop and test again in the cool water. If necessary repeat the procedure until the mercury stands at the correct place on the scale. The initial test may show that there is too little mercury in the thermometer bulb (mercury below scale); in this case warm up the thermometer until drops form at the capillary, invert it, and tap gently to make mercury from the reservoir join with mercury in the capillary. Re-invert, and place the thermometer in a water bath held 2° above the freezing point of the solvent; the cooling draws mercury from the reservoir into the bulb. When this process is complete, tap the upper part of the thermometer gently to make excess mercury separate from the end of the capillary. Test the thermometer in water at the freezing point of the solvent to make sure that the setting is correct, i.e. the mercury is on the scale. If it is a little above, remove the excess dropwise to the reservoir as described above. Beckmann thermometers require careful handling as they are expensive, and easily broken.

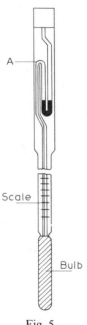

Fig. 5.
Beckmann
thermometer

Determination of freezing point. Remove the tube B from the apparatus. Clean, dry, and weigh the tube. Introduce dry, crystallisable benzene (15–20 g), re-weigh and insert the Beckmann thermometer and the stirrer. Pack the bath with a freezing mixture of ice and water controlled at about 2°. Place the tube in the freezing mixture, and when solid separates, remove it, dry the outside and place it in the air jacket. Stir the solution about once every second (faster stirring can generate undue heat), and note the constant (highest) temperature on the Beckmann thermometer. The first reading is usually approximate. Remove the tube from the apparatus, and warm gently with stirring, until the temperature is about 1° above the freezing point. Replace it in the apparatus, and stir; the supercooling should not be more than 0.5°, crystallisation of solvent being induced by a bout of vigorous stirring, or by adding a crystal of benzene. Follow the temperature rise accompanying crystallisation, noting the highest temperature, and tapping the thermometer with the fingernail to prevent the mercury from sticking in the stem.

Repeat the procedure until three consistent readings for the freezing point of pure solvent have been obtained. Introduce a weighed tablet of naphthalene (about 0.12 g) through the side arm, dissolve it in the benzene with gentle warming, and determine the freezing point of the solution. When consistent results have been obtained, a second weighed tablet should be introduced to obtain a second reading.

Treatment of results. Calculate the molality of the solution in both parts of the experiment using equation (17) and the observed ΔT_f values. Calculate M_2 from eq. (5).

Experiment 7. *Detection of association in solution*

Many carboxylic acids associate in non-polar solvents, generally to form dimers. Acetic, benzoic, and phenylacetic acids are amongst those showing this phenomenon.

Method. Weigh 1, 2, and 3 g of dry phenylacetic acid into each of three graduated flasks (50 ml), make up approximately to volume with dry crystallisable benzene, stopper, and reweigh (weight concentrations are required; the graduated flasks are merely convenient containers). Determine the freezing point of each solution, and of a sample of the solvent, by the technique of the preceding experiment.

Treatment of results. 1. Calculate the apparent molecular weight, M, for phenylacetic acid in each solution. Compare this with the monomer molecular weight of 136.

2. Calculate the degree of association, α, of the phenylacetic acid in each solution, assuming that dimers are formed, e.g. $2C_6H_5CH_2COOH \rightleftharpoons (C_6H_5CH_2COOH)_2$.

For one mole of solute $(1-\alpha)$ moles are unassociated, and $\frac{1}{2}\alpha$ are associated. The total number of gram molecules is therefore $\frac{1}{2}\alpha + (1-\alpha)$. Using the usual freezing point equation, calculate the freezing point depression $(\Delta T_f)_0$ assuming that no association took place, i.e., for the monomer of molecular weight, M_0.

Now,

$$(\Delta T_f)_{exp} = \left[(1-\alpha)+\frac{\alpha}{2}\right](\Delta T_f)_0$$

$$\therefore \quad (1-\alpha/2) = \frac{(\Delta T_f)_{exp}}{(\Delta T_f)_0}$$

$$= \frac{M_0}{M}$$

Hence,

$$\alpha = \frac{2(M-M_0)}{M} \tag{22}$$

Experiment 8. *Micro method for molecular weight determination* (*Rast*)

A number of terpenoid substances which have very large K_f values, e.g. camphor 40, camphene 35, are useful for determining molecular weights by freezing point depression, using small quantities of material. Generally about 10% solutions in camphor are prepared; the method assumes that the freezing point depression equation is obeyed, and that pure camphor separates on freezing. K_f for camphor varies from sample to sample, so either a sample of known K_f value must be used, or it must be measured using naphthalene as solute (10%–15% by weight) in camphor for the method given below.

Method. Seal one end of a clean, thin walled glass tube (1 × 10 cm), allow to cool, and weigh to 0.1 mg. Introduce about 15 mg of acetanilide, ensuring that the sample is all placed at the bottom of the tube, weigh accurately, add about 150 mg camphor and reweigh. Carefully seal the end of the tube, place it in an oil bath, and heat until the contents melt. Rotate the tube to mix the materials thoroughly. For a second determination prepare another tube with acetanilide (20 mg) and camphor (150 mg). Allow the tubes to cool in the oil bath.

Determine the melting points of the two acetanilide camphor mixtures and of camphor itself in capillary tubes using an electrical melting point apparatus; note the temperature at which the last crystal of each sample disappears. Slow raising of the temperature is essential in the region of the melting point.

Treatment of results. 1. Calculate K_f from the naphthalene experiment, if this is necessary.

2. From the melting point difference in the acetanilide experiment, calculate the molecular weight using both results.

Mass spectrometry

Precise molecular weights may now be determined by mass spectrometry, as described in Chapter 13.

REFRACTOMETRY

Index of Refraction

When a ray of light passes from one medium (1), Fig. 6, into another medium (2) it undergoes refraction. The ray travels at a lower velocity in the optically

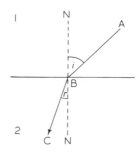

Fig. 6. Index of refraction

more dense medium (2) than in medium (1) which is less optically dense. The angle of incidence (i) and angle of refraction (r) are related by Snell's Law

$$\frac{\sin i}{\sin r} = {}_1n_2 \tag{23}$$

where n is the refractive index of medium (2) relative to medium (1), and the subscripts indicate the direction of the ray. The refractive indices of liquids are nearly always referred to that of air.

The critical angle is used extensively in refractometry. For a narrow cone of rays, a–b, placed close to the boundary between media (1) and (2) (Fig. 7),

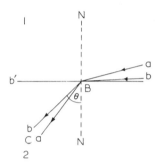

Fig. 7. Critical angle

and observed at C, a band of light is seen. This band has a sharp edge at b, where the actual ray which is incident along the surface of medium (2) (b–b) is observed. No rays are present in the b–b′ region. Now

$$_1n_2 = \frac{\sin i}{\sin r} = \frac{\sin 90°}{\sin \theta} = \frac{1}{\sin \theta} \tag{24}$$

A measurement of the critical angle θ can therefore give the refractive index of medium (2).

Effect of Wavelength

As refractive index generally decreases with increase of the wavelength of light used, a monochromatic source must be used for precision work. Alternatively the refractometer must be capable of compensating for the use of white light. The Cauchy formula

$$n = A + B/\lambda^2 + C/\lambda^4$$

where A, B and C are constants, expresses the variation of refractive index with wavelength when the medium does not absorb light. When the medium and the wavelength are such that absorption occurs, the variation of refractive index with wavelength is 'anomalous' (Fig. 8).

Fig. 8. Refractive index dispersion in the region of an absorption band

Effect of Temperature

The refractive index of a liquid decreases as temperature increases, the percentage decrease ranging from 0.01 to 0.1 % per degree, depending on the nature of the liquid. Since refractive indices are easily measurable to within $\pm 2 \times 10^{-4}$, good temperature control is essential. The temperature of measurement is specified as a superscript numeral on n, and the wavelength of light used as a subscript capital, e.g. n_D^{20}, where D refers to the sodium D line.

Specific Refractive Index Increment

The refractive index of a solvent generally changes if a solute is dissolved in it, and this variation of refractive index can be very useful in following chromatographic separations. The specific refractive index increment (dn/dc) is the change in refractive index of a solvent with change of solute concentration

$$\frac{dn}{dc} = \frac{n_{solution} - n_{solvent}}{\text{concentration of solute}}$$

the concentration being expressed in g/ml, giving dn/dc in ml/g.

A knowledge of dn/dc permits the concentration of a solution to be determined by measuring its refractive index.

Molar Refraction

Lorentz and Lorenz found that the specific refraction

$$[n] = \frac{n^2 - 1}{n^2 + 2} \cdot \frac{1}{\rho}$$

where ρ is the density of a liquid of refractive index n, was almost independent of temperature. The molar refraction:

$$R = \frac{n^2 - 1}{n^2 + 2} \cdot \frac{M}{\rho} \tag{25}$$

has the dimensions of molar volume, and is a property of the number and type of atoms and groupings in the molecule. This additivity arises from the property of polarisability of molecules when subjected to electromagnetic radiation. The incident radiation sets the electrons vibrating and induces dipoles in the molecule. The total polarisability of the molecule is related to the sum of all the induced dipoles, and hence to the number and type of constituent atoms. Polarisability is in turn related to refractive index.

Some atomic refractivities evaluated at the sodium D line are given in Table 3. When the empirical formula of a compound is known and its molar refraction

Table 3. Atomic Refractivities

C	2.418	O (−OH)	1.525	Br	8.748
H	1.100	O (−OR)	1.643	N (−NH$_2$)	2.322
C (C=C)	1.733	O (=O)	2.211	N (−NHR)	2.502
C (C≡C)	2.398	Cl	5.967	N (−NR$_2$)	2.840

measured, the additivity of values for atoms in the molecule can be a useful guide in structure determinations.

Refractive Index of Water

Some values for the refractive index of water relative to dry air at 760 mm mercury pressure are given in Table 4.

Table 4

15°	20°	25°	30°
1.33339	1.33299	1.33250	1.33194

Experiment 9. *Determination of refractive index using the Abbé refractometer*

The optical system of the Abbé refractometer, illustrated in Fig. 9, employs the critical angle principle. The liquid under examination is placed between the two prisms. The top face of the lower prism is ground, to diffuse the light rays in all directions. Rays passing from the liquid to the upper prism (which must have a higher refractive index than the liquid) may be refracted in the normal way, giving a bright field in the eyepiece. Rays which strike the liquid/glass interface at grazing incidence give rise to the critical ray. These effects result in the field of view containing a dark and a light area, with a sharp dividing line.

The optical diagram for the upper prism is drawn out in Fig. 10. Let the liquid have a refractive index n (air to liquid) and let N be the refractive index of the prism compared to air,

then,

$$N = \sin \alpha / \sin \beta$$

$$\sin \theta = {}_{glass}n_{liquid} = n/N$$

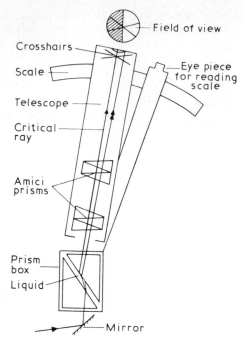

Fig. 9. Optical system of the Abbé refractometer

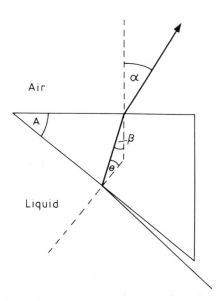

Fig. 10. Optical diagram for upper prism of Abbé refractometer

and

$$A = \beta + \theta$$

Now

$$n = N \sin (A - \beta)$$
$$= N \sin A \cos \beta - N \cos A \sin \beta$$

As

$$\sin \alpha = N \sin \beta$$

$$\cos \beta = \sqrt{\left\{ 1 - \frac{\sin^2 \alpha}{N^2} \right\}}$$

Substituting:

$$n = N \sin A \sqrt{\left\{ 1 - \frac{\sin^2 \alpha}{N^2} \right\}} - \cos A \sin \alpha$$

$$n - \{ \sin A \sqrt{(N^2 - \sin^2 \alpha)} \} - \sin \alpha \cos A \qquad (26)$$

Equation 26 gives the refractive index of the liquid relative to air in terms of two constants for a particular prism (A and N), and a measurable angle, α. The Abbé refractometer measures the angle α between the critical ray emergent from the upper side of the prism and the normal. Using the constants A and N this angle has been converted directly into refractive index, which is printed on the calibrated scale of the instrument.

The telescope of the instrument is fixed, and the prism box is attached to the scale. The prism box is rotated until the critical ray falls on the cross hairs of the telescope. The value of n at this setting is read from the scale.

Method. For accurate measurements attach the prism box to a thermostat bath regulated at 25°. Open the prism box, and place a few drops of liquid on the lower prism; close the box. Adjust the mirror to give a bright illumination of the field. Turn the knurled knob until the field has a light and a dark section. If there is a coloured fringe between the two areas, adjust the Amici prisms until the boundary is sharp and black, set it on the cross hairs, and note the reading of refractive index. The Amici prisms deviate all wavelengths except those of the sodium D lines, and may also be used to determine the dispersion of the liquid.

Open the prism box, and wipe off the liquid with cotton wool moistened with acetone. Use first *n*-octane, and then *n*-octene as the liquid, and determine the refractive index of each.

Treatment of results. 1. Octane has $M = 114.2$, $\rho_{25} = 0.699$ g/ml, and octene has $M = 112.2$, $\rho_{25} = 0.715$ g ml. Calculate the molar refraction of each liquid, using the observed refractive indices.

2. Compare these values with those calculated from the atomic refractivities on p. 17, and note the contribution of the double bond.

Note. A simple immersion model of the Abbé refractometer is available consisting of the upper prism fixed to a telescope, and a scale. The prism is dipped in a beaker of liquid, the scale reading observed, and converted to refractive index using tables.

Experiment 10. *Determination of critical micelle concentration by refractometry*

Substances which form micelles in water generally have two distinct regions in their molecules, the hydrophobic portion (hydrocarbon chain) and the hydrophilic portion (polar group). The formation of micelles from a number of monomers places all the hydrocarbon chains together in the centre of the micelle, thus lowering the free energy of the system. The concentration at which micelles are first detectable is known as the critical micelle concentration (CMC). In general, the physical properties of solutions of soaps (or of other substances forming micelles) undergo sharp changes at the CMC. A plot of refractive index against concentration should show a change in slope at the CMC.

Method. Prepare 200 ml of a 25 % w/v solution of butyric acid in water. Accurately prepare 2.5, 5, 7.5, 10, 15 and 20% solutions of butyric acid in water by measuring appropriate volumes from a burette into 50 ml volumetric flasks, and making up to volume with water. Measure the refractive index of each solution, and of the original 25% solution at 25°, using the Abbé refractometer. Both dilutions and measurements must be very carefully made. Measure also the refractive index of water.

Treatment of results. Plot refractive index as ordinate and concentration as abscissa. Two straight lines should be obtained, intersecting at the CMC.

Note. Butyric acid forms micelles in aqueous solution in a concentration region suitable for measurements with the Abbé refractometer. The majority of soaps form micelles at low concentrations, and the precision of the Abbé is insufficient to detect the CMC. In these cases an interference refractometer, such as the Rayleigh, must be used.

OPTICAL ACTIVITY

The electric fields associated with a beam of monochromatic light vibrate in all directions perpendicular to the direction of propagation of the light. Certain crystalline materials such as Iceland spar have different refractive indices for light whose field vibrates parallel or perpendicular to the principal plane of the crystal. As a result a Nicol prism constructed of this material transmits only light whose electric field oscillates in one plane (Fig. 11). Optical activity concerns the interaction of such plane-polarised light with certain materials, particularly solutions of some organic compounds.

The electric field of plane-polarised light can be considered to be composed of two components of fixed magnitude rotating in opposite sense to one another (left circularly polarised and right circularly polarised) and the plane-polarised beam is the vector sum of these components (Fig. 11).

When plane-polarised light passes through a medium it is retarded to an extent which is indicated by the refractive index of the medium. When the latter is optically inactive both circularly polarised components are retarded to the same extent and the beam emerges from the medium polarised in the same plane

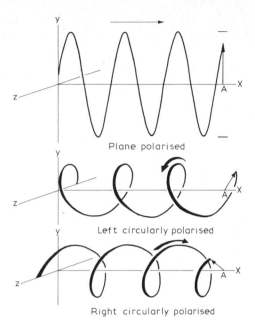

Plane polarised

Left circularly polarised

Right circularly polarised

Fig. 11. The electric field of plane-polarised light and its two circularly polarised components. The electric field of the plane-polarised beam at point A is the vector sum of the electric fields of the circularly polarised beams (indicated by arrows)

as the incident beam. If the medium is optically active the components are retarded to different extents because the refractive indices of the medium for left circularly polarised light (n_L) and right circularly polarised light (n_R) differ. As a result of this circular birefringence ($\Delta n = n_L - n_R$) the beam emerges from the medium still plane-polarised but with the plane of polarisation inclined at an angle α degrees to the plane of polarisation of the incident beam, given by:

$$\alpha = \frac{1800}{\lambda} l\,\Delta n \qquad (27)$$

where l is the light-path in decimetres and λ the wavelength in cm. α is the optical rotation of the medium and is positive when the plane of polarisation is rotated clockwise relative to that of the incident beam when viewed looking towards the light source.

If the wavelength of the light is such that the medium absorbs a fraction of the radiation, an optically active medium may show a second physical effect arising from unequal absorption of left and right circularly polarised light. As a result of this circular dichroism the beam emerges from the medium elliptically polarised. The ellipticity, ψ degrees, of the emergent beam is given by:

$$\psi = \frac{1800}{\lambda} l\,\Delta\kappa \qquad (28)$$

where l and λ have the same meaning as in eq. 27, for optical rotation, and $\Delta\kappa = \kappa_L - \kappa_R$, the difference between the absorption indices* for left and right circularly polarised light. Using the more familiar absorbance, A (Chapter 8)

$$\psi = \frac{2.303 \times 1800}{4\pi} \Delta A \qquad (29)$$

Any medium that shows circular dichroism must at the same time show circular birefringence and hence optical rotation. Both result from unequal interaction of the medium with left and right circularly polarised light and hence are closely related phenomena.

A more detailed account of the relationship between optical rotation and circular dichroism is given by Foss (1963).

Effect of Concentration, Solvent and Temperature on Optical Activity

The dependence of optical rotation and circular dichroism of a solution of an optically active substance on concentration may be taken into account by calculating the specific optical rotation and ellipticity or the molecular rotation and ellipticity as shown in Table 5.

Table 5

Specific rotation	Specific ellipticity
$[\alpha]^t_\lambda = \dfrac{\alpha^t_\lambda}{lc}$	$[\psi]^t_\lambda = \dfrac{\psi^t_\lambda}{lc}$
units $-$ deg $\cdot$ cm^2 $\cdot$ decagram^{-1}	

Molecular rotation	Molecular ellipticity
$[\Phi]^t_\lambda = \dfrac{M[\alpha]^t_\lambda}{100}$	$[\Theta]^t_\lambda = \dfrac{M[\psi]^t_\lambda}{100}$
units $-$ deg $\cdot$ cm^2 $\cdot$ dmol^{-1}	

α^t_λ = optical rotation $\}$ in degrees at temperature $t°$C
ψ^t_λ = ellipticity $\qquad$ and wavelength λ nm
$\quad c$ = concentration, g cm^{-3}, of solute
$\quad l$ = light-path, dm
M = molecular weight of solute

* κ is defined by the equation:

$$I = I_0 \, e^{-4\pi\kappa/\lambda}$$

where I and I_0 are the intensities of the transmitted and incident light

$$\therefore \quad \kappa = \frac{2.303\lambda}{4\pi l} A$$

It may also be shown that

$$[\Theta] = 3300\,\Delta\varepsilon$$

where $\Delta\epsilon = \epsilon_L - \epsilon_R$, the difference between the molar absorption coefficients of left and right circularly polarised light.

Measurements of optical rotation are frequently made with sodium D light and usually, though not necessarily, at 20° and specific rotations based on such measurements are reported as $[\alpha]_D^{20}$. Specific rotation is generally concentration-dependent and in dilute solution an equation of the type

$$[\alpha] = A + Bc + Cc^2 \tag{30}$$

can be used to describe results, A, B and C being constants. It is necessary always to state the concentration at which a specific rotation was measured unless a procedure for extrapolating $[\alpha]$ to zero concentration and quoting $[\alpha]_{c=0}$ is followed.

Temperature may have a pronounced effect on specific rotation, and suitable temperature control is required for precise work. Effects may arise due to changes of intermolecular interactions with temperature, or to changes in equilibria between configurations. Tartaric acid provides an example of the latter effect, the variation in $[\alpha]$ being about $10\%/°C$, owing to the equilibrium between two forms with different optical rotatory power varying with temperature.

The particular solvent used can have a large effect on the results, e.g. a 20% w/w solution of nicotine in chloroform has $[\alpha]_D^{20} \simeq +4°$, while the same concentration in water gives $[\alpha]_D^{20} \simeq +10°$. Chloramphenicol gives a change in sign, for example $[\alpha]_D^{25} = +19°$ in ethanol changes to $-25°$ in ethyl acetate. These effects, which can be very large, indicate that a statement of the solvent used must always accompany any report of $[\alpha]$.

Optical Rotatory Dispersion (ORD) and Circular Dichroism (CD) Spectra

Information about the structure of organic compounds and their optical purity can be gained from measurements of optical rotation at a single wavelength. More detailed and precise information can be gained by recording the variation of α with wavelength (ORD spectra) or of ψ with wavelength (CD spectra),

The ORD, CD and absorption spectra of D-10-camphorsulphonic acid (Fig. 12) illustrate many of the general features of optical activity. At wavelengths well clear of the absorption bands of the carbonyl group ($\lambda > 450$ nm) $[\Phi]$ increases in magnitude as λ decreases and as in many other examples, the spectrum can be fitted by an equation of the form

$$[\Phi] = \frac{K}{\lambda^2 - \lambda_0^2} \qquad \text{(one-term Drude equation)} \tag{31}$$

where K and λ_0 are constants, as shown by the linearity of the $1/[\Phi]$ vs λ^2 plot. For some substances more complicated Drude equations are required. The ORD spectra of many optically active aliphatic compounds are of this

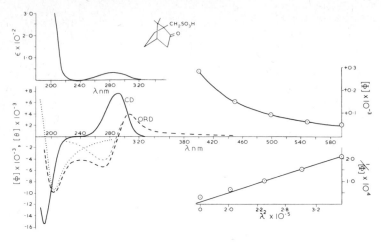

Fig. 12. Absorption, ORD and CD spectra of D-10-camphorsulphonic acid in water at 27°

'plain' type which may be positive, as in this case, or negative. One obvious advantage in analytical work of measuring α at wavelengths lower than the sodium D line (589.6 nm) is the gain in sensitivity; the molecular rotation of D-10-camphorsulphonic acid at 370 nm is about ten times its value at 589.6 nm.

In the region of the absorption band ($\lambda_{max} = 287$ nm) D-10-camphorsulphonic acid shows circular dichroism, the shape of the CD spectrum between 240 nm and 320 nm being similar to that of the absorption spectrum. The ORD spectrum in this region takes a sigmoid course (cf the variation of refractive index with wavelengths, Fig. 8) with a distinct peak and trough (extrema) at 306 nm and 270 nm and change of sign at 290 nm. This is known as a Cotton effect and by convention, when the sign of the CD is positive and the higher wavelength extremum of the ORD spectrum is positive, the Cotton effect itself is said to be positive.

Below 240 nm D-10-camphorsulphonic acid has a second optically active absorption band ($\lambda_{max} \simeq 190$ nm) but this time the Cotton effect is negative. The CD and the higher wavelength extremum of this part of the ORD spectrum (204 nm) are both negative (the lower wavelength extremum is inaccessible).

In this relatively simple example the two absorption bands associated with the single carbonyl chromophore are well resolved, as are the corresponding CD peaks. Circular dichroism and optical rotation both result from the unequal interaction of left and right circularly polarised light with chromophores in dissymmetric molecules and are closely related phenomena. The ORD spectrum corresponding to a particular CD band can be calculated from the CD spectrum by means of a Kronig–Kramers transform:

$$[\Phi]_\lambda = \frac{2}{\pi} \int_0^\infty [\Theta]_{\lambda'} \frac{\lambda'}{\lambda^2 - \lambda'^2} d\lambda' \tag{32}$$

where $[\Phi]_\lambda$ is the molecular rotation at wavelength λ, $[\Theta]_{\lambda'}$ is the molecular ellipticity at wavelength λ' and λ and λ' are the main variable and parameter of integration respectively. The dotted lines in Fig. 12 are the approximate contributions to the ORD spectrum calculated from the two CD bands of D-10-camphorsulphonic acid. The ORD Cotton effects are incompletely resolved since they spread on either side of the corresponding CD bands. In more complex molecules, possibly with several optically-active chromophores, the higher resolving power of CD is advantageous in determining the contribution of each chromophore to optical activity. On the other hand many optically-active compounds have absorption bands at such low wavelengths that their solutions show no CD in the accessible spectral region; the ORD Cotton effects associated with these absorption bands extend into the accessible spectral region so that the optical activity of such substances can be studied by means of optical rotation. CD and ORD are thus complementary techniques.

More detailed descriptions of the relationship between ORD and CD spectra and the stereochemistry of organic compounds are given in a number of recent reviews (e.g. Snatzke, 1967).

Instrumentation

(a) *Visual polarimetry.* The optical system for a simple visual polarimeter is shown in Fig. 13. The light source is usually a sodium vapour lamp whose light is made parallel by lens L_1 and polarised by the Nicol prism, P_1. In the absence of optically active material in the sample tube a second Nicol prism,

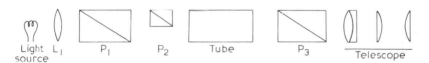

Light source L_1 P_1 P_2 Tube P_3 Telescope

Fig. 13. Polarimeter optical system

P_3, can be rotated about the axis of the instrument so that the principal planes of P_1 and P_3 are mutually at right angles; as a result no light is transmitted by P_3 and the field of view in the telescope is dark—P_1 and P_3 are then said to be crossed. If now an optically active sample is placed in the tube, light will once more emerge from P_3. Extinction can be restored by rotating P_3 through an angle, measured on a calibrated circular scale, which will be equal in magnitude and sign to the rotation of the sample. It is difficult to determine precisely at which angular position of P_3 the illumination of the field is at a minimum because the transmitted intensity change per unit angular rotation is very small near the extinction point. A third prism, P_2 (or a half wave plate) is therefore inserted to cover half the field. This causes two plane-polarised rays to pass through the tube, differing in phase by half a wavelength.

If the analysing prism is crossed with the unobstructed part of the field, the part is darkened, as in A, Fig. 14, but some light is present in the obstructed half. Rotation of the analyser in the correct direction gives a field as in B, in which both beams are almost completely extinguished, and both areas of the field are of equal intensity. Further rotation gives C, in which light from the

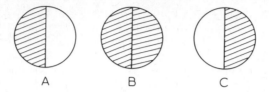

Fig. 14. Fields of view of polarimeter

obstructed half of the field is eliminated, but some rays from the unobstructed half pass through. This device of obstructing half the field gives a means of finding the balance point (B), by judging when the illumination of the two areas is of equal intensity.

(b) *Photoelectric spectropolarimeters.* The light source is usually a xenon arc lamp which emits continuously between 190 nm and 700 nm. The light passes through a monochromator and the monochromatic beam then passes through the polariser, sample cell and analyser as in the visual polarimeter.

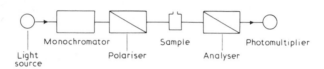

Fig. 15. Diagram of the optical components of a spectropolarimeter

Two methods have been used to determine the position of the analyser which gives extinction, both employing a photomultiplier to detect the transmitted light. In the simpler of these the polariser is caused mechanically to oscillate through a small angle (say 2°). The plane of polarisation of the beam passing through the sample is thus modulated through the same angle. To position the analyser at its extinction angle it is rotated until the intensity of the light reaching the photomultiplier (and hence the photocurrent) is the same for both extreme positions of the polariser.

In the second class of instruments the modulation of the plane of polarisation is achieved not mechanically but electrically by interposing in the lightpath between the polariser and the analyser a Faraday cell. This consists of a silica cylinder surrounded by a coil through which is passed an alternating current (typically 60 Hz). The magnetic field of the coil induces in the silica optical rotation whose sign depends on the direction of the current through the coil. The plane of polarisation of the beam which leaves the Faraday cell is thus modulated. When the analyser is not in its extinction position this causes an asymmetric a.c. current in the photomultiplier which is detected and used to drive the analyser to its extinction position. ORD spectra may be recorded automatically.

(c) *Circular Dichroism.* One of several methods for measuring circular dichroism is shown diagrammatically in Fig. 16. A beam of plane-polarised light is passed through an electro-optic modulator (EOM) or Pockels cell.

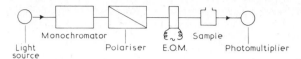

Fig. 16. Diagram of the optical components of a circular dichrometer

This is a 0°-Z-cut plate of potassium dideuterium phosphate oriented so that the incident light is parallel to either its X or Y crystallographic axes. When an alternating electric field (typically 325 Hz) is applied to transparent conducting films on the opposing Z-surfaces of the plate it transmits alternately the left and the right circularly polarised components of the plane-polarised beam. When these are passed through a circularly dichroic sample the photo-current from the photomultiplier contains a steady component proportional to the average transmittance of the sample and a component, alternating with the frequency of EOM voltage, proportional to the difference in sample transmittance for left and right circularly polarised light. The electrical system processes the alternating component so as to record the circular dichroism of the sample.

More detailed descriptions of ORD and CD instrumentation are given in several reviews (Tinoco, 1970; Chignell and Chignell, 1972).

Experiment 11. *Determination of concentration dependence of specific rotation*

Method. Prepare accurately 25 ml each of solutions of 1, 2, 3, 4, and 5% w/v quinidine sulphate in *0.1N* H_2SO_4. Check the polarimeter to see that the field is evenly illuminated (otherwise adjust the light source), and that the telescope is focussed on the dividing line of the field of view. Determine the zero of the instrument as follows.

Clean the polarimeter tube, A (Fig. 17) and screw on one end plate; set the thread gently finger tight otherwise some polarisation may arise due to strain in the glass. Holding the capped end downwards, fill the tube with *0.1N* H_2SO_4, and screw up the second cap, Dry the outside of the tube, and clean the end plates with lens tissue. Set the tube in the instrument, so that any air bubbles lie in the enlargement at the side of the tube. [Some

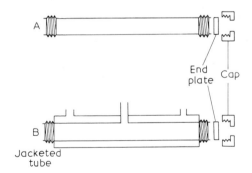

Fig. 17. Two types of Polarimeter tubes

tubes have no accommodation for bubbles. Fill these with one cap in place until a meniscus is built up at the open end of the tube. Slide the glass end plate artfully through the meniscus and screw on the cap.]

Rotate the analyser until the two areas of the field are illuminated with equal intensity, and read both verniers. Take three readings of the balance point approaching from one side, and three approaching from the other. Rest the eyes now and again, otherwise some sensitivity is lost. Calculate the mean readings, i.e. one mean of approximately 0°, and one of approximately 180°.

Repeat this procedure with each of the five solutions of quinidine sulphate in $0.1N$ H_2SO_4 in turn. Note also the sign of the rotation.

Treatment of results. 1. Correct the observed rotations of the solutions for the rotation of the solvent, and calculate $[\alpha]_D$ for each solution.

2. Plot $[\alpha]_D$ against concentration, and note the shape of the graph. Use the general equation (26) to obtain a straight line graph, and find the value of $[\alpha]_D$ at zero concentration.

Experiment 12. *The optical activity of Levodopa*

The specific rotation of Levodopa in the visible region is low ($[\alpha]_D^{20} = -12°$ in N HCl). Assay of optical purity is facilitated by formation of a complex with aluminium sulphate which has enhanced optical rotation; this is the basis of the BP assay.

Method. Prepare a 2.5% (w/v) solution of Levodopa in N HCl and also a 1.0% (w/v) solution of Levodopa-aluminium sulphate complex: to 10 ml of a 5% (w/v) solution of Levodopa in $0.5N$ HCl add 10 ml of a 21.5% (w/v) solution of aluminium sulphate previously filtered and 20 ml of a 21.8% (w/v) solution of sodium acetate and make up to 50 ml with water. Using a photoelectric polarimeter, determine the ORD spectra of both solutions between 600 nm and 350 nm paying particular attention to values at 589 nm and 365 nm.

Comment on the relative magnitudes of the spectra in relation to assay of optical purity. Does your material meet the BP specification?

Dilute the 2.5% solution of Levodopa to 0.01% with N HCl and the 1% solution of the complex to 0.005% with $0.1N$ HCl containing 8.7% (w/v) sodium acetate. Record the absorption, ORD and CD spectra of the two dilutions between 350 nm and 220 nm.

Comment on the relationship between the three spectra for each solution. What advantages might be gained by utilising observations in this spectral region for the assay of optical purity of Levodopa?

Experiment 13. *Polarimetric study of the transformation of the gelatin helix in solution*

In many macromolecules optical activity stems, not only from asymmetric carbon atoms but also from coiling the polypeptide chain into a helix, which may be right or left handed and optically asymmetric. The helix is generally stabilised by hydrogen bonding. On heating, increased thermal agitations can break down the hydrogen bonds, giving a change in optical rotation as the helical form is destroyed.

Method. Prepare a 5% w/v solution of gelatin in water. Cool the solution, and pour it into a jacketed polarimeter tube, B (Fig. 17). Leave the tube in a refrigerator overnight.

Set up a thermostat bath with an external circulating device, e.g. 'Circotherm', adjacent to the polarimeter. Fill the thermostat bath with cold tap water and also sufficient ice to make the temperature 10°C. Place the jacketed tube in the polarimeter, and circulate the cold water through it. When temperature equilibrium has been reached (after about 20 min) take readings of the rotation as described in Experiment 11. Raise the temperature of the bath by five degree intervals up to 30°, and measure the rotation at each temperature. Take a final reading at 40°. The zero of the tube should strictly be measured at each temperature, but it is sufficiently accurate to determine a single zero reading at 25°, and use this for correcting the other measurements.

Treatment of results. Calculate the specific rotation at each temperature, and plot a graph of $[\alpha]_D$ against temperature. The curve should show a decrease in specific rotation as temperature rises. Comment on the changes in slope.

With a spectropolarimeter this experiment can be done with more dilute gelatin solutions since the specific rotation of gelatin increases markedly as the wavelength is reduced. Thus $[\alpha]_{313}$ and $[\alpha]_{208}$ are respectively about $7\times$ and $100\times[\alpha]_D$. Whereas a 5% solution is required to give reasonable results, using a 0.5 dm tube in a visual polarimeter, as described above, a 0.1% solution is adequate with a 1 cm cell in a spectropolarimeter at 313 nm.

ORD and CD are used extensively in studying the conformation of protein molecules in solution.

Further Applications

The recent commercial availability of precise circular dichrometers has made it possible to exploit the higher resolving power of circular dichroism in the assay of optical purity of some compounds whose structure is such that there is little difference between the ORD spectra of their enantiomers. One example is the determination of the proportions of the two diastereoisomers in Phenethicillin Potassium (Stenlake, Wood, Mital and Stewart, 1972). The strong association of some small optically inactive organic molecules with dissymmetric macromolecules has been shown to induce optical activity in the small molecule. Such effects have proved useful in studying interactions of drugs with nucleic acids (Perrin and Hart 1970) and plasma proteins, particularly albumin (Chignell and Chignell, 1972; Perrin and Hart, 1970; Wood and Stewart, 1971; Rosen, 1970).

VISCOSITY

The internal friction of liquids, due to intermolecular attractions, is known as viscosity. In a flowing liquid each layer of molecules exerts a drag on the next, and to cause the liquid to flow, work must be done to push the layers past one another. The lines in Fig. 18 represent the moving layers, in contact

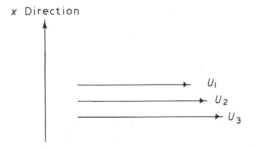

x Direction

U_1
U_2
U_3

Fig. 18. Velocity gradient in fluid flow. (Horizontal arrows represent velocities of layers of liquid)

with one another over an area A; the velocity gradient is du/dx, x being chosen at right angles to the direction of flow. Newton showed that the applied force, F, was proportional to A and to du/dx, the proportionality constant being the coefficient of viscosity, η; hence,

$$F = \eta \frac{du}{dx} A \tag{33}$$

Liquids which obey this equation when flowing are said to give streamlined or Newtonian flow. Very rapid flow can cause Newtonian flow to become turbulent flow, in which the energy causing the liquid to flow is no longer used exclusively to slide the planes of molecules over one another, but part is dissipated as eddies and turbulence.

Poiseuille's Law. *The capillary viscometer*
When a liquid flows in a tube (Fig. 19) the layer adjacent to the wall is stationary,

Fig. 19. Flow of liquid in a tube. (Length of horizontal lines represent velocities of flow)

whilst that in the centre flows fastest. The flow can be considered to consist of concentric rings of molecules moving over one another. The velocity of flow, v, at a distance, r from the centre of the tube of radius, a is given by

$$v = \frac{P}{4\eta l}(a^2 - r^2)$$

in which P is the pressure difference maintained between the ends of the tube, of length, l. Integration of this equation to find the total volume, V, flowing in unit time through the tube gives Poiseuille's law

$$V = \frac{Pa^4}{8\eta l} \tag{34}$$

For a liquid flowing under its own pressure head, the pressure may be expressed in terms of density, ρ, acceleration due to gravity, g, and height; remembering V was defined per unit time, we can rewrite eq. (34) as:

$$\eta = k\rho t \tag{35}$$

This equation applies to the capillary viscometer, which is shown in Fig. 20. The simple Ostwald viscometer has the form of a U-tube, with bulbs at D and E, a capillary tube C–C' and marks at A, B, and F. Liquid is forced up bulb D to above mark A, and the time taken for the meniscus

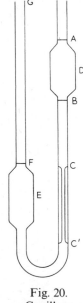

Fig. 20. Capillary viscometer

to fall from A to B noted. If a liquid (1) of known viscosity is compared with another (2) of unknown viscosity, in the same apparatus under identical experimental conditions, then:

$$\text{Known liquid:} \quad \eta_1 = k\rho_1 t_1$$

$$\text{Unknown liquid:} \quad \eta_2 = k\rho_2 t_2$$

$$\therefore \quad \eta_2 = \frac{\eta_1 \rho_2 t_2}{\rho_1 t_1} \tag{36}$$

The constant (k) for a particular viscometer need not be determined, as it cancels (eq. (36)). To obtain readily measured flow times, viscometers with narrow capillary tubes are used for liquids of low viscosity, while wider capillary tubes are used in viscometers for high viscosity liquids.

Units of Viscosity

The SI unit of viscosity (dynamic viscosity) is $1 \text{ m}^{-1} \text{ kg s}^{-1}$ but dynamic viscosity is often expressed in cgs units, i.e. 1 poise = $1 \text{ g cm}^{-1} \text{ sec}^{-1} = 10^{-1} \text{ m}^{-1} \text{ kg s}^{-1}$. For water at 20° $\eta = 0.01002$ poise = 1.002 centipose. Kinematic viscosity, v, is dynamic viscosity divided by density. The SI unit is $\text{m}^2 \text{ s}^{-1}$ but again it is frequently expressed in cgs units, i.e., 1 stoke = $1 \text{ cm}^2 \text{ s}^{-1} = 10^{-4} \text{ m}^2 \text{ s}^{-1}$. For water at 20° $v = 0.01004$ stokes = 1.004 centistokes. Another term sometimes used is the fluidity, which is the reciprocal of dynamic viscosity.

Other viscosity functions which are used, particularly in studying macromolecules in solution are:

$$\text{relative viscosity} = \eta_{rel} = \frac{\eta}{\eta_0} = \frac{t\rho}{t_0 \rho_0}$$
(viscosity ratio)

$$\simeq \frac{t}{t_0} \quad \text{(dilute solutions)}$$

$$\text{specific viscosity} = \eta_{sp.} = \eta_{rel}^{-1}$$

$$\text{reduced specific viscosity} = \frac{\eta_{sp}}{c}$$
(viscosity number)

$$\text{intrinsic viscosity} = [\eta] = \lim_{c \to 0} \left(\frac{\eta_{sp}}{c} \right) \tag{37}$$

where η and ρ are the viscosity and density of a solution of concentration $c \text{ g cm}^{-3}$ and η_0 and ρ_0 are the viscosity and density of the solvent. $[\eta]$ is a function of the size and shape of macromolecules in solution.

Viscosity and Temperature

Viscosity decreases by about 2% per degree rise in temperature. Strict control of temperature during measurements is essential. The relationship between viscosity and temperature is expressed by an equation of the form:

$$\eta = A \, e^{B/RT} \tag{38}$$

A and B being constants for a particular liquid. A plot of $\log \eta$ against $1/T$ should therefore give a straight line.

The Couette Viscometer

With non-Newtonian fluids, whose viscosity depends on shear rate, capillary viscometers can give misleading results. In the Couette viscometer the liquid is contained in the annular space between two coaxial cylinders (Fig. 21) and the rotational force on the inner cylinder is measured when the outer one is rotated at known angular velocity, ω. By making measurements at different values of ω the viscosity of the liquid may be determined at different shear rates.

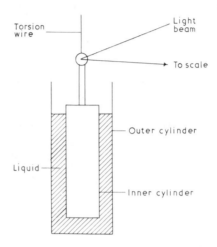

Fig. 21. Scheme of Couette viscometer

The Falling Ball Viscometer

By allowing a solid sphere to fall through a liquid, and measuring its limiting velocity, v, the gravitational force pulling the sphere down is equal to $mg = \frac{4}{3}\pi r^3 (\rho_2 - \rho_1)g$, where r is the radius of the sphere and $(\rho_2 - \rho_1)$ the density difference between sphere and liquid. The gravitational force is balanced by the viscous drag, which, from Stokes' Law, is $6\pi \eta r v$. Equating and solving for η, gives the expression:

$$\eta = \frac{2(\rho_2 - \rho_1)r^2 g}{9v} \tag{39}$$

Thus a measurement of the limiting velocity can be used to determine the viscosity of the liquid, η.

Provided that the velocity of fall is not too rapid, the main theoretical difficulty is to account for wall effects. In narrow tubes additional resistance effects are present. and v is smaller than expected. This makes the apparent viscosity calculated from eq. (39) too high. A number of equations are available for correction for wall effects; they generally involve terms in $2r/D$ where D is the diameter of the tube. Such a correction is given in the Pharmacopoeial formula for the falling ball viscometer which is derived from eq. (39) using the diameter of the sphere in place of the radius, and expressing the result as the kinematic viscosity (see p. 31) in centistokes.

$$v = \frac{d^2(\rho_2 - \rho_1)g \times 0.867}{0.18v\rho_1}$$

Experiment 14. *Calibration and use of capillary viscometers*

(a) *Determination of the intrinsic viscosity of dextran in Dextran Injection BP*

Method. Clean and dry a size A Ostwald viscometer (capillary bore $\simeq 0.5$ mm). Prepare dilutions containing 2.0, 1.0, 0.5 and 0.25% w/v of dextrans from the Dextran Injection provided. Using a plumb line, set the viscometer upright in a thermostat set at 37°. Fill the viscometer through tube G with saline solution to mark F (Fig. 20). When the apparatus has come to temperature, adjust the liquid level exactly to this mark. Slip a short length of clean rubber tubing on to A and apply suction until the liquid rises above mark at A. Place the finger on the end of this tube to prevent the liquid level falling while removing the rubber tubing. Find the time taken for the liquid level to fall between marks A and B. Repeat until results agreeing to 0.1 second are obtained. Remove the viscometer from the bath, wash it out, dry it and repeat the procedure using each of the dextran solutions in turn. Calculate the specific viscosity of each solution and by plotting η_{sp}/c vs c (eq. (37)) determine the intrinsic viscosity of the dextran. Compare it with the BP requirement.

(b) *Determination of viscosity of Methylcellulose*

Method. Prepare a 2% solution of Methylcellulose in water by weighing 2·0 g into a 250 ml conical flask (containing a magnetic follower) fitted with a ground glass stopper, adding 100 ml water heated to 85–90°, inserting the stopper, and stir for 10 min. Place in ice and continue stirring until the solution is of uniform consistency. Allow the solution to attain room temperature.

The suspended level viscometer (Fig. 22) is used in this experiment. Introduce sufficient liquid through tube A to ensure that the bulb D will be full. Set the viscometer on the thermostat at 20°. After temperature equilibrium has been attained, fit a clean piece of rubber tubing on to B. Close the tube C with the finger and suck liquid up above mark E. Close B by 'nipping' the rubber tubing, open C, quickly detach the tubing, and time the fall of the meniscus between marks E and F. This procedure leaves a hanging column of liquid in BG. This type of capillary viscometer has the advantages that precise filling is not required, errors in setting the viscometer exactly upright are minimised, and dilutions may be made within the viscometer.

Determine reproducible flow times for the methylcellulose solution and for a solution of glycerol of known concentration (between 90–95% w/w). If necessary determine the density at 20° of the glycerol solution used for calibration.

Treatment of results. Find the viscosity of the glycerol solution used from the data in Table 6.

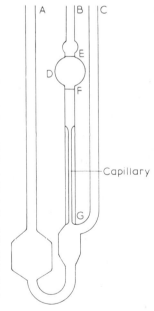

Fig. 22. Suspended level viscometer

Table 6

% glycerol w/w	90	91	92	93	94	95
Viscosity (cp, 20°)	234.6	278.4	328.4	387.7	457.7	545
ρ_{20}, g/ml	1.2347	1.2374	1.2401	1.2428	1.2455	1.2482

(A convenient means of interpolation is from a graph of $\log \eta$ against ρ_{20}.) Calculate both the absolute and kinematic viscosities of the methylcellulose solution, using $\rho_{20} = 1.005$ g/ml for methyl cellulose solution; compare the answer with the BP limits.

Experiment 15. *The falling ball viscometer—wall effects*

A convenient form of the falling ball viscometer is shown in Fig. 23, having a tube graduated at lengths of 0, 25, 100, 175, 200, and 220 mm.

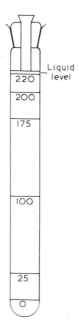

Liquid level

220
200
175
100
25
0

Fig. 23. Falling ball viscometer

Method. Fill the tube with glycerin (>99.5 % w/w). Place the tube in a thermostat at 20°, and allow 20–30 min for temperature equilibrium. Use $\frac{1}{16}$ in. (1.59 mm) diameter steel spheres, pre-wetted with the liquid whose viscosity is to be determined, also at 20°. Introduce a sphere into the funnel at the top of the apparatus, and time its fall between the 200 mm and 0 mm marks. Repeat until consistent results are obtained. As the limiting velocity, v, is used in the equation for determining viscosity by this method, it should be ascertained that this velocity has been obtained, by timing spheres between the 200 mm and 100 mm marks, then between the 100 mm and 0 mm marks. Repeat the experiment with $\frac{1}{8}$ in. diameter (3.176 mm) spheres. Measure the internal diameter of the fall tube; for the BP apparatus this is 2.5 cm.

Repeat the whole experiment using a large measuring cylinder (500 ml, internal diameter 4–5 cm) fitted with a large rubber bung and a funnel. Calibrate the cylinder to allow a 20 cm fall to be used in the first part of the experiment.

The falling ball method is used to determine the viscosity of a pyroxylin solution, prepared as BP using 1.59 mm diameter balls in the apparatus used in the first part of the experiment, and timing the fall between 175 mm and 25 mm marks. The viscosity in centistokes is calculated from the basic formula (39) with the substitution of diameter for radius, and the inclusion of a correction factor for wall effects.

Treatment of results. 1. Calculate the apparent viscosities of the glycerin from eq. (39) for the four experiments (two sizes of sphere in two cylinders). Note that the largest ball in the smallest cylinder gives the highest viscosity (Take $\rho_2 - \rho_1 = 6.49$ g/ml).

2. Plot the apparent viscosity against the ratio $2r/D$ and extrapolate to $2r/D = 0$. This is a hypothetical case where the diameter of the cylinder is infinite and wall effects are negligible. Hence the correct viscosity should be obtained from the intercept on the viscosity axis. The apparent viscosity may be multiplied by the correction factor F,

$$F = 1 - 2.104 d/D + 2.09 d^3/D^3$$

in which d is the diameter of the sphere, and D that of the cylinder. Calculate F for each of the four cases mentioned above, and calculate the viscosity in each case.

SURFACE TENSION

In the bulk of a liquid, a molecule is, on average, subjected to attractions in all directions. At a liquid surface, a molecule is not completely surrounded by others, and is subject to a resultant attraction acting inwards towards the bulk of the liquid. Because of this inward force, the surface has a tendency to contract. To increase the area of the surface, work must be done against the inwardly directed attractive forces.

Surface energy is the work required for unit increase in surface area.

Surface tension (γ) is the force acting at right angles to a line of unit length, present in the surface. As work is force × distance, extension of surface area by 1 cm² can be considered as moving a line of length 1 cm through a distance of 1 cm (Fig. 24). The surface energy therefore has the same numerical value as the surface tension.

Fig. 24. Diagrammatic representation of extension of surface area by 1 cm²

Following similar arguments, the work necessary to extend the interface between two immiscible liquids by unit amount is the interfacial energy, and the interfacial tension is the force acting at this interface. Surface and interfacial tension are important in pharmacy in studies of surfactants, the properties of emulsions, and to provide information on liquid surfaces.

The SI unit of surface tension is $N\,m^{-1}$.

$$1\ mN\,m^{-1} = 1\ dyne\,cm^{-1}$$

Effect of Temperature

In general both surface and interfacial tension decrease with a rise in temperature. The surface tension of water at various temperatures is given in Table 7 (Gross, Young, and Harkins).

Table 7. Surface Tension of Water at Various Temperatures

$t°C$	0.20	4.99	10.02	20.00	40.20	60.00
γ (mN m^{-1})	75.66	74.96	74.26	72.79	69.57	66.23

The surface tension decreases by approximately $0.15\ mN\,m^{-1}\,deg^{-1}$.

Effect of Solute

While it is difficult to generalise on the effect of solutes on surface or interfacial tension, two classes of substances give specific behaviour. Inorganic salts, like sodium chloride or calcium chloride, in fairly high concentrations, raise the surface tension of water. This is probably due to the surface layer being of pure water, and there being a lack of solute in it. Gibbs' equation relates the change of surface tension, $d\gamma$, to the surface excess of solute, Γ, and change in chemical potential, $d\mu_i$, so for a solute species i

$$-d\gamma = \Gamma_i\,d\mu_i$$

If the surface tension increases with concentration, Γ_i is negative, i.e. the solute has a smaller concentration in the surface than in the bulk of the solution.

The second important class of substances is the surfactants. These are strongly adsorbed at surfaces and decrease the surface tension. For a dilute solution of a non-electrolyte surfactant, giving a fairly high adsorption at the surface, Gibbs' equation reduces to

$$-\mathrm{d}\gamma = \Gamma_2\,2.303RT\,\mathrm{d}\log C_2 \tag{40}$$

where Γ_2 and C_2 are the surface excess and bulk concentrations of solute respectively, concentrations having been substituted for activities. This form of equation is also applicable to ionised surfactants in salt solutions, but for ionised surfactants in pure water the additional factor 2 is required on the right hand side.

The progressive addition of a surfactant to water lowers the surface tension (A to B, Fig. 25), the solute distributing itself between the bulk and surface of

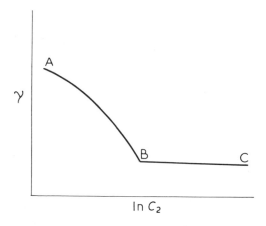

Fig. 25. Change in surface tension by addition of surfactant to water

the solution. At B, the solute molecules in the bulk begin to aggregate to form micelles, and the addition of further solute (B to C) has little effect on the surface tension. The surface excess, Γ_2, may be determined from the slope of the γ vs $\log C_2$ curve in the A–B region; the concentration at which micelles first form (critical micelle concentration = CMC) can be found from the break in the curve at point B.

Application of Gibbs' equation is one of the most important methods of studying adsorption of soluble solutes at liquid surfaces.

Dynamic and Static Surface Tension

Often, when measurements are made of the surface tension of a surfactant solution, the values obtained change with time, owing to slow adsorption at the surface or the presence of impurities. These values are dynamic surface tensions, while that obtained after full equilibration of surface and bulk phases, which no longer changes with time, is the static value.

In pharmacy, there is more interest in the surface tension of solutions than that of pure liquids. Hence a method like capillary rise, which is excellent for pure liquids but not for solutions, is not considered. The Wilhelmy plate method, and to a lesser extent the drop weight method, can be used for dynamic as well as static measurements.

Experiment 16. *Surface tension of Water by the Wilhelmy Plate Method*

When a thin platinum plate dips vertically into the surface of a liquid, the surface tension pulls the plate downwards, acting all round its perimeter. If l is the perimeter of the plate, the downward pull due to surface tension $= l\gamma$. To counteract this downward pull an upward force of mg dynes must be applied

$$l\gamma = mg$$

or

$$\gamma = mg/l \tag{41}$$

To avoid having to apply buoyancy corrections for the volume of plate submerged in the liquid, a technique is used which ensures that the bottom of the plate just touches the surface. The apparatus is shown in Fig. 26.

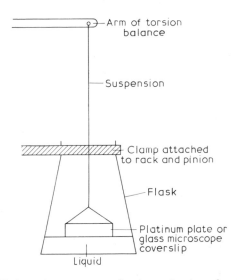

Fig. 26. Wilhelmy plate apparatus for determination of surface tension

Method. Clean the depolished platinum plate by dipping it in chromic acid for 2–3 min, and washing it in six changes of water* in a beaker. Dry the plate with filter paper. Do not touch it with the fingers; it can be lifted by means of the nylon suspension. Attach the suspension to the arm of the (0–500 mg) torsion balance, unclamp the beam, set the pointer to a reading of zero mg, and adjust the zero control until the indicating pointer comes to the zero mark. Clamp the beam, ensuring that it remains at the zero position.

* Water must be freshly redistilled from permanganate solution, and stored in a well seasoned glass container.

Clean a 250 ml Erlenmayer flask with chromic acid, wash well with water, and place about 50 ml water in the flask. Clamp the flask to the rack and pinion device, and lift it until the plate hangs about 0.5 cm above the surface of the water. Rack up the flask very slowly. This is a convenient time to ensure that the bottom of the plate is parallel to the surface, by observing the degree of parallelism of the bottom of the plate and its image in the surface. If it is not parallel, remove it from the flask, gently bend the wire part of the suspension. Continue racking up the flask until the plate just touches the surface. Unclamp the balance arm.

Rotate the handle of the balance, until the indicating pointer is exactly on the zero mark. Note the reading in mg. Turn the handle back to zero mg, so that the plate sinks in the water, and obtain two agreeing readings for the pull. By this method the plate is always raised in the surface to take a reading, which ensures that the contact angle between water and plate is correct (0°). At the end of the experiment measure the temperature of the water.

Treatment of results. Calculate* the surface tension of water from eq. (41).

Experiment 17. *Determination of area/molecule and CMC from surface tension measurements*

Fill a burette with a 5% solution of the nonionic detergent polysorbate 80 ('Tween' 80) in *0.1N* KCl. Prepare 50 ml of the following dilutions: 2%, 0.1%, 0.05%, 0.005%, 0.0005%, and 0.0001%, making the flasks up to volume with *0.1N* KCl solution.

Clean the plate and flask as in Experiment 16. Rinse out the flask with two 5 ml portions of the most dilute solution, place the remainder in the flask, and measure its surface tension. With some samples of sodium caprylate, changes of surface tension with time may be noted. Continue to take readings until a constant value is obtained within 2 mg. Clean the flask and plate as before, and measure the surface tension of each solution.

Treatment of results. 1. Calculate the equilibrium surface tension of each solution from eq. (41), and plot γ against log (per cent concentration).

2. Determine the CMC.

3. Determine $d\gamma/d \log C$ from the slope of the graph in the concentration region just below the CMC. Taking $R = 8.31$ J mol^{-1} deg^{-1}, calculate Γ_2 from eq. (40) (moles/cm^2). Calculate the area (Å^2) per molecule.

Experiment 18. *Determination of the surface tension of water by the drop weight method*

A hanging drop of the liquid is slowly formed on a circular glass tip. The force causing the drop to adhere to the tip is:

$$\text{circumference} \times \text{surface tension} = 2\pi r\gamma$$

where r is the radius of the tip. The force causing the drop to fall is *mg*, where *m* is the weight of the drop. At the instant when the drop falls:

$$2\pi r\gamma = mg$$

$$\gamma = \frac{mg}{2\pi r}$$

This simple theory is complicated by the fact that, during detachment, some of the drop remains on the tip, so the simple formula is modified by the intro-

---

* If the perimeter of the plate is not given, it must be measured with a cathetometer.

duction of a correction factor, F.

$$\gamma = \frac{mgF}{r} \tag{42}$$

(2π has been taken into the correction factor). F was originally determined by measuring the drop weight of liquids whose surface tensions were known, falling from tips of known radius. F is found by calculating the volume of the drop, V, from its weight, working out V/r^3, and reading off F from Table 8.

Table 8. Correction Factors for the Drop Weight Method (Harkins and Brown)

V/r^3	4.65	3.98	3.43	3.00	2.64	2.34
F	0.254	0.257	0.259	0.261	0.262	0.264
V/r^3	2.09	1.88	1.55	1.31	1.21	1.12
F	0.265	0.265	0.266	0.265	0.264	0.263

Method. The apparatus is shown in Fig. 27. The polythene tubing connecting the reservoir and tip must be well extracted with water. The end of the tip must be ground flat, and its radius determined with a cathetometer.

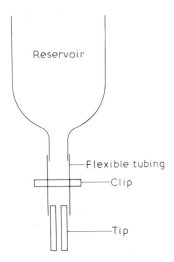

Reservoir

Flexible tubing

Clip

Tip

Fig. 27. Drop weight apparatus for determination of surface tension

Immerse the tip in chromic acid for 2–3 min. Wash down the outside of the tip, and run about 50 ml water through the apparatus. Close the clip, and put about 20 ml water into the reservoir. Adjust the clip until drops are falling at about 2 min intervals. Collect three drops in a tared weighing bottle, and reweigh. Repeat for a further three drops. Avoid vibrations of the apparatus while drops are being collected. At the end of the experiment note the temperature of the water in the reservoir.

Treatment of results. Using the density data in Table 8, and the mean drop weight of each set collected, calculate V/r^3. Calculate the surface tension from eq. (42).

METRICATION IN SCIENTIFIC LITERATURE

The Royal Society Conference of Editors of Scientific Journals has made the following recommendations to promote the general adoption of the metric system in the United Kingdom:

1. That the system of units known as SI should be adopted in all scientific and technical journals.

2. That, in order to keep to a minimum the difficulties that will inevitably arise during the period of transition, the change-over should be effected as quickly as possible.

SI (which is the abbreviation in many languages for Système International d'Unités) is an extension and refinement of the traditional metric system.

A number of these recommendations are now in process of being adopted and two, nm (nanometre) and Hz (hertz), are used throughout this text as they are already in general use. It is likely that other SI units will become more widely adopted during the life of this edition. Students who are interested are referred to the pamphlet on this subject entitled 'Quantities, Units and Symbols' which was published by the Royal Society in 1971.

References

Chignell, C. F. and Chignell, D. A. in *Methods in Pharmacology*, ed. C. F. Chignell, Appleton-Century-Crofts, New York, 1972, Vol. 2, p. 111.

Foss, J. G., *J. chem. Ed.* (1963) **40**, 592.

Perrin, J. H. and Hart, P. A., *J. Pharm. Sci.* (1970) **59**, 431.

Rosen, A., *Biochem. Pharmacol.* (1970) **19**, 2075.

Snatzke, G., *ORD and CD in Organic Chemistry*, Heyden, 1967.

Stenlake, J. B., Wood, G. C., Mital, H. C. and Stewart, S., *Analyst* (1972) **97**, 639.

Tinoco, I., *Methods of Biochemical Analysis* (1970) **18**, 81.

Wood, G. C. and Stewart, S., *J. Pharm. Pharmac.* (1971) **23**, 248S.

2 Analysis of Drugs in the Solid State

A. T. FLORENCE

Modern concepts of quality differ considerably from those of a decade or so ago. We are now concerned not only with chemical purity but also with those other characteristics of a drug which may influence its safety or efficacy. Many physico-chemical properties of drugs in their solid state have thus assumed importance. For example, particle size and crystal form of a poorly soluble drug such as digoxin may influence the rate and extent of drug absorption and this may be reflected in differences in clinical response or toxicity.

Gastro-intestinal absorption of a drug from a solid dosage form involves the following steps:

Drug in dosage $\xrightarrow{\text{rate of solution}}$ drug in solution $\xrightarrow{\text{rate of absorption}}$ drug in blood
form   in gastro-intestinal
                       lumen

If the rate of solution is slow as with poorly soluble drugs then this may become the rate-limiting step in the adsorption process. The rate of solution of a solid is directly proportional to exposed surface area. The rate of solution may also be determined by the crystal form and if the drug exists in a number of polymorphic forms—and it has been estimated that some 20% of all organic compounds exhibit polymorphism—then the properties of the crystal have to be properly defined.

A complete profile of the solid state properties of a drug would include determination of melting point, particle size distribution, rate of solution, equilibrium solubility and the investigation of possible polymorphism if the drug substance has been found to have a low aqueous solubility. The parameters determined for each drug will depend on its subsequent formulation or use. For example, the particle size distribution and shape of solids determines powder flow characteristics, and their degree of fineness will influence the physical stability of suspensions. Formulation of a drug in an inhalation aerosol poses several problems of physical chemistry and analysis. Particles for inhalation above 5 μm are not satisfactorily transported to the required destination in the respiratory tract; particles below 1 μm are exhaled.

PARTICLE SIZE ANALYSIS

Pharmacopoeial Standards on Particle Size

An indication of the scope of particle size requirements in official monographs is given in Table 9. This lists instances where the particle size characteristics

Table 9. Particle Size Control of Drugs and Adjuvants in Official Compendia

Substance or Preparation		Remarks
Aspirin	BP	In fine powder form for preparation of Soluble Aspirin Tablets and Soluble Aspirin, Phenacetin and Codeine Tablets
Bentonite	BPC	Consists of particles about 50–150 μm with numerous smaller particles about 1–2 μm
Bephenium Hydroxynaphthoate	BPC	Surface area of not less than 7000 cm^2 g^{-1} determined by air permeability method.
Betamethasone	BPC	Ultra-fine powder* to be used for preparation of solid dosage forms to achieve a satisfactory rate of solution
Cellulose, Microcrystalline	BPC	Colloidal water-dispersible type differentiated from non-dispersible form by size
Cortisone Acetate	BPC	Ultra-fine powder to be used for preparation of solid dosage forms
Cortisone Injection	BP	Maximum diameter of drug 30 μm
Dithranol Ointment	BP	Prepared from dithranol in fine powder form
Ergotamine Aerosol Inhalation	BPC	Most of the individual particles have a diameter not greater than 5 μm; no individual particle has a length greater than 20 μm
Fusidic Acid Mixture	BPC	95% of particles have a maximum diameter of not more than 5 μm
Gold (^{198}Au) Injection	BPC	About 80% of radioactivity is present in particles between 5 and 50 nm in diameter
Griseofulvin Tablets	BP	Particle size determined from distintegrated tablet generally up to 5 μm in maximum dimension although larger particles may occasionally be greater than 30 μm
Hydrocortisone preparations	BP & BPC	All subject to limit on particle size of Hydrocortisone or Hydrocortisone Acetate. See Hydrocortisone Acetate Ointment BP, Hydrocortisone Cream BPC, Hydrocortisone and Neomycin Cream BPC, Hydrocortisone and Neomycin Ear drops and Eye drops BPC, Hydrocortisone Eye Ointment BPC, Hydrocortisone Lotion BPC and Hydrocortisone Suppositories BPC
Insulin preparations	BP	See Insulin Zinc Suspension (Crystalline) BP, Insulin Zinc Suspension (Amorphous) BP, Biphasic Insulin Injection BP, Biphasic Insulin Injection BP
Isoprenaline Inhalation Aerosol		As for Ergotamine Aerosol Inhalation BPC
Kaolin, Light	BP	Size and shape of particles stated
Kaolin, Heavy	BPC	Size and shape of particles stated
Macrisalb (^{131}I) Injection	BP	Sterile suspension of denatured albumin iodinated with ^{131}I: particle size in range 10–100 μm

Table 9 *continued*

Substance or Preparation		Remarks
Magnesium Carbonate, Heavy	BPC	Subspherical particles, 10–20 μm in diameter, usually arranged in clumps of 4 to 20 particles
Magnesium Carbonate, Light	BPC	Small acicular crystals 7 μm long, 1–2 μm thick in clumps of about 10–100 crystals
Magnesium Trisilicate	BPC	Rounded particles and thin flat lamellae, plus smaller particles
† Methisazone Mixture	BP	Weight median diameter of particles not greater than 15 μm
Novobiocin Mixture	BPC	75% of particles less than 10 μm; not less than 99% less than 20 μm
Nystatin Ointment	BPC	No particle of nystatin has a maximum diameter greater than 75 μm
Orciprenaline Aerosol Inhalation	BPC	As for Ergotamine Aerosol Inhalation BPC
Phenolphthalein	BPC	Microcrystalline phenolphthalein** to be used in Liquid Paraffin Emulsion with Phenolphthalein BPC to prevent sedimentation of the phenolphthalein.
Prednisolone	BPC	Ultra-fine particles to be used in solid dosage form to achieve satisfactory solution rate
Prednisolone Pivalate Injection	BP	Particles rarely exceed 20 μm in diameter
Prednisone	BPC	As Prednisolone.
Salbutamol Aerosol Inhalation	BPC	As for Ergotamine Aerosol Inhalation BPC
Sodium Cromoglycate Cartridges	BP	When examined immediately after dispersion in pentan-1-ol by exposure for 20 sec to low intensity ultrasonic waves exhibit 2 types of particles: small rounded particles of maximum diameter 10 μm, together with larger angular particles of length 10–150 μm but mostly within the range 20–80 μm
Sulphur Precipitated	BP & BPC	Grouped amorphous subglobular particles free from crystals (BP); spherules 1.5–11 μm in diameter
Talc purified	BP	Irregularly shaped angular particles, either as flakes about 3–5 μm long or fragments about 14 μm long with jagged and laminated ends.

* The following terms are used *inter alia* in the description of powders in the British Pharmacopoeia and the British Pharmaceutical Codex.
Coarse Powder: a powder of which all the particles pass through a No. 170 sieve and not more than 40% pass through a No. 355 sieve.
Fine Powder: a powder of which all the particles pass through a No. 180 sieve.
Ultra-fine powder: a powder of which the maximum diameter of 90% of the particles is not greater than 5 μm and of which the diameter of none is greater than 50 μm.
   ** BPC 1973, p. 679.   † BP Addendum 1975.

are essential either to the adequate dissolution of the drug (e.g. Betamethasone BPC) or deposition in the respiratory tract (Ergotamine Aerosol Inhalation BPC) or release (Hydrocortisone Suppositories and Ointment BP). Other examples which may be found in this compilation include adjuvants such as bentonite, cellulose and talc.

The papers of Lees (1963), Fincher (1968) and Heywood (1963) deal with pharmaceutical and analytical aspects of powders. A comprehensive review of methods for determining particle size has been published (Analytical Methods Committee Society for Analytical Chemistry, 1963). These should be consulted for further information.

Concepts of Particle Size

There is no difficulty in defining the size of a spherical particle since only one parameter is necessary, the diameter. Powders, however, consist of irregularly shaped particles which may vary from flakes to elongated particles and the problem is how to define the size uniquely by one parameter. The apparent size of a particle depends to a large extent upon the method of measurement and the conventions used in defining particle size. The principles involved in the classification of particles and the corresponding methods of measurement are:

Principle of Similarity	*Method*
Geometrical	Microscopic examination, sieving
Hydrodynamic	Sedimentation, elutriation
Volume	Coulter counter
Surface properties	Gas permeability, adsorption

The results obtained by the above methods will be the same for spherical particles, because of their symmetry. An irregular particle cannot be defined in geometrical terms, although it has a definite volume and surface area, and one cannot assign to it a diameter (or length, breadth and thickness) that enables its volume or surface area to be accurately determined. The 'size' of an irregular particle as measured by the methods listed is represented by the diameter of a sphere that exhibits the same properties as the particle. Thus sieving refers the particles to imaginary spheres which will just pass through the sieve apertures; sedimentation refers the particles to spheres of the same density and having the same terminal velocity in a particular fluid. The particle diameter can thus be defined in the following terms:

Sieve size: the nominal sieve aperture through which the particle just passes. In the BP and BPC sieves were formerly designated by mesh number (number of meshes per inch). Sieves are now designated by aperture size and the numbers indicate the nominal aperture size, measured in mm for aperture sizes of 1 mm or greater and in μm for aperture sizes of less than 1 mm.

Projected area: diameter of a circle of area equal to the projected area of the particle resting in its most stable position.

Stokes': diameter of sphere of equal density having the same settling velocity as the particle in a fluid medium.

An irregular particle will be assigned different sizes by the above methods and the discrepancy between the results will increase as the particle departs further from sphericity. It is possible to combine the 'sizes' from different methods to give non-dimensional parameters known as 'Shape Factors', which serve to characterise the shape of the particle. Shape factors are essentially the ratio of the mean sizes of the powders obtained by two different methods. Thus there are as many shape factors as there are pairs of methods. The shape factors given in Table 10 can be used when it is necessary to combine size distributions obtained by different methods. However, if the particles are of extreme shapes, it is preferable to use factors specifically determined for the powder. A possible method of establishing shape factors is to overlap the methods of size determination so that one or more size classes are assessed by both methods. It is assumed that the factor for an individual powder is independent of particle size.

Table 10. Correlation of Particle Diameters

A. Known particle size	Sieve	Sieve	Projected area	Projected area	Stokes	Stokes
B. Particle size required	Projected area	Stokes	Sieve	Stokes	Sieve	Projected area
Factor to obtain B. Multiply A by:	1.4	0.94	0.71	0.67	1.07	1.5

Size Distribution

Consider a sample of powder examined under a microscope from which 500 particles are sized by comparing them with some suitable scale so that the particles can be grouped into classes of different sizes. The number of particles lying in each interval of 2 μm is shown in Table 11. The particles in this particular example all lie between 2 and 22 μm. The data can be expressed in the form of a histogram (Fig. 28) where the abscissa represents the particle size and the ordinate the frequency per class interval. The histogram does not provide a unique pattern for a given particle size distribution, since its shape varies if the scale of particle size intervals is changed.

Table 11. Example of Size Distribution in a Sample consisting of 500 Particles

Size interval in μm	2–4	4–6	6–8	8–10	10–12	12–14	14–16	16–18	18–20	20–22
Number of particles	25	88	107	110	55	45	30	18	12	10
% in each interval	5	17.6	21.4	22	11	9	6	3.6	2.4	2
% undersize	5	22.6	44	66	77	86	92	95.6	98	100

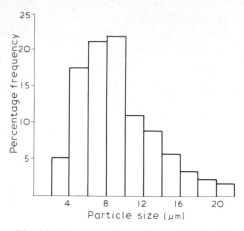

Fig. 28. Histogram (from data in Table 11)

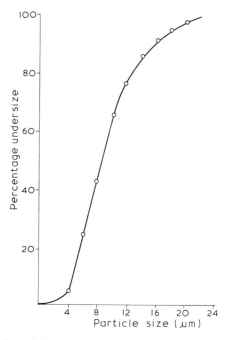

Fig. 29. Cumulative undersize curve (from data in Table 11)

An alternative method is to express the data in the form of a cumulative percentage curve where the class frequencies are calculated as a percentage of the total number of observations. The cumulative percentage which may be 'oversize' or 'undersize' is plotted against the particle size as shown in Fig. 29. The values for undersize percentage appear in Table 10.

The example just discussed was a size distribution based on numbers of particles. In sieving and sedimentation methods the distribution is calculated on a weight basis. If it is necessary to convert from a size distribution by number to one by weight, the conversion is usually made by assuming that all the particles have the same shape factor, i.e. the volume of the particle is proportional to the cube of the measured 'equivalent diameter'.

Mean Size of a Particulate System

The concept of mean size requires some explanation since this may be calculated on the basis of numbers of particles, weight of particles or surface area of particles. This acquires importance in correlation problems.

The simplest average diameter is the arithmetic mean diameter $\bar{d}_n$ which we define for two diameters d_1 and d_2 as,

$$\bar{d}_n = (d_1 + d_2)/2$$

In general terms we define the *numerical mean diameter* by

$$\bar{d}_n = \frac{\Sigma n \cdot d}{\Sigma n} \tag{1}$$

where n is the number of particles with equivalent diameter d. In many pharmaceutical applications the surface area of the particles is important and the *surface mean diameter* can be employed. This is defined as

$$\bar{d}_s = \frac{\Sigma nd^3}{\Sigma nd^2} \tag{2}$$

because the surface area is proportional to nd^2.

The significance of surface mean diameter $\bar{d}_s$ is that the specific surface (surface area per unit weight) can be calculated from it by the following equation:

$$S = \frac{6}{\bar{d}_s \sigma_s} \tag{3}$$

where σ_s is the density of the solid and S is the specific surface ($m^2 \, kg^{-1}$) where $\bar{d}_s$ is measured in m. This equation is only strictly correct for spherical or cubical particles but may be used without correction for a shape factor if the particles are not too asymmetric. If the weight of each fraction, rather than number or surface area, is noted then the *weight mean diameter* is obtained from

$$\bar{d}_w = \frac{\Sigma nd^4}{\Sigma nd^3} \tag{4}$$

If the particle density does not vary with particle size then this quantity is identical to the volume mean diameter. It is important that the terminology of mean diameters used above is strictly adhered to. For example, the weight mean diameter is not the same as the diameter of a particle of mean weight. The latter quantity is termed the *mean weight diameter* and is equal to

$$\sqrt[3]{\left(\frac{\Sigma nd^2}{\Sigma n}\right)} \tag{5}$$

a quantity used when one is concerned with the number of particles per g of material.

From the data in Table 10 the following mean diameters are obtained

$$\bar{d}_n = 9.28\ \mu m$$
$$\bar{d}_s = 12.9\ \mu m$$
$$\bar{d}_w = 14\cdot3\ \mu m$$

Specific surface $186.1\ m^2\ kg^{-1}$ (density $\sigma_s = 2.5 \times 10^3\ kg\ m^{-3}$).

METHODS OF PARTICLE SIZE ANALYSIS

1. Sieving

Sieves are generally used for grading coarser powders although they are capable of separating fractions as fine as 76 μm. Such fine powders, however, may clog the sieve apertures and wet sieving methods may be necessary (e.g. Phenothiazine Dispersible Powder B.Vet.C. 1965). The lower limit of 76 μm for dry sieving may be reduced if special sieves prepared by the electro-deposition of metals are used in conjunction with an air flow through the sieves.

Specifications for test sieves appear in British Standards to which reference should be made for further details. To obtain reproducible particle size analyses the detailed procedure laid down in British Standard 410 (1962) should be carefully adhered to. The powder is passed through a number of sieves of successively smaller mesh size and the weight remaining on each sieve is determined. The method of shaking the sieves is important to obtain rapid sieving and it is usual to use a mechanical shaker that imparts gyratory and vibratory movement to spread the material over the whole mesh. Errors can arise from overloading the sieves and not allowing sufficient time for the passage of particles.

2. Microscope Method

Direct observations of the particles would appear to be very reliable, but unless the technique of measurement is carefully standardised considerable errors may be introduced. The sources of error include sampling, technique of slide preparation and choice of diameter. Under the microscope a particle is seen as a projected area whose dimensions depend on the orientation of the particle on the slide. The diameter that is frequently used is the diameter of a circle whose area is equivalent to the projected area of the particle. The diameter is estimated by using a graticule placed in the microscope eyepiece. British Standard 3406, Part 4 describes in detail a graticule technique. A less tedious method is double-image microscopy where a beam-splitting device between the objective and the eyepiece produces two identical images in the field of view. The images can be moved apart and a scale indicates the amount of displacement. When the images are in edge to edge contact the 'diameter' can be read off from an electrical meter in the Timbrell instrument which is commercially available.

3. Sedimentation Methods

These methods which have been critically reviewed by the Analytical Methods Committee (1968) are based on the free settling of particles individually dispersed in a suitable fluid. A sphere, of diameter d, falling slowly in a fluid reaches a terminal velocity given by Stokes' Law:

$$v = \frac{d^2 g(\sigma - \rho)}{18\eta} \qquad (6)$$

where

v = terminal velocity (m s^{-1})

g = acceleration due to gravity (9.81 m s^{-2})

σ = density of the solid (kg m^{-3})

ρ = density of the dispersion medium (kg m^{-3})

and

η = viscosity of the dispersion medium (poise or 10^{-1} N s m^{-2})

The equation is only applicable to streamline motion of the particles, i.e. Reynolds Number (Re) must not exceed 0.2 calculated from the equation $d\rho v/\eta$ = Re. For large dense particles it may be necessary to increase either the density or the viscosity of the suspending fluid to maintain streamline flow. This limits the maximum size of particle to about 60 μm for a solid of density 3×10^3 kg m^{-3}, settling in water.

To determine the size distribution, the powder is dispersed uniformly through the fluid and at suitable time intervals the concentration of powder at a given depth in the suspension is determined. This can be done by the use of a suitable pipette, e.g. Andreasen Pipette, and from the weight of solid in the sample the percentage of 'undersize' particles can be calculated. Pipette sampling is an incremental method and gives a direct measure of the particle concentration. In cumulative methods, e.g. sedimentation balance, the proportion of material falling out on the pan as a function of time, must be subjected to a differentiation process which can be performed in a tabular method (B.S.3406, Part 2) or graphically in order to calculate the percentage of oversize material. To avoid particle/particle interaction the concentration of solid in the suspension should not exceed 1 % v/v.

4. Conductivity Method (*Coulter counter*)

The instrument is shown diagrammatically in Fig. 30. The change in resistance of a solution of electrolyte caused by the presence of a particle is used to calculate the volume of the particle. The suspension flows through a small aperture having an immersed electrode on either side, with particle concentration such that the particles traverse the aperture substantially one at a time. The passage of each particle displaces electrolyte within the aperture, momentarily changing the resistance between the electrodes and producing a voltage pulse of magnitude proportional to particle volume.

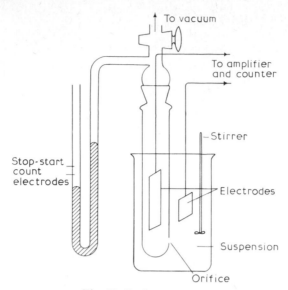

Fig. 30. Coulter counter

When the controlled external vacuum is applied via the stopcock, the mercury column in the manometer is drawn past the 'start-stop-count electrodes'. Closure of the stopcock permits the mercury to rebalance, thus drawing the sample suspension through the aperture and permitting a count to be taken on a known volume of suspension by actuation of the 'start-stop-count electrodes'. The series of pulses so obtained are amplified and fed to a threshold circuit which discriminates against particles below, and permits counting of those which exceed, the threshold value. Successive counts at different threshold levels give the data for determining the cumulative frequency oversize curve. The diameter of the particle is expressed as the cube root of the particle volume.

Orifice tubes are available with diameters ranging from 30 μm to 200 μm. Each aperture is able to measure particles of equivalent diameters between 2% and 40% of its own stipulated diameter. The instrument has to be calibrated by using a reference powder of known particle size as described in the experimental section. The instrument constant will be different for each orifice tube/electrolyte combination for Coulter Model A.

5. Permeability to Gas Flow

The resistance offered by a powdered material to the passage of air flow is a function of the surface area of the powder. The relationship that is assumed to hold between the rate of gas flow and the specific surface is given by the Kozeny equation:

$$V = \frac{\epsilon^3}{(1-\epsilon)^2} \cdot \frac{\Delta P A}{k\eta LS^2\sigma^2} \tag{7}$$

where

$$V = \text{volume of gas flowing in } m^3 s^{-1}$$
$$\epsilon = \text{porosity of the bed}$$
$$\Delta P = \text{pressure drop across bed}$$
$$A = \text{cross-sectional area of bed } (m^2)$$
$$\sigma = \text{true density of the powder kg } m^{-3}$$
$$\eta = \text{viscosity (Poise, } 10^{-1} \text{ N s } m^{-2})$$
$$L = \text{thickness of the bed (m)}$$
$$S = \text{specific surface } (m^2 \text{ kg}^{-1})$$

and

$$k = \text{Kozeny constant} = 5.0 \pm 0.5$$

In deriving this equation it is assumed that a bed formed from granular material is equivalent to a group of similar parallel channels such that the total internal surface of the pore channels is equal to the total external surface of the powder and the total internal volume is equal to the pore space in the powder bed. The mean equivalent radius R of the pore channels can be expressed in terms of the porosity of the powder:

$$R = \frac{2\epsilon}{S\sigma(1-\epsilon)} \tag{8}$$

The substitution of this equivalent radius into a modified form of the Poiseuille equation enables the Kozeny equation to be derived.

Scope of the Methods

Most methods of particle size analysis have limitations in the range of sizes which they can cover. While sieving is capable of assessing powders over a wide range of sizes (16 mm to 10 μm) other methods such as visible light microscopy are more limited. In Table 12 the approximate limits are quoted as a guide.

Sampling Procedures

As with all methods of analysis it is essential that the sample taken is representative of the bulk material under examination. The test samples must be truly representative of the bulk quantity. In the case of aerosol formulations, for example, the sample taken when the can is full may not necessarily be identical to that taken when the can is nearly exhausted.

Sampling procedures to obtain unbiased samples for sizing are described in detail in B.S.3406, Part 1. In the experiments described in this chapter it is likely that the samples taken for analysis will be sufficiently large so that sampling errors should not be significant. However, if very small samples are required as for example in microscopical examination, the method of sampling

Table 12. Scope of Methods of Particle Size Analysis

Methods of analysis	Lower limit (microns)
Sieving:	
B.S. sieves	75
Electroformed holes	20
Elutriation:	
Gravitational: air or water	5–10
Centrifugal: air	5
Sedimentation:	
Gravitational: liquid	2
Centrifugal: liquid	0.05
Microscope:	
Optical (practical limit)	1
Electron	0.01
Coulter counter	0.6–0.8
(30 μm orifice tube)	
Air permeability	0.1–1

is critical and the reader is referred to the precise procedure laid down in B.S.3406, Part 4.

There is also a useful discussion of sampling in Mullin (1972, Chapter 11).

It is also essential that the sampling technique or indeed measuring technique does not alter particle size as may occasionally occur with emulsions on dilution into the saline required for the Coulter Counter.

Some pharmaceutical preparations, e.g. aerosol inhalations, ointments and creams require special methods of sampling (see below).

Practical Experiments

Experiment 1. *Size analysis of calcium carbonate by sedimentation*

Selection of suitable dispersion medium. The following tests can be applied to detect whether the suspension is deflocculated.

(*a*) *Rheological behaviour.* Gradually add the dispersion medium to the powder on an ointment tile, and work the mass with a spatula to a pasty consistency.

A good dispersion flows off the spatula in long syrupy threads and has dilatant properties, i.e. stiffens up on being worked. A flocculated paste does not flow off the spatula even on tapping and the mass appears dull and pasty in contrast to the dilatant paste. This test is useful in making a preliminary selection of dispersing agent.

(*b*) *Microscopical examination.* Place a small drop of dilute suspension on a microscope slide; carefully cover with a cover slip and examine at a suitable magnification.

A flocculated suspension will show clumping of the particles whilst a deflocculated suspension will show the particles individually dispersed.

(c) *Sedimentation volume*. This method can be used to determine quantitatively the optimum amount of dispersing agent. Suspensions are prepared using different amounts of dispersing agent and the sediment volume is measured after the suspensions have been allowed to stand for a suitable time. The minimum sedimentation volume is obtained with the best dispersing agent. Prepare 100 ml quantities of the following solutions in distilled water:

Sodium pyrophosphate 10^{-2}, 10^{-3}, 10^{-4}, 10^{-5} molar
Cetrimide 2%, 0.1%, 0.01%, 0.001%.

Use these solutions to prepare 2% w/v suspensions of calcium carbonate. Transfer the suspensions to 100 ml stoppered graduated cylinders. Shake the suspensions and then allow them to sediment for several hours or overnight. From the appearance of the suspensions determine the optimum concentration of dispersing agent. Confirm your results by applying tests (a) and (b).

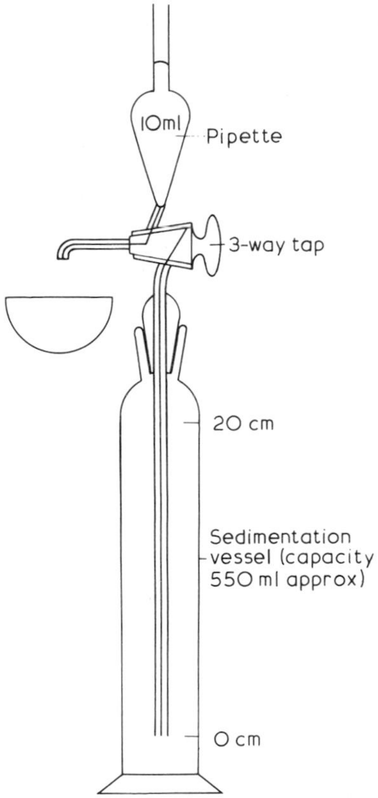

Andreasen pipette method. The pipette is shown in Fig. 31. The tip of the pipette is 20 cm below the zero mark. Prepare 600 ml of sodium pyrophosphate solution of concentration found to be most effective in Experiment 1. Mix calcium carbonate (5 g) with a little of the solution to form a smooth paste in a mortar using a brush to break up agglomerated particles. Make the suspension up to 600 ml with more solution and transfer it to the Andreasen sedimentation vessel up to the 20 cm mark. Reserve the excess suspension for determination of dissolved carbonate. The temperature of the suspension should be kept constant for the whole period of the experiment, and, if the room temperature is likely to vary by more than a few degrees, the sedimentation vessel should be placed in a thermostat bath. Note the temperature of the suspension. Agitate the suspension with a plunger type stirrer, replace the pipette head and start a stopclock. Immediately take the first sample by sucking the suspension into the pipette and then discharge it into a conical flask. Rinse the pipette with a little water and add the rinsings to the flask. Take subsequent samples at 2, 4, 8, 16, 32, 64, 128 and 256 min. Take further samples the following day if necessary. Determine the amount of calcium carbonate in each sample by adding 0.2N HCl (10 ml) to each flask, boil to remove carbon dioxide, and back titrate with 0.1N NaOH (methyl orange as indicator). Owing to the slight water solubility of the carbonate, a blank should be carried out on the aqueous phase after removal of the solid by centrifuging. The excess suspension previously reserved can be used for this.

Fig. 31. Andreasen pipette

The titration figure for the first sample is proportional to the initial powder weight whilst the subsequent samples give the cumulative weight undersize corresponding to the values of the Stokes' diameters.

Calculation of Stokes' diameters. These diameters are calculated from the Stokes' equation most conveniently expressed as follows from eq. (6):

$$d_{st} = 29.17 \left[\frac{h\eta}{(\sigma - \rho)t} \right]^{\frac{1}{2}}$$
(9)

where

d_{st} = Stokes' diameter in μm
η = the viscosity of the dispersion medium (poise) (water = 0.01 poise at 20°C)
σ = the density of the solid (kg m^{-3})
ρ = the density of the dispersion medium (kg m^{-3}) (water = 998 at 20°C)
h = mean depth at which sample is taken (m)
t = time of sedimentation for each sample (s)

Since the level of the suspension falls as samples are taken, a mean depth is used in the above calculation. For instance if the level falls 0.6 cm on removing the first sample, the mean depth of sampling is 19.7 cm. The mean depths for the subsequent samples are 19.1, 18.5 cm and so on.

Determination of density of solid. Introduce about 10 g of calcium carbonate into a 25 ml density bottle. Weight the bottle and contents together with stopper. Add sufficient dispersing fluid so that the bottle is about half filled. Place the bottle without the stopper in a vacuum desiccator and gradually evacuate it. When no further air is released from the powder, add sufficient dispersing fluid to fill the bottle. Replace the stopper, dry and weigh. If a liquid other than water is used its density must also be determined. The density of the powder is calculated from the formula:

$$\text{Density} = \frac{S(W_2 - W_1)}{VS - (W_3 - W_2)}$$
(10)

where

W_1 = weight of dry density bottle
W_2 = weight of density bottle + powder
W_3 = weight of density bottle + suspension
S = specific gravity of dispersion medium
V = volume of density bottle

Results. Plot the percentage of solid in each sample against the corresponding Stokes' diameter to obtain a cumulative curve (undersize), and construct a histogram to show the size frequency distribution. An approximate value for the specific surface of the powder can be obtained from the histogram assuming that the particles are cubical. The surface area contributed by the particles in each block of the histogram is given by $600 \, n/\sigma D$ where n is the percentage by weight of the particles in each block, D is their mean size and σ is the density. The specific surface m^2 kg^{-1} is obtained by the summation of the surface areas so obtained.

Experiment 2. *Determination of size distribution by microscopy*

As an introductory exercise it is suggested that a fractionated sample of powder is examined, e.g. $-200 + 300$ mesh, so that the sizing can be done at one magnification. Where the size range is large it is necessary to count fields of different sizes using appropriate magnifications to ensure that sufficient large particles have been counted in relation to the smaller ones. The 'diameter' of the particles can be measured as mean projected diameters by the use of graticules mounted in the eyepiece of the microscope, so that the particles can be matched up with a circle of known diameter on the graticule. If a projection microscope is available, the image of the graticule, together with image of the particles can be viewed together. Alternatively, the graticule may be omitted from the eyepiece and the projected particles matched up with circles drawn on transparent sheet, which can be moved over the projection screen. Operator fatigue can be reduced by the use of an image splitting eyepiece fitted on to any conventional microscope. This device produces two co-planar images which can be made to move across each other by rotating a micrometer dial (calibrated in μm). The amount of shear is directly proportional to the micrometer reading. Where the microscope field consists of an assortment of differently sized particles, the proportion of those larger than a given diameter can be conveniently estimated by setting the micrometer dial to the appropriate value. The images that overlap (i.e. oversize particles) are easily distinguished from those that are completely sheared apart (i.e. undersize particles).

In the following exercise, a bench microscope fitted with a graticule is used to measure the size distribution of a fractionated sample of lactose or magnesium carbonate mesh. About 600 particles should be counted.

Apparatus. (*a*) *Bench microscope.* This should be provided with (i) coarse and fine focussing; (ii) focussing and centring substage condenser; (iii) an adjustable substage condenser diaphragm. A mechanical stage giving two graduated movements at right angles, each capable of being read to 0.1 mm by scale or micrometer screw, is desirable but not essential. It is convenient to have a graduated draw tube so that the magnification can be varied in order to match up with an exact distance on the stage micrometer (see B.S.3406, Part 4) but this is not essential. Eye piece $\times$ 20 fitted with standard graticule; objective $\times$ 4 or $\times$ 5.

(*b*) *Standard graticule.* The graticule (Fig. 32) conforms to B.S.2625. The relative dimensions of the graticule are such that the reference circles increase in geometrical progression with the constant ratio of $\sqrt{2}$. Two calibration marks are inscribed on the longer axis of the grid to enable easy calibration of the reference circles.

(*c*) *Stage micrometer.* The micrometer, 0.1 mm scale length is subdivided into 100 μm and 10 μm divisions.

Method. (*a*) *Microscope.* Adjust the condenser and light source to obtain even illumination of suitable intensity.

Place the stage micrometer on the microscope stand and align it with the longer axis of the grid on the graticule. Adjust the magnification by altering the tube length until there is an exact correspondence between the distance separating the calibration marks on the graticule and the length on the stage micrometer. For example, if the distance between the calibration marks is 800 μm then the sizes of the reference circles are:

Circle	7	6	5	4	3	2	1
Micrometer	106	75	53	37	27	29	13

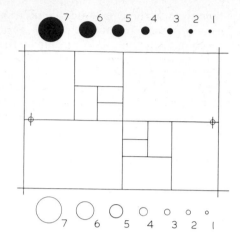

Fig. 32. Graticule (BS.2625)
Distance between calibration marks = 60.4 units
Diameter of circle: 1, 1.00; 2, 1.41; 3, 2.00; 4, 2.83; 5, 4.00; 6, 5.66; 7, 8.00

(b) *Preparation of sample.* Mix a little of the lactose sample with liquid paraffin on a watch glass using a glass rod. When satisfactory mixing has been achieved, a drop of the suspension should be removed with a dropping rod and transferred to a clean microscope slide. Gently lower a cover slip on to the sample. The drop size should be such that no surplus liquid spreads outside the edges of the cover slip.

(c) *Counting procedure.* The graticule is subdivided into four large rectangles. Count the particles within these areas. If the sample on the slide is examined over an area of 20 mm × 20 mm it is possible to examine say 25 fields by moving the slide 4 mm at a time by reference to the scales of the micrometer screws on the mechanical stage. The number of particles within each rectangle should on average not be more than 6.

A particle is recorded as belonging to the field area if it lies wholly within the boundary lines of the field area and also if it lies on either of two adjacent sides or the corner formed by these sides. Particles touching the other two sides or the other three corners do not belong to the field area and are not recorded.

Examine every particle present in each field area individually and compare its area mentally within the areas of the graticule circles. A particle whose area is estimated to be smaller than that of say circle 6 but larger than that of circle 5, is assigned to the class interval defined by the diamters of circles 5 and 6 and so on for other size classes.

(d) *Calculation of size distribution.* The counts for each size class of particle have been observed over the same area of slide so that the percentage of number in each class can be calculated directly (B.S.3406, Part 4 gives the method of calculating if different magnifications have been used for different size classes).

(i) Draw a graph showing the percentage undersize by number as a function of size as shown in Fig. 29.

(ii) Calculate the weight size distribution by converting the numbers of particles to weight, in each size range, by cubing the diameter of the particle d, and multiplying by the number of particles n. $\Sigma n d^3$ will be the relative total weight of the sample, and hence the percentage weight of each fraction will be obtained by $n d^3 / \Sigma n d^3 \times 100$. This calculation assumes that the weight and shape of particles is independent of size.

Alternative Experiments Illustrating Different Methods of Sampling for Microscopy

Experiment 3. *Particle size distribution in Isoprenaline Aerosol Inhalation*

Method. With the oral adaptor in position to mimic conditions of usage, fire 1 spray on to a clean dry microscope slide held about 5 cm from the end of the mouthpiece of the adaptor perpendicular to the direction of the spray. Remove the slide and wash with 2 ml of carbon tetrachloride taking care not to remove the deposit. Allow to dry and examine the residue under the microscope.

The reproducibility of the distribution on repeated sampling can be tested.

Experiment 4. *Particle distribution of hydrocortisone in Hydrocortisone Eye Ointment*

Liquid paraffin is used to dilute this eye preparation. The particle size distribution is determined by microscopy, as in Experiment 2.

Experiment 5. *Particle size of hydrocortisone in Hydrocortisone Suppositories*

Melt one suppository with the aid of gentle heat, dilute the melt in a suitable volume of liquid paraffin and determine the particle size distribution.

Experiment 6. *Determination of the specific surface of calcium carbonate by air permeability measurement*

The Rigden apparatus used for this determination is shown in Fig. 33. The two ends of the cell E containing the powder are connected to the two ends of

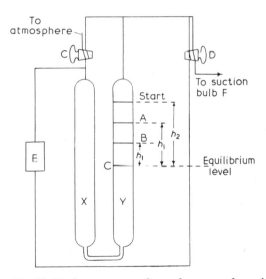

Fig. 33. Rigden apparatus for surface area of powders

a U tube containing an oil of low vapour pressure. The equilibrium level of the oil is at C. The oil is sucked into one arm Y of the manometer by means of bulb F and as it returns to equilibrium level it forces air through the powder bed. The time taken for a given volume of oil to travel between the marks

A and B on the manometer is measured and the specific surface calculated from a modification of the Kozeny equation.

Method. Cut a circle of filter paper to cover the perforated disc at the base of the permeability cell. Determine the weight of the cell plus paper. Insert the plunger into the cell and obtain the zero reading for the depth of the cell by measuring the overall length of the cell + plunger with vernier callipers. Weigh about 5 g calcium carbonate and pack it into the standard one-inch cell in four or five increments. After each increment gently tap the cell on the bench with the plunger inserted and hold with the thumb on the cap of the plunger. Determine the final depth of the bed by measuring the overall length of the cell + plunger. Remove the plunger carefully taking care not to disturb the bed. Next weigh the cell and contents to obtain the actual weight of powder used and connect the cell to the manometer as shown in Fig. 33.

With taps C and D in the appropriate positions, draw the manometer fluid into arm Y to above the starting line. Open taps C and D so that the manometer fluid falls in Y and forces air through the cell. Note the time taken for the liquid to fall to A or B and apply the following modification eq. (7) to obtain the specific surface (S).

$$S^2 = \frac{\epsilon^3}{(1-\epsilon)^2} \cdot \frac{2g\rho t \cdot A}{k\eta La \ln (h_1/h_2)\sigma^2} \tag{11}$$

where

A = cross-sectional area of the permeability cell (m^2)
ϵ = porosity calculated from $\epsilon = 1 - W/A\sigma_s d$
W = weight of powder in the bed (kg)
σ = solid density of the powder (kg m^{-3})
L = depth of the bed (m)
g = acceleration due to gravity (m s^{-2})
ρ = density of U-tube fluid
t = time of flow (s)
k = Kozeny's constant which has a value of about 5.0 for granular solids
η = viscosity of air (1.81×10^{-4} poise at 20°C)
a = cross-sectional area of each limb of the U-tube (m^2)
h_1 = distance from mark A or B to equilibrium level C
h_2 = distance from start to equilibrium level C.

A fluid suitable for use in the manometer is dibutyl phthalate of density 1.045 g cm^{-3} (1.045×10^3 kg m^{-3}).

Repeat the determination several times using the same amount of calcium carbonate but pack the bed more firmly each time in successive experiments so that the effect of porosity on the calculated specific surface can be observed. Although the Kozeny equation states that the specific surface is independent of the porosity of the bed, in practice this is only true over a limited range of porosity. Compare the results you obtain with the specific surface estimated from the Andreasen pipette method.

Experiment 7. *Calibration of the Coulter counter. Model A*

There are three basic calibrations required for a particular aperture/electrolyte combination. These are as follows:
  (*a*) Primary Coincidence Factor 'P'.
  (*b*) Aperture Current Expansion Factors 'F'.
  (*c*) Calibration Factor 'K'.

(a) *Primary Coincidence Factor 'P'*. Primary coincidence is defined as the loss of count at a particular threshold setting, due to two or more particles entering the aperture simultaneously and being counted as one particle. The count loss n'' is proportional to the square of the observed count:

$$n'' = p\left(\frac{\bar{n}}{1000}\right)^2 \qquad (12)$$

where

  $\bar{n} = $ average count at each threshold setting

  $p = $ a constant factor dependent on aperture diameter, d, in μm, and manometer volume V, in ml, and is given by:

$$p = 2.5\left(\frac{d}{100}\right)^3\left(\frac{0.5}{V}\right)$$

The relationship between n'' and instrument count is shown in Fig. 34.

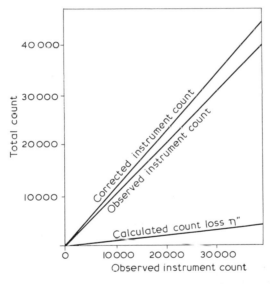

Fig. 34. Coulter counter. Graphs of calculated count loss and corrected instrument count against observed instrument count for aperture of 100 nm (manometer volume = 0.5 ml)

A Secondary Coincidence Factor may also have to be considered where two or more particles are recorded as one larger particle but under normal conditions this factor can be neglected. Primary coincidence must be kept down to 10% of the instrument count.

(b) *Aperture Current Expansion Factors 'F'*. The threshold setting is employed as the basic scale of particle size measurement, but the aperture current setting expands the threshold scale by a factor of approximately two, for each adjacent setting. The aperture current setting determines the amount of resistance introduced into the circuit and so alters the sensitivity of the threshold setting by changing the pulse height produced by each particle. Actual values of the factors are determined from a knowledge of the resistance between the electrodes and then reading off from the appropriate tables of Aperture Current Expansion Factors provided with the instrument.

(c) *Calibration Factor 'K'*. Threshold settings are employed as a measure of particle size and the Calibration Factor 'K' describes the relationship between the equivalent volume diameter and the threshold setting t'

$$d = K\sqrt[3]{(F \cdot t')} \tag{13}$$

where F is the aperture current expansion factor.

The relationship between d and $\sqrt[3]{(F \cdot t')}$ is shown in Fig. 35 where different values of threshold and aperture current settings have been used. This shows that these two settings are inter-related and threshold values can be expanded by the use of different aperture current settings without any loss in linearity.

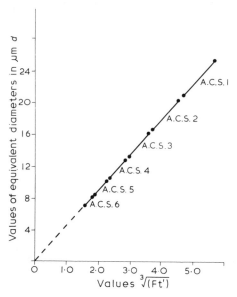

Fig. 35. Coulter counter. Scale expansion of threshold t' by use of increasing aperture current settings (ACS)

One method for evaluating the Calibration Factor is to use mono-sized particles of known size. Certain pollen grains are substantially spherical and mono-sized so these are frequently used, e.g. hazel pollen (20 μm). These pollens have to be accurately sized by microscope methods since errors arising here will introduce a corresponding error in K. The particles are dispersed in the electrolyte solution to produce a maximum count with a primary coincidence of 1 to 2% only. The pollen grains or other suitable particles must be chosen so that their diameters do not exceed 5 to 20% of the aperture diameter.

One aperture current setting is chosen and counts are taken at small increments of the threshold over the whole range of threshold settings. Special care is required to obtain the maximum count and to perform a higher density of counting where the count drop is greatest. The procedure is repeated at as many aperture settings as possible; see Fig. 36. The threshold settings at 50%

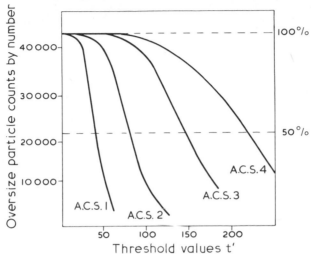

Fig. 36. Coulter counter. Graphs of oversize particle counts by number against threshold values at four aperture current settings

counts are read off, multiplied by the appropriate aperture current expansion factors F and the value of K is determined from eq. (13). The actual values for F can also be determined from the data in Fig. 35 since

$$t'_2/t'_1 = F_{1,2}; \quad t'_3/t'_1 = F_{1,3}; \quad t'_4/t'_1 = F_{1,4} \quad \text{and so on,}$$

where F is the aperture current expansion factor corresponding to threshold setting t'. At aperture current setting 1, F is equal to 1.

An average value of K can be obtained if the data are plotted as shown in Fig. 37 where values of F are taken from the tables supplied with the instrument, or obtained by the method just described.

Experiment 8. *Measurement of particle size using the Coulter counter*

The following method is suitable for water-insoluble powders that disperse readily in aqueous solutions of sodium chloride, e.g. powdered silica. A small amount of surface active agent is generally added to ensure adequate dispersion. The approximate particle size range of the sample should be known since the size of the largest particles should not exceed 40% of the orifice diameter (see Methisazone Mixture BP, Addendum 1975; method specifies 280 μm orifice tube).

Method. (1) Prepare 500 ml 0.9% sodium chloride solution and clarify the solution by filtration through a No. 3 sintered glass filter under gravity, or through a membrane filter using negative pressure. This clarification is necessary to reduce the 'background'

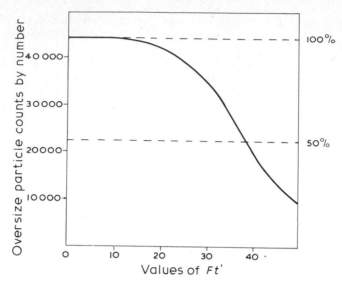

Fig. 37. Coulter counter. Graphs of oversize particle counts by number against threshold values at four aperture current settings

count to a small proportion of the particle count. If the 30 or 50 μm orifice is to be used then filtration through a membrane filter capable of retaining particles of 0.5 μm and above will be necessary. Filtration through a No. 3 glass filter is often sufficient when using an orifice of 100 μm or above.

(2) Fill the orifice tube and the beaker with filtered vehicle. Start the stirrer immersed in the vehicle and adjust the electrodes, if necessary, so that the orifice is clearly defined in the microscope mounted alongside the beaker.

(3) Open the vacuum stopcock carefully and adjust the vacuum supply so that the mercury in the manometer is drawn clear of the start-stop electrodes. When the stopcock is closed the sample is sucked through the orifice and the count is automatically made as the mercury passes the electrodes.

(4) Prepare a concentrated dispersed suspension of the powder in the vehicle and add a little of this to the beaker. Set the Gain Index at 3 or 4, Aperture Current Setting at 10 and Threshold at 10. Note the count and repeat the count using a Threshold of 20. Set the Aperture Current at 1 and the Threshold at 250 and note this count. If the first two counts are large and in reasonable agreement, and the third count negligibly small then the orifice tube/electrolyte is suitable for this particular sample of powder since all the particles have been accounted for.

(5) Remove the beaker of suspension and rinse the aperture tube with fresh vehicle. Determine the background count of the vehicle in a clean beaker by setting the Aperture Current and Threshold to obtain the maximum count as described in the previous paragraph, and then alter the Aperture Current and Threshold so that progressive counts are obtained on particles of larger size. Repeat each count.

(6) Calculate the values of equivalent diameters to correspond with the settings chosen (eq. 13).

(7) Calculate coincidence corrections to find the maximum count which should not be exceeded for this combination of orifice and manometer volume.

(8) Add sufficient of the concentrated suspension of powder (see 4) to the vehicle until the maximum count is slightly less than the permissible maximum count.

(9) Count at the same Aperture Current and Threshold settings as used for the background counting, taking at least two consecutive readings at each setting.

(10) During each count observe the orifice through the microscope to check that it does not become obstructed.

(11) Calculate the results by the deduction of the appropriate background counts and addition of the count loss values to produce a cumulative percentage oversize plot by number or by weight. Data sheets provided by Coulter Electronics Ltd. enable the data to be conveniently organised for determining the weight cumulative plot. Reference should be made to these sheets for further information.

References

Analytical Methods Committee, Society for Analytical Chemistry: *Classification of Methods for Determining Particle Size*, Analyst (1963) **88**, 156–87.

Analytical Methods Committee: Determination of Particle Size, Part 1: *A Critical Review of Sedimentation Methods*, Society for Analytical Chemistry, London, 1968.

B.S.410. *Test Sieves*, British Standards Institution, 1962.

B.S.3406. *Methods for the Determination of Particle Size of Powders*, Parts I–IV, British Standards Institution, 1963.

Fincher, J. H., *J. Pharm. Sci.* (1968), **57**, 1825.

Heywood, H. *J. Pharm. Pharmac.* (1963) **15**, 56T.

Lees, K. A., ibid. (1963) **15**, 43T.

Mullin, J., *Crystallisation* (2nd edn.), Butterworths, London, 1972.

Other Sources

B.S.1796. *Methods for the Use of B.S. Fine Mesh Test Sieves*, British Standards Institution.

Edmundsen, I. C., *Advances in Pharmaceutical Sciences*, Vol. 2, eds. Bean, H., Beckett, A. H. and Carless, J. E., 1967, pp. 95–179.

Herdan, G., *Small Particle Statistics* (2nd edn.), Butterworths, London, 1960.

THERMAL METHODS OF ANALYSIS

Crystalline drugs may exist in several different but chemically identical forms, or *polymorphs*. Where the different forms arise because of inclusion of small amounts of solvent of crystallisation, these forms are termed *pseudopolymorphs*. Some methods of distinguishing between the crystalline forms not only of drugs but of fats, waxes and polymers are discussed in this section. Different polymorphs of a compound, although they may have different rates of solution and different equilibrium solubilities produce identical species in solution. All methods of value in differentiating between crystal forms therefore depend on measurement of properties in the solid state.

Grinding of drugs or recrystallisation, isolation and drying procedures may alter the form in which the drug exists in the dosage form. Thus methods of sample preparation are of great importance in studying polymorphism.

Infra-red spectra often exhibit differences between crystal forms. The BPC 1974 has an infra-red limit test for the inactive form of Chloramphenicol Palmitate. Infra-red spectroscopy is dealt with elsewhere in this book (Chapter 11). In this section we will deal with two methods of thermal analysis, namely differential thermal analysis (DTA) and differential scanning calorimetry (DSC).

Differential Thermal Analysis and Differential Scanning Calorimetry

In differential thermal analysis a record is made of the temperature difference (ΔT) between the sample and a reference material against time or temperature as the two specimens are subjected to identical temperature regimes in an environment being heated or cooled at a controlled rate. The reference material, e.g. alumina is a substance which does not undergo any physical or chemical change in the temperature range of interest, and ideally should have thermal conductivity properties similar to that of the sample.

Differentials canning calorimetry is a related technique in which the sample and reference materials are subjected to linear heating, but the two materials are maintained at the same temperature. What is recorded is not ΔT, but the heat flow into the sample to maintain isothermal conditions.

Because phase changes occur in the samples during heating and cooling DTA and DSC are used to detect polymorphic changes. Specific heats, heats of fusion and heats of dehydration and decomposition of samples can also be determined. In some cases, as the degree of purity of the sample affects melting behaviour and hence the shape of the thermogram, both techniques can be used also to assess sample purity but generally only in substances of high purity.

DTA and DSC are specially useful in the study of fats, oils, waxes and polymers. The investigation of crystallisation and melting behaviour in single or multi-component fats and waxes is one such application. Thermal analysis is also useful in deriving phase diagrams of solid mixtures which form eutectics and solid solutions. As these systems are finding increasing use in pharmaceutical formulations, DTA and DSC techniques will provide a useful method for their analysis and control.

Differential Thermal Analysis

Apparatus and Methodology

Apparatus for differential thermal analysis consists of sample and reference holders, a furnace for programmed heating, a detector of thermal differences created in the differential thermocouple, an amplifier and a recorder. Commercial instruments capable of working from $-150°$ to $+1600°C$ are available. A typical arrangement of the sample chamber is shown in Fig. 38. One heater supplies heat at a constant rate to sample S and reference R. The samples are placed in metal cups and the thermocouple junction detects the difference in temperature ΔT (i.e. $T_S - T_R$) as the two samples are heated. Until the sample exhibits a phase change or chemical change $\Delta T = 0$. The regularity in the arrangement and orientation of the molecules in a crystal begins to fall with rise in temperature long before the melting point is reached; as the melting point is approached the process is accelerated so that a further decrease of the degree of order in the system requires less and less energy.

A direct heating curve and a differential heating curve are shown in Fig. 39. The direct trace shows an endothermic change in the sample temperature, resulting in the typical trace shown in the differential form. Melting transitions are always endothermic. An idealised DTA curve is shown in Fig. 40. BCD is the endothermic peak: EFG is an exothermic peak.

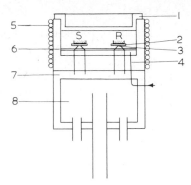

Fig. 38. Specimen platforms and heating and cooling assembly in the Stanton–Redcroft Model 671 Differential Thermal Analysis Instrument (by courtesy of Stanton–Redcroft, London)

1 tight-fitting lid with Pyrex window
2 metal platform
3 chromel-alumel thermocouple welded to platform
4 gas inlet
5 heating coils, specially insulated
6 metal disc through which thermocouple assemblies protrude
7 cooling assembly
8 cooling chamber flushed with liquid nitrogen which enters through smaller pipes and leaves by the central pipe

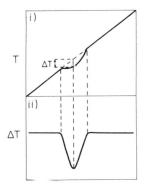

Fig. 39. (i) A plot of sample temperature (T) versus temperature of chamber showing an endothermic change in sample; (ii) the corresponding differential curve (ΔT) of this sample compared with a reference substance treated in the same way

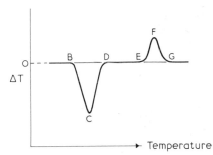

Fig. 40. A typical DTA curve showing an endotherm (BCD) and exotherm (EFG) with increasing temperature. The important features of the DTA curve are the points of deviation from base line B and E and the temperature maxima C and F. C represents the conventional melting points of the substance

Factors Affecting DTA Results

In order to standardise the DTA technique it is essential that regard is paid to a number of factors which can affect the results. These include: (1) sample weight, (2) particle size, (3) heating rate, (4) atmospheric conditions surrounding the sample, and (5) conditions of sample packing into the dishes.

Form of Samples in DTA and DSC

Samples may be studied in the form of powders, fibres, single crystals, polymer films, hydrated solids or as liquids. It is essential for meaningful results that good contact is obtained between the sample and the metal sample holder. In some cases it is desirable to control the atmosphere surrounding the sample and reference material and most commercial instruments have facilities for running samples under reduced pressure or in inert atmospheres. When decomposition of the sample occurs it is advisable to crimp the specimen dishes, although in some cases a pin-hole should be left to avoid build-up of pressure. It is best to investigate the effect of crimping on the results. When the sample is fibrous or fluffy, white kieselguhr may be a suitable reference substance as an alternative to alumina.

The number of experiments which can be carried out using DTA and DSC instrumentation is extensive. A comprehensive account of the pharmaceutical applications of the techniques is given by Ferrari and Inoŭe (1972) and further details of the uses of DTA and DSC will be found in the bibliography at the end of this chapter. Some examples of the uses and results are shown in Figs. 41 to 43. Figure 41 shows the DSC traces for the forms of Phenobarbitone. Figure 42 illustrates DSC results for Carnauba Wax showing three main peaks indicative of the complex nature of the wax. Waxes melt over a considerable range of temperatures and can give highly characteristic peak shapes. This points to the use of thermal analysis in both qualitative and quantitative analysis. Figure 43 illustrates a method of possible value in the analysis of polymers for pharmaceutical use. As the physical properties rather than the chemical properties of polymeric substances frequently determine their utility, DTA and DSC can provide a means of characterisation. In this case a polymer blend of high and low density polyethylene exhibits two distinct endotherms on heating, from which can be obtained the total crystallinity of the sample.

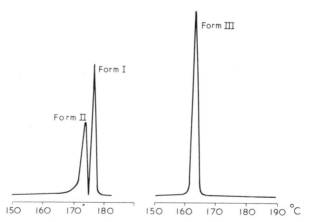

Fig. 41. Polymorphism of Phenobarbitone studied by DSC at a heating rate of 5°/min. The low temperature peak is the melting of a metastable crystal form which reverts to the most stable form melting at 176°C (by permission, Perkin-Elmer Ltd)

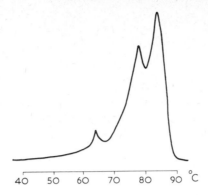

Fig. 42. DSC trace of Carnauba Wax run at 10°/min showing three peaks at 64°, 78° and 84°C. Redrawn by permission from Perkin-Elmer literature on Differential Scanning Calorimeter, Model DSC-1B

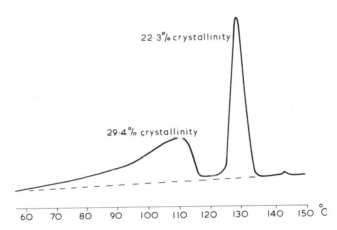

Fig. 43. Illustration of use of DSC in analysis of polymer blends which are often difficult to distinguish by usual analytical techniques from copolymers. The figure shows the melting region of a blend of 25 % linear and 75 % branched polyethlene. The high melting crystallites produce a peak well-resolved from that of the crystallites consisting of shorter linear and branched molecules.

Total crystallinity 51.7%. Redrawn, by permission, from Perkin-Elmer literature on the Model DSC-1B Differential Scanning Calorimeter

From the percentage crystallinity other pharmaceutically relevant parameters such as relative water vapour permeability may be calculated (Florence, 1974).

Quantitative DTA

One of the simplest expressions for the area under the ΔT-time (t) DTA curve is the equation of Speil *et al.* (1945), modified by Kerr and Kulp (1948). The final form of their equation is

$$\frac{m \, \Delta H}{g \lambda_s} = \int_{t_1}^{t_2} \Delta T \cdot dt \tag{14}$$

m is the mass of sample in g; ΔH is the enthalpy change per g of substance; g is a constant dealing with the effect of the geometry of the sample arrangement and construction on heat transfer; λ_s is the coefficient of thermal conductivity of the sample. If A is the peak area, $\Delta H = A/K$ where K is $(g\lambda_s/m)$ and is temperature-dependent (Cunningham and Wilburn, 1970). K should ideally be determined over a range of temperature, e.g. using indium, tin and lead. Areas under the $\Delta T - t$ curve can be obtained by cutting and weighing Xerox copies of the trace.

The problems of heat transfer in the system render application of this equation difficult and not universally valid.

The effect of heating rate is obviously of importance as the rate of reaction depends on the temperature of the sample. If the temperature rise from the external heater is considerable during the period of the reaction then the DTA curve will differ from that obtained at a slower rate of heating. Cooling cycles are often of use in studying phase transitions. Rates of cooling can be critical if supercooling effects are to be avoided. Where supercooling occurs a shift in the position of the exotherm may occur. Instrumental differences and operating conditions are such that the consistency of temperature measurement in DTA is ± 5 to $6°$. Individual workers on a single instrument should, however, achieve a precision better than $\pm 2.5°C$. In order to achieve accuracy in recording temperature calibration materials have to be used preferably over a range of temperatures. Table 13 includes some compounds which can be used for this purpose.

Table 13. Materials for Temperature
Calibration of DTA and DSC Instruments

Reference Compound	Literature T_m (°C)
p-Chlortoluene	7.5
p-Nitrotoluene	51.5
Cyclododecane	60.8
Naphthalene	80.3
Benzoic acid	122.4
Potassium nitrate	127.7
Indium	156.4
Salicyclic acid	159.5
o-Iodobenzoic acid	163.0
Anthracene	216.2
Bismuth	271.3
Potassium Perchlorate	299.5

Sources. B. Wunderlich, *Differential Thermal Analysis in Techniques of Chemistry*, Vol. 1, Part V, ed. A. Weissberger, Ch. 8, Wiley-Interscience, New York, 1971.

A. Blăzek, *Thermal Analysis*, Van Nostrand Reinhold, London, 1972.

Interpretation of Results

In the analysis of pharmaceuticals several thermal effects may be noted, the most important of these and their influence on the DTA curve being shown in Table 14.

Table 14. Qualitative Interpretation of DTA and DSC Curves

Phenomenon	Thermal Effect		Peak Type
	Endothermic	Exothermic	
Dehydration	1		Broad, large
Desorption	1		Broad with no pronounced peak
Crystalline transition	1		Sharp, small
Melting	1		Sharp, medium to large
Sublimation	1		No typical peak
Decomposition	1	1	Generally large
Oxidation degradation	1	1	Generally broad, large

From Instruction Manual for DTA Model 671, Stanton Redcroft Ltd. London by permission.

Differential Scanning Calorimetry

In this technique sample and reference are heated separately and the sample and reference temperatures are monitored separately and continuously maintained at the same level (Fig. 44). When the sample absorbs and evolves energy,

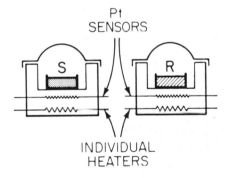

Fig. 44. Schematic representation of sample holders, sensors and heaters of the Perkin-Elmer Differential Scanning Calorimeter by courtesy of Perkin-Elmer Ltd.

Isothermal conditions ($T_S = T_R$) are maintained by comparing the signal from a platinum resistance thermometer in the sample holder with an identical sensor in a reference holder. The continuous and automatic adjustment of heater power (energy per unit time) to achieve this provides a varying electrical signal which is recorded.

more or less power is required by the sample to allow it to be maintained at the same temperature as the reference. This differential power is recorded as a function of temperature and the peak area represents the energy of the transition. DSC is thus more readily quantified than DTA although both techniques require standardisation and calibration. For most pharmaceutical purposes the techniques are of equal utility.

Agreement on a convention for the presentation of DSC and DTA data is obviously required. It has been suggested that DSC scans should have the format shown in Fig. 45 where endothermic transitions are represented by an

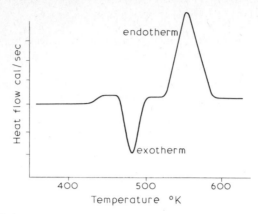

Fig. 45. Characteristic DSC curve (thermogram) Heat flow cal (or J) sec^{-1} as a function of temperature (in °K). Endotherms are above the origin, exotherms below, (*cf.* DTA traces) the form suggested by Perkin-Elmer (Thermal Analysis Newsletter, 1965, No. 2)

increase in the ordinate from the baseline position, exotherms by a decrease in the ordinate.

The same factors which affect DTA results also affect the validity of DSC results. For quantification of DSC with reference to heats of fusion, Wunderlich (1971) recommends the following substances (Table 15).

Table 15. Heat of Fusion Reference Materials

Reference Compound	Melting Temperature °C	Heat of Fusion kJ kg^{-1}
Benzoic acid	122.4	147.28
Urea	132.7	241.84
Indium	156.4	28.47
Anthracene	216.2	161.92

Determination of Purity by DSC

The presence of minute amounts of impurity in a substance broadens its melting range and lowers its melting point by an amount ΔT, which may be related to the mole fraction (X_2) of the impurity by the Van't Hoff relation

$$\Delta T = \frac{RT_0^2 X_2}{\Delta H_f} \tag{15}$$

where ΔH_f is the heat of fusion of the sample in J mole^{-1}, T_0 is the melting point of a pure sample (°K), T_m is the melting point of the sample and $\Delta T = T_0 - T_m$. For compounds which are 99.5 mole % pure ($X_2 = 0.005$) or more, the melting point depression is very small but even at this level of purity the melting range will have increased considerably. It is thus the range combined with the melting depression that is used in the assessment of high purity by DSC.

The differential scanning calorimeter measures the thermal energy per unit time, dq/dt, transferred to or from a sample as the sample holder is heated at a linear rate, dT/dt. Ideally the read-out of the system is proportional to dq/dT, then

$$\frac{dq}{dt} = \frac{dT}{dt} \cdot \frac{dq}{dT} \tag{16}$$

where dT/dt, the scanning rate, is the proportionality constant.

Since eq. (15) shows the melting point depression is directly proportional to X_2, we find that the fraction of sample melted at any temperatures T_s is given by

$$f = \frac{(T_0 - T_m)}{(T_0 - T_s)} \tag{17}$$

and as $(T_0 - T_m) = RT_0^2 X_2 / \Delta H_f$

$$f = \frac{RT_0^2 X_2}{\Delta H_f (T_0 - T_s)} \tag{18}$$

where $f = q/\Delta q$. The object is therefore to obtain at various sample temperatures the percentage of the sample which has melted. On rearrangement and substitution we obtain

$$T_s = T_0 - \frac{1}{f} \frac{RT_0^2 X_2}{\Delta H_f}$$

The sample is heated slowly through the melting range. The fraction melted (f) at any temperature T_s is obtained from the curves. A plot of sample temperature versus $1/f$ should be a straight line with a slope equal to $-RT_0^2 X_2 / \Delta H_f$ and an intercept of T_0. ΔH_f is obtained from the area under the DSC curve so X_2 can be calculated. The procedure is illustrated in Fig. 46. In this example applied to testosterone the slope $-(T_0 - T_m)$ is 0.227. For the DSC curve $\Delta H = 31.77$ kJ mole^{-1}. Substituting the values in eq. (15)

$$0.227 = \frac{8.28 \times (428.14)^2}{31.772 \times 1000} X_2$$

we obtain

$$X_2 = 0.0042$$

$$\therefore \quad \text{purity is } \underline{99.6\%}$$

Several limitations of the method are discussed by Wragg (1974).

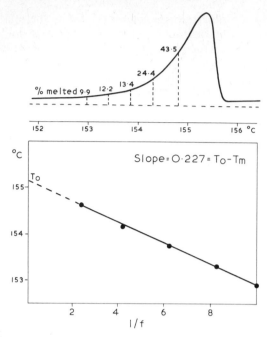

Fig. 46. DSC curve for testosterone from which the lower plot was constructed. The lower diagram is a plot of the reciprocal of the fraction melted $(1/f)$ as a function of the temperature of the sample for a sample of testosterone heated at $1.25°/\text{min}$. The slope of the line is $0.227 = (T_0 - T_m) = RT_0^2/\Delta H_f \cdot X_2$. ΔH_f is $6.66\,\text{kcal mole}^{-1}$, $\therefore$ $X_2 = 0.004$. Purity 99.6% (by courtesy of Perkin-Elmer Ltd)

Practical Experiments

Experiment 9. *Calibration of instrument and familiarisation runs*

The instruction manuals for commerical instruments must be followed. Careful sample preparation and loading of the specimen dish is essential. Reproducibility will only be obtained if precisely the same packing procedure is carried out each time. Contact of the specimen dish with the bench or hands should be avoided.

Method. With sample (2–10 mg) and reference material dishes in place check that there is no contact between the two dishes (possible in some instruments) and ensure satisfactory contact between pan and baseplate (pans can become distorted). Attach the head-unit and allow the system to equilibrate for 5–10 min.

Adjust the recorder to give a baseline such that small exotherms can be accommodated on scale. A heating rate of 10°/min is suggested for most experiments. Start the heating programme and record the DSC or DTA trace. In some instruments (e.g. Stanton–Redcroft DTA) temperature is recorded as a function of time and ΔT-temperature traces have to be obtained from the original trace.

Samples of the standard materials (indium, etc.) in Table 13 should be run at a heating rate of 10°/min from ambient to 100°C above the melting point of each substance, to check or calibrate the temperature scale of the instrument. Use alumina (BDH Ltd. analytical quality, calcined material which has been stored over magnesium perchlorate) as reference material.

To investigate the effects of rate of heating on the DSC or DTA trace, identical weights of benzoic acid should be studied at 2°, 10° and 25°C/min from ambient to 200°C.

Experiment 10. *Demonstration of a pre-melting crystalline transition (DSC/DTA)*

Method. Run a sample of potassium nitrate (weight in mg will depend on the instrument available) unground in static air at 10 °C/min over the temperature range ambient to 350°C. After the temperature has reached about 350°C, allow the sample chamber to cool to about 100°C by passing liquid nitrogen through the cooling assembly and repeat the run up to 400°C.

Note the peak melting temperature and compare the value with that in Table 13. Note the crystalline transition temperature and comment on the effect of the cooling and reheating cycle.

Experiments 11 and 12. *Detection of polymorphism and pseudopolymorphism in pharmaceuticals by DSC or DTA*

11. *Triamcinolone Forms A and B.* A large number of pharmaceuticals exhibit polymorphism and most can be studied by DSC or DTA techniques. Triamcinolone exists in two forms, A and B. Form A is obtained by recrystallisation of triamcinolone from 60% aqueous isopropanol. Form B is obtained by recrystallisation from dimethylacetamide water mixtures.

Method. Run samples of Form A and Form B from ambient to 350°C.

With a new sample of Form A programme the instrument to stop the heating cycle at 270° and cool to ambient. Start the heating cycle to 350°C and compare the traces with those of Form A and B. The favoured form of the compound is the form to which the melt reverts on heating and cooling.

12. *Ampicillin trihydrate*

Method. Run a sample of ampicillin trihydrate from ambient to 200°C. Endotherms due to the desorption of solvating water and to the melting transition should be observed.

Experiment 13. *Determination of the purity of a sample of testosterone by DSC*

Method. Run the drug sample (about 1.5 mg) from 150° to 156°C at a slow heating rate (1.25°/min). Treat the results as described in the text, and make a Xerox copy of the chart paper. The complete curve can be cut out to the base line (obtained in the absence of sample) and the cut-out weighed (x mg). Cut out the area under the curve from the start of the trace up to 153° and weigh this (y mg). The fraction (f) melted is obtained from y/x. Cut out the area between 153° and 153.4°, weigh this adding the weight to that of the first portion ($y + y^1$ mg). Repeat this procedure at 153.8, 154.4 and 155°, and calculate successive f values. ($\Sigma y/x$). Plot T_s versus $1/f$. Determine T_0 by extrapolation as shown in Fig. 46 and calculate the slope of the line and carry out the calculation of mole fraction of the impurity (X_2) as explained in the text.

In work of the highest precision account has to be taken of the thermal characteristics of the apparatus, in particular the thermal lag which results in the indicated temperature being greater than the temperature that the sample has reached. At sufficiently slow heating rates this temperature differential ($\sim 0.2°$K) can be neglected. For a full discussion of the effect and how to correct for it, the account by Gray (1966) should be studied.

Cognate Determinations

In order to facilitate the investigation of the purity of other materials and to check experimental ΔH_f values a partial list of enthalpies of fusion of some pharmaceuticals is appended in Table 16.

Table 16. ΔH_f Values for Selected Drugs

Drug	ΔH_f (kcals mole^{-1})	Source
Chlordiazepoxide	7.3	MacDonald, Michaelis and Senkowski (1972)
Chlorprothixene	7.4	Rudy and Senkowski (1973)
Diazepam	5.9	MacDonald, Michaelis and Senkowski (1972)
Levallorphan Tartrate	10.4	Rudy and Senkowski (1973)
Paracetamol	6.8	Fairbrother (1974)
Sulphamethoxazole	7.5	Rudy and Senkowski (1973)
Testosterone	6.6	Perkin-Elmer (1966)

References

Cunningham, A. D. and Wilburn, F. W., in *Differential Thermal Analysis*, Vol. 1, ed. R. C. Mackenzie, Academic Press, London, 1970.

Fairbrother, J. E., *Analytical Profiles of Drug Substances*, Vol. 3, ed. K. Florey, New York, 1974.

Ferrari, H. J. and Inoŭe, M., in *Differential Thermal Analysis*, Vol. 2, ed. R. C. Mackenzie, Academic Press, London, 1972.

Florence, A. T., *Pharm. J.* (1974) **213**, 36.

Gray, A. P., *Thermal Analysis Newsletter*, Perkin-Elmer, 1966, No. 5.

Kerr, F. P. and Kulp, J. L., *Am. Minerologist* (148) **33**, 387.

MacDonald, A., Michaelis, A. F. and Sekowski, B. Z., in *Analytical Profiles of Drug Substances*, Vol. 1, ed. K. Florey, Academic Press, New York, 1972, pp. 15–37 and 79–99.

Rudy, B. C. and Senkowski, B. Z., in *Analytical Profiles of Drug Substances*, Vol. 2, ed. K. Florey, Academic Press, New York, 1973.

Speil, S., Beukelhammer, L. H., Pask, J. A., and Davis, B., *Bur. Mines Tech. Papers*, 664, 1945.

Wragg, J. S., *Pharm. J.* (1974) **212**, 587.

Wunderlich, B., in *Physical Methods of Chemistry*, Part V, Vol. 1, ed. A. Weissberger and B. W. Rossiter, Wiley–Interscience, New York, 1971.

Other Sources

Blăzek, A., *Thermal Analysis*, Van Nostrand Reinhold, London, 1972.

Daniels, T., *Thermal Analysis*, Kogan Page, London, 1973.

Mackenzie, R. C. (ed.), *Differential Thermal Analysis*, Academic Press, London, Vol. 1, 1970; Vol. 2, 1972.

Nomenclature in Thermal Analysis I and II, *Talanta* (1969) **16**, 1227; ibid. (1972) **19**, 1079.

3 Chromatography

W. D. WILLIAMS

Chromatography is essentially a technique for the separation of the components of mixtures by a continuous distribution of the components between two phases, one of which is moving past the other. The systems associated with this definition are (a) a solid stationary phase with a liquid or gaseous mobile phase and (b) a liquid stationary phase with a liquid or gaseous mobile phase; the former (a) gives rise to adsorption chromatography and the latter (b) to partition chromatography. The theoretical treatment of these divisions gives results which are not entirely satisfactory and the experimental procedures are diverse. The latter, however, have followed from the development of chromatography from its early days and it is convenient to consider the procedures separately.

COLUMN CHROMATOGRAPHY

Adsorption Chromatography

The technique was originally developed by the Russian botanist Tswett in 1906 during the course of an investigation into the nature of leaf pigments. He found that leaf pigments extracted with light petroleum were adsorbed on the top of a column of calcium carbonate supported in a glass tube. As more solvent was allowed to percolate through the column the region of pigmentation became broader and finally separated into distinct and differently coloured bands. Prolonged washing with solvent caused complete separation of the bands which could be eluted separately. It is one of the simplest laboratory exercises to illustrate the use of column chromatography. Tswett's work attracted little attention and it was not until 1931 when polyene pigments were investigated by Kuhn and Lederer that interest in chromatography was renewed.

The principle underlying the separation of the components is adsorption at the solid liquid interface. For successful separation, the compounds of a mixture must show different degrees of affinity for the solid support (or adsorbent) and the interaction between adsorbent and component must be reversible. As the adsorbent is washed with fresh solvent the various components will therefore move down the column until, ultimately, they are arranged in order of their affinity for the adsorbent. Those with least affinity move down the column at a faster rate than those with greatest affinity for the adsorbent.

An adsorbent which is already saturated with respect to one substance may take up a small quantity of a second. The latter displaces the former and consequently if a solution of a mixture is percolated continuously through the column and the eluate is examined for the presence of substances, a plot of amount of substance (per ml of eluate) against volume of eluate will appear as in Fig. 47. This technique was redeveloped by Tiselius in 1940 and is known as *frontal analysis*. It is convenient for the determination of the number of components in a mixture as each is represented by a step in the chromatogram;

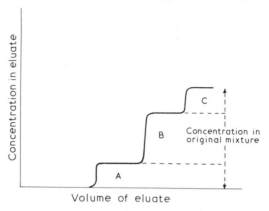

Fig. 47. Frontal analysis: curve for a mixture of three substances, A, B and C

the height of each step is proportional to the concentration of that component in the original mixture. In this method some of the first component only (A in Fig. 47) is obtained pure. Closely allied to this technique is *displacement analysis* in which a small volume of the mixture is added to the column which is developed by a solution of a substance which is capable of displacing all the components of the mixture. When the result of the experiment is plotted as described for Fig. 47 a diagram similar to that in Fig. 48 is obtained. Although the division between components is sharp there are no intervening fractions of solvent only, and some mixing is bound to occur between pairs unless more displacing agents are used to separate A from B, B from C and C from D. The technique is useful

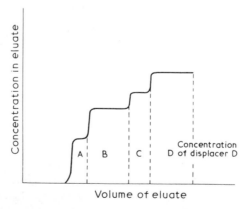

Fig. 48. Displacement analysis: curve for a mixture of three substances, A, B and C displaced by a fourth, D

for preparative work but the investigation of conditions for carrying out this method requires considerable time.

Partition Chromatography

The adsorbent used in the method described above can also be used as a support for a liquid phase which forms a thin layer around each particle. Separation of the components of a mixture occurs because of differences in the partition coefficients of the components between two liquid phases of which one is stationary and the other mobile. Martin and Synge in 1941 developed the concept of the 'theoretical plate' in order to establish a satisfactory theory for partition chromatography. The column is considered as being made up of a large number of parallel layers or 'theoretical plates', and when the mobile phase passes down the column the components of a mixture on the column distribute themselves between the stationary and mobile phases in accordance with their partition coefficients. The rate of movement of the mobile phase is assumed to be such that equilibrium is established within each plate. The equilibrium, however, is dynamic and the components move down the column at a definite rate depending on the rate of movement of the mobile phase. The R value of a component is

$$R = \frac{\text{rate of movement of component}}{\text{rate of movement of mobile phase}}$$

$$= \frac{\text{distance moved by component}}{\text{distance moved by front of mobile phase}} \tag{1}$$

The inter-relationship between R and the partition coefficient α can be shown to be

$$R = \frac{A_m}{A_m + \alpha A_s} \tag{2}$$

where

A_m = average area of cross-section of mobile phase

A_s = average area of cross-section of stationary phase

$$\alpha = \frac{\text{concentration of component in stationary phase}}{\text{concentration of component in mobile phase}}$$

Although the formula is of potential value in devising optimum conditions for chromatography, these are more frequently determined empirically.

Practical Considerations

Adsorption

Adsorbents and solvents. Many adsorbents of varying degrees of activity are available and they should preferably conform to the following requirements. They should be insoluble in solvents, chemically inert, active but not so active that no movement of components occurs, colourless to facilitate observation of zones, allow suitable flow of mobile phases, and have reproducible properties from batch to batch. It is not always possible for adsorbents to comply with all these requirements as the mobile phases and particle size also play a part.

The amount of a substance adsorbed from solution by an adsorbent can be determined by shaking a known weight of adsorbent with a known volume of solution at a fixed temperature until equilibrium is attained. The adsorbent is filtered off and the concentration of the substance in the filtrate is determined by any suitable means. If this procedure is carried out with solutions of different concentrations the results can be expressed graphically by plotting the amount of substance adsorbed per gramme of adsorbent against the concentration of solution. The curves so obtained represent adsorption isotherms, which are important in explaining the appearance of chromatograms.

The curves may take any one of the three forms (Fig. 49), each of which explains a characteristic appearance of peaks in a chromatogram.

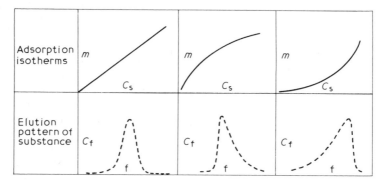

Fig. 49. Adsorption isotherms and related elution patterns (diagrammatic) of substances from a column of adsorbent; m, weight of substance adsorbed per g of adsorbent; C_s, concentration of solution; C_f, concentration of each fraction; f, number of each fraction

(a) *Linear adsorption isotherms* are obtained when the amount of substance adsorbed per gramme of adsorbent is proportional to the concentration of solution. When a substance moves as a band through a column of adsorbent there is no tendency for any portion of the band to be adsorbed more strongly than another. Therefore, a symmetrical peak is obtained as the eluate from the column is examined.

(b) *Convex adsorption isotherms* are obtained when adsorption from weak solutions is greater than from strong solutions. Therefore, even if the pattern of the band of substance is initially symmetrical, the substance, in low concentration, at the front of the band is held more strongly by adsorption than is the centre of the band as the band of substance moves down the column. Therefore the centre of the band 'catches up' with the front and a sharp leading edge to the band is obtained. By a similar argument, the tail of the band becomes longer and longer. This appearance (Fig. 49) is one which is met frequently in practice.

(c) *Concave adsorption isotherms* on the other hand, are obtained when adsorption from strong solutions is greater than from weak solutions so that the appearance of an initially symmetrical band of substance after passage through the column is once more characteristic (Fig. 49). It is not often that this type occurs.

Although the adsorption forces involved in chromatography are weak, cognisance must be taken of undesirable chemical changes that might occur because of the properties of the column material itself. Thus, an alkaline grade of alumina may cause hydrolysis of esters or lactones. Other changes associated with a poor choice of column are isomerisation, neutralisation of acids or bases and decomposition of compounds. The last may be put to good use in certain preparations, e.g. cadalene from oil of cade forms a crystalline picrate which is decomposed on a short column of alumina. The pure cadalene is eluted leaving the picric acid fixed on the column.

The degree of activity of an adsorbent may be modified by chemical or physical treatment, e.g. alumina may be activated by treating with acids or bases followed by washing with water and heating strongly. Careful addition of water to the treated alumina allows different degrees of activity to be obtained and reproduced from batch to batch. Preliminary treatment of the material in this way often overcomes the property which leads to the undesirable effects noted above. It may, indeed, introduce increased specificity of the column for certain compounds, e.g. when silica gel is freshly prepared in the presence of propyl orange and the dye is finally removed by elution, the column has a greater affinity for propyl orange than it has for the methyl, ethyl and butyl analogues. A similar situation obtains when the silica gel is prepared in the presence of one enantiomorph of an optically active compound, e.g. laevorotatory quinine.

Adsorption is most powerful from non-polar solvents such as petroleum ether or benzene and a single solvent may often be effective in developing the chromatogram. The rate of movement of the components down the column can be increased by the addition of a second solvent to the mobile phase; the second solvent is usually more polar than the first. Strain (1942) has arranged both adsorbents and solvents in order of adsorptive and eluting power respectively and Table 17 lists the series. It is usual to redistil all solvents before use, so that traces of non-volatile matter, e.g. grease, are completely absent.

Silica gel and cellulose are also used frequently but mixtures of adsorbents are generally used for some purpose additional to that of adsorption, e.g. charcoal in fine powder in admixture with a light coarser powder to assist passage of solvent.

For solvents, the change from one to another should be gradual, e.g. the change-over from petroleum ether to benzene should be done in proportions such as the following (petroleum ether first) 100, 95:5, 90:10, 80:20, 60:40, 40:60, 10:90, 100. Such a procedure is time-consuming and a more rapid elution may be achieved by the addition of about 0.5 to 1.0% of ethanol to the first non-polar solvent used.

Alteration in the composition of the eluting solvent may also be achieved by adding the second solvent gradually to a reservoir of the first with efficient mixing; the solvent entering the column therefore becomes gradually and continuously richer in the second solvent. This technique is known as gradient elution which, with proper choice of adsorbent and gradient, often reduces tailing of the compounds on the column.

Preparation of the column. A typical arrangement for column chromatography is shown in Fig. 50.

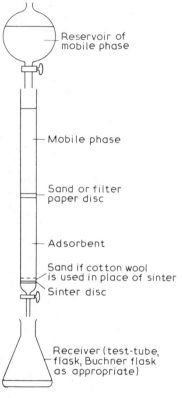

Fig. 50. Apparatus for column chromatography

Method. Prepare the chromatographic column by mixing the adsorbent into a slurry with the solvent and pouring the mixture into the glass tube which contains solvent. The sand serves to give a flat base to the column of adsorbent when cotton wool is used instead of a sinter disc. After the adsorbent has settled, add a filter paper disc and sand then run off the supernatant liquid until the level falls to about 1 cm above the top layer of sand. The filter paper disc and sand is one means of avoiding disturbance of the absorbent as fresh mobile phase is added to the column in the initial stages of development. The level of solvent must never be allowed to fall below the level of adsorbent, otherwise the latter develops cracks and becomes useless for chromatography because the solvent runs through the cracks rather than between the particles of adsorbent.

The preparation of the slurry may prove difficult with dense adsorbents and it is convenient to pour the powder directly into the solvent in the tube. Frequent tapping of the tube and stirring of the mixture assists in even packing and removal of air bubbles or pockets. Alternatively, the tube may be packed with the dry powder and the solvent allowed to percolate through with the stopcock open until the level falls to about 1 cm above the adsorbent.

The dimensions of the column and quantity of adsorbent depend upon the nature and amount of the substance to be chromatographed but a rough guide is given in Table 18.

Use of column. Wash the column with about 50 ml of the mobile phase used to prepare it, which should be the least polar solvent in which the mixture will dissolve. Add the mixture dissolved in a small volume of solvent and carefully allow it to run into the sandy layer by opening the stopcock. Add a small volume of solvent and wash in the mixture. Repeat with gradually increasing quantities of solvent and develop the chromatogram, collecting the eluate in appropriate receivers if the components are to be eluted from the column.

Detection and recovery of components. For those mixtures which are coloured, visual examination of the column is usually sufficient to locate the coloured components. Colourless components may also be detected visually if they

Table 17. Adsorbents and Solvents

	Adsorbent		Solvent
Weak	Sucrose Starch Inulin Talc Sodium carbonate		Petroleum ether Carbon tetrachloride Cyclohexane Carbon disulphide
Medium	Calcium carbonate Calcium phosphate Magnesium carbonate Magnesium oxide Calcium hydroxide	increasing eluting power	Ether (ethanol-free) Acetone Benzene Toluene Esters
Strong	Activated magnesium silicate Activated alumina Activated charcoal Activated magnesia Fullers earth		Chloroform Alcohols Water Pyridine ↓ Organic acids Mixtures of acids or bases with ethanol or pyridine

Table 18. Column Characteristics

Adsorbent/adsorbate Weight Ratio*	30:1
Length/diameter Ratio†	10–15:1

Column length
    (a) multi-component system
    (b) components with similar affinities for adsorbent
    (c) components with different affinities for adsorbent
	
	long column
	long column
	short column

* The 30:1 weight ratio is suitable for preparative separations. For analytical purposes the ratio (30:1) is much too small, and often mg quantities of substance are chromatographed on 20 g or more of adsorbent (see Experiment 3).

† In general, narrow columns give better separations than wide columns.

fluoresce, e.g. quinine and ergometrine. Recovery of the components after detection on the column requires extrusion of the column of adsorbent and isolation of each zone for extraction with solvents. If plastic tubing is used instead of glass tubes the zones are conveniently isolated by cutting the tubing into sections.

It is, however, more convenient to complete the chromatogram by eluting the various components with solvents. For colourless compounds the eluate is collected as a large number of fractions, each of small volume.

The advent of automatic fraction collectors has enabled large numbers of fractions to be obtained without the tedium associated with manual collection.

The large number of fractions also assists in obtaining better separation of components providing attention is directed to correct choice of flow rate. Each fraction is examined appropriately for the presence of a compound. The examination may be by evaporation of the solvent from each fraction and weighing the residue, by simple spot tests, by examination of the fraction by paper or thin layer chromatography or by spectrophotometry, either directly or after addition of reagents.

By comparison with detectors used in gas chromatography (p.114) the monitoring systems discussed above are relatively insensitive. The eluate from the column can, however, be examined by an argon or flame ionisation detector provided that a phase transformation is carried out. The basis of the method described by Maggs and Young (1967) is to sample the eluate continuously by means of a thin wire moving through the eluate. The liquid phase is removed from the wire by evaporation in a stream of argon and the solute (if any) remains on the wire which now enters a pyrolysis chamber adjacent to an argon ionisation detector (p.117). The pyrolysis products are swept into the detector by a stream of argon and the response is recorded in the normal way. Water, and solvents which boil above 180° are not suitable as eluting solvents when this method is used.

Partition

In this type of chromatography the solid material of the column functions solely as a support for a thin layer of liquid phase which is ordinarily polar in character, e.g. water or aqueous buffer solutions, the mobile phase being one of those normally used in adsorption chromatography. The two liquid phases must be in equilibrium and this is accomplished by shaking them together in a separator and allowing them to separate. The aqueous layer is used to coat the support which is afterwards transferred to the chromatographic tube and packed firmly. The other layer is used as developing solvent. A change from one solvent to another during development is inappropriate here (compare adsorption chromatography) as the equilibrium would be disturbed. Of the solid supports, silica gel is probably most convenient for aqueous systems as it is capable of holding a considerable volume of water whilst retaining its powder form. Cellulose powder is also frequently used.

The technique has been extended to enable organic liquids to be used as the stationary phase with kieselguhr commonly used as a support. It is made water-repellent by treating the dry powder with an ethereal solution of dichlorodimethylsilane (CARE), allowing the ether to evaporate in a fume cupboard and washing the residue with methanol until free from acid. Dry the powder and store in a well-stoppered jar. The column is prepared in the way described above but the equilibrated organic phase is used as the stationary phase. This method is known as reversed phase chromatography.

The procedure for partition column chromatography differs little from that of adsorption chromatography and a typical elution pattern from a mixture of diphenylamine (0.1 mg) and phenothiazine (0.1 mg) is shown in Fig. 51. Each fraction (10 ml) in this experiment (Experiment 3) was examined at 253 nm.

Fractions 10–14 contained diphenylamine.

Fractions 20–29 contained phenothiazine.

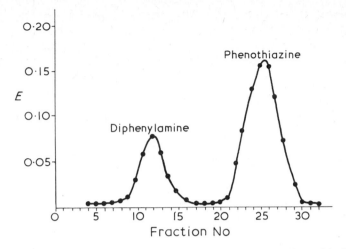

Fig. 51. Chromatogram for a mixture of diphenylamine and phenothiazine

On a larger scale the appropriate fractions may be evaporated for recovery of the components.

Practical Experiments

Experiment 1. *Determination of the percentage w/w of strychnine in syrup of ferrous phosphate with quinine and strychnine*

The assay depends upon the observation that strychnine hydrochloride is soluble in chloroform.

Method. Introduce into a chromatograph tube a layer of Celite 545 (1 g) previously mixed with water (1.0 ml). Pack well with the aid of a flat-ended glass rod and add a second layer of Celite 545 (3 g) well mixed with $2N$ hydrochloric acid (3.0 ml). Pack firmly. Mix the sample (about 1 g, accurately weighed) with Celite 545 (5 g), $2N$ hydrochloric acid (2.0 ml), and water (1.0 ml) and transfer quantitatively to the column using more Celite 545 (1 g) to clean the beaker. Transfer the Celite 545 to the column, clean the beaker with a small wad of cotton wool and use the latter to push down any traces of Celite 545 adhering to the sides of the chromatograph column.

Wash the column with anaesthetic ether saturated with water (125 ml in portions) and elute the strychnine with chloroform saturated with water (150 ml) in portions. Combine the chloroform eluates and remove the solvent with the aid of gentle heat and a stream of air. Dissolve the residue in $0.1N$ hydrochloric acid with the aid of heat, cool, transfer to a 25 ml volumetric flask and dilute to the mark with acid.

Measure the extinction at the peak absorption (254 nm) and calculate the percentage w/w of strychnine, using E_1^1 (p. 236) for strychnine at 254 nm, as 375. For this calculation the weight per ml of the syrup must also be determined.

Garratt (1964) refers to the use of a 3-point correction procedure to allow for irrelevant absorption and gives the equation

$$E_{\text{corrected}} = 6.9(E_{254} - 0.533E_{247} - 0.467E_{262})$$

Experiment 2. *Determination of the primary and secondary glycosides in digitalis leaf, both calculated as digitoxin*

Method. Weigh the powdered leaf (4 g), add ethanol (70%, 40 ml), shake gently for 1 hour and filter. To the filtrate (5 ml), add water (35 ml), solution of lead acetate (15%, 5 ml) shake well and filter (No. 54 paper). Transfer the filtrate (40 ml) to a separator, add ethanol (95%, 20 ml) and extract with chloroform (3 × 25 ml). Bulk the extracts, dry with Na_2SO_4, filter and adjust the filtrate to 100 ml with chloroform. Evaporate 25 ml of the solution to dryness at a temperature not exceeding 65°. Dissolve the residue as completely as possible in ethanol-free chloroform (2 ml) and transfer to a column of cellulose powder (3 g) previously prepared with the aid of ethanol-free chloroform. Elute with ethanol-free chloroform (30 ml), using the first 10 ml to wash out the flask used for evaporating the chloroform solution. The eluate is fraction A. Continue the elution with chloroform: ethanol (90:10, 30 ml), collecting the eluate in the flask used for evaporating the 25 ml of chloroform solution. This is fraction B. Evaporate each fraction to dryness.

Dissolve each residue in separate portions of methanol (each of 5 ml) and add to each, sodium picrate reagent (1% picric acid in 0.5% sodium hydroxide freshly prepared, 5 ml). Measure the extinction at 485 nm after 25 min using a blank of reagents and calculate the amount of secondary glycosides (fraction A) and primary glycosides (fraction B) both calculated as digitoxin. E_1^1 for digitoxin is 190 in this reaction.

Note. The rather large quantity of leaf and ethanol is to ensure that a representative sample is obtained and to allow for adequate filtrate. Similarly, 40 ml of filtrate from the reaction with lead acetate (a decolourising agent) is used to avoid the difficult washing procedure which would be required if the whole of the filtrate were to be used. The chloroform used in preparing the column and for eluting fraction A must be free from ethanol otherwise some of the primary glycosides will appear in that fraction. To confirm that separation is complete, examine fractions A and B obtained in a separate experiment, by paper or thin-layer chromatography (Experiment 4, p. 94).

Cognate Experiments

Digitoxin tablets. The column is prepared with acid and water-washed kieselguhr using formamide as the stationary phase and benzene: chloroform (3:1) as eluant. After completion of the assay the sample may be tested for *Other Digitoxosides* by eluting the column with chloroform and applying the Keller Kiliani reaction to the residue obtained on evaporation of the eluate.

Experiment 3. *Determination of phenothiazine in the presence of diphenylamine and carbazole*

Crude phenothiazine may contain diphenylamine, carbazole and phenothiazone as likely impurities and Bailey, Barlow and Holbrook (1963) have developed an elegant chromatographic method to determine phenothiazine in the presence of these impurities.

Solvent system. Shake acetonitrile (1 volume) with hexane (10 volumes); use the lower layer as stationary phase and the upper layer to develop the chromatogram.

Column. Mix Celite 545 (25 g) with stationary phase (12.5 ml) and transfer to the chromatographic tube (70 × 2 cm) in small portions. Pack each portion firmly with a flat-ended glass rod. The prepared Celite 545 is bulky and a large tube is required.

Method. Dissolve the sample (about 100 mg accurately weighed) in methanol and dilute accurately to 100 ml with methanol. Transfer the solution (10 ml) to a 100 ml volumetric flask and dilute to volume with methanol. Transfer 1.0 ml of the dilution (equivalent to 0.1 mg of phenothiazine) to a small beaker and evaporate to dryness. Dissolve the residue in stationary phase (1.0 ml), add Celite 545 (2 g), mix well and transfer to the column. Clean the beaker with more Celite (1 g) and add to the column. Add the developing solvent carefully to the column and adjust the flow rate to about 2 drops per second. Collect about 36 fractions each of 10 ml and measure the extinction of each at 253 nm against developing solvent as reference. Repeat the procedure using a carefully purified phenothiazine as standard.

Calculate the percentage of phenothiazine in the sample by means of the formula:

$$\% = \frac{E_T \times W_S \times 100}{E_S \times W_T}$$

where

E_T = sum of extinctions (p. 236) obtained from chromatogram band for sample
E_S = sum of extinctions obtained from chromatogram band for standard
W_T = weight of sample (mg)
W_S = weight of standard (mg)

Plot the readings against fraction number and identify any impurities by the position of the peaks:

e.g. diphenylamine is eluted first (fraction numbers 8–12 approximately) followed by phenothiazine (13–17) and carbazole (20–30).

Note. The authors of this method used acid-washed Celite 545 for the separation. In the absence of this material, silane-treated Celite (40–60 mesh) proved very satisfactory, the separation of the components being slightly better than that quoted in the original paper.

Cognate Experiment

Fluocinolone Acetonide in formulated products. Use the system *n*-hexane: dioxan:water (100:40:5). The lower layer is used to prepare the column and the upper layer is used as eluant phase. Measure the extinctions at 238 nm. This method, developed by Bailey, Holbrook and Miller (1966) is of wide applicability.

Various preparations of Betamethasone 17-Valerate are examined in a similar fashion using the system *n*-hexane:dioxan:methanol:water (100:30:5:5). It is important that *n*-hexane of spectroscopic grade be used as the ordinary grade contains much aromatic material which gives rise to considerable ultraviolet absorption.

Experiment 4. *Determination of Phytomenadione*

Chromatography is carried out on alumina containing water (7%), to control adsorption, with precautions against exposure to bright light.

Method. Prepare a column (5 mm internal diameter) in the normal manner with alumina (4.5 g) using trimethylpentane as eluant. Weight accurately about 0.1 g of sample and dissolve in trimethylpentane in a 200 ml volumetric flask. Dilute to volume and mix well. Apply the solution (2 ml) to the column and elute with a mixture (20 ml) of trimethylpentane: anaesthetic ether (49:1). Dilute the eluate to 50 ml with trimethylpentane and measure the extinction at about 249 nm in a 1 cm cell using 420 as the E_1^1 at 249.

Cognate Determinations

Phytomenadione Injection
Phytomenadione Tablets

PAPER CHROMATOGRAPHY

Paper partition chromatography was developed by Consden, Gordon and Martin (1944) as a technique for the separation of amino acids. Paper is used as the support or adsorbent but partition probably plays a greater part than adsorption in the separation of the components of mixtures, as the cellulose fibres have a film of moisture round them even in the air-dry state. The technique is therefore closely allied to column partition chromatography, but whereas the latter is capable of dealing with a gramme or more of material, the former requires microgrammes. It is, therefore, an extremely sensitive technique of enormous value in chemical and biological fields.

The properties of chromatographic paper may be modified to suit various requirements, and Table 19 lists the grades available in Whatman products.

As in column chromatography, aqueous buffer solutions may be incorporated in ordinary papers to assist in the separation process, or they may be further modified by incorporating organic liquids such as liquid paraffin and hydrocarbon grease to give supports for reversed phase chromatography.

The movement of components on the paper depends on the amount and nature of stationary phase compared with the amount of mobile phase in the same part of the paper, and also on the partition coefficient. R values (p. 77) are frequently called R_F values in paper chromatography and the equation (3) is commonly employed for R_F. Strictly, however, because the rate of movement of the mobile phase at the solvent front tends to be faster than at the position of the component on the paper, it is better to define R_F as

$$R_F = \frac{\text{distance travelled by centre of component}}{\text{distance travelled by solvent front}} \qquad (3)$$

R_F values are of considerable importance in paper chromatography as much information is available concerning the movement of compounds on paper with various solvent systems. They offer a means of tentative identification of components and simplify experimental work in reducing the number of reference substances used in any one experiment, e.g. if examination of a trial chromatogram of a protein hydrolysate reveals no spot corresponding to an R_F value of about 0.2 with liquefied phenol as mobile phase, it is unlikely that aspartic acid is present. Reliance should not be placed upon one solvent only but the behaviour of the sample in several such systems should be correlated with that of reference compounds in the same systems. Some differences in R_F values between those found and those quoted in the literature may well occur, as the conditions used are probably slightly different, e.g. in temperature, stationary phase, size of spots, quality of paper and equilibration times. In order to eliminate or minimise these variables the R_F values are sometimes related to that for a standard substance, e.g. glucose in the investigation of carbohydrates.

Table 19. Papers for Chromatography

Grade No.	Solvent Flow Rate	Applications	Remarks
1	Medium	General	Most widely used paper
2	Slow	Amino acids, peptides proteins	Similar to No. 1 but with a slower solvent flow rate
3	Medium	Inorganic separations	Thick paper, also used for electrophoresis
3 MM	Medium	As for No. 3 but is useful for larger volumes of solution than those required for other papers	As for No. 3
4	Fast	Amino acids and sugars	Rapid resolution of simple mixtures
17	Very fast	Preparative work and electrophoresis	Very thick, soft paper
20	Very slow	Amino acids	Produces good resolution and compact spots for most compounds
31 ET	Very fast	Electrophoresis	Thick paper, acid-washed
54	Fast	Amino acids and sugars	Hardened paper, can be impregnated with adsorbents
540	Slow	General purposes	Double acid-washed, hardened and of minimum metallic impurity
541	Fast	General purposes	Double acid-washed, hardened and of minimum metallic impurity
542	Very slow	General purposes	Double acid-washed, hardened and of minimum metallic impurity
SG 81	Medium	Water-insoluble substances, e.g. lipids, sterols, terpenes, insecticides, plant pigments, inorganic substances; also dyes	Paper loaded with silica gel adsorbent; used with nonpolar solvents
AH 81	Slow	Water-insoluble substances; uses as for SG 81	Paper loaded with aluminium hydroxide
AC 81		Reverse phase chromatography	An acetylated paper of acetyl content 12% by weight
AC 82		As for AC 81	25% Acetyl content
GF 81–83	Very fast		Glass fibre material for use where cellulose is prohibited; better results obtained by impregnating with water-soluble salts, silica gel or aluminium hydroxide
ST 81	Medium	Amines, lipids, steroids, vitamins	Impregnated with a suitable silicone; absorbs non-polar solvent preferentially
ST 82	Slow	Amines, lipids, steroids, vitamins	Impregnated with a suitable silicone; absorbs non-polar solvent preferentially
ST 84	Fast	Amines, lipids, steroids, vitamins	Impregnated with a suitable silicone; absorbs non-polar solvent preferentially

The ratio:

$$R_X = \frac{\text{distance substance moves from origin}}{\text{distance reference substance moves from origin}} \qquad (4)$$

is known as the R_X value where X refers to the reference substance, e.g. when glucose is used as a reference substance the term would be R_g value. The ratio is particularly useful where the chromatogram must be allowed to continue for such a time that solvent reaches the end of the paper and drips off the end. R_F values are clearly not possible under these conditions.

R_F values are also used in the determination of structure of certain groups of compounds; this is done indirectly via R_M values. Martin (1949) has stated that the differences between the logarithms of the partition coefficients for adjacent members of a homologous series is constant. Therefore by applying the relationship

$$R_F = \frac{A_m}{A_m + \alpha A_s} \qquad \text{(substituting in equation 2)}$$

$$\alpha = \frac{A_m}{A_s}\left(\frac{1}{R_F} - 1\right)$$

$$\therefore \quad \log \alpha = \log\frac{A_m}{A_s} + \log\left(\frac{1}{R_F} - 1\right)$$

The term $\log(1/R_F - 1)$, is known as R_M. The change in this value (ΔR_M), when a substituent or functional group is added to compounds, should be the same for all the compounds in the same solvent system. Mikes (1966) and Lederer and Lederer (1957) give a table of ΔR_M values for various solvent systems.

Practical Considerations

Apparatus

The paper functions in the same way as a column but since evaporation of the mobile phase would occur as development progressed it must be enclosed in a chamber to prevent such loss. Chromatography tanks are therefore required and the assembly for the two types of paper chromatography are shown in Fig. 52 (descending technique) and Fig. 53 (ascending technique). Both these methods use vertical papers but horizontal chromatography is also used for rapid separation of mixtures into their components. In one form with specially cut papers, up to five mixtures may be examined at one time (Fig. 54). The solvent rises to the centre of the paper and spreads radially by capillary action to develop the chromatogram.

Methods

Descending chromatography. Draw a pencil line about 12 cm from one end of the selected paper and apply 2–10 μl volumes of solutions of the mixtures and reference samples to this line. The spots should not be larger than 5 mm and spaced at about 5 cm intervals on the line or, if a larger number of samples is examined, not closer than 2 cm. Prepare the tank by putting some mobile phase or equilibrated stationary phase (as appropriate) in the bottom. Place the paper in the tank so that the end is held firmly in the *empty* trough

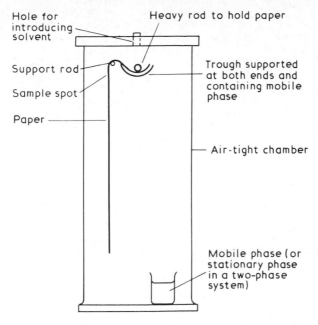

Hole for introducing solvent

Heavy rod to hold paper

Support rod

Sample spot

Paper

Trough supported at both ends and containing mobile phase

Air-tight chamber

Mobile phase (or stationary phase in a two-phase system)

Fig. 52. Apparatus for descending paper chromatography

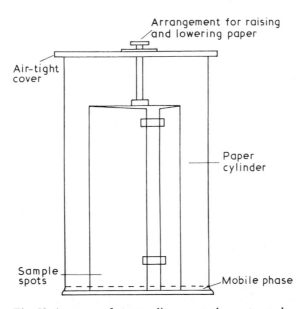

Arrangement for raising and lowering paper

Air-tight cover

Paper cylinder

Sample spots

Mobile phase

Fig. 53. Apparatus for ascending paper chromatography

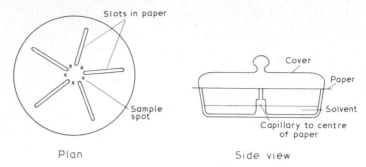

Fig. 54. Apparatus for circular chromatography

and the line on which the spots are located is about 3 cm below the support rod (Fig. 52). Seal the tank and allow to stand for several hours, or overnight, for equilibrium between paper and volatile phases to be attained. Add the remainder of the mobile phase to the trough as quickly as possible via the stoppered hole immediately above the trough (Fig. 52) and allow development to proceed until the solvent front is about 5–10 cm from the end of the paper. If the R_F value is not required, and separation of slow moving components is the main interest, the solvent may be allowed to run to the end of the paper and to drip from the end if necessary.

Remove the paper, taking care not to splash mobile phase from the trough on to the paper, and allow the paper to dry in a current of air or in an oven if the solvent boils at a high temperature. Locate the compounds by a suitable means (see below).

Ascending chromatography. Draw a pencil line about 3 cm from one end of the selected paper and apply 2–5 μl volumes of solutions of sample and reference compounds at about 2 cm intervals on the line. Fold the paper, at right angles to the line, into cylindrical form and hold the edges together with clips.

Prepare the tank (Fig. 53) by placing the mobile phase in the bottom to a depth of about 1 cm, and line the tank with filter paper. The filter paper allows a rapid saturation of the atmosphere of the tank with solvent vapour. Unlined tanks may also be used; the solvent tends to run slightly slower under these conditions. Place the cylinder in the tank so that the lower end (with spots) is above the surface of the liquid and allow to equilibrate for 2 or 3 hours. At the end of this time lower the paper into the mobile phase and allow development to take place until the solvent front has reached a suitable height (15–20 cm). Remove the paper, dry as described above and locate the components (see below).

Elaborate apparatus is not essential for this method and it can be conveniently carried out in readily available laboratory apparatus, e.g. strips of paper can be attached to corks and hence can be suspended with the ends dipping into solvent contained in test-tubes or measuring cylinders. On this small scale, development times are short. Although not more than about two spots can be placed on the paper, several assemblies can be prepared and used at the same time, so that a fairly rapid assessment of solvent systems for best separation of components can be made.

Two-dimensional chromatography. Draw a pencil line along one side of the chromatography paper and place the sample solution (2–5 μl) on the line about 3 cm from its end. Develop the chromatogram in one solvent system by the method of ascending chromatography. Dry the paper and refold it into a cylinder at right angles to the first. Develop with a second solvent system, dry and locate the components of the mixture. Identify by treating reference compounds in the same way. The separation of a five component mixture might appear as shown in Fig. 55 and Experiment 1 (Polymyxin B Sulphate) is a typical example.

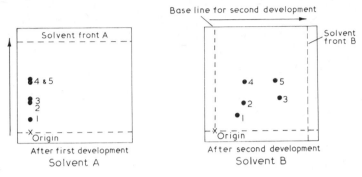

Fig. 55. Two-dimensional paper chromatography (diagrammatic)

Detection of Components

Components of mixtures are located more readily in paper chromatography than they are in column chromatography as it is easier to apply both chemical and physical methods if no visible colour is present.

Irradiation with light of 254 nm often reveals those compounds which absorb in the ultraviolet region, as dark areas on a pale purple background. The irradiation itself may also induce fluorescence either immediately, e.g. quinine and ergot alkaloids, or after several minutes, e.g. p-chloroacetanilide. Fluorescence is usually better observed with light of wavelength 366 nm either before or after irradiation with light of shorter wavelength.

Chemical reagents for the development of colour are numerous as they react either with a specific group in the molecule to be detected or they are non-specific. Typical common reagents are listed in Table 20.

The reagents are applied most conveniently as a fine spray by means of an atomiser, except where vapour is employed, as with iodine. They are unpleasant in the mist form and should be used in a fume cupboard with a good draught. If a portion only of the chromatogram is to be developed, e.g. to detect the position of the required component on the chromatogram of the mixture, the position of the reference spot can be determined by laying a strip of the filter paper impregnated with the reagent along the development path of the reference substance. Alternatively, this path only may be sprayed if the remainder of the chromatogram is protected by means of stiff paper or cardboard.

Specialised means of detection may be used for some compounds, e.g. radioactive materials, by autoradiography or Geiger–Müller counter; and antibiotics by laying the chromatogram on nutrient agar inoculated with an appropriate strain of micro-organism—the zones of inhibition of growth indicate regions of active material.

Quantitative Measurements

It is not possible to standardise the conditions of an experiment to such an extent that the results from one chromatogram can be applied to another

Table 20. Common Developing Reagents for Paper Chromatography

Reagent	Remarks
Iodine vapour	Organic bases give brown spots
Potassium permanganate (1 %) in acetone/water	General reagent: yellow or pale spots on pink background
Bromocresol green (0.05 %)	Acids give yellow spots on green background
Ninhydrin 0.1 % in water-saturated butanol	Aminoacids give purple spots on heating the paper for 5–10 minutes at 100°
Sodium nitrite (0.5 %) followed by 0.1 % N-(1-naphthyl) ethylene diamine dihydrochloride in N HCl	Sulphonamides give purple or red spots if a primary aromatic amine is present
3,5-Dinitrobenzoic acid (1 %) in N sodium hydroxide solution	Cardiac glycosides give red spots
2,4-Dinitrophenylhydrazine (saturated) in $2N$ HCl	Aldehydes and ketones give yellow spots
Dragendorff's reagent diluted with glacial acetic acid (1:5)	Alkaloids give brown or orange spots

carried out later. Therefore, the sample to be examined, and the standards, must be applied to the same paper taking the utmost care to maintain the size of spots as uniform as possible. After development, the concentration of the required component can be determined by making use of a suitable test, e.g. ergometrine can be determined in extracts of ergot by means of its fluorescence in ultraviolet light. Visual comparison of sample with standard spots enabled Foster, Macdonald and Jones (1949) to determine ergometrine with sufficient accuracy (20 %) for it to serve as a useful screening test for its presence in samples of ergot.

Determination of the component can also be carried out on the paper when it forms a coloured spot with a suitable reagent. This method generally involves measuring the area of the coloured spot and comparing it with the areas of standard spots. A graph is constructed by plotting the logarithm of the weight of component against the area of the spot, as described by Fisher, Parsons and Morrison (1948). The measurement of the area of spots can be carried out directly by means of a planimeter, or indirectly by cutting out the areas of paper and weighing them. Alternatively, squared paper can be laid over the spots and the area determined by counting the squares.

A more sensitive method to a simple visual comparison of the intensity of colour of spots (cf. fluorescence), is the measurement of the transmission of light through the spot by means of a photodensitometer. In order to reduce the general scattering of light by the paper in this measurement, it is convenient to treat the paper with an oil, providing it has no untoward effect on the spot, such as causing the colour to run. Reflectance measurements would also appear to offer a means of measurement.

For substances having characteristic absorption, the paper containing the required component can be cut out and the component eluted with a polar solvent for spectrophotometric examination. A blank should be carried out on a similar area of paper adjacent to the spot to allow for non-specific absorbing materials that may be eluted by the solvent. The solvent should preferably be added drop by drop to the paper and allowed to run into a suitable receiver. This avoids introducing much fibrous material into the solvent. Alternatively the paper containing the component is cut into small pieces and a known volume of solvent added. Vigorous shaking, followed by centrifugation yields a clear supernatent layer for spectrophotometric examination. This method is adopted for Methotrexate and its preparations.

Practical Experiments

Experiment 1. *To identify the amino acids obtained on hydrolysis of Polymyxin B Sulphate*

Method. Dissolve the sample (5 mg) in water (0.5 ml) and hydrochloric acid (0.5 ml) in a 1 ml ampoule. Seal the ampoule, wrap in a thin layer of cotton wool or copper foil and heat at 135° for 5 h. Allow the ampoule to cool, open carefully and transfer the contents to a small crucible. Evaporate to dryness on a water-bath and continue to heat until the odour of hydrochloric acid is no longer detectable. Dissolve the residue in 0.5 ml of water and use the solution (0.005 ml) as a spot on the chromatography paper. On the same paper place separate spots (0.005 ml) of an aqueous solution containing 10 μg of leucine, phenylalanine, threonine, serine and α,γ-diaminobutyric acid respectively, and one spot containing 50 μg of a mixture of all five amino acids in equal amounts. Proceed by the general method for paper chromatography (ascending or descending technique, p. 88) using the upper layer of the mixed solvent system n-butanol, glacial acetic acid, water (4:1:5) as the mobile phase (*Note*) with the lower layer of the system placed in the chromatography tank. Dry and spray the paper with 0.1 % ninhydrin in n-butanol saturated with water and heat at about 90° for 5–10 min.

Identify the amino acids in the hydrolysed sample by comparison of the spots obtained from the sample with those from the known amino acids.

Note. A certain amount of butyl acetate is formed slowly, and the mixture should be allowed to stand for at least 48 hours before use in order that the equilibrium quantity of ester be present.

Cognate Experiments

Bacitracin Zinc. Heat the sample and diluted acid in an ampoule for 5 hr at 135°, use the amino acids cysteine, glutamic acid, histidine, isoleucine, leucine, lysine, ornithine, phenylalanine and aspartic acid as controls and apply the two-dimensional technique, using n-butanol, glacial acetic acid, water (4:1:5) and liquefied phenol as the solvent systems. See also TLC page 103.

Liquefied phenol should be used first, with the paper arranged so that the solvent runs along the machine direction of the paper. This direction is normally marked on the box containing the paper, or may be taken as being along the longer side of rectangular sheets. Solvents run faster parallel to this direction than they do at right angles to it.

Colistin Sulphate. Heat the sample and diluted acid in an ampoule for 5 hr at 135°, use the same amino acids as described for Polymyxin B and note the distinction. The mobile phase should be that from a mixture of t-butanol, methanol, water (4:5:1).

Experiment 2. *To separate a mixture of potassium iodide and potassium iodate*

Method. Prepare a 1–2 % solution of the sample in sodium bicarbonate solution (1 %) and place a spot on paper. On the same paper, place at well spaced intervals, spots of potassium iodide (1%) and potassium iodate (2%). Proceed by the general method for ascending chromatography using the mobile phase, methanol: water (3:1). Allow to dry and detect the potassium iodide spots with filter paper impregnated with acetic acid and potassium iodate, and the potassium iodate spots with filter paper impregnated with acetic acid and potassium iodide.

The experiment is simple but it forms the basis of a test for the radiochemical purity of Sodium Iodide (^{131}I) Solution. The radioactivity of the chromatogram obtained by adding potassium iodide, potassium iodate and sodium bicarbonate to the solution and treating as described above, should be located in one spot only—that corresponding to potassium iodide.

The addition of sodium bicarbonate is necessary to neutralise any trace of acid in the paper, otherwise iodine would be liberated.

Cognate Experiment

Sodium Phosphate (^{32}P) *Injection.* Phosphoric acid solution is added to the injection and the solvent system is *t*-butanol, water, formic acid (8:4:1). All the activity must be associated with the phosphate spot, the position of which may be determined chemically by using an acid solution of ammonium molybdate.

Experiment 3. *To separate a mixture of oxalic, succinic and glutaric acids*

Method. Dissolve the sample in water to give an approximately 3% solution and apply about 0.01 ml to paper using solutions of the acids (3%) as controls. Proceed as for the general method for ascending chromatography using the solvent system liquefied phenol: formic acid (99:1). Detect the acids by spraying with bromocresol green indicator. If the spots are indistinct, carefully expose the paper to ammonia vapor to give bright yellow spots on a blue background.

Experiment 4. *To examine extracts or tinctures of Digitalis Leaf*

Method. Impregnate chromatographic paper with formamide by passing the paper through a shallow tray containing a solution of formamide (30%) in acetone. Allow the acetone to evaporate and apply 0.1 ml aliquots of the tincture, extract or fraction A (Experiment 2, p. 84) in the form of a streak (2 cm) along the starting line. Apply in the same way reference solution of Purpurea glycosides A and B, digitoxin, gitoxin, strospeside and digitalanum verum (or as appropriate for the species of digitalis examined) and develop with chloroform saturated with formamide (descending technique) or xylene: methyl ethyl ketone (1:1) (ascending technique). Detect the zones by spraying with a mixture of 25% solution of trichloroacetic acid in ethanol: 3% aqueous chloramine (14:1) followed by heating at 120° for 4 min. Alternatively, a spray of xanthydrol (0.125% in glacial acetic acid) to which is added 1% v/v hydrochloric acid immediately before use, can be used.

To separate the primary glycosides the developing solvent is chloroform: tetrahydrofuran: formamide (50:50:6.5).

The colour developed is not intense and a streak, rather than a spot, gives more readily observed zones. The method is, in part, that of Cowley and Rowson (1963) and it can be applied as a thin-layer technique using cellulose powder (compare barbiturates, TLC Experiment 4, p. 107).

Official Examples

A number of substances and preparations of the British Pharmacopoeia are examined by paper chromatography and these are listed in Table 21.

Table 21. Paper Chromatography of Pharmacopoeial Products

Substance	Method	Tested for	Mobile Phase	Visualisation
Capreomycin Sulphate	D	Capreomycin I ($\prec 90\%$)	Propanol:water: glacial acetic acid:triethyl- amine (75:33: 8:8)	254 nm radiation followed by quantitative spectrophoto- metry
Chlormerodrin (^{197}Hg) Injection	D	Radiochemical purity $\prec 95\%$	Methanol:n- butanol:strong ammonia (7:5:3)	Diphenyl- carbazone (1%) in ethanol (95%) Radioactivity
		Mercuric chloride $\succ 2\%$		
Ergotamine Injection	D	Ergot alkaloids and related substances	Special conditions	
Erythromcyin Estolate	A	Identity and distinction from erythromycin	Isobutyl methyl ketone (alkali- washed)	Bacillus pumilus
Gentamycin Sulphate	D	Identity	Chloroform: methanol:strong ammonia:water (10:5:3:2)	Ninhydrin (0.25%) in pyridine– acetone mixture and heat at 105°
Hydroxocobalamin	D	Coloured impurities	Special conditions	
Liothyronine Sodium	C	Di-iodothyronine (2%)	Amyl alcohol:t- amyl alcohol: strong ammonia:water (5:5:3:3) Upper layer	Ninhydrin (0.25%) in acetone containing glacial acetic acid (1%)
		Thyroxine sodium (5%)		
Methotrexate Injection Tablets	D	Assay	Special conditions	
Phenformin Hydrochloride	D	Related biguanides	Ethyl acetate: ethanol (95%) water	Special conditions
L-Selenomethionine (^{75}Se) Injection	D	Radiochemical purity ($\prec 90\%$)	n-Butanol:glacial acetic acid: water (60:15:25)	Ninhydrin (0.5%) in n-butanol Radioactivity
Sodium (^{125}I) Iodide and Sodium (^{131}I) Iodide Preparations	A	Radiochemical purity	Methanol:water (3:1)	See Experiment 2 Radioactivity
Sodium Iodohippurate (^{131}I) Injection	A	Radiochemical purity ($\prec 95\%$)	n-Butanol:water: glacial acetic acid (120:50:30)	366 nm radiation Radioactivity
Sodium Pertechnetate	D	Radiochemical purity ($\prec 95\%$)	Methanol:water: (4:1)	Radioactivity
Sodium (^{32}P) Phosphate Injection	D	Radiochemical purity	t-Butanol:water formic acid (40:20:5)	Acid molybdate Radioactivity
Vancomycin Hydrochloride	D	Identity	t-Amyl alcohol: acetone:water (2:1:2)	Bacillus subtilis

continued on page 96

Table 21—*continued*

Substance	Method	Tested for	Mobile Phase	Visualisation
Vitamin A Ester Concentrate	D	Retinol	Dioxan : methanol : water containing butylated hydroxyanisole (1 %) (70 : 15 : 5) Reversed phase system	366 nm radiation

The Codex applies ascending paper chromatography to a few preparations for identification of the components italicized in Table 22. Equilibration times of 24 hr for the chromatography tank and 1.5 hr for the paper are specified before lowering the paper into the mobile phase.

Table 22

Preparation	Mobile Phase and Conditions	Visualisation
Aerosol Inhalation *Isoprenaline* *Orciprenaline* *Salbutamol*	Citric acid (0.48 %) in *n*-butanol : water (87 : 13) Paper saturated with sodium dihydrogen citrate (5 %) and air dried	Alkaline lithium and sodium molybdo-phospho-tungstate solution
Eye-Drops *Physostigmine* and *Pilocarpine*	Citric acid (0.48 %) in *n*-butanol : water (87 : 13) Paper saturated with sodium dihydrogen citrate (5 %) and air dried	Iodoplatinate solution
Paint *Brilliant Green* and *Crystal Violet*	Ethyl acetate : pyridine : water (55 : 5 : 40). Use upper layer	Self-indicating

THIN-LAYER CHROMATOGRAPHY

The techniques of paper and column chromatography are widely different and it is difficult to apply the results of the former to formulate conditions for the latter. This possibility of prediction would be of considerable value in proceeding from levels of less than a mg (paper chromatography) to levels that only column chromatography can handle. This is realised when adsorbents for column chromatography are used in the form of thin layers. They must, however, be modified, e.g. in particle size, so that they adhere to a suitable support. Although the use of thin layers of adsorbent on glass plates was described as early as 1938, many difficulties were encountered and the technique was developed much later as suitable materials and apparatus became available. It has achieved phenomenal success, not only in investigating column methods, but also in its rapidity, excellent resolving power, and in its application to analytical problems (μg scale) and preparative work for which 'thicker' layers can be used.

Although the term adsorbent is frequently used in the following text it must be remembered that adsorption may not always be the principle on which the separation of components of a mixture may be achieved; the correct principle, viz. adsorption or partition, is generally self-evident from the nature of the experiment or preparation of the plate.

The substances most frequently used as adsorbents are silica gel, alumina and cellulose and to give stable layers they often contain binders such as calcium sulphate (gypsum) or starch. A test for *Adhesive Power* in the European Pharmacopoeia specifies that when a vertical jet of air (1 mm diameter at 2 atmospheres pressure) is directed onto the prepared plate (p. 98) in a horizontal position no particles should be displaced until the jet is not greater than 5 cm from the surface (Silica gel H, Table 23) or 3 cm from the surface (Silica gel G, Table 23).

Table 23 lists a selection of commercially available materials.

Table 23. Materials for Thin-Layer Chromatography

Name	Composition	Use with Adsorbent/Water Ratio
Silica gel H	Silica gel without binder	1:1.5
Silica gel G	Silica gel with calcium sulphate	1:2
Silica gel GF	Silica gel with calcium sulphate and fluorescent indicator	1:2
Alumina Neutral Basic Acidic	Aluminium oxide without binder	1:1.1
Aluminium Oxide G	Aluminium oxide with calcium sulphate	1:2
Cellulose Powder MN 300	Cellulose powder without binder	1:5
Cellulose Powder MN 300 G	Cellulose powder with calcium sulphate	1:6
Kieselguhr G	Diatomaceous earth with calcium sulphate	1:2
Polyamide Powder	Polyamide	1:9 (Chloroform:methanol 2:3)

Various combinations of these materials with binders and inorganic fluorescent indicators are available and they may be used as in adsorption, partition or reversed phase chromatography. Additions to materials for thin-layer work are dextran gels, e.g. Sephadex of various grades, and ion-exchange compounds which are based on cellulose derivatives.

Practical Considerations

Preparation of the Plate

General method. The sizes of glass plates for use with commercially available spreaders are usually 20×20, 20×10 or 20×5 cm.

Mix the adsorbent (30 g) in a mortar to a smooth consistency with the requisite amount of water or solvent (Table 23) and transfer the slurry quickly to the spreader. Spread the mixture over 4–5 plates (20×20 cm) or a proportionately larger number of smaller plates and allow the thin layers to set (about 4 min when calcium sulphate is present). Transfer the plates carefully to a suitable holder and after a further 30 min, dry at 100–120° for 1 hr to activate the adsorbent. Cool and store the plates in a desiccator over silica gel. The thickness of the moist thin layer should be about 0.25 mm.

Special methods. (i) The adsorbent may be mixed with aqueous solution of salts, acids, bases, buffer solutions and water-soluble organic phases, e.g. formamide, for special purposes. The plate is prepared as described under the general method above except that the activation process is omitted. Partition rather than adsorption is the means whereby separation of the components of mixtures is achieved on these plates.

(ii) Reversed phase systems. After preparation as described under the general method, dip the plate into a solution of the appropriate phase, e.g. silicone oil or hydrocarbon grease in light petroleum, and allow the solvent to evaporate. Alternatively, develop the plate with a solution of the appropriate phase and allow the solvent to evaporate.

(iii) Preparative thin layer. The layers are 0.5–2 mm thick, prepared as described under the general method, but using a smaller quantity of water and allowing a longer time for the initial drying off of the plate.

(iv) In the absence of a commercial spreader, usable plates can be prepared by using strips of adhesive tape as guides for thickness of layer and for holding several clean plates in a row. The slurry is poured onto the row of plates and spread quickly over them by means of a glass rod pushed along the adhesive tapes. The tapes are removed and the plates dried and activated in the normal manner.

(v) Microscope slides are conveniently coated by a dipping technique in the following way: prepare a slurry of the adsorbent by shaking with chloroform or chloroform-methanol (2:1) and insert two microscope slides (back to back) into the slurry. Withdraw the slides, allow to drain, separate the slides and dry.

(vi) The same technique is adopted for coating the outside of test-tubes except that the test-tube is inverted after removing it from the slurry.

(vii) The slurry, prepared in the normal way, is sprayed onto the surface of glass plates, using a laboratory spray gun.

(viii) The adsorbent, mixed with an organic solvent, e.g. chloroform or ethyl acetate, is distributed evenly over a glass plate by careful tilting, and, after evaporation of solvent, is dried in the normal way.

(ix) Layers of loose adsorbent can be spread on plates using tapes as guides for thickness and a glass rod as scraper to smooth the powder. Alternatively, the glass rod can be lifted clear of the glass plate to the required degree, by means of thin rubber tubing at each end of the rod. The final plates must be developed in the horizontal or slightly inclined position.

In all methods the plates should be tidied before use by cleaning the edges and backs (microscope slides).

Preparation of Chromatogram

Remove about 2 mm of adsorbent from each edge of the plate to give a sharply defined edge.

Apply 2–5 μl volumes of a 1% solution of the mixture and of reference substances in an organic solvent to the plate (Fig. 56) with the aid of a template; the spot size should

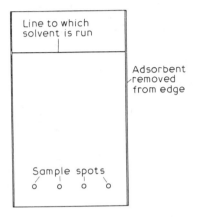

Fig. 56. Preparation of plate for thin-layer chromatography

be about 0.3 cm. For weak solutions (see Experiment 1), several applications may be necessary and each spot should be allowed to dry before applying another volume of solution to the same spot. This technique may also be adopted so as to obtain different concentrations of sample but it is not recommended; it is better to use single spots of solutions of different strength. Spot size is thus more uniform and results are found to be more consistent. Allow the solvent to evaporate and transfer the plate to a previously prepared developing tank; this preparation is done about 30 min before insertion of the plate. The tank is similar to that used for paper chromatography (ascending technique) (Fig. 53) or may be one specially designed to allow of a reproducible angle of inclination of the plate. It should be lined with filter paper dipping into the developing solvent so as to maintain an atmosphere saturated with the solvent vapour in the tank. This method helps to eliminate the 'edge effect' which sometimes occurs, particularly when mixed solvents are used. In this effect a substance moves faster when near the edge than when in the centre of the plate. Allow the solvent to rise a distance of about 10 to 15 cm, remove the plate and allow to dry using heat or a current of air as appropriate.

Detection of Compounds

The use of highly corrosive spray reagents is possible with silica gel, alumina and kieselguhr thin-layers. Therefore, in addition to the physical methods and those reagents for paper chromatography (p. 92) the following corrosive reagents may also be used:

(i) sulphuric acid, concentrated or 70–80%: causes charring of organic compounds on heating to 100°.

(ii) mixtures of sulphuric acid with sodium or potassium dichromate or with other acids such as nitric acid.

Thin layers of cellulose or those containing starch suffer the same disadvantages of paper in the detection of substances.

The recording of the results after detection of the components is no more difficult than with paper chromatography in that the chromatogram may be traced on paper, photographed or recorded with a photodensitometer. The preservation of the plates, however, poses problems but it can be achieved by spraying with a solution of plastic material such as collodion or a polymer material. When set, the film is peeled off and mounted on paper or card. Transparent adhesive Cellophane tape is also used.

Single Compounds. There are over 130 substances in the British Pharmacopoeia for which examination by TLC on silica gel G is specified and the list in Table 24 is purely representative to illustrate type of examination, solvent systems and detection methods.

Table 24. Thin-layer Chromatography of Pharmacopoeial Products

Substance	Tested for	Mobile Phase	Spray Reagents and Conditions
Amitriptyline Hydrochloride	Ketone (0.05%)	Carbon tetrachloride: benzene (3:7)	4% formaldehyde solution in sulphuric acid and examine with 366 nm radiation
Atropine Sulphate Eye Ointment Injection	Identity	Chloroform:acetone: diethylamine (5:4:1)	Potassium iodobismuthate
Bisacodyl Suppositories Tablets	Foreign substances (1%)	Xylene:ethyl methyl ketone (1:1)	254 radiation
Carbenoxolone Sodium	Related compounds (2%)	Diethylamine:ethyl acetate:methanol (3:14:10) for plate prepared with phosphoric acid (0.25%)	254 nm
Cascara Tablets	Frangula	Ethyl acetate: methanol:water (100:17:30)	Nitrosodimethylaniline followed by potassium hydroxide in ethanol (50%) and heating at 105°
Chlorcyclizine Hydrochloride Tablets	N-Methylpiperazine	Chloroform:methanol: strong ammonia (90:8:2)	Platinic chloride (0.16%): potassium iodide (2%)
Chlorpropamide Tablets	p-Chlorobenzene sulphonamide and NN'-dipropylurea (0.33%)	Isopropanol:cyclo-hexane:strong ammonia:water (15:3:1:1)	Sodium hypochlorite followed by potassium iodide in starch mucilage
Desipramine Hydrochloride	Iminodibenzyl (0.2%)	Toluene:methanol (95:5)	Potassium dichromate (0.5%) in sulphuric acid:water (4:4)

Table 24 *continued*

Substance	Tested for	Mobile Phase	Spray Reagents and Conditions
Dichlorophen	4-Chlorophenol (0.1%)	Toluene	Ferric chloride: potassium ferri-cyanide (10%:0.5%)
Diethylcarbamazine Citrate	N-Methylpiperazine (0.1%)	Ethanol (95%):glacial acetic acid:water (6:3:1)	Platinic chloride: potassium iodide
Dodecyl Gallate	Identity	Petroleum:toluene: glacial acetic acid (2:2:1)	Phosphomolybdic acid and ammonia fumes
Doxycyline Hydrochloride (on Kieselguhr G specially prepared)	Related compounds 1, 2 and 2% respectively	Ethyl acetate shaken with disodium edetate solution at pH 7	Ammonia fumes
Emetine Hydrochloride	Other alkaloids (2.5%)	Chloroform:methanol (85:15)	Iodine in chloroform, heat and examine with 366 nm radiation
Ethambutol Hydrochloride Tablets	Identity and (+)-2-amino-butanol (1%)	Ethyl acetate:glacial acetic acid:hydrochloric acid:water (11:7:1:1)	Cadmium and ninhydrin solution and heat at 90°
Liquorice Liquid Extract (plate made with 0.25% phosphoric acid)	Identity	Chloroform:methanol (95:5)	254 nm radiation
Nitrazepam Tablets	Decomposition and and related substances 0.5%	Nitromethane:ethyl (85:15)	254 nm radiation
	2-Amino-5-nitro-benzophenone (0.1%)		Method 1
Pentagastrin	Identity	(1) s-Butanol:ammonia (1%) (3:1) (2) n-Butanol:glacial acetic acid:water (4:1:5) (Upper layer)	10N Sulphuric acid to hot plate and heat at 105°
	Foreign substances (2%)	(3) Ether (anaesthetic):glacial acetic acid:water (6:2:1)	
Phthalylsulpha-thiazole	Related substances other than sulphathiazole (0.5%)	n-Butanol:N-ammonia (15:3)	Dimethylamino-benzaldehyde in hydrochloric acid (1%) in ethanol (95%)
Promethazine Hydrochloride Injection Tablets	Isopromethazine hydrochloride (2%) Other substances (0.5%) Identity as for Chlorpromazine Hydrochloride	Acetone:strong ammonia (100:2)	254 nm radiation

continued on page 102

Table 24 *continued*

Substance	Tested for	Mobile Phase	Spray Reagents and Conditions
Quinidine Sulphate	Other Cinchona alkaloids (2.5%)	Benzene:ether: diethylamine (20:12:5)	Potassium iodobismuthate
Quinine Bisulphate Quinine Dihydro- chloride Quinine Sulphate			
Saccharin	4-Sulphamoyl-benzoic acid (1%)	Chloroform:methanol: strong ammonia (100:50:11.5)	Sodium hypochlorite solution (0.5% Cl$_2$) then potassium iodide in starch mucilage
Saccharin Sodium	Toluene-2-sulphon- amide (1%)		
Sulphadiazine	Related substance (0.5%)	n-Butanol:N ammonia	Dimethylaminobenz- aldehyde
Tranylcypromine Sulphate	Cinnamylamine acid sulphate (0.6%)	Methanol	Ninhydrin in n-butanol and heat at 105°

There is only one example of the use of alumina, viz. for Colchicine but cellulose powder is used for those substances in Table 25. Many of the tests involve comparison with results of tests on reference substances and are, therefore, limit tests.

Mixtures. Most assays are non-specific, and as an aid to complete identification of components in formulations, TLC is a particularly useful adjunct. Moreover it can be applied to those preparations where difficulty would be encountered in quantitative work but where, in quality control, it must be shown that the active substance is present, e.g. where tinctures of belladonna, colchicine, gelsemium, ipecacuanha, nux vomica and opium are incorporated. The preparations cannot be applied directly to TLC plates but extraction procedures based on the principles discussed in Part 1, Chapter 13, allow the alkaloids to be concentrated in an organic solvent for examination. This aspect of TLC has been considerably developed in Appendix 5 of the British Pharmaceutical Codex.

Sensitivity of Detection Methods

The development of new drugs requires that considerable attention must be paid to the level of impurities and to the method of detection. The latter must be shown to be adequate for the limits quoted in specifications and if the impurities are known, the sensitivity of the method is readily determined. Standard volumes of gradually decreasing concentration of impurity are applied to the plate, and after development in the normal manner, various methods of detection are used to determine the most sensitive. Having thus determined the limit of detection, the drug itself may now be examined *at suitable loadings.* It is useless to apply 50 μg of drug to the plate if a specification states '0.5% of impurity A' and the limit of detection is 0.5 μg of A. At least 100 μg of drug must be applied in this example and the procedure should be

Table 25. Thin-layer Chromatography of Pharmacopoeial Products

Substance	Tested for	Mobile Phase	Spray Reagents and Conditions
Allopurinol	3-Aminopyrazole-4-carboxamide (0.1%)	n-Butanol saturated with ammonia	254 nm
Bacitracin Zinc	Constituent amino acids after hydrolysis	Isopropanol:ethyl methyl ketone: N Hydrochloric acid (60:15:25)	Cadmium and ninhydrin solution and heat at 100°
Histamine Acid Phosphate	DL-Histidine mono-hydrochloride (1%)	Isopropanol:ethyl methyl ketone: strong ammonia (75:20:5)	Cadmium and ninhydrin solution and heat at 100°
Levodopa Capsules Tablets	Identity	n-Butanol:glacial acetic acid:water (50:25:25)	Ferric chloride (5%) and Potassium ferricyanide (2.5%) mixture
	Foreign amino acids (0.5%)	Isopropanol:ethyl methyl ketone: N Hydrochloric acid (16:15:25)	Cadmium and ninhydrin solution and heat at 100°
Metformin Hydrochloride	Dicyandiamide after solvent extraction (0.04%)	Isobutyl methyl ketone:methoxy-ethanol:glacial acetic acid:water (6:4:0.6:1)	Sodium nitroprusside (10%) Potassium ferricyanide (10%) Sodium Hydroxide (10%) Equal volumes

checked by adding the impurity to the sample spot to note any effect on separation of spots.

If the impurities in the drug have not been identified *an assumption* is made that they will react towards a visualising reagent in the same way as the drug itself because they are likely to be related to it. Therefore the drug is applied to the plate at say, 200 and 100 μg and also at 1, 0.5 and 0.1 μg levels. Any spots due to impurities can then be compared in size and intensity (but not in position) with those from low loadings of the drug.

Recovery of Components

Recovery of the components of mixtures from the plates is conveniently done by means of a Craig tube (Fig. 57) to remove the adsorbent with component, followed by solvent extraction of the powder. It is particularly useful for preparative thin-layer chromatography in which the mixture is applied as a streak along the starting line and fairly large quantities of powder are therefore removed by suction.

Alternatively, the zones are removed by means of a spatula and extracted with a polar solvent such as ethanol. Glass sinters should be used to filter off the adsorbent to avoid contamination with traces of grease or fibres; the quantities of material involved are so small that it must be emphasised that all apparatus must be scrupulously clean.

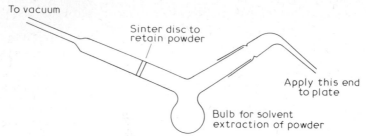

Fig. 57. Craig tube for removal of powder from thin-layer plates

Comparison of Paper and Thin-layer Chromatography

Two-dimensional chromatography can also be applied to a plate as well as to paper, and multidevelopment in one direction to separate slow moving components is very easily carried out if care is exercised in drying off the plate between each treatment with solvent.

Thin-layer chromatography has a considerable advantage over paper chromatography in that plates can be made with mixtures of adsorbents so as to give several effects:

(a) a uniform mixture of adsorbents
(b) different adsorbents succeeding each other on the same plate
(c) a gradient of two adsorbents from 100 % of the first and 0 % of the second, to 0 % of the first and 100 % of the second.

The results (b) and (c) are obtained by insertion of suitable dividers in the slurry holder, e.g. for result (b), parallel dividers are used so as to give two or more sections for the two or more different adsorbents; for result (c), a diagonal divider is used with two adsorbents, one in each compartment.

Probably the only disadvantage of thin-layer chromatography as compared with paper chromatography lies in the initial preparation of the plates. In other aspects it appears in a very favourable light, because both adsorptive and partition effects can be used, whereas in paper chromatography partition effects play the predominant role; development time is much shorter with thin-layer methods than with paper chromatography, the areas of the spots are compact and a wide range of methods of detection are available. Even the preparative difficulty is being solved to a certain extent by the provision of precoated glass, plastic and aluminium sheets. With supports other than glass, care must be exercised in the choice of solvent as the supports must be impervious and inert.

However, it is difficult to envisage the very high uniformity of chromatography paper from batch to batch being achieved from batch to batch of plates, so that the inherent value of paper chromatography (for R_F, R_X and R_M values) is not likely to be superseded.

Vapour-Programmed Thin-layer Chromatography

De Zeeuw (1968) compared the results obtained with saturated and unsaturated chambers when using multicomponent solvent systems. He concluded that better separation of the components of mixtures of chemically related substances was obtained in unsaturated chambers, i.e. those in which equilibrium between atmosphere and plate was *not* established before commencing development. He attributed this result to a concentration gradient of adsorbed vapour from the bottom to the top of the plate. The more polar constituents of the solvent system are adsorbed preferentially so that spots of solute pass continually into areas enriched in polar solvent. The migration of fast-running spots is thus increased whereas slow-running spots remain in an area of less polarity; the migration rate of the latter is, therefore, less affected.

De Zeeuw confirmed this explanation by means of a specially constructed apparatus allowing full control of vapour over the whole plate. Essentially, the plate is placed horizontally 0.5 mm over a series of twenty-one troughs, each containing the solvent system or one or more of its constituents in a carefully arranged pattern. The vapour from each trough equilibrates with the adsorbent directly above each trough thus affording precise control over the concentration gradient of adsorbed vapour. The development solvent is led onto the plate by means of filter paper.

By contrast sandwich chambers are also used for the express purpose of attaining rapid equilibrium between plate and atmosphere. In this method the plate is separated from a glass plate of similar size by a thin mask of about 3 mm thickness. The two are clamped together and placed in a small solvent trough for development. The glass plate may also be clamped directly to the chromatographic plate. A sandwich chamber is prescribed for the identification of, and test for foreign substances in, Pentagastrin.

Quantitative Measurements

Normal spectrophotometric methods may be applied to the solvent extracts obtained as described above, provided that they take the form of reaction with reagents to produce a colour. Direct examination of extracts in the ultraviolet region of the spectrum gives rise to some error because of irrelevant absorption by substances extracted from the adsorbents.

The relationship between the spot area on the plate and the concentration of the component has been investigated and several methods of correlation have been suggested. Purdy and Truter (1962) found a linear relationship between the logarithm of the weight of substance and the square root of the spot area and have provided formulae for general use.

The density of the spot may also be used and this method is likely to be of increasing value with the advent of sensitive densitometers.

Practical Experiments

Experiment 1. *To examine a petroleum ether extract of an umbelliferous fruit*

Plate. Use kieselguhr G prepared with 0.05% aqueous fluorescein instead of water and chloroform:benzene, 1:1 by volume as the developing solvent.

Method. Shake the coarsely powdered fruit (0.5 g) with petroleum ether (b.p. 60–80°, 5 ml) and allow the suspension to settle. Transfer approximately 0.03 ml of the supernatant to the plate by successive applications to produce a spot as small as possible. Develop the plate until the solvent has risen about 15 cm.

Allow the plate to dry, examine in ultraviolet light (366 nm) and note the presence of any dark fluorescence-quenching spots against the bright yellow fluorescent background. Expose the plate briefly to bromine vapour, re-examine under ultraviolet light and note any persistent fluorescein fluorescence against a dull background. Spray the plate with a saturated solution of 2,4-dinitrophenylhydrazine in N hydrochloric acid and note any orange spots. Finally, after air-drying the plate, spray with sulphuric acid containing vanillin (1%).

The results of the various methods of examination of the plate serves to identify many umbelliferous fruits, but does not distinguish caraway from dill nor anise from fennel. Table 26 correlates the results obtained with some fruits and for a complete account the original paper must be consulted (Betts, 1964).

Table 26. Thin-layer Characteristics of Petroleum Ether Extracts of Umbelliferous Fruits

Fruit	Fluorescence	R_F	Compound	Br$_2$	2,4D	Vanillin
Anise or Fennel	Quenched	0.70	Anethole	+		
		0.40	Anethole		+	
Parsley	Quenched	0.65	Apiole (?)	+	+	
Indian Dill	Quenched	0.60	Dillapiole (?)	+		
		0.40	Carvone	+	+	
Ajowan	Quenched	0.40	Thymol	+		red-mauve
Cumin		0.60	Cuminadlehyde	−	+	
		0.55				
	No quenching but visible after Br$_2$					
Caraway		0.4	Carvone		+	
Dill						mauve-
Coriander		0.25	Linalol			brown

The reasons underlying the adoption of the identification procedure are as follows. (i) Many substances are capable of quenching fluorescence and any such components of the oil show up as dark areas against the bright yellow fluorescence of fluorescein. (ii) This test alone is not very informative, but additional information is obtained by exposing the plate *briefly* to bromine vapour. The sodium salt of fluorescein is converted to that of tetrabromo-fluorescein (eosin) which does not fluoresce, so that if any of the components absorb bromine preferentially, e.g. unsaturated systems, the fluorescein in the areas of those components remains unaffected and a persistent fluorescein fluorescence is observed. Over-exposure of the plate to bromine will render this test useless; compare, in this respect, the use of ammonia vapour to render acid components visible as yellow spots against a blue background of indicator in Experiment 3 (paper chromatography, p. 94). (iii) Dinitrophenylhydrazine reacts with aldehydes and ketones to yield dinitrophenylhydrazones which vary in colour from yellow to orange-red or very dark red depending upon the carbonyl compound. Such compounds are therefore readily detected by the appearance of pale yellow to red spots on the plate. (iv) 1% Vanillin in sulphuric acid is not a specific reagent for one class of compound but is useful in the present instance in giving a variety of colours, e.g. limonene (yellow to brown colour), linalol (mauve/brown), thymol (red/mauve) and khellin (bright yellow).

Experiment 2. *To detect* p-*chloroacetanilide in Phenacetin and Paracetamol*

Plates. Kieselgel G, 250 μm thick on 20 × 20 cm glass plates.

Solvent. Cyclohexane, acetone, diisobutylketone, methanol and water (100:80:30:5:1).

Standard solutions. (*a*) Dissolve p-chloroacetanilide-free phenacetin (0.3 g) in dichloromethane (8 ml), add 0.009 % w/v solution of p-chloroacetanilide in dichloromethane (1 ml) and dilute to 10 ml with dichloromethane.

(*b*) Dissolve p-chloroacetanilide-free paracetamol (1.50 g) in methanol (8 ml) add 0.045 % solution of p-chloroacetanilide in methanol (1 ml) and dilute to 10 ml with methanol.

Sample solutions. (*a*) Dissolve phenacetin (0.3 g) in dichloromethane and make up to 10 ml.

(*b*) Dissolve paracetamol (1.50 g) in methanol and dilute to 10 ml.

Method. Apply 2 × 5 μl portions of sample solution (*a*) as a single spot. On the same line, but not closer than 1.5 cm, apply 2 × 5 μl portions of standard solution (*a*). On the same plate apply sample and standard solutions (*b*) using 1 × 2 μl of each solution.

Develop the plate until the solvent has risen 15 cm past the spots. Dry the plate in a stream of cold air and hold it within 2–3 cm of a source of ultraviolet light (253.7 nm) for 10 min and then examine under ultraviolet light (365 nm).

Compare the intensity of fluorescence of the spots due to p-chloroacetanilide in the standard with those (if any) in the samples.

The method gives an approximate estimate of the amount of p-chloro-acetanilide in phenacetin and paracetamol as such or in tablets. Savidge and Wragg (1965) suggest, however, that it is better suited for use as a limit test.

Experiment 3. *To separate a mixture of sulphonamides*

Method. Prepare a 0.5 % solution of the sulphonamides in the solvent system acetone : diethylamine:methanol (20:4:3) or chloroform:methanol:water (32:8:5) and 0.1 % solutions of reference sulphonamides in the same solvent. Apply 5–10 μl aliquots on a thin layer of silica gel (250 μm) previously activated by heating at 100–105° for 1 hr and cooling in a desiccator. Develop in the solvent system above, and after air-drying, examine under ultraviolet light to locate and identify the spots as dark areas. Alternatively, detect the sulphonamides by spraying with 0.05 % sodium nitrite followed by a solution of N-(1-naphthyl) ethylenediamine dihydrochloride (0.1 g) in water (30 ml) and hydrochloric acid (10 ml).

Experiment 4. *To separate a mixture of barbiturates*

Method. Prepare a 0.5 % solution of the barbiturates in chloroform and 0.4 % solutions of reference barbiturates. Apply 5–10 μl aliquots to a plate prepared with cellulose (without binder, 25 g) made into a slurry with *1M* potassium nitrate (50 ml) and spread to a depth of 250 μm. Dry overnight. Develop with n-butanol:pentanol (1:1) which has been saturated with solution of ammonia.

Spray with a solution of mercurous nitrate in *0.1 N* nitric acid and identify the barbiturates as white spots.

The method is that of Hjelt, Leppanen and Tamminin (1955) applied as a thin-layer technique rather than as a paper chromatographic method. Chatten and Morrison (1965) have separated barbiturates on a silica gel plate made up with *0.1N* sodium hydroxide and have proceeded to determine the barbiturates quantitatively by isolation as the mercuric salts, treatment with dithizone and measurement of the extinction of the mercuric-dithizone complex.

Experiment 5. *To show the presence of strychnine and brucine in Nux Vomica Mixture (Acid or Alkaline)*

Plate. Silica Gel G.

Solvent. Benzene, ethyl acetate, diethylamine (70 : 20 : 10).

Standard solutions. 0.1% Strychnine hydrochloride and 0.1% brucine sulphate, both in ethanol.

Method. Mix the preparation (10 ml) with dilute sulphuric acid (10 ml) and extract with two successive portions of chloroform each of 10 ml. Reject the chloroform extracts and make the aqueous solution alkaline with dilute ammonia solution. Extract with four successive portions of chloroform each of 10 ml, evaporate the combined extracts to dryness, cool and dissolve the residue in ethanol (0.5 ml). Apply a portion (10 μl) to the prepared plate and, separately, 10 μl portions of the standard solutions. Develop the chromatogram, heat the plate at 105° for 30 min and spray with dilute potassium iodobismuthate solution. The spots in the test chromatogram should correspond in position and colour to the spots in the chromatograms of the standard solutions. Subsidiary spots should be ignored.

Cognate Experiment

Ipecacuanha Mixture, Paediatric. Use the solvent system chloroform, diethylamine (90 : 10) along with cephaeline hydrochloride (0.1%) and emetine hydrochloride (0.1%), both in ethanol, as standards.

Experiment 6. *To examine Dexamethasone for the presence of related foreign steroids*

Method. Prepare a silica gel G plate and lined tank as previously described using as the mobile phase dichloromethane, solvent ether, methanol and water (77 : 15 : 8 : 1.2). Apply separately to the plate, 1 μl of each of three solutions (1) 1.5% Dexamethasone, (2) 1.5% dexamethasone BRCS and (3) 0.03% of each of prednisolone BCRS, prednisone BCRS and cortisone acetate BCRS, all solutions being in chloroform: methanol (9 : 1). Allow the mobile phase to rise 15 cm from the line of application, remove the plate from the tank, allow to dry and detect the compounds with alkaline tetrazolium blue solution (Note 1).

The principal spot from solution (1) corresponds with that from solution (2) in position, colour and intensity (identity) and any other spot from solution (1) is not more intense than the proximate spot from solution (3) (limit on related foreign steroids) (Note 2).

Note 1. The reagent is prepared immediately before use by mixing tetrazolium blue (0.2% in methanol, 1 volume) with sodium hydroxide (12% in methanol, 3 volumes).

Note 2. The method described for Dexamethasone is Method A and this also applies to some others as indicated under cognate determinations. Method B, also indicated therein, requires a mobile phase of ethylene chloride, methanol and water (95 : 5 : 0.2) and for solution (3) a mixture (0.03%) of each of prednisone BCRS, prednisolone acetate BCRS, cortisone acetate BCRS and deoxycortone acetate BCRS.

Cognate Determinations

Beclomethasone Dipropionate	(B)
Deoxycortone Acetate Implants	(B)
Fludrocortisone Acetate	(B)
Fluocinolone Acetonide	(A)

Fluocortolone Hexanoate	**(B)**
Fluocortolone Pivalate	**(B)**
Hydrocortisone Hydrogen Succinate	**(A)**

For solution (3) use hydrocortisone BCRS and hydrocortisone acetate BCRS (0.03 % of each).

Hydrocortisone Sodium Succinate   **(A)**

Solution (3) is hydrocortisone acetate BCRS (0.03 %) and a fourth solution is also specified as hydrocortisone BCRS (0.075 %).

Hydroxyprogesterone Hexanoate. Silica gel H/UV254 is specified with mobile phase cyclohexane, ethyl acetate (1:1). This is an example of the substance itself being used at a lower concentration to serve as a limit on the steroids. The sample is applied at 1 % and 0.01 % levels and detected by quenching of fluorescence.

Megesterol Acetate. The mobile phase is ethylene chloride, methanol and water (92:8:0.5) and detection is by development of a fluorescence (366 nm excitation) after spraying with ethanolic sulphuric acid (10 %) and heating at 110° (10 min). Megestrol AS is used as a control at a limit corresponding to 0.5 %.

Methylprednisolone	**(A)**
Prednisolone Pivalate	**(B)**

Experiment 7. *To determine the limit of detection of methimazole in Carbimazole*

Method. Prepare a 1 % solution of Carbimazole in chloroform and 0.05, 0.01, 0.005 and 0.0025 % solutions of methimazole in chloroform (Note). Use plates of silica gel G UV/254 and silica gel G, a mobile phase of chloroform, acetone (9:1) and apply 10 μl of each solution to the plates. Develop in the normal manner, dry the plates in air and use the following methods of detection:

(a) by examination under radiation of 254 nm—both components show up as purple areas.
(b) dilute potassium iodobismuthate solution—pink spots.
(c) iodine vapour—brown spots.
(d) dichlorobenzoquinonechlorimine—yellow spots.

Comment on the results and the specification limit of 0.5 %.

Note. Methimazole is readily prepared from Carbimazole by acid hydrolysis and extraction into chloroform, a reaction which serves as an identity test.

GAS CHROMATOGRAPHY

The phenomenal advances made in gas chromatographic instrumentation and application to analytical problems over a very short period of years, warrants a separate section to describe what is essentially a development of column chromatography (p. 75) in which the mobile phase is a gas instead of a liquid. The technique requires the vaporisation of the sample which is carried through a prepared column, at a suitable temperature, by a stream of carrier gas (the mobile phase). During the passage of the vapour of the sample through the column, separation of the components of the sample occurs by adsorption effects (p. 77) if the prepared column consists of particles of adsorbent only, or by partition effects (p. 82) if the particles of adsorbent are coated with a liquid which forms a stationary phase. In the latter instance it is better to use

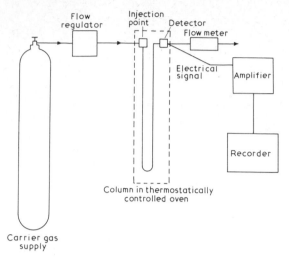

Fig. 58. Basic gas chromatography apparatus (diagrammatic)

the term *support* for the liquid phase rather than *adsorbent*, as adsorption effects are undesirable in partition columns, and supports are normally treated to eliminate, as far as possible, such effects. The two types of gas chromatography are, therefore, gas-solid and gas-liquid chromatography respectively.

It is essential that the sample is stable when vaporised and during its passage through the prepared, or *packed* column, in order to avoid decomposition products and the production of a complex chromatogram as the carrier gas elutes the products from the column. If unstable or non-volatile compounds must be examined, a reasonable approach to the problem is the preparation of derivatives of the compounds, e.g. trisilyl ethers of carbohydrates.

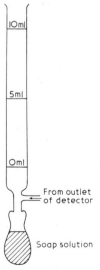

Fig. 59. Soap bubble meter

James and Martin (1952) first operated gas chromatography successfully, and the basic apparatus for the technique is shown diagrammatically in Fig. 58.

The carrier gas, at a suitable pressure, passes through the injection point which is heated to the temperature of the column or, if a flash heater or specially heated injection block is used, to about 50° above that of the column. The sample is instantaneously vaporised on injection and passes down the column with the carrier gas. Ideally, the various components of the sample are separated in so doing, and a detector at the outlet of the column produces an electrical output proportional to the amount of compound emerging from the column. The response of the detector is amplified and recorded to provide a chromatogram. Some means of measuring the gas flow is desirable at the outlet of the detector and the simplest form is a soap bubble flow meter (Fig. 59).

The soap solution provides soap films, the rate of ascent of which through the calibrated tube is a measure of the gas flow.

Carrier Gas

The gases used are argon, helium, nitrogen and occasionally hydrogen, but the selection of a carrier gas also depends on the type of detector, and it is convenient to discuss gases in association with detectors (p. 114).

Columns

Packed columns are conveniently described as analytical columns when sample solutions of a few microlitres or less are used, and preparative columns when larger volumes, e.g. 0.5 ml and more, are injected. In the latter instance, the components of a mixture can be collected at the outlet of the detector in a pure form in relatively large quantities, i.e. on a preparative scale. Analytical columns are generally from 1–1.5 m long and about 3–6 mm outside diameter. The tubes from which the columns are made may be of glass or metal but the former is advisable if there is the possibility of interaction between the metal and sample, e.g. steroids are better examined in glass columns. Preparative columns are about 3–6 m long and about 6–9 mm outside diameter but, clearly, these values, and those for analytical columns are representative only, as satisfactory results may be obtained with shorter or longer columns depending upon the types of compound to be separated and the packed columns available.

The selection of a packed column depends upon the type of analysis to be performed, and gas-solid chromatography is the method of choice for gases. The low solubility of gases such as oxygen, nitrogen, carbon dioxide, hydrogen and methane in liquids makes partition columns of little use. Reliance must be placed on adsorption effects and typical adsorbents are silica gel, alumina, charcoal and molecular sieves. They are affected by moisture and adsorbed gases, and must therefore be prepared for the column by heating and cooling in an appropriate carrier gas before use.

The appearance of the chromatograms often reflect the non-linearity of the adsorption isotherms (p. 78, Fig. 49), but the disadvantage of tailing peaks can be overcome to a certain extent by 'poisoning' the adsorbent with a small amount of liquid stationary phase. *Active sites*, which are those local parts of a packed column where adsorption effects are more pronounced than elsewhere in the column, are reduced in activity. The columns prepared in this way may appear to be somewhat similar to partition columns (below), but the latter differ in that great care is taken to reduce the natural adsorption effect of the support by the method of manufacture and by preliminary treatment with a silicone compound. Further, with gases, as in this context, partition plays little or no part in the separation procedure.

In recent years, columns packed with highly porous organic polymer beads have proved very suitable for the analysis of water, alcohols and low molecular weight gas and liquid mixtures. Hollis (1966) suggests that adsorption plays little part in the separation of the components of the mixtures.

The polymers in the Porapak series are based on ethylvinylbenzene–styrene–divinylbenzene (Porapak P), ethylvinylbenzene–divinylbenzene (Porapak Q) and ethylvinylbenzene–divinylbenzene modified with polar monomers

(Porapak R, S and T). Janâk (1967) has pointed out the new possibilities when these materials are applied as lipophilic stationary phases in thin-layer and column chromatography.

Gas-liquid chromatography offers considerably more scope for the analysis of mixtures than does gas-solid chromatography because many stationary phases are now available. (Table 27.) See also Table 29, p. 124.

Table 27. Stationary Phases for Gas-liquid Chromatography

Stationary Phase	Solvent for Stationary Phase	Stabilisation	Temperature Maximum in Use	Uses
Apiezon L $\frac{1}{2}$%	Light Petroleum	250	245	Hydrocarbons, steroids and esters
Dinonyl phthalate 10%	Acetone	125	115	Esters, carbonyl compounds
Carbowax 1000 20%	Chloroform	150	135	Alcohols, chloroform, camphor, essential oils
Carbowax 20M 10% with KOH 5%	Methanol	200	175	Volatile bases, e.g. amphetamines
Polyethylene glycol adipate 10%	Methylene chloride	225	190	Methyl esters of fatty acids nitriles essential oils
S.E. 30 1.5% and 5%	Chloroform	250	250	Hydrocarbons, Me compounds of barbiturates, general purposes
Versamide 10%	Chloroform	200	190	Phenols

The supports for the liquid phase in partition columns are generally based on silica, e.g. kieselguhr or glass beads of suitable mesh size. Kieselguhr is most useful, as liquid phases may be incorporated to the extent of 0.5 to 25% without the mixture becoming too soft or lumpy to pour easily into the tube. Glass beads are capable of maintaining a thin film only and the percentage of liquid phase must, therefore, be low (0.2–0.5%).

The preparation of the packed column is conveniently described by reference to a column of dinonyl phthalate on Celite.

Method. Coating of support. Dissolve dinonyl phthalate (1.4 g) in acetone (40 ml) in a 250 ml round-bottom flask and add prepared Celite (100–120 mesh : 12.6 g) to the solution, slowly, and with gentle mixing. Place the flask on a warm water-bath and apply a moderate vacuum to assist evaporation of the solvent. Rotate the flask continually and gently during this stage to avoid depositing the dinonyl phthalate as a film on the glass and to avoid production of fine particles by mechanical abrasion. After removal of the solvent, heat the residue under vacuum on a boiling water-bath for 1 hr.

Packing of column. For straight tubes, and for those with U-bends, attach a small funnel to the tube or limb of a U-tube, and pour in sufficient of the coated support to fill about 2.5 cm of tube. Settle the packing by tapping the tube carefully on a rubber bung, and also by means of a mechanical vibrator. Repeat this procedure until the tube or U-tube is filled. Plug the top of the tube or limbs with glass or asbestos yarn.

Preheating of column. Insert the packed column in the oven or heating jacket of the chromatograph and pass a slow stream of argon or nitrogen through the column for 5 min; *do not let the issuing gas pass through the detector.* Raise the temperature to 125° whilst continuing to pass inert gas, and allow to stand for at least 24 hours under these conditions. Allow the column to cool, still with gas flowing, and connect the detector. The apparatus is now ready for use at an appropriate temperature.

The quantities given are sufficient for about 2 metres of approximately 6 mm O.D. tubing, and the mesh size is suitable for fairly rapid rates of gas flow without excessive pressure at the injection head. For long columns, and high concentrations of stationary phase, larger mesh sizes, e.g. 80–100 or 40–60, are more convenient. For lower percentages of stationary phase, e.g. 0.5% Apiezon L, the evaporation of solvent requires the utmost care to avoid depositing a rim of the grease on the flask. For these small percentages of stationary phase another technique may be adopted, viz. add the support to a strong solution of the stationary phase, filter rapidly on a Buchner funnel, allow to drain, and dry the residue. The exact amount of stationary phase on the support must be determined by a suitable assay procedure, e.g. solvent extraction.

Packing a straight tube or limbs of U-tubes offers no particular difficulty but coiled tubes are slightly more difficult to fill. However, by applying a vacuum to one end, the coated support is sucked into the coil which is ultimately filled. If a sufficient degree of vacuum is not obtained the coil must be filled by careful manipulation.

The pretreatment of the column is essential as volatile impurities are removed without contaminating the detector. The value of this treatment may not be particularly evident if non-polar samples are examined, but if polar materials, e.g. esters, are injected, tailing of the peaks may occur. This is often due to adsorption at active sites, but after continual use, the column shows a marked improvement in performance. The temperature at which stabilisation of the column is carried out, is about 10–20°C above that at which the column is normally operated. Table 27 (p. 112) lists some common stationary phases and their operating conditions.

If no improvement in performance appears after several days use, several injections of ethereal solutions of dichlorodimethylsilane or hexamethyl-disilazane at the operating conditions may be tried, the column being disconnected from the detector whilst this treatment is carried out.

In contrast to packed columns, Golay in 1956 prepared open tubular columns 4 m by 1.5 mm coated internally with ethylene glycol. A remarkable increase in performance was obtained over that of ordinary packed columns and present day Golay capillary columns range from thirty to many hundreds of metres in length. The internal diameter is normally 0.25 mm but both smaller and larger diameters are used. The ability to separate the components of a mixture is increased 10–40-fold over that of ordinary packed columns. Golay columns are prepared for use by passing a solution of the stationary phase through the capillary and evaporating the solvent, thus leaving the phase as a film on the wall of the capillary.

Injection Systems

The injection system is an extremely important part of the apparatus and must be such that the sample enters the column as a small 'plug' of vapour. Hamilton syringes (1–10 μl volumes) with needles long enough to inject the sample directly onto the column without interruption of gas flow are ideal. For quantitative results, the rubber septum through which the needle is inserted must be stout enough to remain leak-proof over several injections and whilst the sample is being injected. If necessary, stout plugs of clear silicone rubber can be made quite easily using the monomer and a catalyst. They remain serviceable over a large number of injections. Capillary pipettes of 0.025, 0.05 and 0.1 μl volumes can also be used but the gas flow must be interrupted for the insertion of the pipette.

Golay columns require extremely small loads and therefore a stream splitter is required at the injection point to divert a small proportion only of the sample on to the column.

Detectors

The gas from the column is monitored constantly and any differences between it and the normal gas is recorded, i.e. detectors are, in most apparatus, differential. The chromatogram appears as a record of detector response against time, or, as the gas flow is known, against volume of carrier gas. Fig. 60 illustrates the trace obtained with a typical detector and includes the appearance of a corresponding curve for an integral detector such as was used by James and Martin in their first experiment with fatty acids using gas chromatography. The trace for the latter is a summation of some property related to the mass of the components passing through the column, e.g. volume of alkali required to neutralise the acid.

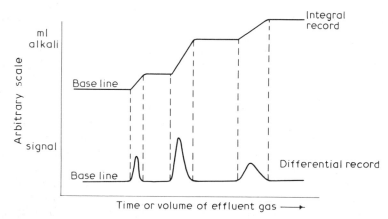

Fig. 60. Gas chromatogram of a mixture of three acids using differential and integral detectors

Few detectors are selective, i.e. when a peak is recorded no information is normally obtained on the type of compound which causes that peak (but see electron capture detector below). The detectors are therefore universal which is an advantage. Among other desirable properties for a detector are:

(i) the sensitivity should be high and without instability at high sensitivities
(ii) the volume should be low so that the compound eluted from the column in a small 'plug' of carrier gas is not diluted further within the detector itself
(iii) the response should be rapid and linear with concentration of compound. In practice the detector is calibrated to determine the optimum range
(iv) the response should be unaffected by flow rate of carrier gas and temperature.

To accommodate all these requirements, many types of detector have been made and those in most frequent use are based upon thermal conductivity or thermal effects, and ionisation phenomena.

Katharometer. This is based upon the alteration of the thermal conductivity of the carrier gas in the presence of an organic compound. The principle of the detector is illustrated diagrammatically in Fig. 61. The platinum wires

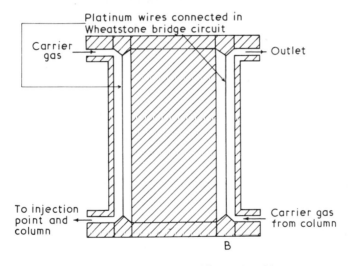

Fig. 61. Katharometer (diagrammatic)

are heated electrically and assume equilibrium conditions of temperature and resistance when carrier gas alone passes over them. They are mounted in a Wheatstone bridge arrangement and when a compound emerges, the thermal conductivity of the gas surrounding wire **B** alters, hence the temperature and resistance of the wire change with a concomitant out-of-balance signal which is amplified and recorded.

The sensitivity of this detector is low by comparison with the sensitivities of other detectors and is affected by fluctuations in temperature and flow rate. Nevertheless it is of particular value for preparative work where a high concentration of a substance occurs in the gas phase. Hydrogen or helium are the best carrier gases for this detector as nitrogen sometimes gives rise to peak reversal in the chromatogram. The sensitivity of the detector depends upon the difference between the thermal conductivity of the carrier gas and that of the compound eluted from the column. With nitrogen as carrier gas, compounds may have a greater or smaller thermal conductivity than that of nitrogen so that the response of the detector may be either positive or negative. Peak reversal is not likely when using hydrogen or helium as the figures for thermal conductivity given in Table 28 show.

Table 28. Thermal Conductivity of Carrier Gases

Carrier gas or compound	Hydrogen	Helium	Nitrogen	Methane	Hexane
Thermal Conductivity $\times 10^5$	32.7	33.9	5.2	6.5	3.0

Helium is safer than hydrogen because there is no explosion hazard.

Flame Detector. When hydrogen is used as the carrier gas it may be burnt at the outlet of the column, as a small flame about 3–6 mm high. When a compound emerges a considerable alteration of the flame temperature occurs. This change is readily detected by a thermocouple placed over the flame. The detector is sensitive but is affected by alterations in the gas flow.

Flame ionisation detector. This detector is relatively simple in design (Fig. 62) and operates by change in conductivity of the flame as the compound is burnt.

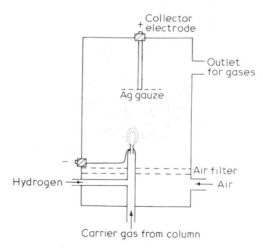

Fig. 62. Flame ionisation detector (diagrammatic)

The change in conductivity of the flame does not arise by simple ionisation of the compounds emerging from the detector. Partial or complete stripping of the molecule appears to occur to give charged hydrogen-deficient polymers or aggregates of carbon of low ionisation potential; the exact mechanism, however, is still in doubt. The electrical resistance, and hence conductivity of the flame, therefore alters appreciably as a component emerges from the column. The carrier gas is usually nitrogen or argon and is mixed with hydrogen before passing to the burner tip which is often made of a platinum capillary. This forms one electrode, the other being a silver gauze about 1 cm above the top of the flame. Designs vary, and some have the collector electrode as a cylinder surrounding the flame. The potential across the electrodes is quite low, e.g. Pye Panchromatograph: 50 V; Perkin-Elmer F 11: 150 V; laboratory-made detector: 100–400 V. The detector is extremely sensitive, has a low background signal, is insensitive to small changes in carrier gas flow and water vapour (but see Experiment 5) and responds to most organic compounds. The sample is, of course, lost in the flame but the advent of stream splitters at the column outlet combined with the high sensitivity of the detector enables a portion only of the effluent to be used for detection. The remainder passes on for recovery of samples if necessary.

Argon ionisation detector. This was designed by Lovelock in 1958 for use with argon as carrier gas and depends on the excitation of argon atoms to a metastable state about 11.7 eV above the ground state. The excitation is achieved in the detector by irradiating the carrier gas stream with either α-particles from radium-D or with β-particles from ^{90}Sr or tritium. The high energy particles ionise argon atoms and the electrons which are released are accelerated by the applied field of about 750–1500 V. On collision with argon atoms their energy is transferred to the argon atoms which are thereby raised to the metastable state. Collision of the metastable atoms with substances of lower ionisation potential than 11.7 eV causes ionisation of those substances and, therefore, an increase in current through the detector.

The detector is extremely sensitive to most organic compounds but the sensitivity is affected by water and is much reduced for halogenated compounds. The response varies with the temperature of the detector, and for high temperatures (240°) voltages of 1000 or less are usually necessary.

In a modified detector it is possible to use helium as carrier gas and, as its metastable state lies 19.8 eV above the ground state the detector is sensitive to those compounds for which a poor response is given by the argon ionisation detector, e.g. methane, carbon dioxide, carbon monoxide and oxygen among many.

The *electron capture detector* ionises the carrier gas by means of a radioactive source. The potential across two electrodes is adjusted to collect all the ions and a steady saturation current is, therefore, recorded; the applied potential is normally about 20 V. If a compound containing an electronegative element such as oxygen or, in particular, a halogen, enters the detector, electrons are 'captured' by the compound and a negative ion is produced. Combination between positive and negative ions is very rapid as compared with recombination of positive ions and electrons so that a reduction in current occurs in the detector.

For compounds of high electron-affinity argon may be used as carrier gas and, for those of lower affinity, nitrogen, hydrogen or carbon dioxide. The reason for this is that compounds of high electron-affinity are able to capture electrons of high energy whereas those of low affinity, e.g. ethers and hydrocarbons require electrons of low energy. The order of electron energy produced with various carrier gases is:

argon > nitrogen > carbon dioxide

Comparison of sensitivity of detectors. The performance characteristics of detectors depends not only on the detectors themselves but also on the associated electronic equipment. Further, for katharometers, the carrier gas and temperature of filaments play a significant role, whilst for the electron capture detector sensitivity, is strictly applicable only to electron capturing compounds. Table 28, therefore, lists *approximate* values only for the sensitivities of the detectors discussed in this chapter.

Table 28. Sensitivity of Detectors

Detector	Katharometer	Flame	Argon	Electron Capture
Minimum detectable concentration (v/v)	10^{-6}	10^{-11}	10^{-11}	10^{-12}*

Nomenclature

A number of terms relevant to gas chromatography have arisen and among them the following are of frequent occurrence in describing the chromatographic behaviour of a compound or column.

Retention time is the time of emergence of the peak maximum after injection of the compound; it is represented by x in Fig. 63.

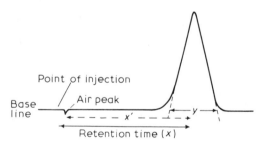

Fig. 63. Gas chromatogram (diagrammatic) for a pure compound injected along with a trace of air; detector-argon ionisation

Retention volume (V_R) is the volume of carrier gas required to elute one half of the compound from the column as indicated by the peak maximum and is given by

$$V_R = x \times f$$

where

f = flow rate of the carrier gas at the outlet pressure of the column and the temperature of the column

* For CCl_4

Adjusted retention volume (V'_R) allows for the gas hold-up of the column which is due to the interstitial volume of the column and the volume of injector and detector systems. It is given by:

$$V'_R = x' \times f$$

Net retention volume (V_N). The average flow rate of the carrier gas differs from the outlet flow rate because the gas is compressible and a pressure gradient exists down the column. A factor (j) must be used to correct the adjusted retention volume and therefore

$$V_N = V'_R \times j$$

$$V_N = V'_R \times \frac{3}{2}\left[\frac{(P_i/P_o)^2 - 1}{(P_i/P_o)^3 - 1}\right]$$

where

$$P_i = \text{pressure of carrier gas at column inlet}$$

$$P_o = \text{pressure of carrier gas at column outlet}$$

Specific retention volume, V_g, is the net retention volume per g of liquid phase at $0°$

$$V_g = \frac{V_N 273}{T W_L}$$

where

$$W_L = \text{weight of liquid phase}$$

$$T = \text{temperature of column (degrees absolute)}$$

Specific retention volumes are convenient data from which experimental retention volumes may be calculated for the particular column conditions in use. The other retention volumes described above apply to columns having particular gas hold-up volumes and amounts of liquid phase and are, therefore, applicable only to those columns.

Relative retention volume. The determination of the various retention volumes requires a considerable amount of experimental work which is essential if the information is to be meaningful to other workers. A simpler treatment which eliminates the need for calculating or measuring many of the parameters whilst retaining information of value is to determine relative retention volumes. Retention volumes for compounds are expressed relative to the retention volume of a standard compound on the same column under the same conditions as the compound examined. Therefore, this ratio is given by:

$$\frac{V_N \text{ (sample)}}{V_N \text{ (standard)}} = \frac{V'_R \text{ (sample)} \times j}{V'_R \text{ (standard)} \times j} = \frac{x' \text{ sample}}{x' \text{ standard}}$$

Relative retention volumes can be represented, therefore, by ratios of the distances on the recorder chart and are the same as relative retention times.

Retention Index. The retention index devised by Kováts (1958, 1965) makes use of the linear relationship between the logarithm of the adjusted retention volume V'_R and carbon number in n-hydrocarbons. For any hydrocarbon C_zH_{2z+2} the retention index, I, is *defined* as $100z$. Thus for n-decane $I = 1000$.

The retention index for any drug is obtained from a chromatogram in which the drug is eluted between two n-hydrocarbons C_zH_{2z+2} and $C_{z+1}H_{2z+4}$. Thus, if the compound appears between decane and undecane, the retention index lies between 1000 and 1100. The exact value is calculated from the general formula:

$$I^{\text{St.phase}}_{\text{temp}} = 100N \frac{\log x'(\text{drug}) - \log x'(A_1)}{\log x'(A_2) - \log x'(A_1)} + 100Z$$

where

$x' =$ adjusted retention time (Fig. 62)
$A_1 =$ alkane with z carbon atoms eluting before drug
$A_2 =$ alkane with $(z+N)$ carbon atoms eluting after drug
$N =$ difference in carbon number between the two alkanes

Note that it is the adjusted retention time which is used in the calculation and this leads to some difficulty in that the gas hold-up time is required when examining the chromatogram. Conditions must therefore be identical for drug and hydrocarbon. For rapid calculation of retention indices, Caddy, Fish and Scott (1973) used the plot of $\log x'$ vs carbon number as a nomogram. Multiplication of the carbon number found for $\log x'$ (drug) by 100 yielded a retention index of sufficient accuracy for tentative identification purposes in forensic science. The values were obtained on SCOT columns (support coated open tubular columns) of Apiezon L/KOH and Carbowax 20M/KOH. A plot of the retention indices obtained on the former column against those obtained on the latter gave a graph for use in the identification of certain amines.

ΔI values, e.g. $(I^{\text{Carbowax}}_{150^\circ} - I^{\text{Apiezon L}}_{150^\circ})$ have proved useful in the determination of the structure of compounds but the topic is beyond the scope of this chapter. Reference should be made to Kováts (1965) and references cited therein.

Height equivalent to a theoretical plate. A column may be considered as being made up of a large number of theoretical plates where distribution of sample between liquid and gas phase occurs. The number of theoretical plates (n) in a column is given by the relationship:

$$n = 16\frac{x^2}{y^2}$$

As y and x are proportional to the distances on the recorder chart, the correction for pressure variations in the column is not necessary for this calculation. The height equivalent to a theoretical plate (HETP) is given by

$$\text{HETP} = \frac{\text{Length of column}}{n}$$

Temperature programming. It is possible for a mixture to contain compounds which boil over a very wide range of temperature, e.g. a mixture containing hydrocarbons from C_{20} to C_{35}. Consequently, the higher members of the mixture may require several hours to be eluted if a fixed temperature (isothermal operation) consistent with resolving the lower members is used. Under such circumstances the temperature of the column may be programmed so that a gradual rise in temperature occurs and elution of the compounds is complete within a reasonable time. Various types of programme may be used, e.g. temperature increasing linearly with time or a linear increase followed by a period of isothermal operation.

To obtain a steady base line, i.e. a base line free from drift, it is usual to operate temperature programming with two columns, one of which serves as a reference column so that the effect of such factors as the following are compensated:

   (i) alteration in viscosity of mobile phase with increase in temperature
  (ii) alteration of gas flow with increase in temperature
 (iii) increase in 'bleed' rate of liquid stationary phase, i.e. the elution of traces of the liquid phase as the temperature increases. Ideally, this would be nil if the column is operated well below the limit for the stationary phase.

Quantitative Analysis

The accuracy and precision of a gas-chromatographic method depend on a number of factors most of which are within the control of the analyst.

Column performance. The number of theoretical plates (p. 120) in a column is a convenient criterion for assessing the performance or efficiency of a column. A value of 2500 (or more) for a 1.5 m column is indicative of good performance but in the determination of this value an optimum time for the emergence of the compound occurs: generally, 10–15 min is a useful guide. Also, the value will vary according to the compound chosen for the experiment and the flow rate of the carrier gas. The experiment relates only to a single peak and therefore does not strictly refer to the ability of the column to separate mixtures of compounds; usually, however, the separating ability of the column increases with increase in n.

In pharmacopoeial assays using gas-chromatography a minimum value of 600 theoretical plates per metre is required when a column other than the recommended one is used. The performance of any column is also assessed by reference to the *symmetry factor* of a peak calculated from the expression

$$\frac{y_x}{2A}$$

where y_x is the width of the peak at one-twentieth of the peak height; A is the distance between the perpendicular dropped from the peak maximum and the leading edge of the peak at one-twentieth of the peak height.

The factor is important in that when its value lies between 0.95 and 1.05 the peak heights of sample and internal standard may be used in calculations (see Experiment 5).

For satisfactory results in quantitative work the *resolution* between measured peaks on the chromatogram must be greater than 1.0. The British Pharmacopoeia states that the resolution should be calculated from the expression

$$\frac{2(t_{R_b} - t_{R_a})}{y_a + y_b}$$

where t_{R_a} and t_{R_b} are the distances along the base line between the point of injection and perpendiculars dropped from the maxima of two adjacent peaks; y_a and y_b are the respective peak widths.

Internal Standard. An internal standard (or marker) is a compound added in constant amount to solutions in order to compensate for the small variations in volume of sample injected. These variations are almost inevitable when using volumes of 0.2 to 1 μl and could well exceed the tolerances allowed in pharmaceutical formulations. In calculations or preparation of calibration curves the ratio

$$\frac{\text{peak height of component}}{\text{peak height of internal standard}}$$

is used instead of peak height of component. Thus, as an extreme example, if a volume α μl were injected followed by a volume 2α μl, the peak heights for component and internal standard would be greater for the latter injection than for the former but the ratio should, ideally, be the same. If, however, a mixture consists of two components of similar physical characteristics (see Experiment 4) the composition may be determined by direct injection of the mixture and measurement of peak areas without resort to the use of an internal standard.

The areas of peaks can be determined by integrator, planimeter or by the following geometrical methods

(i) Draw lines to the two sides of the peak so that they intersect both with each other and the base line. Take the area of the peak as that of the triangle abc (Fig. 64), i.e. area $= \dfrac{hs}{2}$.

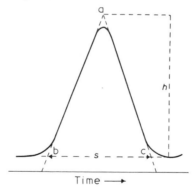

Fig. 64. Peak area by triangulation

(ii) Area = peak height × width of peak at half peak height.

The method takes into account the fact that detectors do not necessarily respond to different substances to the same extent. The relative response can be found from the equation

$$\frac{A_1}{A_2} = \alpha \frac{W_1}{W_2}$$

where α is the response factor

$$A_1 = \text{area for component 1}$$
$$A_2 = \text{area for component 2}$$
$$W_1 = \text{weight for component 1}$$
$$W_2 = \text{weight for component 2}$$

Unfortunately, α often varies with different proportions of W_1 and W_2. Therefore, several such proportions are required to ensure either that α remains constant or, if not, that a calibration curve may be constructed from which the composition of the unknown may be determined. In the extreme case the result may have to be used to assist in the preparation of a standard mixture almost identical in composition with that of the unknown.

Detection. The most useful routine method in quantitative analysis is flame ionisation. With aqueous solutions, a flame ionisation detector is essential as an argon ionisation detector is markedly affected by the presence of moisture. Although the flame detector is normally regarded as being little affected by water, for quantitative results care must be taken to ensure that the water does not emerge at the same time as a component of whatever mixture is examined. For a discussion of the effect, see the reports by Foster and Murfin (1965) and Singer (1966). To determine the exact time at which the water emerges, place a small pellet of calcium carbide at the exit of the column and inject a small volume of water. Note the point of injection and the time required for a peak to emerge. The peak is due to acetylene produced by the action of water on the carbide. This interference occurs when Carbowax columns are used because water is eluted after the lower alcohols. In the official assay procedure on a Porapak column, water is eluted first.

Derivatives. Some classes of drugs such as barbiturates, basic nitrogenous compounds and phenols, even when examined on various columns under different conditions of temperature and gas flow, still exhibit non-linearity of response and tailing of peaks. This is attributable to adsorption of the compounds on the column. A derivative may solve the problem by reducing the polarity of the compound and by rendering it more volatile.

Trimethylsilyl derivatives are very useful for phenols (see Experiment 9) but their decomposition in the flame leads to deposits of silica in the detector; consequently, loss of sensitivity and lack of precision may be expected. These effects vary from instrument to instrument, depending upon the design of the detector. The only remedy to maintain precise and accurate results is to clean the collector electrode and burner tips frequently.

Trifluoroacetyl derivatives do not suffer from this disadvantage but note that silica-based liquid stationary phases may bleed over a period of time and deposit silica when the flame is left on. The use of halogenated derivatives enables an electron-capture detector to be used for increased sensitivity but this is rarely necessary in pharmaceutical analysis where relatively large amounts of active ingredient are available from formulations.

Barbiturates are generally methylated with dimethyl sulphate to give *N*-methyl derivatives of much better chromatographic characteristics than those of the parent compounds.

The use of such derivatives requires an initial period of investigation to determine the time required for complete reaction and, possibly, to check the stability of the product. It is wise to adopt a standard procedure at the chromatographic stage such that a check is obtained on incomplete reaction or instability. In any one assay, duplicate standard solutions (S_1 and S_2) and duplicate test solutions (T_1 and T_2) should be prepared and injected in a definite sequence as follows:

$$S_1, T_1, T_1, S_2, T_2, T_2, S_1$$

Use the mean ratio of compound/internal standard for S_1, S_2 and the mean ratio for T_1, T_1 to calculate the first result. Similarly, calculate a second result using the means for S_2, S_1 and T_2, T_2 respectively. Depending upon the time of elution of components, the period between the first injection of S_1 and the last could be up to several hours. If the ratios are in reasonable agreement, it can be concluded that nothing abnormal has occurred in the solutions during the assay.

Pharmaceutical Applications

Specifications in the pharmaceutical industry are concerned not only with crude drugs and chemicals but also with intermediates used in the manufacture of the latter and formulations. Many of the intermediates are relatively simple compounds and amenable to analysis by GLC. The principle underlying the analysis is that *all* the sample injected into the GLC instrument is eluted and detected. Therefore, the total area of the main and subsidiary peaks represent 100% and a measure of the purity of the sample is readily determined by comparison of relative areas *on the assumption* that detector response is similar for all components. Volatile substances such as Ethanol, Halothane and Methoxyflurane can be examined for related compounds in the same way (see Table 29).

Table 29 Some Official Applications of Gas Chromatography

Substance	Tested for	Column	Temp.	Internal Standard
Alcohol (95%)	Other alcohols	Porapak Q	150°	Ethyl methyl ketone (0.02%)
Allopurinol	Formamide (0.1%)	Porapak Q	200°	Dimethylformamide (0.013%)
Ampicillin Sodium	Dichloromethane (0.2%)	Diglycerol (20%)	60°	Ethyl methyl ketone (0.02%)
Cephaloridine	Residual solvent	Macrogol (10%) 1000	120°	Ethyl methyl ketone (0.25%) and di- methylformamide (0.375%)
Clioquinol (as silyl derivative)	C_9H_5ClINO (≮ 90.0%)	Silicone gum rubber (SE 30, 5%)	220°	Octadecane (0.2%)
Clofibrate	Volatile related substances	Silicone gum rubber (SE 301, 30%)	185°	Ethyl 2-methyl-2- phenoxy- propionate
Colchicine	Ethyl acetate or chloroform	Macrogol 1000 (10%)	75°	Ethanol (0.1%) or Ethanol (0.02%)

Table 29 *continued*

Substance	Tested for	Column	Temp.	Internal Standard
Doxycycline Hydrochloride	Ethanol (4.3–6.0 % w/w)	Porapak Q	135°	Propanol (0.05 %)
Fenfluramine Hydrochloride Tablets	Foreign substances	Carbowax 20M (10 %) Potassium hydroxide (2 %)	135°	N,N-Diethylaniline (0.01 %)
Halothane	Volatile related compounds	Polyethylene glycol 400 (30 %, 1.8 m) followed by dinonyl phthalate (30 %, 1 m)	50°	1,1,2-Trichloro-1,2,2-trifluoro-ethane (0.005 %)
Lincomycin Hydrochloride Capsules Injection (Silyl derivative)	Lincomycin (≮ 82.5 %) Lincomycin B (≯ 5 % of area of Lincomycin peak)	Silicone gum rubber (SE 30, 3 %)	260°	Tetraphenylcyclo-pentadienone
Methoxyflurane	Volatile related compounds	Polyethylene glycol (30 %, 1.8 m) followed by dinonyl phthalate (30 %, 1 m)	80°	Chloroform (0.1 %)
Novobiocin Calcium (sodium edetate used to give clear solution) Novobiocin Sodium	Residual solvent (cf. Experiment Chapter 11)	Polyethylene glycol (10 %)	100°	Propanol (0.1 %)
Orciprenaline Sulphate	Methanol	Polyethylene glycol 1500 (10 %)	75°	Propanol (0.4 %)
Primidone Tablets	Ethyl phenyl malondiamide (1 %)	OV 17 (2 %)	240°	—
Tetracosactrin	Acetic acid (8–13 % calc. with reference to peptide content)	Porapak Q	150°	Dioxan (0.1 %)
Tranylcypromine Sulphate Tablets	Cis-isomer	OS-124 (10 %) (Polyphenylether-5 rings)	220°	4-Bromoaniline hydrochloride
Warfarin Sodium (clathrate)	Isopropanol (4.3–8.3 w/w)	Polyethylene glycol 1500 (10 %)	70°	Propanol (0.5 %)
Wool Alcohols Wool Fat	Antioxidants	Silicone gum rubber (SE 30, 10 %) (Use pre-column of silanised glass wool)	150°	Methyl n-decanoate (0.005 %)

Bulk preparation of crystalline chemicals generally involves crystallisation from a solvent and the detection and determination of solvent residues or solvent of crystallisation should form part of specifications. Normal drying processes do not always remove solvent quantitatively from the solid, and GLC and sometimes infra-red absorption (Chapter 11) are alternative methods. Compounds capable of geometrical or optical isomerism can in certain instances be examined to determine the relative proportions of each form, e.g. Tranylcypromine Sulphate. Generally this type of examination requires the formation of derivatives.

There are relatively few applications of GLC to formulations in Pharmacopoeias but the reverse is the case in industry. Not only is quantitative information directly available but GLC is very often stability-indicating in that decomposition products are separated from the parent compound. In the British Pharmacopoeia, examples relate in the main to the determination of residual solvent, to volatile related compounds as indicated in Table 29 and to the determination of ethanol in galenicals.

Combination of Gas Chromatography with Other Techniques

The identification of fractions in gas chromatography is essentially comparative in that the characteristics of the unknown are compared with those of known compounds. By correct choice of column, the fractions consist of single substances only, so that if each is examined by other methods for identification, a powerful analytical tool becomes available. This may be done in several ways, and gas chromatography is now used in conjunction with infrared spectrophotometry (p 331) and with mass spectrometry (p 424).

Practical Experiments

Experiment 1. *To determine the efficiency of a column*

Method. Select a sample appropriate to the column to be tested, e.g. ethyl acetate or propionate for a dinonyl phthalate column, and prepare a 10% solution of the ester in cyclohexane or other low boiling point non-polar solvent. With the chromatograph at about 50° and a gas flow of about 40 ml per min, inject 0.2–0.4 μl of the solution. After the components are eluted, repeat the injection: the result should be identical with that from the first injection with regard to the time required for elution of the components. Measure the distance (x) and the peak width (y) (Fig. 62) and calculate the number of theoretical plates (n) from the formula

$$n = 16\frac{x^2}{y^2}$$

where x and y are expressed in the same units.

Experiment 2. *To determine the optimum flow rate of carrier gas*

Method. As for Experiment 1 but with flow rates of 10, 20, 30, 50, 60, 70, and 80 ml per min. Plot a graph of flow rate against height equivalent to a theoretical plate (HETP) where

$$\text{HETP} = \frac{\text{length of column}}{\text{number of theoretical plates}}$$

An optimum gas flow exists for each substance and the Van Deemter equation relates band broadening to three factors. In a simplified form the equation is

$$\text{HETP} = A + \frac{B}{u} + Cu$$

where

$$u = \text{linear gas velocity in cm/sec}$$

$$= \frac{\text{length of column (cm)}}{\text{retention time for air peak}}$$

A = eddy diffusion term caused by different rates of flow around different sized particles. Normally its value is small.

B = molecular diffusion term caused by diffusion of sample in gas phase

C = mass transfer term caused by a finite time being required for exchange of solutes between gas and liquid phases

Therefore a minimum in the experimental curve is to be expected.

Experiment 3. *To examine a homologous series of compounds and the relationship between retention time and number of carbon atoms*

Method. Prepare a mixture of approximately equal amounts of one of the following series:

methyl esters of lower fatty acids	(dinonyl phthalate)
hydrocarbons	(Apiezon L)
ethers	(dinonyl phthalate)
dialkylaminoethanols	(Carbowax 10%: KOH 5%)

Inject the sample (0.2–0.3 μl) on to a suitable column as indicated in brackets above and measure the retention time for each component in a series. Plot the logarithm of the retention time against number of carbon atoms in the molecule.

The components are eluted in order of molecular weight and there is no need to inject samples of individual compounds for identification of the peaks. The graph should indicate a straight line relationship between carbon number and the logarithm of the retention time.

The examples chosen are largely artificial in that for these it is a simple matter to identify any particular member of a series by means of an authentic sample. The real value of the linear relationship shown in the experiment lies in the examination of complex mixtures such as normal and branched chain hydrocarbons obtained from plant waxes or from decomposition of quaternary ammonium compounds such as Cetrimide, or of fatty acids from fixed oils. Examination of the hydrocarbons, for example, may reveal compounds of carbon number C_{20} to C_{35} or more. Moreover, isoparaffins as well as normal paraffins may be present. Providing a few reference samples are available, the line corresponding to the normal paraffins, and that for the isoparaffins, may be drawn and identification of the individual peaks carried out with a minimum of reference material.

Experiment 4. *To determine the composition of a mixture of tertiary butyl alcohol and isobutyl alcohol*

Conditions. Dinonyl phthalate	10%
Temperature	50°

Method. Prepare a series of mixtures of the alcohols to contain from 25% w/w tertiary butyl alcohol with 75% w/w isobutyl alcohol to 50% by weight of each component.

Inject each mixture (0.1 μl) in turn into the column and measure the areas of the peaks for the two alcohols. Plot a graph of

$$\frac{A_1}{A_1 + A_2} \quad \text{against} \quad \frac{W_1}{W_1 + W_2}$$

where

A_1 = area of peak for tertiary butyl alcohol
A_2 = area of peak for isobutyl alcohol
W_1 = weight of tertiary butyl alcohol
W_2 = weight of isobutyl alcohol

Determine A_1 and A_2 for the unknown mixture and read off the value of $W_1/(W_1 + W_2)$ from the graph. Calculate the percentage composition of the mixture.

Experiment 5. *To determine the percentage v/v of ethanol in Stramonium Tincture*

Conditions. Porous polymer beads (Porapak Q or Chromosorb 101), (100–120 mesh), 1.5 m (Note 1). Temperature 160°.

Method. Prepare a standard solution containing dehydrated ethanol (5% v/v) and propanol (5% v/v). Dilute the sample (10 ml) in a 100 ml volumetric flask with propanol (5 ml) and dilute with water. Inject the standard solution and sample solution alternately (3 of each) and calculate the ratio

$$\frac{\text{height of ethanol peak}}{\text{height of propanol peak}}$$

for standard and sample.

Average the results for each and calculate by proportion the percentage of ethanol in the tincture (Note 2).

Note 1. A column 1 m long is adequate for separating all the alcohols.

Note 2. For preparations containing industrial methylated spirit the presence of methanol should be confirmed and the amount determined.

Extension. Obtain 12 replicates for the standard or sample solution and calculate the coefficient of variation of the results. Hence calculate for the result of the determination of ethanol the range within which the true value should lie (95% probability).

Cognate Experiments

Chloroform in aqueous pharmaceutical preparations. Many of these preparations contain 0.1 to 0.2% of chloroform and dilution is therefore not possible. A special injection system is required and for details see the paper by Brealey, Elvidge and Proctor (1959).

Less difficulty is experienced with preparations such as Aconite, Belladonna and Chloroform Liniment where the chloroform is extracted with carbon tetrachloride containing 1.0% v/v of methylene chloride as internal standard and the solution is compared directly with a 1.66% solution of chloroform in carbon tetrachloride with 1.0% methylene chloride as internal standard.

To allow for possible interference by constituents of the liniment with the peak for the internal standard, a separate sample is extracted with carbon tetrachloride alone and examined.

A suitable column is silicone gum rubber SE 30 (10% w/w) on white diatomaceous earth (80–120 mesh) at 100° with flame ionisation detector.

Experiment 6. *To determine the percentage of menthol in Peppermint Oil*

    Conditions. Column: Carbowax 1000 (10%)

               Temperature: 130°

Method. Prepare standard solutions of menthol in ethanol to contain 0.5, 1.0, 1.5 and 2.0% w/v including in each, 2.0% of camphor. Dilute the Peppermint Oil with ethanol (w/v) so that the expected menthol content of the dilution is about 1.5–2.0%, and include 2.0% of camphor. Inject 1 μl aliquots of each solution and calculate the menthol content (w/w) of the oil by reference to a calibration curve prepared in the normal way (Experiment 5).

The menthol content determined by this method is generally below that determined chemically.

Cognate Experiment

Camphor Liniments. The presence of fixed oil is a disadvantage if the dilution of the sample is not sufficiently great. Therefore, prepare dilutions to contain 0.1, 0.2, 0.3, 0.4% of camphor with 0.3% of menthol as internal standard. Under these conditions the detector is ^{90}Sr with argon as carrier gas.

In the absence of fixed oil, the camphor is conveniently extracted with carbon tetrachloride, nitrobenzene being used as internal standard. In the methods described in the British Pharmaceutical Codex, a calibration curve is not used and a direct comparison of the standard solution containing 0.4% of camphor and 0.4% nitrobenzene in carbon tetrachloride is made with an extract of the liniment obtained by extraction with 0.4% nitrobenzene in carbon tetrachloride. To check that miscellaneous extracted material does not appear with the nitrobenzene peak, a third solution consisting of an extract of the liniment with carbon tetrachloride is also examined. If a peak appears in the same position as that for nitrobenzene allowance is made for this in the final calculation.

The column suggested for determination of Camphor in Aconite Liniment; Aconite, Belladonna and Chloroform Liniment; Ammoniated Belladonna and Camphor Liniment, is silicone gum rubber 5% w/w SE 30 on 60–80 mesh white diatomaceous earth. A temperature of 140° and a flame ionisation detector are satisfactory.

Methyl Salicylate in Liniment and Ointments. The standard solution contains 1.0% w/v of benzyl alcohol (internal standard) and 1.0% of methyl salicylate in light petroleum (b.p. 80–100°). The sample is diluted with light petroleum to give a concentration of methyl salicylate of about 1%, and 1% of benzyl alcohol is included. The third solution required for reasons given above (for camphor) is one of the sample but without the internal standard.

Suitable conditions are 10% w/w of Macrogol 1500 on white diatomaceous earth (60–80 mesh) at 110° with flame ionisation detector.

Experiment 7. *To determine the composition of the fatty acids of Arachis Oil*

Method. Saponify the oil (2 g) with *0.5N* ethanolic potassium hydroxide (25 ml) for 30 min, cool the mixture, add water (50 ml) and extract the mixture with ether (3 × 50 ml), using gentle shaking to avoid emulsions. Wash each extract with water (20 ml) and reject the ether extracts.

Bulk the aqueous layers, acidify with *N* hydrochloric acid (40 ml) and extract the fatty acids with ether (3 × 25 ml). Wash the extracts with water (20 ml), dry the bulked extracts with Na_2SO_4 (1 g) and evaporate the ether using a current of nitrogen to remove the last traces of solvent.

Esterify the residue with a methanolic solution of boron trifluoride (5 ml) by boiling the mixture for 2 min. Cool the mixture, dilute with water (20 ml) and extract the esters with petroleum ether b.p. 40–60°, (3 × 20 ml) washing each extract with water (20 ml). Bulk the extracts, dry with Na_2SO_4 (1 g) and use the solution (0.5 μl) for injection into a chromatograph.

The conditions for separation of the esters are:

Columns: (i) Apiezon L 0.5%
              (ii) Polyethyleneglycol adipate 10%
Temperature: 150° for column (i)
              190° for column (ii)
Detector: β-ionisation or flame ionisation

Measure the area of the peaks and calculate the proportion which each peak bears to the sum of all the areas. Identify the components by the use of authentic samples of methyl esters of lauric, myristic, palmitic, stearic, oleic and linoleic acids.

When Apiezon L is used as the stationary phase the esters emerge in order of molecular weight, but with the adipate column unsaturated esters emerge after the corresponding saturated ester, e.g. methyl oleate emerges after methyl stearate.

In determining the composition of the esters the assumption is made that the peaks obtained represent all the esters present in the sample injected.

The method of esterification is an elegant one applicable to quantities of acid of less than 1 mg to very much more.

Experiment 8. *To determine volatile bases in urine*
Conditions. Column: 10% Carbowax 6000 and 5% KOH
            Temperature: 140°
            Detector: Flame ionisation

Method. Pipette urine (2–5 ml) into a glass-stoppered centrifuge tube, neutralise with dilute hydrochloric acid or sodium hydroxide solution, as appropriate, and add 0.1 ml *5N* hydrochloric acid. Extract the urine with freshly distilled ether, centrifuge and reject the ether layer. Add *5N* sodium hydroxide (0.5 ml) to the urine and extract with ether (3 × 2.5 ml) centrifuging between each extraction. Transfer the ether extracts to a 15 ml test tube, the base of which is drawn out to a fine taper. Add *N,N*-dimethylaniline solution (5 μg base/ml, 1 ml) and concentrate the solution on a water bath at 40° to about 50 μl volume. Inject 2 μl into the column. Calculate the ratio of peak heights for *N,N*-dimethylaniline to bases and read off the concentration of the bases from a calibration curve prepared by treating known quantities of the appropriate base in the same way.

The method was developed by Beckett and Rowland (1965) with the object of determining amphetamine in urine. Methylamphetamine and β-phenylethylamine are separated easily from amphetamine and the internal standard. The presence or absence of amphetamine in a treated patient's urine is linked closely with the pH of the urine so that a negative result is not conclusive evidence that amphetamine has not been taken. The method, however, introduces a confirmatory test for amphetamine when a peak corresponding to amphetamine appears in the trace—add acetone to the residue and inject the solution. Amphetamine forms an acetonide which appears from the column several minutes later than does amphetamine itself.

The inclusion of potassium hydroxide in the column is essential when examining amines as considerable tailing will otherwise occur.

Experiment 9. *To determine the percentage of adrenaline in Adrenaline Injection*

The method is that described by Boon and Mace (1969) who prepared the tri-O-trimethylsilyl ether of adrenaline for examination by gas-liquid chromatography. In the preliminary isolation of adrenaline from the injection (or any of its formulations) use is made of ion-pair formation with di(2-ethylhexyl)-phosphoric acid at pH 7.4 thus avoiding possible oxidation such as is likely to occur at higher pH values if normal isolation techniques are used.

Reagents. Phosphate buffer solution pH 7.4 made by adding *0.2N* NaOH to *0.2 M* potassium dihydrogen phosphate (250 ml) until the pH reaches 7.4 and diluting to 500 ml.

Di-(2-ethylhexyl) phosphoric acid (1 %) in chloroform.

Silanising reagent. NO-bis(trimethylsilyl) acetamide (1 vol) in dry pyridine (1 vol). The solution should contain methyl myristate (5 mg/ml) as internal standard.

Column. 5 % OV 17, 1.5 m, temperature programmed from 190° to 250° at 8° per min. In the absence of a temperature programmer a temperature of about 210–220° should be selected.

Method. To the injection (10 ml) add phosphate buffer pH 7.4 (15 ml) and extract with the di-(2-ethylhexyl) phosphoric acid solution (4 × 10 ml). Filter each extract through a small plug of cotton wool supporting a little Na_2SO_4, bulk the chloroform extracts and remove the solvent by evaporation preferably by use of a rotary evaporator until all chloroform is removed (Note 1). To the residue add silanising reagent (2 ml) and allow the mixture to stand for 2 hr (Note 2). Inject the solution (1–2 μl) and compare the ratio of peak heights for the sample with that for a standard treated in the same way.

Note 1. It is essential to remove suspended water which reacts with the silanising agent. A residue of adrenaline in di-(2-ethylhexyl) phosphoric acid remains.

Note 2. The silanising agent contains the internal standard and must, therefore, be accurately measured. The reaction rate is increased by the di(2-ethylhexyl) phosphoric acid and the standard must be treated in the same way; even so, 2 hr should be allowed.

HIGH PRESSURE LIQUID CHROMATOGRAPHY

Theory

The general principles applicable to high pressure liquid chromatography have already been discussed under its four basic modes *viz.* column chromatography (adsorption and partition), ion exchange (Part 1, Chapter 6) and gel permeation

(Part 1, Chapter 6). HPLC is also closely akin to GLC in nomenclature descriptive of the performance of columns, e.g. efficiency in terms of theoretical plates (pp 120 and 126) and resolution. With regard to optimum flow rate (p 126), however, the diffusion term B/U in the van Deemter equation and its modifications is negligible because diffusion in the liquid mobile phase is considerably smaller than in the gas phase. Consequently, no minimum appears in the graph relating HETP to flow rate under practical working conditions. Flow rate can, therefore, be increased, within reason, with not too great an increase in HETP.

As in GLC, a packed column must have a capacity to retain samples and to separate the components of a mixture. The *capacity factor* k^1 is defined as the ratio of net retention volume (V_N) to void volume (V_0). If the flow rate is a ml per min the respective volumes are

$$V_N = a(t_r - t_0) \text{ (Compare GLC, p 118)}$$

$$V_0 = at_0$$

and

$$k^1 = \frac{V_N}{V_0} = \frac{t_r - t_0}{t_0} \text{ (Fig. 65).}$$

The chromatogram provides the data for calculation as illustrated in Fig. 65.

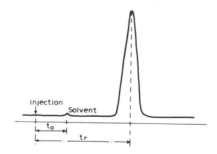

Fig. 65. Chromatogram of a pure compound, the solvent for which differs slightly from the mobile phase to give the 'solvent peak'

Fundamentally,

$$k' = K \frac{V_s}{V_0}$$

where

$$K = \text{distribution coefficient}$$
$$V_s = \text{volume of stationary phase}$$
$$V_0 = \text{volume of mobile phase (void volume).}$$

When $k' = 0$, components elute in volume V_0 and no separation is possible as they elute by analogy with TLC, with the 'solvent front'.

The analogy may be carried further in that compounds of R_F 0.2 and 0.5 in TLC would lead to k' values of 4 and 1 respectively in HPLC using similar packing and solvent systems. Unfortunately, direct conversion of TLC data to HPLC is a gross oversimplification at present. Even so, the wealth of information available for TLC on silica gel should prove invaluable in developing adsorption systems for HPLC

The separation of two components depends upon their relative capacity factors and a *separation factor* α which is defined as

$$\alpha = \frac{V_2 - V_0}{V_1 - V_0} = \frac{V_{N_2}}{V_{N_1}} = \frac{k'_2}{k'_1}$$

A general formula for the resolution (R_s) of two closely spaced peaks is given by

$$R_s = \frac{1}{4}\left(\frac{\alpha - 1}{\alpha}\right)\left(\frac{k'_2}{k'_2 + 1}\right)\sqrt{N}$$

where N = number of theoretical plates of the column. An assumption is made that, as the two peaks are closely spaced, the peak widths are the same. The capacity factor can be readily altered to effect changes in resolution by altering solvent composition rather than by changing the column packing. If, however, this approach fails with, e.g. a particular adsorbant packing, partition or ion-exchange columns may be tried. It is not easy to predict the effect on resolution because, although the alteration on k'_2 may be anticipated, the alteration on α may be uncertain as both components are involved in the separation factor.

Apparatus

The basic system is illustrated diagrammatically in Fig. 66.

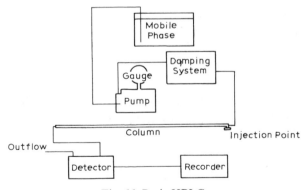

Fig. 66. Basic HPLC

The mode of operation of this system is *isocratic*, i.e. one particular solvent or mixture is pumped throughout the analysis. For some determinations the solvent composition may be altered gradually in a predetermined manner to give *gradient elution*. Qualitatively this may be done by using 100 ml quantities of mixtures of varying proportions, e.g. non-polar solvent with increasing amounts of polar solvent, but clearly this is a crude method. Precision systems giving accurate reproducible solvent profiles are required for quantitative work and for separation of 'difficult' mixtures, e.g. aminoacids.

Pumps

Pumps are required to deliver a constant volume of mobile phase at pressures ranging from 200 to 8000 lb/sq. in. Although not essential, it is useful to have a pump capable of pressures up to 8000 lb/sq. in because advantage may be taken of the increased flow rate and also of new developments in column technology now leading to short columns of packing material of particle size less than 10 μm.

Mechanical pumps of the reciprocating piston type give a pulsating supply of mobile phase. A damping device is therefore required to smooth out the pulses so that excessive noise at high levels of sensitivity or low pressures does not detract from detection of small quantities of sample. This type of pump is extremely useful however in that a constant volume of liquid is delivered, the actual value being set by adjustment of piston stroke. This means that the pressure shown on a gauge acts as an indicator of working conditions. Thus, if the column becomes partially blocked, a rise in pressure occurs until ultimately the relief valve (essential in this type of pump) operates. Similarly, leakage from column connections or pump valves show up as lower pressures. In both cases suitable maintenance measures can be put into operation immediately.

Pulse-free operation is given by pumps operating a screw-driven syringe and by pneumatic pumps based on gas pressures. The latter type operate as constant pressure pumps, the flow rate depending upon the pressure selected. Developments in reciprocating pumps have led to pulse-free operation by using twin pistons carefully phased in operation.

Detectors

Essential requirements are high sensitivity, good stability and low dead volume. Two commonly used detectors involve ultraviolet absorption and refractive index. Absorbance may be measured at 254 nm (fixed wavelength); at up to five preset wavelengths; or over the ultraviolet-visible region by means of a spectrophotometer designed specially for HPLC. The advantage of the last detector lies in the increased sensitivity for those compounds which absorb strongly at about 220 or 320 nm but for which the absorption curves rapidly decrease towards 254 nm. Digoxin is an example of such a compound which has absorption at 220 nm about 70 times that at 254 nm. In general, however, bearing in mind the nature of mobile phases and the usually broad absorption bands encountered, the 254 nm detector is surprisingly adequate. With regard to solvents, the cut-off points are a guide to the usefulness of any solvent with ultraviolet absorption detectors. If a solvent has significant absorption at 254 nm, e.g. ethyl acetate has an E_1 value of about 0.6, the injection of samples containing small amounts of transparent solvents, e.g. water, methanol or ethanol onto adsorption columns, gives rise to negative peaks in the chromatogram as the solvents are eluted. Similar remarks may also apply when solutions of ammonium nitrate are used, depending upon their ionic strength.

The refractive index detector is a universal detector in that some solvent is usually available in which the sample gives rise to a measurable difference in refractive index between solvent and solution. The two detectors can be coupled in sequence providing narrow bore tubing and low dead volume fittings are used thus allowing some conclusions to be drawn from chromatograms on the properties of eluted compounds.

Of comparable importance in detection systems is the method described on p 82 now modified, however, to convert all carbon-containing compounds to methane which is the actual compound measured. Detectors, both general and for particular purposes, are continually being developed and for information on such systems as polarography, fluorescence, heat of adsorption, capacity and conductivity suggested reading at the end of the chapter should be consulted.

Columns

Columns are made of stainless steel of about 2.3 mm internal diameter and Swagelock or Simplex fittings are required for connection to injection and detector tubing. It is essential to use a fitting incorporating a 2 μm filter or frit at the outlet of the column to avoid particles of packing material blocking the narrow bore tubing leading to the detector. Connection to the detector should be made with low dead volume fittings.

Considerable attention is being paid to the internal surface of columns and although the simple treatment described in Experiment 1 may seem excessive it is by no means the case. Scrubbing and polishing of the internal surface and more recently, drilling, have been adopted to remove the longitudinal striations which contribute to erratic flow of solvent and consequent poor efficiency of the packed column.

Injection

A pressure syringe can safely be used through a septum up to pressures of 1000 lb/sq. in but above that pressure special injection systems are desirable. In one such type the sample is placed in a sample loading loop which is then turned so that the sample is swept out onto the column by the mobile phase. This system operates at all pressures.

Packing Materials

The efficiency of high pressure liquid chromatography depends considerably on the particles of packing material being porous and of small diameter (about 40 μm). The latter factor reduces band-broadening caused by slow mass-transfer within the particles which occur in several forms. *Porous silica beads* are marketed under trade names such as Porasil and Sil-X-1 both of which have adsorbent properties. Columns of these materials are used in liquid/solid chromatography for the analysis of non-polar samples. Elution problems can be largely overcome by increasing the polarity of mobile phases, such as trimethylpentane or chloroform, with acetonitrile or methanol. The samples need not necessarily be entirely non-polar as is shown by the success of TLC using silica gel plates.

Porous silica beads of controlled porosities may also be used to fractionate materials of high molecular weight. The principle is similar to that described for gel filtration (Part 1, Chapter 6) but Sephadex cannot withstand the pressures required in HPLC. Porous silica beads of the required properties are Porasil and Vit-X for use with aqueous solutions. Hydrophobic porous polystyrene or polystyrene acetate beads for use with non-aqueous solvents are Poragel and Styragel.

Increased efficiency is obtained if the particle consists of a *porous silica layer* surrounding a solid core. The layer is approximately 1 μm thick and materials of this type are Corasil, Sil-X-II and Zipax. They require smaller samples sizes than those for porous beads to avoid overloading the column but this is no real disadvantage with sensitive detectors now available.

Zipax is a specially inert material and is normally used as a support for liquid stationary phases rather than for liquid/solid chromatography. The other materials quoted above may also be used with a liquid film and the surface so modified is the basis of liquid/liquid chromatography. Providing the stationary phase does not dissolve in the mobile phase no special precautions are necessary in using the treated column. If some solubility is possible, e.g. by modification of the mobile phase with polar solvents, a precolumn may be necessary to avoid stripping of the stationary phase. Liquids used in coating porous layer and porous beads are Carbowax, β,β-oxidipropionitrile (BOP), trimethyleneglycol and XE60 though there appears no reason why many other GLC stationary phases should not be used. The loadings are generally quite small at about 0.5 %.

Stripping of the stationary phase can be avoided to a large extent by means of *bonded supports* (Table 30). Chemical reaction of the Si—OH group of silica beads or layer with reagents leads to films bonded by means of ester, ether or Si—C bonds and hence to different degrees of resistance when hydrolytic solvents are used. Some of these supports are used in the reversed phase mode, i.e. the mobile phase is more polar than the stationary phase.

Table 30. Bonded Support Materials

Material	Type	Supplier
*Octadecyl Sil-X-1	C_{18}-hydrocarbon bonded to porous bead	Perkin-Elmer
*Octadecyl Sil-X-11	C_{18}-hydrocarbon bonded to porous layer bead	Perkin-Elmer
*Phenyl Sil-X-1	Bonded phenyl groups	Perkin-Elmer
Fluoroethyl Sil-X-1	Ether	Perkin-Elmer
Durapak		Waters Associates
OPN/Porasil C	Ester	
CW400/Porasil C	Ester	
n-C_8/Porasil C	Ester	
CW400/Corasil 1	Ester	
Bondapak		Waters Associates
*n-C_{18}/Corasil 1	Ether	
*Phenyl/Corasil 1	Ether	
Permaphase		Du Pont
*ODS	Octadecyltrimethyl siloxane	
AAX	Anion exchange	
ETH	Aliphatic ether	

* Reversed phase mode.

Ion exchange resins are available as porous beads, as solid polymer coatings and bonded layers but their mode of action may not be entirely due to ion exchange, e.g. when non-ionic materials are separated.

Practical

The object of the experiments is to illustrate the development of methods for formulations which normally require some time and effort to examine by conventional means. All the determinations can be carried out with particular mobile phases in analytical columns 2 feet by 2.3 mm internal diameter and ultraviolet absorption (254 nm) for detection. Gradient elution, which adds considerably to the cost of instruments, is not required. Dissolved air should be removed from mobile phases before use by warming or applying a vacuum.

Experiment 1. *To prepare an analytical column with packing material of particle size about 40 µm*

Material of particle size 30 µm or over can be dry-packed but to ensure satisfactory results more time must be allowed than for GLC columns. Owing to the small particle size there is little point in applying a vacuum to one end and gentle tapping of the column is sufficient. A mechanical filler which automatically lifts and drops the column about 12 mm is very useful for preparation of reproducible and efficient columns.

Method. Rinse the column with nitric acid (25%), water, methanol and ether and dry at 100°C (Note 1). Plug one end of the column with cotton wool, attach a fitting incorporating a 2 µm frit to this end and add enough material to fill about 1 cm of the column. Tap the column vertically on a hard surface for about 1 minute with an occasional tap on the side (Note 2). Continue in the same way until the column is filled to within 3 mm of the top and plug the space firmly with cotton wool (Note 3). Remove the 2 µm frit, attach the column to the injection point and pump the selected mobile phase (100 ml) through the column which should not be connected to the detector at this stage (Note 4). Attach the outlet of the column to the detector inlet tube with a connector containing a 2 µm frit and continue pumping. The column is now ready for use.

Note 1. This is a cleaning process to remove any salts, dust or fatty substances from the stainless steel tube.

Note 2. This method is reported to give more efficient columns than that adopted for GLC where a vibrator is often used. The fitting of the frit or other connection at this stage avoids damage to the end of the column as it is tapped on a hard surface.

Note 3. It is assumed that on-column injection will be used so that other possible means of plugging the column are not possible, e.g. a porous Teflon plug or metal frit.

Note 4. This is a general cleaning up process.

Experiment 2. *To prepare and pack a liquid coated support*

The example selected is XE 60 (0.5%) on Zipax but the general procedure is the same for other liquids apart from choice of solvent.

Method. Dry the Zipax (5 g) overnight at 100° and allow to cool in a desiccator (Note). Transfer to a 250 ml flask and add toluene (5 ml) and XE 60 (25 mg) dissolved in warm toluene (25 ml). Evaporate to dryness *in vacuo* using a boiling water bath and a rotary evaporator. Add toluene (20 ml) to the residue and again evaporate to dryness. Heat for a further 2 hr.

Table 31. Some Applications of HPLC

Substance or Formulation	Column	Mobile Phase	Type of Chromatography	Reference
Ergot alkaloids	Bondapak Phenyl/Corasil	Acetonitrile : water with *0.1M* ammonium carbonate (35:65)	Reverse phase	Waters Assoc.
Cough preparations (Vasoconstrictors, Antitussives, Antihistamines)	Bondapak Phenyl/Corasil	Acetonitrile : water with 0.1 % $(NH_4)_2 CO_3$ [50:50]	Reverse phase	Waters Assoc.
Phthalate Plasticisers	μ Porasil	Ethylacetate/iso-octane	Normal phase	Waters Assoc.
Polyvinyl Chloride	Styragel	Tetrahydrofuran	Normal phase	Waters Assoc.
Propylene Glycol	Bondapak C_{18}/Corasil	Acetonitrile/water (25:75 v/v)	Reverse phase	Waters Assoc.
Methyl Testosterone Testosterone Propionate Dexamethasone	Bondapak C_{18}/Corasil	Methanol/water (75:25 v/v)	Reverse phase	Waters Assoc.
Penicillin G	Porasil A conditioned with *0.1M* PO_4^{-3} at pH 6.7	Chloroform/methanol/*0.1M* PO_4^{-3} at pH 6.7	Liquid/liquid	Waters Assoc.
Tetracycline	Special Hydrocarbon Phase	Acetonitrile/water/EDTA	Reverse phase	Waters Assoc.
Fat-soluble vitamins (Hexane extract)	μ Bondapak C_{18}	Methanol/water (95:5 v/v)	Reverse phase	Waters Assoc.
Water-soluble vitamins (Methanol extract)	CX/Corasil	Ammonium hydrogen phosphate (*0.05M*)	Reverse phase	Waters Assoc.
Cosmetic preparations (Estrone, Estrogenic Cream)	Bondapak C_{18}/Porasil	Acetonitrile/water (50:50 v/v)	Reverse phase	Waters Assoc.
Librium	C_{18}/Corasil	Acetonitrile/water with 0.1 % $(NH_4)_2 CO_3$ (25:75 v/v)	Reverse phase	Waters Assoc.
Quinones	Bondapak C_{18}/Corasil	Methanol/water (50:50 v/v)	Reverse phase	Waters Assoc.
Herbicides	Bondapak C_{18}/Porasil	Methanol/water (50:50 v/v)	Reverse phase	Waters Assoc.
Mild analgesics (Caffeine, Phenacetin, Aspirin, Salicylic Acid)	Bondapak AX/Corasil	Na Citrate (*0.03M*) pH 3.52	Reverse phase	Waters Assoc.

Sample	Column	Eluent	Manufacturer
Synthetic fibre coatings (Acrylamide, Acrylic Acid, Sulphonic Acid)	Biorad AG1-×10 (200–400 mesh) 20% w/v	Sodium acetate (0.1N) pH 8.2	Perkin-Elmer
Phthalic acid isomers Mono-, di- and tri-phosphates	325 mesh Biosil Silica Gel 'Permaphase' AAX	Methanol Linear gradient 0.005M KH$_2$PO$_4$ (pH 3.0) to 0.5M KH$_2$PO$_4$ (pH 4.0)	Perkin-Elmer Du Pont Instruments
Alcohol mixture (α methyl benzyl, alcohol, α,α'dimethyl benzyl alcohol, 2-phenylethanol, cinnamyl alcohol benzyl alcohol)	'Permaphase' ETH	Hexane	Du Pont Instruments
Nucleic acid	'Zipax' SCX	Nitric acid (0.01M) ammonium nitrate (0.05M)	Du Pont Instruments
Fused-ring aromatics 60% Chlorinated Biphenyls	'Permaphase' ODS 'Permaphase' ODS	Methanol/water (60:40 v/v) Methanol/water (60:40 v/v)	Du Pont Instruments Du Pont Instruments
Methyl prednisolone and related steroids	'Zipax' BOP	Tetrahydrofuran/heptane (20:80)	Du Pont Instruments
Carbaryl and 1-Naphthol from a plant extract	'Zipax' TMG	n-Hexane saturated with T.M.G.	Du Pont Instruments
Tartrazine and reaction products	SAX	Linear gradient 4% min-distilled water pH 9.2 to distilled water pH 9.2 + 0.05M sodium perchlorate	Du Pont Instruments

Dry pack the column as described in Experiment 1 and pump through methanol (10%, 100 ml) to remove traces of toluene.

Note. This is to remove adsorbed water.

Experiment 3. *Determination of Benzocaine and Phenazone in Ear Drops*

The composition of the preparation is stated to be as follows:

$$
\begin{array}{lll}
\text{Benzocaine} & & 1\%\ \text{w/v} \\
\text{Phenazone} & & 5\%\ \text{w/v} \\
\text{Glycerol} & \text{to} & 100\%
\end{array}
$$

Both components can be determined by means of ultraviolet absorption as a two-component system. A wavelength can be selected at which one component does not absorb whereas the other has significant absorption albeit not quite at the maximum. Therefore if HPLC is to be applied to the problem, the simplest possible method must be evolved to offer any real advantage over conventional spectrophotometry. Consequently, as the preparation is water-miscible a reversed phase column (C_{18}/Corasil II) is selected using water with a suitable proportion of acetonitrile to elute the components.

Method. Accurately weigh the sample (about 1.3 g) into a 50 ml volumetric flask and dilute to volume with water. Mix thoroughly and inject 10 μl accurately measured (Note 1) into a chromatograph stabilised with water : acetonitrile (70 : 30) and an ultraviolet detector (254 nm). Elute the components at about 0.8 ml per min at an appropriate sensitivity (Note 2) and compare the peak heights (Note 3) with those for a standard solution (10 μl) containing benzocaine (0.02%) and phenazone (0.1%). Make replicate injections (3) of sample and standard, average the results and calculate the percentage w/v of benzocaine and phenazone using a wt/ml of 1.26 for the preparation.

Note 1. With volumes of 10–20 μl the precision of the results is very good and an internal standard is not strictly necessary.

Note 2. The sensitivity setting for the benzocaine should be increased to obtain a satisfactory peak height for this compound (see Fig. 67A).

Note 3. An integrator would be preferable for measurement purposes as the peaks are rather broad.

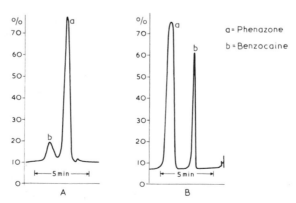

Fig. 67. Elution patterns in reversed phase mode (A) and absorption (B) a: Phenazone b: Benzocaine

Alternative Method

An adsorption technique can be used with chloroform (A.R.) as solvent and a column of porous layer silica beads such as Corasil II. The elution pattern under these conditions is shown in Fig. 67B.

Method. To chloroform (50 ml) in a 50 ml volumetric flask add the sample (about 1.3 g accurately weighed), stopper the flask and shake vigorously for several minutes (Note 1). Allow to stand 5 min for the chloroform to clear. Prepare a standard solution by adding to a chloroform solution (50 ml in a volumetric flask) of benzocaine (0.02 %) and phenazone (0.1 %), glycerol (1.3 g) (Note 2) and shaking vigorously for several minutes. Inject 10 μl volumes of standard and sample and complete the assay described above (Note 3).

Note 1. Glycerol is insoluble in chloroform and vigorous shaking is required to ensure extraction of components into the chloroform.

Note 2. Glycerol is added to allow for any retention of components that might occur in accordance with the distribution coefficient.

Note 3. Benzocaine is eluted first in this method and the peak height is comparable with that of phenazone. Increase of the sensitivity setting is not, therefore, required.

Experiment 4. *Determination of Aspirin and Caffeine in Aspirin and Caffeine Tablets*

The components are soluble in chloroform and it would appear that the system described under Experiment 3 (Alternative Method) would be the method of choice. Under those conditions, however, the peak due to aspirin tails badly. Although aspirin is not appreciably soluble in water it is soluble in acetonitrile and, moreover, is not precipitated on dilution with water. A reversed phase system is possible but accurate control of the acetonitrile content is essential to separate aspirin and caffeine.

Method. Weigh and powder 20 tablets. Accurately weigh a quantity of powder equivalent to one tablet into a porosity-3 sinter-glass crucible. Extract carefully and thoroughly with 5 × 5 ml of acetonitrile, with thorough mixing and stirring. Dilute the filtrate to 200 ml with water, mix well and inject 15 μl samples for analysis. Use a standard solution of aspirin (0.18 %) and caffeine (0.015 %) made from solutions in acetonitrile such that the final concentration of acetonitrile is 10–15 % v/v. Elute with acetonitrile (10 % v/v), measure the peak heights, and calculate the quantity of aspirin and caffeine in a tablet of average weight.

Cognate Experiment

Investigate the effect of acetonitrile concentration and flow rate on the resolution of a mixture of aspirin and paracetamol.

Hint: Paracetamol appears between aspirin and caffeine under the conditions above.

Experiment 5. *Determination of Caffeine in Caffeine Iodide Elixir*

The Codex describes a gravimetric determination for caffeine and an electrometric determination for iodide both methods being satisfactory. The HPLC method is of interest however because an approximate measure of the iodide content can also be made as indicated in the Note.

Standard Solution. Sodium iodide (9 %) and Caffeine (3 %) in water.

Internal Standard Solution. Phenazone (3 %) in water.

Column. Reverse phase C_{18}/Corasil II.
Mobile Phase. Acetonitrile:water (15:85).

Method. Dilute the sample (5 ml) and phenazone (3%, 5 ml) to 500 ml with water. Similarly dilute the standard solutions and inject 10 μl of each solution in the normal manner. Whilst the experiment is proceeding, pass about 60 ml of the sample dilution through a short column (15 cm × 2 cm) of an anion exchange resin (hydroxyl form), reject the first 50 ml and collect the last 10 ml. Use as blank (10 μl) (Note).

Calculate the peak height ratios as described under GLC (p. 122) correcting the iodide peak for the presence of other compounds. Calculate the percentage of caffeine and the approximate concentration of sodium iodide in the mixture.

Note. Iodide ion has an intense absorption at about 228 nm and absorbs sufficiently at 254 nm to give a satisfactory peak on the chromatogram. Unfortunately, however, it is eluted in little over the void volume of the column and the results are not as precise as those for caffeine. Further, colouring matter from the liquorice and coffee (used in the preparation) appears at the same point in the chromatogram. A correction can be applied for this by removing the iodide ion by passage through a short column of an anion-exchange resin. Figure 68 illustrates the chromatograms obtained for the sample and for the sample from which the iodide is removed.

Iodide can also be removed by addition of silver nitrate but the effect of excess silver ions on the column leads to peculiar results on subsequent injections of iodide. In general, the injection of a sample in a solvent slightly different from the mobile phase may have a significant effect. All columns must be in equilibrium with the mobile phase for reproducible results and, if this is disturbed, immediate injection of another sample may give a different result, e.g. in peak height or retention time. If such an effect is observed, adequate time must be allowed to re-establish equilibrium to ensure that it is, or is not, the cause of poor results.

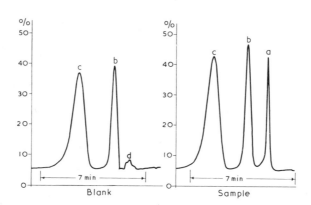

Fig. 68. Chromatograms for Caffeine Iodide Elixir before and after removal of iodide
    a: iodide + other constituents   c: phenazone
    b: caffeine   d: other constituents

Experiment 6. *Determination of benzoic acid in Camphorated Opium Tincture*

No official assay is specified for benzoic acid in this preparation but it can conceivably be carried out by isolation of the acid and titration. A useful exercise may also be developed with this formulation to illustrate the use of isotope dilution analysis. The simplest procedure is, however, HPLC using an anion exchange column and a mobile phase of *0.02M* ammonium nitrate adjusted to pH 9.2 with ammonia. A typical trace is shown in Fig. 69.

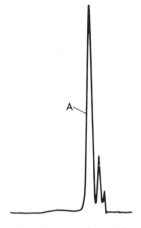

Method. Inject 5 μl of the sample and adjust the sensitivity of the detector to give about 60% full scale deflection for benzoic acid. Compare the height of the benzoic acid peak with that for 0.5% benzoic acid in the buffer solution under the conditions selected.

Fig. 69. Camphorated
Opium Tincture
A: benzoic acid.
Note other constituents

Cognate Determination

Compound Glycerin of Thymol. Determination of Sodium Benzoate and Sodium Salicylate The determination may be used to illustrate the importance of ionic strength in obtaining satisfactory elution of components. If carried out using *0.02M* ammonium nitrate (pH 9.2) as described above the salicylate is eluted slowly (20 min) and the band is very broad. On increasing the ionic strength to *0.2M*, however, the salicylate ion is eluted more rapidly (4 min).

Experiment 7. *Determination of Triprolidine Hydrochloride Codeine Phosphate and Pseudoephedrine Hydrochloride in a linctus*

The preparation is stated to be of the following composition:

Triprolidine Hydrochloride	1.25 mg
Codeine Phosphate	10 mg
Pseudophedrine Hydrochloride	30 mg
Syrup base	to 5 ml

Red colouring is also present.

Cough preparations have been examined on ion-exchange resins and a possible approach to the problem is, therefore, indicated. Waters Associates Ltd used *0.35M* sodium citrate at pH 5.3 to elute epinephrine and various non-ionics from an anion exchange resin but the use of a *0.1M* solution to avoid too high a concentration of salt at pH 5.3 gave the trace shown in Fig. 70 for a mixture of the three compounds in proportions similar to those in the linctus.

The next step is clearly to investigate the effect of pH on the elution pattern and the results are shown in Fig. 70.

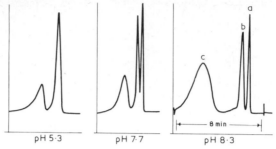

Fig. 70. Chromatogram of standard mixture

a: pseudoephedrine   b: codeine   c: triprolidine

The conditions, therefore, appeared to be *0.1M* sodium citrate which solution had a pH of 8.3. However, injection of diluted sample gave a distorted trace for triprolidine indicating possible overlapping bands. Further investigation at pH 8.8 and 9.2 gave the traces shown in Fig. 71 for the sample.

The influence of pH is clearly evident in the development of the method. What the other components are can only be known to the manufacturer, but note how triprolidine is eluted more slowly with increase in pH. One of the

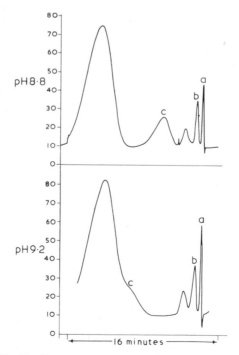

Fig. 71. Chromatogram of sample. a, b, c as in Fig. 70

unknown compounds is eluted more quickly under the same conditions thus leading eventually to overlap of the peaks. The exercise illustrates the importance of examining actual commercial preparations rather than laboratory samples only.

Solutions. (a) Weigh accurately about 2.5 g of sample into a 5 ml volumetric flask and dilute to volume with water. Mix well.

(b) Prepare a solution of a mixture of the components as follows:

Triprolidine Hydrochloride,	about 0.025 g
Codeine Phosphate,	about 0.2 g
Pseudophedrine Hydrochloride	about 0.6 g

all accurately weighed. Dilute to 100 ml with water and further dilute 5 ml of the solution to 10 ml with water. This is the standard for injection.

(c) Sodium Citrate ($0.1M$) adjusted to pH 8.8 with ammonia solution.

Resin. Strong anion exchange (SAX, Du Pont) 2 feet.

Detector. 254 nm radiation.

Method. Inject the standard solution (15 μl) and adjust the sensitivity to give a reasonable peak height for the components (Note) using a flow rate of 0.8 ml/min. Carry out replicate injections for sample and standard solutions using 15 μl volumes accurately measured.

Determine the wt/ml of the linctus and calculate the volume of sample used in preparing the sample solution. Calculate the percentage w/v of each component.

Note. The sensitivity required in determining triprolidine is about four times that for pseudoephedrine and codeine.

Experiment 8. *Single tablet assay of Methyl Testosterone Tablets (10 mg)*

Column. C_{18}/Corasil II.
Mobile Phase. Methanol 60%, 0.5 ml/min.
Standard Solution. Methyl testosterone (0.2 mg/ml).

Method. To one tablet in a 25 ml volumetric flask add water (10 ml) and shake gently until completely dispersed. Add methanol (15 ml), mix well for several minutes and dilute to volume with water. Allow the solids to settle and inject 10 μl of the clear supernatant layer. Compare the peak height with that for the standard solution (10 μl) using the average of two results.

Experiment 9. *Determination of Ethisterone in Ethisterone Tablets (5 mg)*

Steroidal ketones have been determined by HPLC as their 2,4-dinitro-phenylhydrazones. The present experiment uses Isoniazid.

Column. C_{18}/Corasil II.
Reagent. Isoniazid (60 mg), glacial acetic acid (1 ml) in methanol sufficient to produce 25 ml.
Standard Solution. Ethisterone (1 mg/ml) in methanol.
Solvent. Water:methanol (50:50), 1 ml/min.

Method. Weigh and powder 5 tablets and to an accurately weighed quantity of the powder equivalent to one tablet in a 25 ml volumetric flask, add methanol (5 ml) and reagent (2 ml). Dilute to 25 ml with methanol. Allow to stand with occasional shaking for 1 hr and inject 10 μl samples of the clear supernatant liquid. Compare the peak height of the isoniazid derivative with that for the standard solution (5 ml) treated in the same way.

Calculate the amount of ethisterone in a tablet of average weight.

Extensions

Rate of Formation of Derivatives. Examine the standard reaction mixture at intervals of 10 min over a period of 80 min.

Effect of Methanol. Examine the rate of elution of the derivative with 30, 40, 50, 60 and 70% methanol.

Linearity of Detector. Prepare a calibration curve for ethisterone.

Internal Standard. Compare benzophenone (about 1.0 mg in reaction mixture) and ethyl benzoate (about 5 μl in reaction mixture) as internal standards. Select the better of the two and compare the precision using (*a*) the method as described and (*b*) with internal standard.

Experiment 9. *To determine Ampicillin and Flucloxacillin Sodium in capsules containing a mixture of the two substances*

The capsules are stated to contain

Ampicillin trihydrate   250 mg
Flucloxacillin Sodium   250 mg

The determination depends upon the use of an anion exchange resin to separate the two compounds.

Column. Anion exchange resin SAX.

Mobile Phase. 0.2M Ammonium nitrate adjusted to pH 9.2 with ammonia.

Flow rate. 1.2 ml/min.

Method. Dissolve the contents of one capsule in sufficient $0.1M$ NH_4NO_3 at pH 7.5 to produce 50 ml (Note 1). Dilute 5 ml of the opalescent solution to 10 ml with $0.4M$ NH_4NO_3 at pH 9.2 mix and immediately inject 15 μl of the solution (Note 2).

Compare the peak heights of the two components with those from a standard mixture treated in the same way.

Calculate the quantity of Flucloxacillin Sodium and Ampicillin Trihydrate in the capsule.

Note 1. The slightly alkaline solvent is to assist in the solution of capsule contents. It must not be too alkaline at this stage otherwise decomposition of Flucloxacillin occurs as described in Note 2.

Note 2. At pH 9.2 there is a slow decomposition of penicillins in general to the corresponding penicilloic acid.

The latter is eluted more rapidly than the monobasic acid. For a review of the β-lactam antibiotics and their physio-chemical properties (see Hou and Poole, 1971).

Extensions

Capsules of Flucloxacillin and Ampicillin. The decomposition of flucloxacillin may be followed by injecting samples of the solution at pH 9.2 obtained in the experiment above at intervals over a period of days. Ampicillin (Note) gives rise to one peak only during this time so that it may be used as a convenient internal standard. Typical traces during the experiment are shown in Fig. 72.

Note. This does not mean it is stable to alkali and a useful extension of the experiment is to investigate the effect of alkali in buffers of different ionic strengths from *0.01* to *0.1M*.

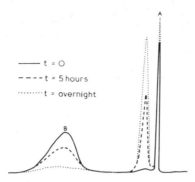

Fig. 72. Effect of time on the chromatogram of a mixture of Ampicillin and Flucloxacillin
Sodium at pH 9.2
A: Ampicillin   B: Flucloxacillin

Other penicillins. Using *0.1M* NH_4NO_3 at pH 9.2 the system is suitable for separating other penicillins from one another except that phenethicillin and potassium penicillin elute together and pyopenicillin and ampicillin are partially resolved (Fig. 73).

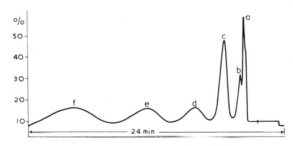

Fig. 73. Chromatogram of penicillins
a, Ampicillin; b, Pyopenicillin; c, Phenethicillin + Phenoxymethylpenicillin; d, Propicillin; e, Cloxacillin; f, Flucloxacillin

Experiment 10. *Determination of methimazole in plasma*

Carbimazole is widely used in the treatment of hyperthyroidism but it has not proved possible to detect it in the plasma of patients. Carbimazole is readily hydrolysed and decarboxylated to methimazole and the determination of the latter in plasma is possible by HPLC (Skellern, Stenlake and Williams, 1974).

Column conditions. C_{18}/Corasil II with water as solvent (0.3 ml/min) or OPN/Porasil C with a solvent of *n*-hexane: tetrahydrofuran (50:50) at a flow rate of 0.75 ml/min.

Method. Dilute plasma (1 ml) with water (3 ml) and extract with chloroform containing 0.5% v/v *n*-octanol (4 × 10 ml). Transfer the extract to evaporating tubes based on a design by Beckett and Rowland (1965) and evaporate to about 100 μl using nitrogen.

Inject 10 μl of the concentrated extract on one of the columns (Note 1). Construct a calibration curve by adding appropriate amounts of methimazole (0.3–1.4 μg) to aliquots (1 ml) of plasma and carrying out the procedure. Calculate the amount of methimazole by reference to the calibration curve (Note 2).

Note 1. In practice the column of OPN/Porasil C proved better as after 15 determinations on the C_{18}/Corasil II the latter had to be purged with methanol to retain the efficiency of the column.

Note 2. The calibration curve is not linear and is indicative of protein binding of methimazole.

Miscellaneous Applications

Table 31 lists a number of applications from manufacturers' literature. (See pp. 137–8).

References

Bailey, F., Barlow, F. S. and Holbrook, A., *J. Pharm. Pharmac.* (1963) **15**, Suppl. 232T.
Bailey, F., Holbrook, A. and Miller, R. J., *J. Pharm. Pharmac.* (1966) **18**, Suppl. 12S.
Beckett, A. H. and Rowland, M., *J. Pharm. Pharmac.* (1965) **17**, 59.
Betts, T. J., *J. Pharm. Pharmac.* (1964) **16**, Suppl. 131T.
Boon, P. F. G. and Mace, A. W., *J. Pharm. Pharmac.* (1969) **21**, Suppl. 49S.
Brealey, L., Elvidge, D. A. and Proctor, K. A., *Analyst* (1959) **84**, 221.
Chatten, L. G. and Morrison, J. C., *J. Pharm. Pharmac.* (1965) **17**, 655.
Consden, R., Gordon, A. H. and Martin, A. J. P., *Biochem. J.* (1944) **38**, 224.
Cowley, P. S. and Rowson, J. M., *J. Pharm. Pharmac.* (1963) **15**, Suppl., 119T.
Deavin, J. C. and Mitchell, O. H. *J. Pharm. Pharmac.* (1965) **17**, Suppl., 56S.
Fisher, R. B., Parsons, P. S. and Morrison, G. A., *Nature, Lond.* (1948) **161**, 764.
Foster, J. S. and Murfin, J. W., *Analyst* (1965) **90**, 118.
Foster, G. E., Macdonald, J. and Jones, T. S. G., *J. Pharm. Pharmac.* (1949) **1**, 802.
Garratt, D. C., *Quantitative Analysis of Drugs*, 3rd edn., Chapman and Hall, London, 1964.
Hjelt, E., Leppanen, K. and Tamminen, V., *Analyst* (1955) **80**, 706.
Hollis, O. L., *Anal. Chem.* (1966) **38**, 309.
Hou, J. P. and Poole, J. W., *J. Pharm. Sci.* (1971) **60**, 503.
James, A. T. and Martin, A. J. P., *Analyst* (1952) **77**, 915; *Biochem. J.* (1952) **50**, 679.
Janâk, J., *Chem. and Ind.* (1967) 1137.
Kuhn, R. and Lederer, E., *Naturwissenschaften* (1931) **19**, 306; *Ber.* (1931) **64**, 1349.
Lovelock, J. E., *J. Chromatog.* (1958) **1**, 35.
Maggs, R. J. and Young, T. E., *Pye Gas Chromatography Bulletin* (1967) **1**, No. 4.
Martin, A. J. P., *Biochem. Soc. Symposium*, No. 3, Cambridge University Press, London, 1949.
Martin, A. J. P. and Synge, R. L. M., *Biochem. J.* (1941) **35**, 1358.
Purdy, S. J. and Truter, E. V., *Analyst* (1962) **87**, 802.
Savidge, R. A. and Wragg, J. S., *J. Pharm. Pharmac.* (1965) **17**, Suppl., 60S.
Singer, D. D., *Analyst* (1966) **91**, 127.
Skellern, G. G., Stenlake, J. B., Williams, W. D. and McLarty, D. G., *Br. J. Clin. Pharmac.* (1974) **1**, 265.
Strain, H. H., *Chromatographic Adsorption Analysis*, Interscience, New York, 1942.
Tiselius, A., *Arkiv. Kemi. Mineral. Geol.* (1941) **14B**, No. 22.
Tswett, M., *Ber. dtsch. botan. Ges.* (1906) **24**, 316, 384; translation, Strain, H. H. and Sherma, J., *J. chem. Educ.* (1967) **44**, 238.

De Zeeuw, R. A., *J. Chromat.* (1968) **32**, 43; *Analyt. Chem.* (1968) **40**, 915; *J. Pharm. Pharmacol.* (1968) **20**, Suppl., 54S.

General Reading

Bobbitt, J. M., *Thin-layer Chromatography*, Reinhold, New York, and Chapman and Hall, London, 1963.

Brown, R. B., *High Pressure Liquid Chromatography; Biochemical and Biomedical Applications*, Academic Press, New York and London, 1973.

Heftmann (ed.), *Chromatography*, Reinhold, New York, and Chapman and Hall, London, 1966.

Kirkland, J. J., (ed.), *Modern Practice of Liquid Chromatography*, Wiley and Interscience, 1971.

Knox, J. H., *Gas Chromatography*, Methuen, London, 1962.

Lederer, E. and Lederer, M., *Chromatography*, Elsevier, 1957, p. 118.

Mikes, O. (ed.), *Chromatographic Methods*, Van Nostrand, London. English edition 1966, p. 35.

Randerath, K., *Thin-layer Chromatography*, 2nd edn., translated by D. D. Libman, Verlag Chemie, Weinheim, with Academic Press, New York, 1966.

Welcher, F. J. (ed.), *Standard Methods of Chemical Analysis*, 6th edn., Vol. III, Part A, pp. 716–37, 738–80.

Liquid Chromatography has been the subject of an International Symposium and the proceedings, including pharmaceutical applications, are reported in the *Journal of Chromatography* (1973) **83**, which volume is devoted entirely to the report.

4 Measurement of EMF and pH

G. O. JOLLIFFE

INTRODUCTION

The acidity and alkalinity of a solution is determined by measuring the concentration of hydrogen and hydroxyl ions in gram equivalents per litre. In many cases the concentrations are better expressed as pH and pOH as defined by the following relationships, suggested by Sørensen (1909):

$$pH = \log_{10} \frac{1}{[H^+]} = -\log_{10} [H^+]$$

$$pOH = \log_{10} \frac{1}{[OH^-]} = -\log_{10} [OH^-]$$

where $[H^+]$ and $[OH^-]$ represent the concentrations, in gram equivalents per litre, of hydrogen and hydroxyl ions respectively. In more recent times, with the realisation of the thermodynamic significance of the activity, these definitions have been modified to:

$$pH = -\log a_{H^+}$$

$$pOH = -\log a_{OH^-}$$

where a is the activity of the ions involved. The more approximate form is still convenient for most purposes and is, therefore, used freely in this chapter.

Even the most highly purified water possesses a small but definite conductivity due to ionisation:

$$2H_2O \rightleftharpoons H_3O^+ + OH^-$$

The hydrogen ion exists in water as the hydroxonium ion, H_3O^+, but for simplicity the following, and more familiar, equation is used:

$$H_2O \rightleftharpoons H^+ + OH^-$$

By applying the law of mass action:

$$\frac{[H^+][OH^-]}{[H_2O]} = \text{a constant}$$

where $[H^+]$, $[OH^-]$ and $[H_2O]$ represent the concentration of the hydrogen ion, hydroxyl ion and water respectively.

In pure water or dilute aqueous solutions, $[H_2O]$ is a constant: hence

$$[H^+][OH^-] = K_w$$

where K_w is known as the *ionic product of water*.

The experimental value of K_w is 1×10^{-14} gram-ions per litre. Therefore, in pure water or in neutral solution where the concentrations of hydrogen and hydroxyl ions are equal:

$$[H^+] = [OH^-] = \sqrt{K_w} = 10^{-7} \text{ g ions per litre (at 25°)}$$

This may also be expressed as:

$$[H^+][OH^-] = K_w = 10^{-14}$$

or

$$\log[H^+] + \log[OH^-] = \log K_w = -14$$

hence

$$pH + pOH = pK_w = 14$$

Thus, a pH value of 7 is obtained for a neutral solution. Acid and alkaline solutions possess pH values less and more than 7 respectively.

The pH of a solution may be determined by measuring the potential difference between a pair of suitable electrodes immersed in the solution. One of the electrodes, the *indicator electrode*, must respond to pH change whilst the other, the *reference electrode*, must remain constant. Each type of electrode itself forms a half-cell which cannot be used on its own.

It can be shown that the potential of a metal electrode (at 25°) immersed in a solution of its own ions is given by the Nernst equation, which can be simplified to:

$$E = E° + \frac{0.0592}{n} \log c$$

where $E°$ is the standard potential of the metal, n the valency of the ions and c the ionic concentration. Thus for a *hydrogen electrode*, where $n = 1$ (the number of electrons involved):

$$E = E_H° + 0.0592 \log[II^+]$$

which may also be written in the form:

$$E = E_H° - 0.0592 \, pH$$

where $E_H°$ is the standard potential of the normal hydrogen electrode, viz. the potential of a hydrogen electrode immersed in a solution of hydrogen ions of unit activity under a pressure of one atmosphere. By convention, the value of $E_H°$ is arbitrarily taken as zero so that the above expression simplifies to:

$$E = -0.0592 \, pH$$

[At any absolute temperature T, the figure 0.0592 should be replaced (in this and similar expressions) by the product $0.000198T$.]

INDICATOR ELECTRODES

Hydrogen Electrode

The hydrogen electrode consists of a small piece of platinum foil, coated electrolytically with platinum black, over which hydrogen gas is passing. The platinum black surface exhibits a strong absorptive power towards hydrogen and, provided that the metal surface remains in continuous contact with the gas, the electrode will act as if it were an electrode of metallic hydrogen. In use, therefore, only a part of the foil is immersed in the solution, the remainder being surrounded by pure hydrogen.

The potential of the hydrogen electrode is used as a reference zero for other potential measurements, as indicated above.

Antimony Electrode

The most satisfactory metal oxide electrode is the antimony-antimonious oxide electrode. For the electrode reaction:

$$Sb + H_2O \rightleftharpoons SbO^+ + 2H^+ + 3e$$

the potential of the antimony electrode is given by the expression:

$$E = E° + \frac{0.0592}{3} \cdot \log [SbO^+][H^+]^2$$

and this can be simplified to the form:

$$E = 0.255 - 0.0592 \, pH$$

In practice, the potential does not vary with pH in a linear manner and, therefore, it is necessary to calibrate the electrode with three or four buffer solutions of known pH before use. The effective range of the antimony electrode is from pH 3 to pH 8. In solutions more acid than pH 3, the oxide becomes appreciably soluble, but if the solution is saturated with oxide, the alkaline limit may be extended to pH 12. The electrode is not suitable in the presence of complexing or strong oxidising agents nor in the presence of metals more noble than antimony. However, the electrode is not easily poisoned, is easily prepared and not readily damaged. It is particularly useful, where applicable, for continuous recording or control of pH and even for measurements involving viscous fluids, heavy sludges or semi-solids.

Glass Electrode

A glass electrode consists of a very thin bulb or membrane of specially prepared, pH responsive glass fused on to a piece of comparatively thick, high resistance glass tube (Fig. 74). In contact with the thin membrane is a suitable solution such as decinormal hydrochloric acid. Electrical contact with this solution is usually made with a silver wire coated with silver chloride, thus acting as an internal reference electrode (i.e. unresponsive to pH change).

The potential of the glass electrode, when immersed in a solution, is given by the expression:

$$E = K + 0.0592 \, (pH_1 - pH_2) \, (\text{at } 25°)$$

where K is a constant, pH_1 is the pH of the solution in the bulb and pH_2 is the pH of the test solution. Now, pH_1 is constant for a given electrode, hence:

$$E = k - 0.0592\,pH_2$$

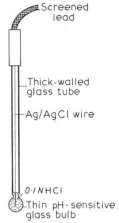

Screened lead

Thick-walled glass tube

Ag/AgCl wire

0.1N HCl

Thin pH-sensitive glass bulb

Fig. 74. Glass electrode

where k, is a constant, known as the asymmetry potential, which depends on several factors such as the existence of strains in the glass, the thickness of the glass bulb and the composition of the solution within.

The advantages of a glass electrode are its rapid response and the fact that it is unaffected by the presence of oxidising or reducing agents, dissolved gases, highly coloured liquids, colloids, suspended matter or moderate concentrations of many salts, with the main exception of sodium salts. The use of modern lithia-silica glasses enables pH measurements to be valid over practically the entire pH range.

The main disadvantage of the glass electrode is that it is extremely fragile. Even a minute scratch on the bulb is often sufficient to render it useless. Most glass electrodes have a very high resistance so that they cannot be used with simple potentiometers. In spite of this, the glass electrode is the most generally useful indicator electrode available for pH measurements.

Rejuvenation of Glass Electrodes

After a long period of use the surface of the glass electrode may deteriorate so that one or more of the following symptons may appear:

(a) slow electrode response
(b) undue sensitivity of the pH reading to physical movement of the electrode
(c) failure of the electrode to check against a pair of buffer solutions
(d) inability to standardise in the range of the meter's asymmetry potential control.

Under favourable circumstances the electrode can be rejuvenated by momentarily immersing the bulb in $0.1N$ hydrochloric acid or by cycling between immersion in acid and alkaline solutions to reduce residual sodium ion effects. If these methods fail to recondition the electrode, then try immersion of the bulb in 20% ammonium fluoride solution for 3 min or in 10% hydrofluoric acid for 15 secs. After this treatment the electrode should be thoroughly rinsed in a stream of tap water and dipped momentarily in $5N$ hydrochloric acid to remove fluorides. After a final rinse in purified water the electrode should be stored in $0.1N$ hydrochloric acid.

Specific Ion Electrodes

Specific ion electrodes respond to the activity of a specific ion in solution rather than to the concentration. For this reason it is sometimes necessary when using them to add a suitable amount of an inert electrolyte to weak solutions to minimise errors due to changes in activity. This approach can, however, produce difficulties when using liquid ion exchange electrodes.

There is an extremely wide range of specific ion electrodes available commercially and details can be obtained quite readily from the manufacturers (e.g. Orion Research Inc., Cambridge, Mass., U.S.A., or Electronic Instruments Ltd., Chertsey, Surrey, England). The theory of operation of each electrode is given in the manufacturer's literature but, by way of example, details of one useful electrode system are given here.

Divalent Cation Activity Electrode. A divalent cation activity electrode (e.g. the EIL Ionanalyser) makes possible the direct measurement of total water hardness with any expanded scale pH meter. The electrode can be used to measure divalent cation levels from saturated solutions down to $10^{-8}M$. The electrode can also be used to obtain data on complex ion equilibria in solution and is a very sensitive end point detector for EDTA titrations of calcium and magnesium ions in aqueous solution.

Activity and concentration measurements. The typical electrode potential *vs* the calcium ion activity or concentration is illustrated in Fig. 75. Note the logarithmic activity/concentration scale.

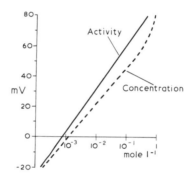

Fig. 75. Electrode potential response *vs* calcium ion activity and concentration

Theory of operation. The electrode detects calcium and magnesium ions by developing a potential across a thin layer of water-immiscible liquid ion exchanger which is selective for divalent ions and is held in place by an inert porous membrane disc (Fig. 76). The aqueous filling solution, containing fixed levels of calcium and chloride ion, contacts the inside surface of the membrane disc which is saturated with the ion exchanger. The calcium ion in the filling solution provides a stable potential between the inside surface of the membrane and the internal filling solution. The chloride ion also provides a constant potential between the filling solution and the internal Ag/AgCl reference element. The electrode thus develops potentials only in response to simple divalent cation activity.

pH Effects. At low pH values the electrode responds to hydrogen ion as well as to divalent cations (Fig. 77). In alkaline solution above about pH 10–11, hydroxide ion complexes a portion of the calcium and magnesium ion, thus reducing

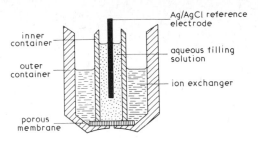

Fig. 76. Specific ion liquid membrane electrode

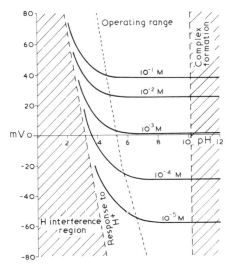

Fig. 77. Electrode potential *vs* pH for calcium or magnesium chloride solutions

the amount of free divalent cations in solution. The electrode does *not* respond to bound or complex ions, e.g. cations bound to citrates, polyphosphates, carbonate or hydroxide ions, or some proteins. The electrode responds only to the activity of free unassociated ions.

Selectivity. The divalent cation activity electrode is designed to respond equally to calcium and magnesium ions, but it also responds to other divalent cations. The electrode does, however, exhibit a limited response to sodium and potassium ions and thus can cause interference with the measurement of divalent cation levels more dilute than about $10^{-4}M$.

Interferences. Direct concentration measurements are not possible when either electrode or method interferences are encountered. *Electrode interferences* are caused when ions are 'mistaken' by the electrode for the ion being measured, e.g. see the shaded area to the right of Fig. 78 where the fluoride ion 'mistakes' OH^- for F^- in alkaline solution. *Method interferences* are problems in solution chemistry which yield incorrect concentration values, e.g. see the shaded area

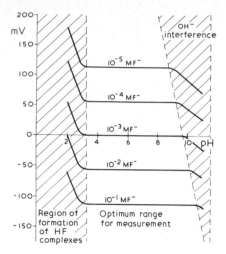

Fig. 78. Electrode potential *vs* pH for fluoride electrode in solutions of varying pH

on the left of Fig. 78 where the fluoride ion forms complexes in acid solutions, thus resulting in a lower level of free fluoride ions to be detected by the fluoride electrode.

Direct electrode measurements are best achieved when using pure solutions of the ion being measured. When other ions are present, maximal accuracy is obtained when the standardising solutions for the electrode also contain these ions at the appropriate concentration levels. These levels should be reasonably constant between the various samples.

Replacement kits. Kits are available for converting to nitrate, chloride, cupric and perchlorate electrodes by replacing the ion exchanger, internal filling solution and membranes.

Standardising solution. Standard calcium chloride solutions are required: *0.1M* and, for water hardness measurements, 100 ppm Ca^{2+} as $CaCO_3$.

Specific ion meters. Some pH meters with an expanded scale also provide a scale which is calibrated in parts per million or moles per litre; this enables direct readings of activity. A calibration curve must be prepared if such a meter is not available (see Fig. 75).

Electrode holder. As the bottom of the electrode is slightly concave, it is essential to tilt the electrode about 20° to ensure no air bubble is trapped.

REFERENCE ELECTRODES

Calomel Electrode

The calomel half-cell is the most widely used reference electrode owing to the constancy of its potential and ease of preparation. The electrode consists of pure mercury in contact with a mixture of mercury and calomel, and a solution of potassium chloride (Fig. 79). The potential of the calomel electrode depends upon the concentration of potassium chloride in the cell. The usual concentra-

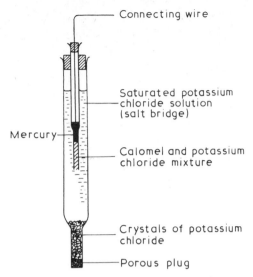

Connecting wire

Saturated potassium
chloride solution
(salt bridge)

Mercury

Calomel and potassium
chloride mixture

Crystals of potassium
chloride

Porous plug

Fig. 79. Saturated calomel electrode

tions of potassium chloride used are decinormal, normal and saturated, the latter being the most convenient. The saturated calomel electrode (S.C.E.) has, however, a high temperature coefficient (-0.76 mV/°C) which is of significance in accurate work. The potentials for the calomel electrode (*vs.* the normal hydrogen electrode) in saturated, *N* and *0.1N* KCl are $+250$, 286 and 338 mV at 20° and $+246$, 285 and 338 mV at 25° respectively.

Silver–Silver Chloride Electrode

The silver–silver chloride reference electrode is more difficult to prepare but as convenient to use as a calomel electrode. It consists of a silver wire, coated electrolytically with silver chloride, dipping into a solution of potassium chloride of definite strength. The potential of the silver–silver chloride electrode (*vs* the normal hydrogen electrode) in saturated, *N* and *0.1N* KCl is $+200$, 235.5 and 288 mV respectively at 25°.

Mercurous Sulphate Electrode

This electrode is similar in construction to the calomel electrode but utilises sulphuric acid (*0.1N*) saturated with mercurous sulphate. It is used, for example, in solutions where silver or lead ions are present, and has a potential of 682 mV.

Salt Bridges

A salt bridge of saturated potassium chloride, potassium nitrate or ammonium nitrate is used to prevent possible contamination of the reference electrodes with the test solutions. Sometimes the salt bridges are designed as a part of the reference electrode (Fig. 79) but are often solidified with a small quantity (3%) of agar.

When two solutions of the same or different electrolytes are brought into contact, a potential difference is set up between them due to the transference of ions across the boundary. This potential difference is known as a diffusion or liquid junction potential. The salt bridge reduces these potentials almost to zero and, therefore, the latter may be neglected for most purposes.

MEASUREMENT OF EMF AND pH

Theory of the Simple Potentiometer

When a resistance wire AB of uniform thickness is connected to a battery or accumulator C, the potential difference between any two points along the wire is proportional to the distance separating these points. Any cell of emf E, lower in value than C, can be connected directly to A and through a galvanometer G to a movable connection D on the wire (Fig. 80).

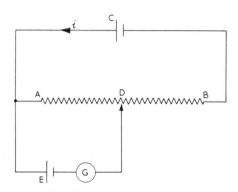

Fig. 80. Simple potentiometer

There is only one position D on the wire such that *no current* passes through the galvanometer. Let the length of wire between A and D be l_1 and the resistance of the wire per unit length be r. Then, if the current flowing through the wire is i, the potential difference between A and D_1 will be equal to irl_1. Thus, at the balance point:

$$E_1 = irl_1$$

In place of E_1, a standard cell of emf E_2 may be used and a corresponding position D_2 can be found on the wire. If $AD_2 = l_2$, then:

$$\frac{E_1}{E_2} = \frac{irl_1}{irl_2}$$

Therefore:

$$E_1 = E_2 l_1/l_2$$

Thus it can be seen that it is possible to measure an unknown emf, E_1, by direct comparison with a standard whose emf is known.

The main advantages of a potentiometer for the measurement of emf are:

(1) it employs a null-point method; therefore, there is no deflection to measure,

(2) at the balance point, no current passes through the cell. This means that the cell is not polarised and that its internal resistance is of no consequence if the null-point detector possesses sufficient sensitivity,

(3) the method is capable of great accuracy.

Commercial Potentiometers

Figure 81 is a simplified circuit diagram of a commercial potentiometer. The primary circuit consists of one or more accumulators or dry cells C connected to slide wire AB via switch S_1 and rheostat R_1. By careful adjustment of R_1

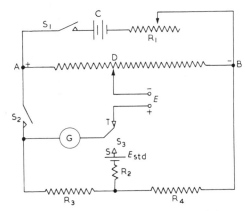

Fig. 81. Simplified circuit diagram of a commercial potentiometer

the potential difference between the ends of the slide wire can be adjusted to a definite value, e.g. 1.4 V. Thus, if the slide wire is divided into 1400 equal parts, each division will represent a potential difference of 1 mV. The distance between A and the numerous possible settings of the sliding contact D is, therefore, calibrated directly in millivolts. In order to obtain the required potential difference between the ends of the slide wire, the secondary circuit incorporates a standard cell E_{std} with its protective series resistance R_2. When switch S_3 is in position S and the galvanometer switch S_2 is closed, R_1 is then adjusted until the galvanometer G is undeflected. At this point, the potential difference across R_3 equals that of the standard cell and the slide wire is thus calibrated. This is because the ratio $(R_3 + R_4)$ to R_3 is chosen such that it equals the ratio of the required potential difference (1.4 V) to the emf of the standard cell. The test cell is connected to the terminals E with S_3 in position T. S_2 is closed and D adjusted until the galvanometer is again undeflected and the calibrated potentiometer reading gives the required emf.

pH Meters

A linear relationship exists between the pH of a solution, at a given temperature, and E, the emf of a cell containing a reference and a suitable indicator electrode, since:

$$E = k - 0.0592 \, pH \text{ (at } 25°)$$

i.e.

$$\Delta E/\Delta pH = -0.0592$$

Thus, a calibration in mV may be converted into pH units by dividing by 0.0592. In practice, however, the resistance values are so chosen that the divisor is 100, thus rendering dual calibration or awkward calculations unnecessary.

Commercial pH meters incorporate a temperature compensating device such as a variable resistance (adjusted manually) or a thermistor (automatic).

In the penultimate equation, k is a constant potential, the asymmetry potential, and is characteristic of the cell used. In a commercial pH meter it is allowed for by adjusting a suitably placed auxiliary potentiometer which injects a potential, $-k$, into the test cell circuit. This adjustment is performed when the test cell contains a standard buffer solution, with the dial of the meter set to the known pH value. This adjustment remains constant over a period of time (often 12–24 hours) but is only valid for a given electrode system.

Most commercial pH meters incorporate a DC amplifier and can measure pH with an accuracy of about ± 0.01 pH units. A switch is usually present so that either a direct reading of potential (mV) or pH may be measured.

Standardisation

Null-point meters are standardised by switching a standard cell into the circuit and adjusting the rheostat to give zero deflection. Direct reading instruments have a zero control potentiometer which must be adjusted before the electrodes are standardised. The standardisation or zero setting should be checked frequently and the controls adjusted, if necessary. Operation of these controls does not vitiate the settings of the asymmetry potentiometer control (q.v.).

For a given electrode system (e.g. glass indicator and calomel reference electrodes) it is necessary to balance out the asymmetry potential and any liquid junction potentials which may be present. The electrodes are immersed in a standard buffer solution of known pH (e.g. *0.05M* potassium hydrogen phthalate, pH 4.00) and the asymmetry control potentiometer is adjusted until the meter reads the desired value.

It is advisable to rinse and immerse the electrode in another buffer solution (e.g. *0.01M* borax solution, pH 9.18) to check that the electrode responds correctly to pH change. The instrument is now calibrated for pH measurement of test solutions but occasional re-standardisation is necessary.

Practical Experiments

Experiment 1. *Determination of* pH *using a glass electrode*

It is not possible to use a simple potentiometer without valve amplification for this experiment. It will be assumed that a commercial potentiometric type pH meter is used, set to read pH units.

(*a*) Connect the meter to its battery or mains supply and switch on. Allow about ten minutes before taking measurements. Earth the instrument case where appropriate.

(*b*) Connect the glass and calomel electrodes to the meter, using the Morton electrode system (Fig. 82). Put saturated potassium chloride in compartment A and water into B.

Note. (i) The tap is pierced in a way such that either compartment may be connected to the drainage tube. The tap should *never* be turned to a position such that the two electrode compartments are connected together, otherwise the potassium chloride will be contaminated by diffusion. (ii) The tap should not be greased, as the film of liquid which surrounds the tap in use provides the necessary electrical contact between the two electrode compartments.

(*c*) Drain compartment B and rinse with *0.05M* potassium hydrogen phthalate, via the filling cup. Use only a sufficient quantity of the buffer solution to immerse the glass bulb to a depth of about 1 cm. Drain and refill with buffer solution to the same depth.

(*d*) Renew the liquid junction by allowing a drop or two of potassium chloride to drain from A. Check that there are no air bubbles in either of the narrow connecting tubes.

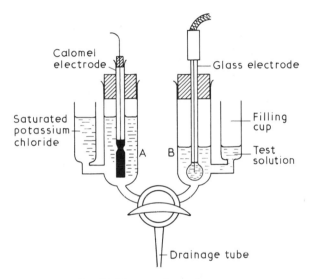

Fig. 82. Morton electrode system

(*e*) With the tap in the neutral position as shown (Fig. 82), the electrodes are ready to be standardised.

(*f*) Standardise the potentiometer—i.e. switch the standard cell into circuit and adjust the battery rheostat until the galvanometer reads zero. It is also necessary, in the first place, to set the manual temperature control to a definite value (often 20°) on some instruments before use.

(*g*) Set the dial readings to 4.00 and adjust the asymmetry control to give no deflection on the meter. The instrument is now ready to measure directly the pH of the test solutions.

(*h*) Drain B and rinse twice with water to a higher level than in (*c*).

(*i*) As (*c*) but using the test solution.

(*j*) As (*d*).

(*k*) Measure the pH of the test solution.

Note. (i) When all measurements are completed, it is advisable to return the electrodes to their usual storage compartments, rinse out the electrode system thoroughly with water and to remove the tap. This prevents any possibility of the tap becoming jammed due to seepage and crystallisation of the potassium chloride.

(ii) Where a manual temperature control is provided, it should always be set to the temperature of the test solution.

(iii) When a thermistor or automatic temperature device is connected to the meter the heat-sensitive tip should be immersed in the test solution.

(iv) The potentiometer should be checked for standardisation before each measurement.

(v) Check periodically that the electrode standardisation has not changed, by using phthalate buffer solution.

(vi) Any suitable buffer solution may be employed in place of potassium hydrogen phthalate. For accurate work, it is a good plan to choose a buffer standard which has a pH value approximately that of the test solution.

Experiment 2. *Determination of the linearity of response of a glass electrode*

(*a*) Prepare 20 ml portions of McIlvaine buffers of pH 2.2, 3.0, 4.0, 5.0, 6.0, 7.0 and 8.0 (Table 32).

(*b*) Set up the pH meter to read potential (mV).

(*c*) Use the Morton electrode system, as in experiment 6, and measure the potential difference between the glass and calomel electrodes for each of the buffer solutions. Wash well with water before each measurement.

(*d*) Plot graphically the potential of the glass electrode (ordinate) against the pH of the buffer solution (abscissa) and measure the slope [$dE/d(pH)$] of the line.

Since:

$$E = -0.0592 \, pH$$

$$\frac{dE}{d(pH)} = -0.0592 \, (25°C)$$

Table 32. McIlvaine Universal Buffer—(pH range 2.2 to 8.0) 20 ml quantities are prepared by mixing X ml disodium hydrogen phosphate (*0.2M*) with Y ml citric acid (*0.1M*).

pH	X	Y	pH	X	Y	pH	X	Y
2.2	0.40	19.60	4.2	8.28	11.72	6.2	13.22	6.78
2.4	1.24	18.76	4.4	8.82	11.18	6.4	13.85	6.15
2.6	2.18	17.82	4.6	9.35	10.65	6.6	14.55	5.45
2.8	3.17	16.83	4.8	9.86	10.14	6.8	15.45	4.55
3.0	4.11	15.89	5.0	10.30	9.70	7.0	16.47	3.53
3.2	4.94	15.06	5.2	10.72	9.28	7.2	17.39	2.61
3.4	5.70	14.30	5.4	11.15	8.85	7.4	18.17	1.83
3.6	6.44	13.56	5.6	11.60	8.40	7.6	18.73	1.27
3.8	7.10	12.90	5.8	12.09	7.91	7.8	19.15	0.85
4.0	7.71	12.29	6.0	12.63	7.37	8.0	19.45	0.55

Note: (i) Sodium ion concentration increases with alkalinity. (ii) Temperature coefficient is small. (iii) Buffer solutions should be freshly prepared using recently boiled and cooled water.

POTENTIOMETRIC TITRATIONS

The end point of most titrations is detected by the use of a visual indicator but the method can be inaccurate in very dilute or coloured solutions. However, under the same conditions, a potentiometric method for the detection of the end point can yield accurate results without difficulty. The electrical apparatus required consists of a potentiometer or pH meter with a suitable indicator and reference electrode. The other apparatus consists of a burette, beaker and stirrer.

The actual potential of the reference electrode need not be known accurately for most purposes and usually any electrode may be used provided its potential remains constant throughout the titration. The indicator electrode must be suitable for the particular type of titration (i.e. a glass electrode for acid–base reactions and a platinum electrode for redox titrations), and should reach equilibrium rapidly. The electrodes are immersed in the solution to be titrated and the potential difference between the electrodes is measured. Measured volumes of titrant are added, with thorough stirring, and the corresponding values of emf or pH recorded. Small increments in volume should be added near the equivalence point which is found graphically by noting the burette reading corresponding to the maximum change of emf or pH per unit change of volume (Fig. 83). When the slope of the curve is more gradual it is not always easy to locate the equivalence point by this method. However, if small increments (0.1 ml or less) of titrant are added near the end point of the titration and a curve of change of emf or pH per unit volume against volume of titrant is plotted, a differential curve is obtained in which the equivalence point is indicated by a peak (Fig. 84).

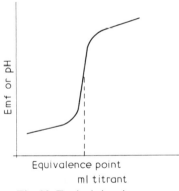

Fig. 83. Typical titration curve

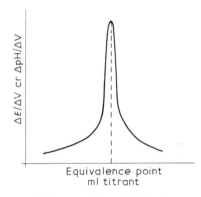

Fig. 84. Typical differential titration curve

Automatic equipment, using a constant flow burette and the pH meter connected to a suitable chart recorder, can be made quite easily and is also commercially available.

Neutralisation Reactions

Any pH responsive indicator electrode may be used, but a glass electrode is usually preferable. The potential at the equivalence point is given by the expression:

$$E = k - 0.0592 \, \text{pH} \ (25°)$$

where k, the asymmetry potential, depends on the electrode system used.

Dibasic and tribasic acids may be titrated with alkali to the intermediate equivalence points provided the dissociation constants of each stage are sufficiently far apart; similarly, mixtures of acids may be titrated satisfactorily (e.g. acetic and sulphuric acids). A sufficiently large inflection is obtained when the difference in pK values exceeds 2.7 pK units. In the titration of a mixture of acids, the first inflection in the titration curve occurs when the stronger acid has been neutralised and the second when neutralisation is complete.

Redox Titrations

The indicator electrode most commonly used is a platinum wire or foil. The potential of the indicator electrode is a function of the *ratio* of the concentrations of oxidised and reduced forms of an ion. For the generalised reaction:

$$\textbf{Red}\text{uctant} + ne \rightleftharpoons \textbf{Ox}\text{idant}$$

the potential of the indicator electrode is given by the expression:

$$E = E° + \frac{0.0592}{n} \cdot \log \frac{[\text{Ox}]}{[\text{Red}]} \quad (25°)$$

where $E°$ is the standard oxidation potential of the system. As before, the equivalence point in a redox titration is indicated by a marked inflection in the titration curve.

Precipitation Reactions

The solubility product of the almost insoluble material formed during a precipitation reaction determines the ionic concentration at the equivalence point. The indicator electrode must readily come into equilibrium with one of the ions. For example, a silver electrode is used for the titration of halides with silver nitrate. The potential of the electrode is given by the expression:

$$E = E° + \frac{0.0592}{n} \cdot \log [M^{n+}] \quad (25°)$$

where $[M^{n+}]$ is the ionic concentration present during the titration and in equilibrium with a slightly soluble precipitate.

Dead-stop End Point Technique

A small potential difference is applied to a pair of identical platinum electrodes immersed in a solution (Fig. 85). Little or no current flows unless the solution is free from polarising substances; this is probably due to adsorbed layers of hydrogen and oxygen on the cathode and anode respectively. Only when both electrodes are depolarised will any current flow. The technique may be applied to any titration when there is a sharp transition at the equivalence point between at least one polarised electrode and the complete depolarisation of both electrodes. The most important example of the dead-stop technique is in the determination of water content using Karl Fischer reagent (see **Aquametry**).

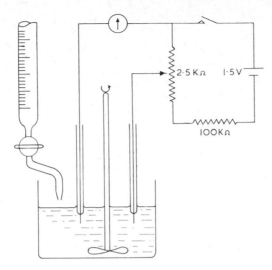

Fig. 85. Dead-stop end point

Practical Experiments

The first two experiments are designed to illustrate the principle of potentiometric titrations and make use of a simple metre bridge. The progress of the titration is followed by measuring the potential difference between the indicator and reference electrodes immersed in the solution being titrated.

Experiment 1. *Titration of hydrochloric acid with sodium hydroxide*

(*a*) Set up the electrode system and connect to the potentiometer as shown (Figs. 86 and 87).

(*b*) Pipette 10 ml hydrochloric acid (*0·5N*) into the beaker (250 ml).

(*c*) Add a sufficient quantity of water to ensure that the indicator and reference electrodes (antimony and calomel respectively) are adequately immersed. The electrodes must, of course, be well clear of the magnet (Fig. 87).

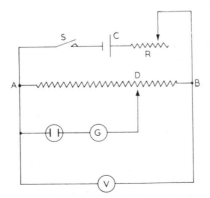

Fig. 86. Circuit diagram of apparatus suitable for use with an antimony/calomel electrode system

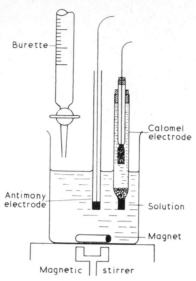

Fig. 87. Electrode system using an antimony indicator electrode

(d) Fill the burette with sodium hydroxide (*0.5N* approx.) and arrange the burette tip to be about 1 cm above the liquid surface so as to avoid splashing.

(e) Close switch S.

(f) Adjust the potential difference between the ends of the metre bridge to 1 V so that the potential drop along the slide-wire is 1 mV/mm.

(g) Place the jockey D vertically on the slide-wire and note the direction the galvanometer needle is deflected.

(h) Release and repeat the procedure successively until the galvanometer needle is undeflected.

(i) Record the length of the wire AD (mm), which equals the emf of the cell (mV).

(j) Switch on the magnetic stirrer and adjust speed to give efficient mixing.

(k) Run in about 2 ml of titrant and allow to mix.

(l) Switch off the stirrer.

(m) Measure the emf of the electrodes as described (g)–(i) above.

(n) As (j)–(m), but reduce the volume of titrant added to 1, 0.5, 0.2 and 0.1 ml quantities as the end point is approached. Always touch off any drops hanging on the burette tip before measuring the emf. Continue the titration until about 5 ml beyond the equivalence point.

(o) Rinse the electrodes, beaker and magnet thoroughly after use.

(p) Plot graphically the emf of the electrodes against the burette readings.

(q) Record the equivalence point of the titration, taken as the point at which the change in emf per unit volume of titrant added is a maximum. Calculate the factor of the sodium hydroxide (*0.5N* solution).

Experiment 2. *Titration of acetic acid with sodium hydroxide*

(a) As experiment 1 (a).

(b) Pipette 10 ml of acetic acid (*0.5N*) into a beaker and add water (approximately 50 ml) as in Experiment 1 (b)–(c).

(c) As Experiment 1 (d)–(q).

Titrations Using a Deflection Type pH Meter

General method

Potentiometric titrations are most readily carried out by using a deflection pH meter. The general method is as follows.

(*a*) Switch on the instrument and allow about 10 min to warm up before use.

(*b*) Connect the saturated calomel electrode to the positive terminal or socket and the glass electrode to the negative. Where applicable, connect the earth terminal to a good earth. Use a beaker, as in Experiments 1 and 2, but a mechanical rather than a magnetic stirrer as great care is needed to avoid damage to the glass electrode. Much accidental damage may be prevented by arranging the calomel electrode and the paddle of the stirrer to be at a lower level than the glass electrode, which should be placed between them (Fig. 88).

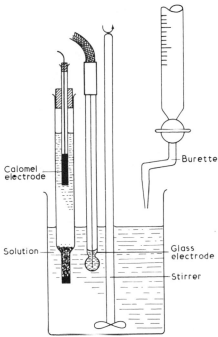

Fig. 88. Electrode system using a glass indicator electrode

(*c*) Adjust the meter to zero and standardise the electrode system by placing *0.05M* potassium hydrogen phthalate solution in the beaker, switch to the 'measure pH' position and adjust the appropriate standardising control to give a pH reading of 4.00. As subsequent readings are to be taken with the stirrer motor running, the motor should be switched on for this operation. [*Note:* the stirrer motor should not be switched on or off while the instrument is switched to the 'measure pH' position].

(*d*) Rinse the beaker and electrodes thoroughly with water and place a measured volume of the solution to be titrated in the beaker. Add sufficient water to cover the bulb of the glass electrode adequately.

(*e*) Switch on the stirrer and measure the pH of the solution.

(*f*) Add about 2 ml of titrant, allow sufficient time to mix and measure the pH of the resultant solution.

(g) Add further quantities of titrant and reduce the volume as the end point is approached to 0.1 ml increments and measure the corresponding values of pH. Periodically, throughout the titration, check the zero setting of the instrument and adjust if necessary.

(h) Obtain further readings of pH for about 5 ml beyond the equivalence point.

(i) Move the 'measure pH' switch to its neutral position, switch off the motor and wash the beaker and electrodes thoroughly with water.

(j) Plot a graph of pH vs ml of titrant added. Read off the equivalence point from the graph and calculate any required data from this value. Calculate the pk_a value of the acid titrated from the Henderson Equation, which is valid over the range pH 4 to pH 10:

$$pH = pk_a + \log \frac{[salt]}{[acid]}$$

Percentage neutralisation	5	10	25	33.3	50	66.6	75	90	95
Value of log [salt]/[acid]	-1.28	-0.95	-0.48	-0.30	0.00	$+0.30$	$+0.48$	$+0.95$	$+1.28$

Calculation of Titration Curves

If the pk value of an acid or base is known, theoretical titration curves can be calculated. Thus in the titration of a weak acid, the initial pH of the solution is given by the expression:

$$pH = \tfrac{1}{2}(pk_a - \log c_a)$$

where c_a is the concentration of acid in gram-equivalents/litre (the calculation must allow for any water added to cover the electrodes).

During the titration, when both free acid and salt are present in the solution, the pH of the solution may be calculated from the Henderson Equation as above, whilst at the end point, when the acid has been completely neutralised, the solution is one of a salt of a weak acid and a strong base, and the pH is given by the expression:

$$pH = \tfrac{1}{2}(pk_a + pK_w + \log c_s)$$

where $pK_w = 14$, and $c_s =$ the final salt concentration in gram-equivalents/litre.

In the titration of polybasic acids, the pH at the intermediate equivalence points is given by the expression:

$$pH = \tfrac{1}{2}(pk_1 + pk_2)$$

The pH beyond the equivalence point can be calculated directly from the excess of alkali added, using the formula:

$$pH = pK_w - pOH$$
$$= 14 - \log [OH^-]$$

where concentrations are expressed in gram-equivalents/litre.

Experiments 3 to 10. *Titrations with sodium hydroxide*

Take 20 ml quantities of the following solutions (all 0.5N) and titrate with sodium hydroxide (0.5N) using the general method outlined above:

3. Hydrochloric acid

4. Acetic acid
5. Hydrochloric acid–acetic acid mixture
6. Boric acid
7. As Experiment 6 but in the presence of 50 ml glycerol
8. Oxalic acid
9. Phosphoric acid
10. Aniline hydrochloride

Experiments 11 to 15. *Titrations with hydrochloric acid*

Take 20 ml quantities of the following solutions (all *0.5N*) and titrate with hydrochloric acid (*0.5N*):
11. Sodium carbonate
12. Sodium nitrite
13. Sodium borate
14. Triethanolamine
15. Triethylamine

Experiments 16 to 19. *Titrations with hydrochloric acid followed by back titration with sodium hydroxide*

Take 20 ml quantities of the following solutions (all *0.5N*) and titrate with hydrochloric acid (*0.5N*) to beyond the equivalence point (e.g. 25 ml). Back titrate with sodium hydroxide (*0.5N*) until excess alkali is present (about 50 ml).
16. Pyridine
17. Piperidine
18. Glycine
19. *p*-Aminobenzoic acid

Experiment 20. *Adjustment of a solution to a given* pH

The stability of many solutions of pharmaceutical substances is governed by the pH of the medium. To obtain maximum stability, therefore, it is necessary to adjust the solution to the required pH by the addition of a suitable quantity of acid or alkali.

(*a*) Measure the total volume of the solution to be adjusted and take a convenient quantity, say 50 ml, and place in the titration cell.

(*b*) Measure the pH [*vs* glass and calomel electrodes] and add sufficient standard acid or alkali until the correct pH is obtained.

(*c*) From the burette reading calculate how much acid or alkali should be added to the remainder of the solution. Add the desired amount and mix well.

(*d*) Again measure the pH, using either the Morton electrode or beaker system and, if necessary, add small quantities of acid or alkali to the bulk solution, with mixing and sampling, until the correct pH is obtained.

REDOX TITRATIONS

These titrations are performed using a bright platinum wire or foil indicator electrode and either a silver–silver chloride or saturated calomel reference electrode. The method is identical with the general method described above except that the meter is set to read millivolts and not pH units. Furthermore, electrode standardisation with buffer solutions is not necessary.

Experiments 21 to 25

General method

Place the solution of the substance to be titrated ($0.1N$; 20 ml) in a 250 ml beaker, add sufficient water to immerse the platinum electrode completely and titrate with the $0.1N$ solutions recommended.

21. Titrate ferrous ammonium sulphate, in the presence of dilute sulphuric acid, with potassium permanganate solution.

22. Titrate ferrous ammonium sulphate, in the presence of dilute sulphuric acid, with potassium dichromate solution.

23. Titrate sodium arsenite, in the presence of 15% hydrochloric acid, with potassium bromate solution.

24. Determine the percentage of antimony in Antimony Sodium Tartrate by titration with potassium bromate in hydrochloric acid solution.

25. Titrate potassium iodide, in the presence of $0.25N$ sulphuric acid with potassium permanganate solution. Note, that near the end point, the electrodes require about a minute to reach equilibrium before their emf can be measured.

26. Use a platinum indicator electrode and a glass reference electrode and titrate 20 ml of a mixture of potassium permanganate ($0.1N$) and ammonium vanadate ($0.1N$) with ferrous sulphate ($0.1N$) in the presence of sulphuric acid ($6N$).

27. Titrate ferrous sulphate ($0.1N$) with ceric sulphate ($0.1N$) in the presence of sulphuric acid ($6N$), as Experiment 26.

PRECIPITATION REACTIONS

Experiments 28 to 32

The indicator electrode consists of polished silver wire and the reference electrode may be either a silver–silver chloride electrode connected to the solution by means of a potassium nitrate salt bridge, or a mercurous sulphate electrode. A glass electrode may also be used as a reference electrode for this type of titration since the hydrogen ion activity remains practically constant throughout the titration.

28. Place 20 ml solution of potassium chloride ($0.1N$) in the beaker and add sufficient nitric acid ($0.05N$) to cover the electrodes. Titrate with silver nitrate solution ($0.1N$), record the emf of the cell as before, and plot a graph of emf *vs* burette reading.

29. Titrate $0.1N$ potassium bromide with silver nitrate.

30. Titrate $0.1N$ potassium iodide with silver nitrate.

31. Titrate 20 ml of a mixture of equal parts of the $0.1N$ potassium chloride, bromide and iodide solutions as in Experiment 28.

32. Titrate 20 ml of a mixture of equal parts of potassium thiocyanate ($0.1N$) and potassium chloride ($0.1N$), as in Experiment 28.

SPECIFIC ION ELECTRODES

Experiments 33 to 37

Specific ion electrodes may be used in conjunction with a reference electrode, and connected to a suitable meter in a similar manner to electrodes used for pH measurements. A conventional pH meter may be used if it has an expanded mV/pH scale, but it is necessary to obtain a suitable calibration curve (Fig. 75). However, it is better to use specific ion meters whereby the concentration is obtained directly from the instrumental readings when ion electrodes are used. These experiments use a divalent cation activity electrode as the indicator electrode.

33. Calibration

Prepare an electrode potential *vs* concentration calibration curve using standard solutions of calcium chloride. Use either:

(*a*) *0.1M* calcium chloride standard solution with dilutions of 1:5, 1:10, 1:50 and 1:100 for experiments involving relatively strong solutions.

(*b*) 100 ppm Ca^{2+} solution with dilutions of 1:5, 1:10, 1:50 and 1:100 for experiments involving the hardness of water, i.e. very low concentrations of calcium.

34. Estimation of calcium chloride

Determine the concentration of calcium chloride in 'unknown' calcium chloride solutions by measuring the potential obtained when the electrodes are immersed in these solutions and by using the calibration curve already prepared.

(The unknown should be about $10^{-3}M$ solution adjusted to pH 10 with ammonium hydroxide.)

35. Electrode response to various ions

Use *0.1M* solutions of various ions at pH 6. Compare the potentials obtained in each case and determine their approximate selectivity constant (*K*). Selectivity constants are expressed as the ratio of electrode response to the ions under consideration to that of the reference ion (Ca^{2+} in this case).

The solutions available should contain the following ions: Zn^{2+}, Fe^{2+}, Cu^{2+}, Ni^{2+}, (Ca^{2+}), Mg^{2+}, Ba^{2+}, Sr^{2+}, Na^+, K^+.

What significance can you place on your results?

36. Determination of the hardness of water

Examine tap water and determine the hardness:
(*a*) straight from the tap
(*b*) after boiling and cooling tap water
(*c*) of laboratory distilled water
(*d*) of the reputed 'soft' water.

37. Determination of calcium by titration

The electrode can be used as the end point detector in titrations with EDTA or other chelators. The end point is where the potential becomes steady on further addition of EDTA.

Determine the concentration of 'unknown' calcium chloride solution by titrating with $10^{-2}M$ NaEDTA. The $CaCl_2$ solution should be adjusted to pH 10.

Compare the result with that obtained in Experiment 34.

5   Conductimetric Titrations

G. O. JOLLIFFE

INTRODUCTION

Solutions of electrolytes normally obey Ohm's law:

$$R = \frac{E}{I}$$

where R, E and I are the resistance of the solution, the applied emf and the current, respectively. The conductance, G, of the solution is the reciprocal of the resistance and is expressed in siemens (S). Hence:

$$G = \frac{1}{R} = \frac{ka}{l}$$

where k, a and l are conductivity, cross-sectional area and length of the conductor, respectively. The conductivity is defined as the conductance of a cube of material having a surface area of 1 m^2 and a length of 1 m. The molar conductivity Λ, is the product of the conductivity and the volume, in cubic metres, containing 1 molecular weight of the solute. Thus:

$$\Lambda = k/c$$

where c is the concentration.

From this equation, it would appear that the molar conductivity approaches an infinite value (Λ_∞) as the concentration approaches zero. In practice it reaches a limiting value.

Ionic Conductivities

Comparison of the molar conductivities of pairs of salts having an ion in common, such as KCl—$NaCl$ and KNO_3—$NaNO_3$, shows that replacement of one ion by another (i.e. replacement of K^+ by Na^+) leads to a constant difference in the conductance values. Therefore, each ion makes a certain definite contribution to the total conductance of the solution, and this is independent of other ions present in the solution. Hence for any salt:

$$\Lambda_\infty = l_+^\circ + l_-^\circ$$

where Λ_∞ is the molar conductivity at infinite dilution, l_+° and l_-° are the ionic conductivities at infinite dilution of the cation and anion respectively.

In conductivity experiments it is important to control the temperature because the ionic conductivity of most ions is increased at least 2 % per °C rise in temperature.

Table 33. Approximate Ionic Conductivity ($Sm^2\,mol^{-1}$) of Ions in Water at Infinite Dilution (25°).

Cation	$l^\circ_+ \times 10^{-4}$	Anion	$l^\circ_- \times 10^{-4}$
H^+	350	OH^-	199
Li^+	39	Cl^-	76
Na^+	50	Br^-	78
K^+	74	I^-	77
NH_4^+	74	NO_3^-	71
Ag^+	62	HCO_3^-	45
$\frac{1}{2}Mg^{2+}$	53	CH_3COO^-	41
$\frac{1}{2}Ca^{2+}$	60	$C_2H_5COO^-$	36
$\frac{1}{2}Ba^{2+}$	64	ClO_4^-	67
$\frac{1}{2}Sr^{2+}$	59	IO_3^-	40
$\frac{1}{2}Zn^{2+}$	53	$\frac{1}{2}CO_3^{2-}$	69
$\frac{1}{2}Cu^{2+}$	54	$\frac{1}{2}C_2O_4^{2-}$	74
$\frac{1}{2}Pb^{2+}$	69	$\frac{1}{2}SO_4^{2-}$	80
$\frac{1}{2}Fe^{2+}$	54	$\frac{1}{2}CrO_4^{2-}$	82
$\frac{1}{3}Fe^{3+}$	68	$\frac{1}{3}PO_4^{3-}$	80

Application of Conductimetric Titrations

During the progress of a titration, changes in the conductivity of the solution usually occur and use is made of this fact to determine, for example, the equivalence point of reactions involving neutralisation or precipitation. The conductance of the solution is measured after each addition of titrant and the results expressed graphically. Ideally, two straight lines are obtained which intersect at the equivalence point, the accuracy under suitable conditions being better than 0.5%. The change in volume throughout the course of a titration should be kept small by employing a titrant which is at least ten times more concentrated than the solution being titrated. In any case a correction for the dilution effect may be applied by multiplying each conductance reading by the ratio total volume to initial volume.

Measurements of conductance obtained in the region of the equivalence point are of little value due to hydrolysis, dissociation or solubility of the reaction product. For this reason, the graph often shows a curvature rather than the ideal clear intersection of two straight lines.

Neutralisation Reactions

In the titration of strong acids and strong bases there is always an initial decrease in conductance followed, beyond the equivalence point, by a rise (Fig. 89). In the titration of sodium hydroxide with hydrochloric acid, for example, the decrease in conductance is due to the replacement of hydroxyl groups with an ionic conductivity of 199 (Table 33) by chloride anions with an ionic conductivity of only 76. Beyond the equivalence point, the increase in conductivity is due to the very high ionic conductivity of the excess hydrogen ions (ionic conductivity 350).

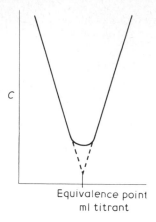

Fig. 89. Conductimetric titration of a strong acid with a strong base

The titration of hydrochloric acid with sodium hydroxide is similarly explained; the initial fall in conductance is due to replacement of the hydrogen
ions (350) with sodium ions (50) and the final rise due to excess hydroxyl ions
(199) being present. The stronger the solutions used the sharper and more
definite is the appearance of the equivalence point, which is invariably obtained
graphically by the intersection of two straight lines (Fig. 90).

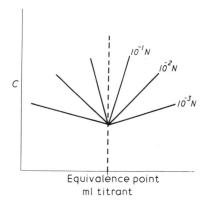

Fig. 90. Conductimetric titration of a strong acid with a strong base showing the effect of
dilution

Titrations of a very weak acid with a strong base or a very weak base with a
strong acid result in a small initial conductance which increases as the concentration of the salt, which is ionised, increases. In the region of the equivalence
point, pronounced hydrolysis often occurs and this is shown by the marked
curvature of the titration curve in this region. The equivalence point must be
located, therefore, by extrapolating the straight line portions of the curve until
they intersect (Fig. 91).

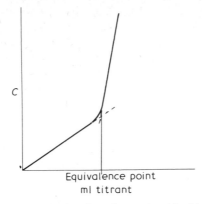

Equivalence point
ml titrant

Fig. 91. Conductimetric titration of a weak acid with a strong base

Conductimetric titrations can be applied to the determination of phenols, weak acids whose salts are coloured, alkaloids, amines, certain dyes, and mixtures of a strong with a weak acid. Phenolsulphonic acid, for example, may be titrated first followed by the phenolic group.

Precipitation and Complex Formation Reactions

Conductimetric titrations of this type may be performed satisfactorily if the reaction product is sparingly soluble or is a stable complex. In order to minimise errors, the intersection of the two branches of the titration curve should be made as acute as possible by suitable selection of titrant. The precipitate should be formed fairly rapidly and should not have strongly adsorbent properties.

EXPERIMENTAL METHODS

Measurement of Conductance

The conductance of a solution is measured by applying a high-frequency (1000 Hz) alternating voltage to a Wheatstone bridge, the conductivity cell (C) being placed in one arm and a standard resistance (R_1) in another (Fig. 92).

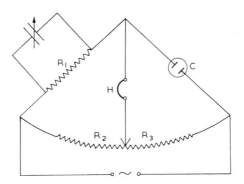

Fig. 92. Schematic diagram of a conductivity bridge

The other two arms of the bridge consist of a slide wire, calibrated in terms of the ratio of the resistances (R_2 and R_3) tapped off at all possible settings of the sliding contact. This contact is adjusted until the sound from the headphones (H) is a minimum; a variable capacitor, approximately equal to the cell capacitance, is sometimes placed across R_1 in order to facilitate the balancing of the bridge. In some commercial conductivity bridges a 'magic eye' tuning indicator is used as the null point detector.

Titration Apparatus

In its simplest form (Figs. 93 and 94), this consists of a Wheatstone bridge, headphones, burette, beaker, stirrer and a pair of platinised platinum electrodes.

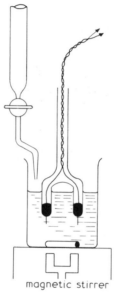

magnetic stirrer

Fig. 93. Electrode system and apparatus for conductimetric titration

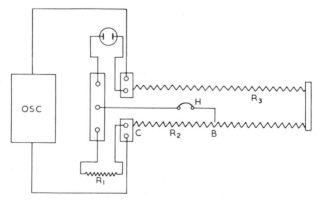

Fig. 94. Conductimetric titrations using a metre bridge

The titration vessel should preferably be immersed in a thermostatically controlled ($\pm 0.1°$) bath.

General Method

(a) *Platinisation of platinum electrodes*

It is convenient to prepare two platinised electrodes simultaneously as follows:

(1) Clean the surface of the platinum wire or foil by careful immersion, for a few seconds, in hot aqua regia, fuming nitric or chromic acid, followed immediately with a thorough rinse with water.

(2) Immerse the electrodes in a solution of chloroplatinic acid (3%) containing lead acetate (0.025%) and connect them, via a reversing switch, milliammeter and rheostat, to a battery as shown in Fig. 95.

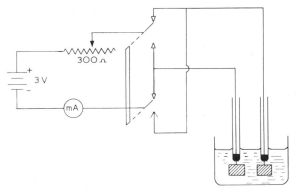

Fig. 95. Apparatus for platinising platinum electrodes

Adjust the current density to about 10 mA cm^{-2} electrode surface and reverse the current through the electrodes every 30 sec until a *thin* black velvety coat is obtained (10–15 min). A thick coating results in an electrode with a very sluggish response.

(3) Remove the electrodes from the solution and rinse with water. Do not throw away or contaminate the platinising solution which can be used repeatedly.

(4) Connect the two electrodes together to the cathode ($-$) and use another platinum electrode as the anode ($+$).

(5) Transfer the electrodes to dilute sulphuric acid, the platinised electrodes still being connected to the cathode, and electrolyse for 5–10 min. This process removes last traces of impurities (chlorine and platinising liquid) adsorbed on to the platinum black and rapidly saturates the electrode surface with electrolytic hydrogen.

(6) Wash the electrodes thoroughly and always store them in water. The platinised electrodes will keep for long periods provided the surface is not allowed to become dry or damaged by scratches.

(b) *Conductimetric titrations using simplified apparatus*

(1) Set up the apparatus, as shown in Figs. 93 and 94, and switch on the 1000 Hz oscillator. The platinum electrodes should be freshly platinised and stored in water when not in use. The surface must not be damaged by scratches or allowed to become dry. Pipette the required amount of the solution (see Table 34 p. 179) to be titrated into the flask and add sufficient water to cover the electrodes.

(2) Measure the conductance by moving the jockey to a position on the slide wire to give minimum sound in the headphones. The balance point should be approached from either side and the mean position noted. Record the length of wire, BC ($=x$ cm).

(3) Add 1 ml portions of titrant, stir well, and find the balance point again. Repeat until about 10 ml of titrant has been added.

(4) Calculate the conductance correspondence to each reading from:

$$G = \frac{x}{R_1(100-x)}$$

where R_1 is the resistance of a value such that the balance point is on the slide wire, x is the length BC of wire in cm. [ABC is a uniform wire, 100 cm in length.]

(5) Plot a graph of conductance (S) against burette reading (ml) and read the end point of the titration from the graph.

(c) *Conductimetric titrations using a commercial conductivity bridge*

(1) Set up the titration cell (Fig. 93) and connect the electrodes to the appropriate terminals of the conductivity bridge, switch on and set to read 'conductance'.

(2) Measure the initial conductance by adjusting the switch which selects R_1 (see Fig. 94), and slide wire, R_2R_3, until the bridge is balanced, as determined by the null-point detector incorporated in the circuit (e.g. headphones, 'magic eye' tuning indicator or AC meter).

(3) Add 1 ml portions of titrant, stir well, and again measure the conductance.

(4) Plot a graph of conductivity against burette reading. If necessary, multiply all the values of conductance by the ratio, total volume/initial volume, in order to correct for volume change.

Experiments

A representative selection of titrations is summarised in Table 34. The equations representing each reaction should be written down and reference made to Table 33. This should enable a student to explain the reason for the shape obtained for each graph.

HIGH-FREQUENCY TITRATIONS

Introduction

Many difficulties are encountered in electrical methods of titration. Conductimetric measurements are complicated by polarisation, difficulty in wetting electrodes with small amounts of liquid, and adsorption at electrode surfaces. Potentiometric and other galvanic methods are usually restricted to ionised solutions and are often impossible where non-aqueous solvents are used, especially if these are poor ionisation media.

By using the field of a high-frequency oscillator it is possible to produce ionic or dipole motion without introducing electrodes into the solution. If such an oscillator is placed in an insulated titration vessel, coupling takes place across the walls. The energy required to produce this ionic or dipole motion causes changes in loading of the oscillator. It is also possible to rectify the radiofrequency current by-passed by the solution and measure the resulting direct current with a microammeter. Such an instrument responds only to changes in conductivity of the solution and is, therefore, suitable only for the titration of electrolytes.

Table 34. Conductimetric Titration Experiments

Experi-ment	Solution	Strength	Vol-ume (ml)	Titrant	Strength	Titration incre-ments (ml)
1	Hydrochloric acid	*0.01N*	50	Sodium hydroxide	*0.1N*	1.0
2	Hydrochloric acid	*0.0001N*	50	Sodium hydroxide	*0.001N*	1.0
3	Acetic acid	*0.1N*	50	Sodium hydroxide	*1.0N*	0.5
4	Acetic acid	*0.001N*	50	Sodium hydroxide	*0.01N*	0.5
5	Hydrochloric acid	*0.01N*	50	Ammonium hydroxide	*0.1N*	1.0
6	Acetic acid	*0.1N*	25	Piperidine	*0.5N*	1.0
7	Phosphoric acid	*0.5M*	50	Sodium hydroxide	*2N*	0.2
8	Boric acid	*0.1N*	50	Sodium hydroxide	*0.5N*	0.5
9	Oxalic acid	*0.2N*	50	Sodium hydroxide	*1.0N*	0.5
10	Oxalic acid	*0.2N*	50	Ammonium hydroxide	*1.0N*	0.5
11	Acetic acid +ammonium hydroxide	*0.1N* *1.0N*	50 4	Sodium hydroxide	*1.0N*	0.5
12	Hydrochloric acid +acetic acid	*0.1N* *0.1N*	10 40	Sodium hydroxide	*1.0N*	0.2
13	Sodium acetate	*0.1N*	50	Hydrochloric acid	*1.0N*	0.5
14	Ammonium sulphate	*0.1N*	50	Sodium hydroxide	*1.0N*	0.5
15	Strychnine hydrochloride +hydrochloric acid +ethanol	*0.01M* *0.01N* —	25 25 50	Sodium hydroxide	*0.1N*	0.2
16	Quinine dihydrochloride +ethanol	*0.01M* —	25 50	Sodium hydroxide	*0.1N*	0.2
17	Sodium sulphate	*0.01M*	50	Barium chloride	*0.1M*	1.0
18	Sodium oxalate	*0.01M*	50	Lead acetate	*0.1M*	1.0
19	Silver nitrate	*0.001N*	50	Potassium chloride	*0.01N*	1.0
20	Silver nitrate	*0.001N*	50	Lithium chloride	*0.01N*	1.0

Mechanism of Loading Effect

(a) *Polar Molecules.* The dipole moment accounts for the greater part of the dielectric constant of a polar molecule and for most of the power loss when such a molecule is subjected to an alternating electric field. The loss which is manifested by rise in temperature of the solution is due to lag in orientation of the dipole with the electric field. This lag, which is known ås the relaxation time (τ), is characteristic of a given molecule and is a function of the molecular dimensions, the viscosity and the absolute temperature.

Maximum loss occurs at a frequency (f) where $f = 1/2\tau$.

(b) *Electrolytes.* In an ionic crystal there is a characteristic resonance due to the forces holding the ions in the lattice. When solution occurs, a loose, quasi-crystalline structure remains. The Debye and Falkenhagen theory accounts for this structure by the suggestion that each ion has a tendency to be surrounded by an 'atmosphere', of ions of opposite charge. In an alternating field of a frequency determined by the relaxation time of the ionic 'atmosphere', a maximum energy absorption occurs in a manner analogous to the absorption in polar molecules.

Instrumentation

A great number of instruments have been described and successfully used for high-frequency titration. The most generally useful ones are those based upon the original designs of Jensen and Parrack and of Blake.

Jensen and Parrack type titrimeter. This titrimeter employs a tuned-anode tuned-grid oscillator, the basic circuit of which is shown in Fig. 96.

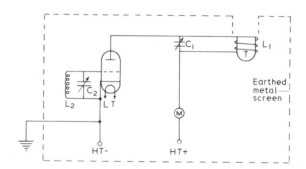

Fig. 96. Jensen and Parrack type titrimeter

Oscillation occurs only when the two tuned circuits L_1C_1 and L_2C_2 are resonant at approximately the same frequency. Capacitor C_1 determines the frequency of oscillation and the insertion of the titration vessel T into coil L_1 loads the oscillator. As the titration proceeds the oscillator loading increases, the anode current alters and is measured by microammeter M. Increase in loading changes the resonant frequency of L_1C_1 and if great enough will stop oscillation. If C_1 is now readjusted, the original resonant frequency may be restored and oscillation restarted.

Blake type titrimeter. This instrument responds to changes in conductivity of the solution and consists of an oscillator, a conductimetric tube, a rectifier and a microammeter, as in Fig. 97.

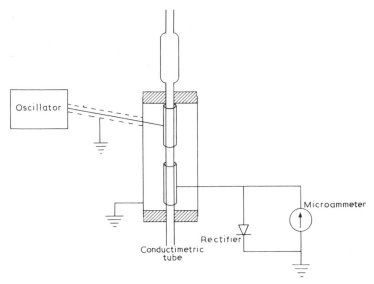

Fig. 97. Blake type titrimeter

The conductimetric tube consists of a thin-walled, narrow-bore Pyrex tube fitted externally with two cylindrical metal electrodes, the whole being surrounded by an earthed metal screen. Samples of solution are drawn up into the tube by a syringe after each addition of titrant. Radio-frequency current is fed, via a screened cable, from the oscillator to one of the electrodes. The current passes across the glass wall of the tube, through the solution and back across the glass wall to the other electrode. The current thus by-passed through the tube is led from the second electrode, rectified and applied to the microammeter.

Applications

The applications of high-frequency titration have not yet been fully explored but are nevertheless extensive and high-frequency titrimeters are commercially available. The method may be used for most determinations but is particularly valuable where chemical methods and other electrometric methods fail.

It is possible to carry out titrations in the presence of very large excesses of indifferent electrolyte. Thus, chloride and sulphate have been determined in very small samples of sediment from ocean beds and silver has been assayed in the presence of a tenfold excess of foreign salts. Sulphur has been determined in petroleum products and in combustible gas mixtures, after conversion to sulphate by suitable pre-treatment.

Titrimetric methods for the determination of thorium are inaccurate and gravimetric methods complicated owing to interference by rare-earth impurities.

The volumetric estimation of beryllium has hitherto been impossible. High-frequency titration has greatly simplified these assays.

The method is very valuable for alkaloidal assays and acid-base titrations in highly coloured solutions where visual indicators cannot be used. It may also be used for non-aqueous titrations which are often limited by the lack of suitable visual indicators and by the lack of adequate electrode systems for potentiometric determinations.

It is possible to analyse complex mixtures by setting up calibration curves. Thus binary and ternary systems such as o-xylene–p-xylene and water-benzene–methyl ethyl ketone may be evaluated and by simple modification the technique may be applied to automatic process control. Another valuable application is the detection of zones and the analysis of the eluate during chromatographic separations.

High-frequency analysis is an elegant and accurate method for following rates of reaction and with the aid of automatic current recorders may be applied to very rapid reactions. It has made possible the direct determination of the products of alkaline hydrolysis of the esters of the chloracetic acids which have half-lives of a few seconds, preventing the use of classical pipetting methods.

6 Polarography and the Elements of Coulometry

G. O. JOLLIFFE

INTRODUCTION

Polarography is an electrochemical method of analysis, devised by Heyrovsky in 1922. The sample, in suitable solution, is placed in a special electrolytic cell which contains a polarisable and a non-polarisable electrode. A gradually increasing voltage is applied to the electrodes of the cell, the corresponding current is measured and the results plotted graphically (either manually or automatically). The current-voltage (cv) curve, which has a characteristic S-shape (Fig. 99), may be used for the *qualitative* and *quantitative* analysis of electro-reducible and electro-oxidisable substances or ions. The curve is known as a *polarogram* and the electrical apparatus (for supplying and measuring the current) as a *polarograph*.

Since the electrolysis current is very small (usually $< 50\ \mu A$) the total amount of reduction due to electrolysis is negligible, so that the same solution may be repeatedly analysed without significant change in the cv curve.

The polarographic method of analysis is applicable to most metals and many organic compounds. Environmental studies are now becoming increasingly important and metals such as lead and cadmium present in samples of chemical or biological origin are readily determined by polarographic analysis. Oxygen (normally removed from solutions) can also be determined polarographically to monitor the extent of river pollution. Additives such as Dimetridazole and Furazolidone to veterinary feedstuffs can be determined polarographically because of the nitro group present. Polarography can often be carried out directly on plasma, urine, bile and saliva. This means that drugs and metabolites may be determined readily in small concentration and volumes.

Some compounds which are polarographically inactive may be studied after preparing a suitable derivative. Chlorpromazine, for example, has to be treated with bromine before polarographic analysis is possible. Similarly, aromatic hydrocarbons require nitration, certain alkaloids and hormones require nitrosation and steroids require reaction with Girard's reagent in order to be determined polarographically.

When several reducible substances are present, the polarogram consists of several 'steps', each of which, under optimum conditions, enables the constituents to be identified and their concentrations determined over the range 10^{-2} to $10^{-6}M$. Polarography is also adaptable to volumes as small as 0.05 ml.

Apparatus

The apparatus used in polarography is basically very simple. It consists of a potentiometer connected to two electrodes via a microammeter. Many types of electrode systems are available but the simplest and most convenient is a *dropping mercury electrode* (dme), which consists of a mercury reservoir connected to a short length of very fine capillary tubing (diameter of aperture 20–80 μm) in conjunction with a *mercury pool anode*.

Fig. 98 illustrates the basic features of the apparatus. R_1 is a shunt for the galvanometer and R_2 is often a calibrated potentiometer, previously standardised with a Weston Standard Cell in the usual way. Alternatively an uncalibrated potentiometer may be used, in which case the applied voltage, V, may be measured by means of a high resistance voltmeter.

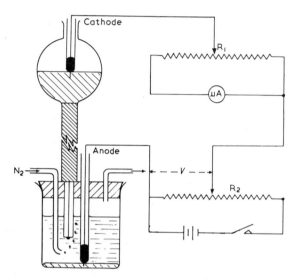

Fig. 98. Apparatus for obtaining polarograms

Method

A suitable solution of the electro-reducible substance under examination is freed from dissolved oxygen by bubbling nitrogen through for several minutes, because oxygen is electro-reducible and would thus interfere with the cv curve obtained. A gradually increasing voltage is applied to the electrodes and the corresponding current is measured and plotted graphically. For accurate work a further polarogram is obtained in the absence of the electro-reducible compound to give a *residual current curve* (broken line in Fig. 99).

A typical cv curve is shown in Fig. 99 where i_r is the *residual current*, i the current at any point on the wave measured with respect to the line AB produced, or (more correctly) with respect to the residual current curve, and i_d the *diffusion current* as defined by the curve and discussed later. The diffusion current is proportional to the concentration of electro-reducible ions present and thus forms the basis of quantitative polarography. $E_{\frac{1}{2}}$, the *half-wave potential* which produces a current (at C) midway along the steeply rising portion of the curve, is characteristic of the particular system being reduced and thus enables qualitative analyses to be performed.

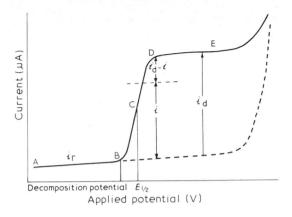

Fig. 99. Typical polarogram

THEORETICAL CONSIDERATIONS

Figure 99 shows a typical polarogram. The gradual rise in current over the portion AB is known as the *residual current*, i_r, and is the sum of the relatively large *condenser current*, i_c, and a very small *faradaic current*, i_f.

$$\text{viz.} \quad i_r = i_c + i_f$$

The condenser current is produced when mercury drops, from the dme, become charged at the mercury-solution interface due to the formation of a Helmholtz double layer of positively and negatively charged ions. The faradaic current is due to traces of various impurities in the solution being reduced. For example, it is very difficult to remove the last traces of oxygen even after bubbling nitrogen through a solution; ordinary distilled water often contains traces of copper and when solutions are deoxygenated in the presence of the pool anode, mercury ions sometimes go into solution with the formation of hydrogen peroxide.

The residual current must *always* be deducted from the total observed current, as shown in Fig. 99, in order to obtain the *diffusion current*, i_d. The minimum detectable concentration of electro-reducible ions depends, to a very large extent, on the accuracy with which this correction is measured.

At the point B, the *decomposition potential*, electrolysis commences and the discharged ions begin to deposit on or amalgamate with the mercury, e.g.:

$$Zn^{2+} + 2e \rightleftharpoons Zn$$

Often only a part of their charge is neutralised and the product is also capable of electro-reduction at a higher applied potential, thus giving a two-step polarogram as for example:

$$Cr^{3+} + e \rightleftharpoons Cr^{2+}$$

$$Cr^{2+} + 2e \rightleftharpoons Cr$$

At point C the corresponding voltage is known as the *half-wave potential*, $E_{\frac{1}{2}}$, and the concentrations of the oxidised and reduced forms at the electrode surface are equal, i.e.

$$[Zn^{2+}] = [Zn]$$

The potential at this point is characteristic of the nature of the reacting material and is independent of the electrode characteristics. Thus, with certain reservations, the half-wave potential may be used for the identification of an unknown substance. As the potential of the pool anode is not constant, a *saturated calomel electrode* (or other suitable reference electrode) is used in place of the mercury pool anode when half-wave potentials are recorded for quantitative use. For accurate work, the potential is corrected for the '*iR*' drop (viz. the potential drop due to the resistance of the cell) where i is the current flowing through the cell and R is its resistance.

The straight portion of the curve, DE, is known as the *limiting current*, i_l, or the *diffusion current*, i_d, in the absence or presence of a *supporting electrolyte* respectively (see p. 185) and is measured (in μA) as the vertical height between AB produced and DE. When the limiting current is reached the electroreducible substance is reduced as rapidly as it reaches the electrode surface and its concentration at the mercury-solution interface remains constant at a value which is negligibly small compared with the concentration in the body of the solution. Under these conditions the current is independent (within certain limits) of the applied potential and in this state the electrode is said to be *concentration polarised*. The extent of polarisation depends on the surface area of the electrode and the current. Hence, for small currents, the minute mercury drops from a dme are easily polarised, whilst the large pool anode is not polarised.

Generally, reducible ions are supplied to the depleted region at the electrode surface by two, more or less independent, forces:

(i) a diffusive force proportional to the concentration gradient at the electrode surface, viz. the *diffusion current*, i_d.
(ii) an electrical force, proportional to the electrical potential gradient at the electrode surface-solution interface, viz. electrical migration or *migration current*, i_m.

Hence,

$$i_l = i_d + i_m$$

The current through an electrolytic solution is carried impartially by all the ions present regardless of whether or not they take part in the electrode reaction. The fraction of the total current carried by any particular species of ion depends primarily on its relative concentration in the solution. Thus, when a large excess of *supporting electrolyte* is present, the current through the solution will be almost entirely carried by the large excess of indifferent ions and the proportion of the current carried by the reducible ions will decrease practically to zero. (A supporting or indifferent electrolyte is a salt, whose ions do not participate in the electrode reaction, which is added to increase the conductance of the solution and to minimise effects of electrical migration.)

$$i_m \simeq 0$$

$$\therefore \quad i_l = i_d$$

Therefore, under normal polarographic conditions, when at least a 50-fold excess of supporting electrolyte is present, the limiting current is, almost solely, a diffusion current.

Ilkovic Equation

Ilkovic, on examination of the various factors which govern the diffusion current in the presence of a large excess of supporting electrolyte, deduced the following equation:

$$i_{max} = 706nD^{\frac{1}{2}}Cm^{\frac{2}{3}}t^{\frac{1}{6}}$$

where i_{max} is the *maximum* diffusion current during the life of a mercury drop, measured in microamperes (1 $\mu A = 10^{-6}$ ampere), n is the number of electrons involved in the reduction of one molecule of reducible substance, D is the diffusion coefficient (cm^2/sec), C is the concentration (millimoles per litre), m is the weight of mercury flowing through the capillary (mg/sec) and t is the time (seconds) necessary for the formation of one drop of mercury, normally between 2 and 7 seconds. The value 706 is a combination of numerical constants, one of which is the density of mercury.

In practice, the *average* current is usually measured with a meter which cannot follow accurately or quickly enough the fluctuations which occur. Under such conditions the average diffusion current, i_d, is given by the expression referred to in future as the Ilkovic equation:

$$i_d = 607\, nD^{\frac{1}{2}}Cm^{\frac{2}{3}}t^{\frac{1}{6}}$$

Hence,

$$i_d \approx \tfrac{6}{7}i_{max}$$

According to the Ilkovic equation, with all other factors constant, it will be noted that the diffusion current is proportional to the concentration:

$$\text{viz.} \quad i_d = kC$$

where k is a constant defined by the Ilkovic equation and i_d has been corrected for the residual current.

This linear relationship may fail when the drop time is too short, owing to the stirring effect disturbing the diffusion layer and producing an abnormally large current. However, the addition to the solution of a small quantity of gelatin, which increases the viscosity, often counteracts this effect.

The term $m^{\frac{2}{3}}t^{\frac{1}{6}}$ in the Ilkovic equation satisfactorily describes the *capillary characteristic* provided the drop time of the capillary is within the usual limits. It will be appreciated that capillaries used in various laboratories will be different with respect to bore, length and reservoir height, thus a knowledge of m and t

will enable comparisons to be made between different capillaries since:

$$i_d = km^{\frac{2}{3}}t^{\frac{1}{6}}$$

Thus

$$i_d/m^{\frac{2}{3}}t^{\frac{1}{6}} = k'$$

Factors affecting the variables in the Ilkovic equation. The factors m and t depend upon the dimensions of the dme and on the pressure due to the mercury column. Drop time depends on the interfacial tension at the mercury-solution interface, the nature of the solution and the applied potential. The rate of flow of mercury also depends on the temperature and, to some extent, on the interfacial tension. It can be shown, by consideration of the rate of flow of liquid through a capillary, that at constant temperature:

$$P \propto m$$

Thus

$$P/m = k''$$

where P is the difference in hydrostatic pressure between the two ends of the tube. The ratio P/m is known as the *capillary constant*.

The height of the mercury column causes changes in the diffusion current, the square of the diffusion current being proportional to the height, after correction for back pressure, of the mercury column above the electrode tip. It can be shown that the value of the back pressure, h_{back}, to be deducted is given by the equation:

$$h_{back} = \frac{3.1}{m^{\frac{1}{3}}t^{\frac{1}{3}}} \text{ cm of mercury}$$

The potential applied to the polarographic cell influences the interfacial tension at the mercury-solution interface. As the applied potential is increased so the interfacial tension, σ, increases, initially to a maximum at about -0.56 volts *vs* Normal Calomel Electrode (NCE) (in the absence of capillary active substances) and then decreases again (Fig. 100). This maximum is known as the

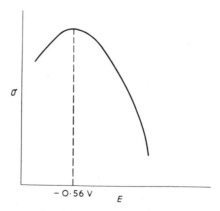

Fig. 100. Electrocapillary curve

electrocapillary maximum, electrocapillary zero or the isoelectric point and at this potential the mercury is uncharged. The position of the electrocapillary zero varies when capillary active ions (e.g. KI, KBr) are adsorbed on to the surface of the mercury drop.

Since it can be shown that drop time is directly proportional to the surface tension, it is to be expected that a drop time *vs* potential curve will be similarly shaped to the electrocapillary curve (Fig. 100).

Every term in the Ilkovic equation, with the sole exception of *n*, varies with temperature and the overall effect is quite complex. The factor most affected by temperature is the diffusion coefficient and, in practice, it is found necessary to control the temperature to $\pm 0.5°$ in order to keep variations in i_d, due to temperature, within $\pm 1\%$.

The solvent viscosity affects the diffusion coefficient to a large extent (the latter is inversely proportional to the viscosity) and the factors *m* and *t* to a relatively small extent.

Polarographic Maxima

Reproducible maxima often occur in cv curves unless eliminated by the addition of a suitable maximum suppressor, such as methylcellulose or gelatin. Figure 101 shows the reduction of oxygen where curve *a* is the unsuppressed

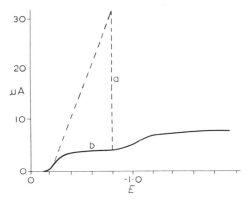

Fig. 101. Reduction of oxygen (a) in absence of maximum suppressor,
(b) in presence of maximum suppressor

oxygen maximum and *b* is the oxygen wave in the presence of gelatin. The height of each maximum depends largely on the concentration of electro-reducible substance in the solution. Maxima are often absent in very low concentrations and become more pronounced as the ion concentration is increased.

No entirely satisfactory explanation for the occurrence of maxima is available although several theories have been advanced by various workers.

Study of the Polarographic Wave

Consider a generalised reversible oxidation-reduction reaction:

$$\text{REDUCTANT} \rightleftharpoons \text{OXIDANT} + ne$$

e.g.

$$Fe^{2+} \rightleftharpoons Fe^{3+} + e$$

From the Nernst equation:

$$E_{\text{applied}} = E^0 + \frac{0.0592}{n} \log \frac{[OX]}{[RED]} \quad \text{(at 25°)}$$

where

$$E^0 = \text{standard electrode potential}$$

Hence, with reference to Fig. 99:

$$E_{\text{applied}} = E^0 + \frac{0.0592}{n} \log \frac{(i_d - i)}{i}$$

Now, at the half-wave potential, the log term becomes zero so that $E^0 = E_{\frac{1}{2}}$. Thus:

$$E_{\text{applied}} = E_{\frac{1}{2}} + \frac{0.0592}{n} \log \frac{(i_d - i)}{i}$$

A graph of $\log i/(i_d - i)$ plotted against E_{applied} will, therefore, give a straight line with a slope of $n/0.0592$. If the line is not straight then the reaction is not truly reversible (Fig. 102).

Thus a 1, 2 or a 3 electron change gives a reciprocal slope 59, 30 or 20 mV (at 25°) respectively.

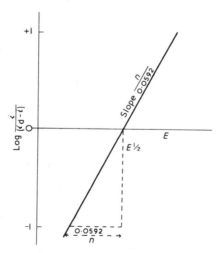

Fig. 102. Graph of $\log i/(i_d - i)$ vs E

Reference Electrodes

When using the dropping mercury cathode, a large non-polarisable reference electrode of mercury is often used. For routine quantitative analysis this is usually satisfactory but, since the electrode does not possess a definite potential, it cannot be used for accurate work unless its potential is measured. The potential of the pool anode varies with the concentration of ions in solution and behaves, for example, as a calomel electrode having the same concentration of chloride ions.

A more useful reference electrode is a saturated calomel electrode (SCE) because its potential is known accurately (246 mV more positive than the normal hydrogen electrode) and does not vary.

The potential (in V) of the polarised dme at the mercury-solution interface, E_m, is given by the expression:

$$E_m = E_{ref} - (V - iR)$$

where E_{ref} is the potential of the reference electrode (viz. 0.246 V in the case of a SCE), V is the applied voltage, i is the current flowing through the cell and R is its resistance (note that by convention, for cathodic reduction, V and i are positive).

In practice, however, it is more usual to quote the *applied potential* and the reference electrode used e.g. -1.25 V *vs* SCE (or NCE etc., as the case may be).

Miscellaneous Phenomena

Hydrogen overvoltage. According to Heyrovsky, hydrogen is reduced at the dme according to the following equations:

$$H + H^+ \rightleftharpoons H_2^+$$

$$H_2^+ + e \rightleftharpoons H_2$$

viz. he assumes that molecular hydrogen is formed by the combination of a hydrogen atom, formed by electrolysis, with a hydrogen ion.

The theoretical potential, E, at which hydrogen is reduced at the dme, at 25°, is given by the equation:

$$E = E_{H^0} + 0.0592 \log [H^+]$$

where E_{H^0} is the standard potential of the normal hydrogen electrode and $[H^+]$ is the concentration of hydrogen ions. In practice, a value (E'), greater than E, is required before hydrogen is evolved. The difference between the two values is known as the hydrogen overvoltage.

The overvoltage depends on the metal, the physical condition of the electrodes, the physical state of the substance deposited, the current density, the hydrogen ion concentration, the presence of other ions in the solution and the temperature.

Catalytic hydrogen waves. Herasymenko and Slendyk (1930) found that the hydrogen overvoltage was greatly increased by cations which deposited at potentials more negative than that of hydrogen (e.g. K^+). Certain cations (e.g. Pt^{4+}) and organic substances had the reverse effect.

Quinine, quinidine, cinchonine and cinchonidine in ammonium chloride; also cystine and cysteine in ammoniacal cobalt buffer solution are found to give well defined catalytic hydrogen waves. Some amino acids (e.g. phenyl-β-alanine) are, however, able to suppress the catalytic cysteine wave.

Oxygen waves. Oxygen is reduced at the dme in two stages, either

$$O_2 + 2H^+ + 2e \rightleftharpoons H_2O_2$$
$$H_2O_2 + 2H^+ 2e \rightleftharpoons 2H_2O$$
(acid conditions)

or,

$$O_2 + 2H_2O + 2e \rightleftharpoons H_2O_2 + 2OH^-$$
$$H_2O_2 + 2e \rightleftharpoons 2OH^-$$
(alkaline conditions)

Oxygen, therefore, must be removed from solutions by displacement with nitrogen or hydrogen otherwise the oxygen wave will be superimposed on the polarograms.

Biological media, e.g. blood serum, sometimes froth badly on deoxygenation with an inert gas. Rapid deoxygenation can be achieved by the addition of glucose, glucose oxidase and catalase.

The removal of oxygen may also be achieved by using carbon dioxide when the solution is acid, or sodium sulphite when neutral or alkaline:

$$2SO_3^{2-} + O_2 \rightleftharpoons 2SO_4^{2-}$$

ORGANIC POLAROGRAPHY

Organic polarography deals with *irreversible* reduction. There are, however, a few exceptions, e.g. the reduction of *p*-quinone or nitrosobenzene:

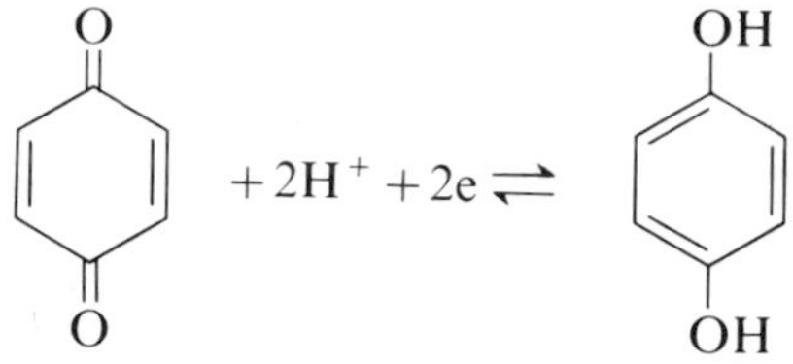

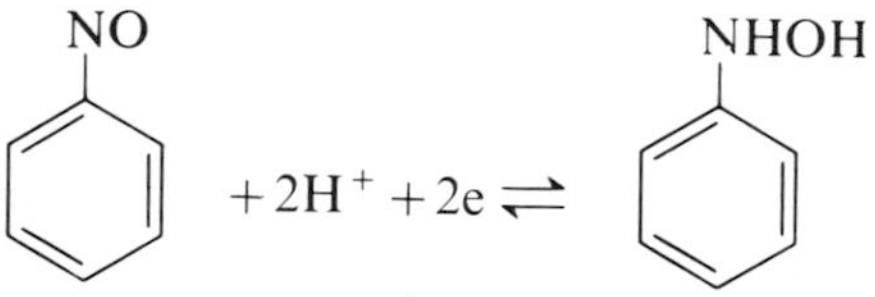

The following are the most important organic groups which are reduced at the dme: aldehydes, ketones, azo, diazo, nitro and nitroso compounds, activated carbon-carbon double bonds, organic peroxides, lactones, organic halogen compounds, disulphides, certain acids and some organo-metallic compounds.

The nature of the reduction product at the dme is often in doubt since it is not necessarily identical with the product produced by chemical reduction. For example, a nitro compound $(R.NO_2)$ may undergo a 2, 4 or a 6 electron change to produce a nitroso compound (R.NO), a hydroxylamine (R.NHOH) or an amine ($R.NH_2$) respectively.

The usual supporting electrolyte (e.g. KCl, KNO_3 etc.) can only be used when the organic substance is water-soluble or is suitably solubilised (e.g. with cetomacrogol 1000). Alternatively, non-aqueous solvents such as dioxan, alcohols, glycerol, glacial acetic acid or liquid ammonia may be used with a soluble lithium compound or quaternary ammonium salt as supporting electrolyte.

The electro-reduction which occurs may be represented in the general form:

$$R + nH^+ + ne \rightleftharpoons RH_n$$

As hydrogen ions are involved, the values of half-wave potential vary with the pH of the solution. Thus, for reproducible results, it is essential for the solutions to be adequately buffered.

The half-wave potential also varies with the supporting electrolyte, the solvent and often with concentration (cf. reversible systems), hence it is necessary to record adequate information when reporting values of $E_{\frac{1}{2}}$. Much early data is incomplete and of little value due to the use of unbuffered solutions or failure to record the solvent used. Furthermore, some early workers reported the *apparent reduction potential*, taken as the point where the cv curve just commences to rise. The value of $E_{\frac{1}{2}}$ may be over 100 mV more negative than this.

There is no good relationship between $E_{\frac{1}{2}}$ and constitution. For example, aromatic nitrogen compounds have values of $E_{\frac{1}{2}}$ varying from -0.1 to -0.7 V (*vs* NCE) over the pH range 1–13; carbonyl compounds have $E_{\frac{1}{2}}$ -0.2 to -1.3 V in acid solutions; aldehydes and ketones have $E_{\frac{1}{2}}$ -1.3 to -2.0 V in alkaline solution.

THE DROPPING MERCURY ELECTRODE

Care of the Dropping Mercury Electrode

1. When setting up the dme for the first time it is essential to use clean, dry and dust-free tubing.

2. Only pure, double or triple distilled mercury should be employed.

3. The conventional dme should be mounted within $\pm 5°$ of the vertical or else the drop time will be erratic.

4. The capillary assembly should be mounted on a heavy stand and on a bench free from vibrations, to prevent premature dislodgement of the mercury drops.

5. Traces of dust cause erratic behaviour of the capillary and should be protected by keeping the tip of the capillary immersed in water when not in use.

6. The mercury reservoir must *not* be lowered unless the capillary tip has been thoroughly washed and immersed in water. When the reservoir is lowered, it should be to a position such that the mercury *just* stops flowing.

7. As each mercury drop falls from the capillary, the mercury thread momentarily retracts slightly into the lumen before the succeeding drop begins to form. This pumping of solution in and out at the end of the capillary tends to make it dirty after long use. The tip can be cleaned by immersing periodically

in 50% v/v nitric acid with the mercury flowing and then washing thoroughly with a jet of water.

Advantages of the Dropping Mercury Electrode

1. It has a smooth and continually renewable surface exposed to the solution being analysed.

2. Each drop formed is unaffected by the reactions which occurred at the surface of earlier drops.

3. Mercury amalgamates readily with most metals.

4. The high hydrogen overvoltage of mercury enables analyses to be carried out in acid solutions.

5. The diffusion equilibrium at the mercury-solution interface is rapidly attained.

Disadvantages of the Dropping Mercury Electrode

1. Mercury has a limited application in the more positive potential range (i.e. when used for anodic polarography), since anodic dissolution of mercury takes place at about $+0.5$ V.

2. The surface area of the drop is never constant.

3. Changes in the applied voltage produce changes in the surface tension of mercury and, therefore, changes in drop size.

4. The addition of surface active agents produce changes in drop size.

5. Mercury may be toxic in certain biological studies.

Other Types of Mercury Electrodes

Several types of mercury electrode have been used in attempts to overcome the disadvantages of the conventional dme due to the ever varying surface area of the drop.

Horizontal mercury electrode. This is a dme, in which the tip of the capillary enters the solution horizontally and not vertically. It has the advantage that current oscillations are reduced to almost zero.

Smoler electrode. This is a dme in which the end of the electrode is bent through 90°. It is equivalent to the horizontal mercury electrode but is mounted vertically (Fig. 103).

Inverted mercury electrode. This electrode, devised by Parker and Griffiths, has the advantage that the surface area is kept constant and, therefore, gives steady current values. It also utilises a dme, as shown (Fig. 104). The heights of the two halves of the apparatus must be carefully adjusted to give an even flow of mercury. The mercury level in the concentric tube on the right hand side should be kept in the position shown.

Fig. 103
Smoler
electrode

Cracked tube electrode. This electrode, invented by Ferrett and Phillips, depends on the large contact angle between mercury and glass. Use is made of a small crack in egg shell tubing (thickness of the wall is about 0.1 mm), as shown in Fig. 105, and the rate of flow of mercury is about 100 mg per minute. The surface area of the exposed mercury surface is constant, and therefore, current fluctuations are absent.

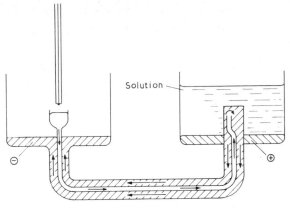

Fig. 104. Inverted mercury electrode

Multiple tip electrode. When several dm electrodes are connected to the same mercury reservoir, the diffusion current is increased. From the Ilkovic equation the diffusion current obtained should be

$$i_d = knD^{\frac{1}{2}}C(m_1^{\frac{2}{3}}t_1^{\frac{1}{6}} + m_2^{\frac{2}{3}}t_2^{\frac{1}{6}} + \ldots + m_n^{\frac{2}{3}}t_n^{\frac{1}{6}})$$

The use of such an electrode has doubtful advantages since the residual current is also increased. At least 25 tips are necessary to ensure that the various drops would always remain out of phase and that the contribution of each to the total current would be small.

DME with electromagnetic tapping device. An electromagnetic tapping device may be used to dislodge drops and so obtain a more uniform drop size than that obtained under normal conditions.

DME with glass hoe. A glass 'hoe' can be attached so that the blade is below the forming drop thus separating the mercury from the capillary tip as soon as

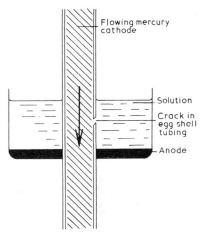

Fig. 105. Cracked tube electrode

the growing drop contacts the blade. This method is reputed to eliminate current oscillations and maxima in the cv curves.

Streaming mercury electrode. This electrode, contrived by Heyrovsky and Forejt, eliminates the effect of the periodic change in area of the dme. The capillary orifice is increased to about 0.1 mm and under a pressure of about 50 cm of mercury, a jet of mercury is sent upwards through the solution under examination at an angle of 45°. The jet of mercury becomes equivalent to a cylindrical electrode with a constant surface area whose surface is being continually renewed (Fig. 106).

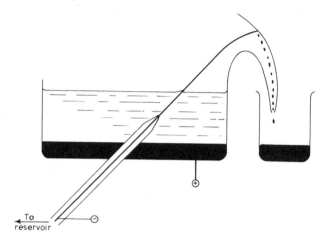

Fig. 106. Streaming mercury electrode (simplified)

Platinum Electrodes

Microelectrodes of platinum, or other noble metals, are useful for studies in the more positive potential range and in biological studies.

Stationary electrodes suffer from the disadvantage that it takes some minutes before a steady reading can be obtained but this disadvantage may be overcome by using a rotating microelectrode (*ca* 1800 rpm). Heyrovsky has suggested that the use of such electrodes should not be classified under the term *polarography* because they do not give highly reproducible results. For example, Cd^{2+} is reduced and deposited on the platinum to give a normal 'polarogram' with increasing negative voltage. However, if the voltage is now slowly reduced to zero a differently shaped curve is obtained since the electrode is now acting as a cadmium instead of a platinum electrode.

BASIC PRINCIPLES OF POLAROGRAPHIC INSTRUMENTATION

Simple Manual Polarograph

Figure 107 shows the circuit diagram of an extremely simple, though practical, polarograph. The polarising circuit consists of a battery (or accumulator) B

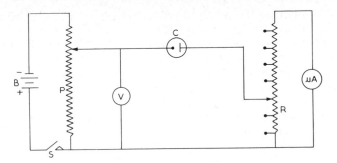

Fig. 107. Circuit diagram of a simple manual polarograph

(3–4 V) connected to an uncalibrated potentiometer (preferably of the helical type) and a switch.

To plot a polarogram, assuming the electrodes of the polarographic cell C have been connected, the meter is first protected by setting it to its minimum sensitivity range by use of the shunt R, and the apparatus switched on by means of switch S. A suitable low potential, e.g. about 0.1 V, as indicated on the high resistance voltmeter V, is applied to the electrodes by suitable adjustment of the potentiometer P. The meter sensitivity is increased, if necessary, and the corresponding current is measured and recorded. The polarising potential is adjusted to a higher value (say 0.2 V) and again the corresponding current is recorded. This process is repeated until a sufficient number of readings enables a polarogram to be plotted graphically.

The use of such a simple instrument illustrates all the essentials of the more elaborate commercial polarographs. One obvious disadvantage for accurate work is the use of an uncalibrated potentiometer and in most simple polarographs a standardising circuit is incorporated (Fig. 108). An additional refine-

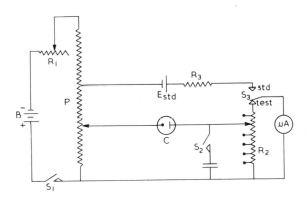

Fig. 108. Circuit diagram of a manual polarograph

ment, to facilitate measurements, is the optional use of a large capacity capacitor (1000–4000 μF) across the microammeter in order to damp the oscillations of the current. E_{std} is the standard cell, one end of which is connected to the appropriate position of the potentiometer, and the other end, protected by a resistance R_3, to the standardising switch S_3. When S_3 is in the 'Std' position, the potentiometer is standardised by adjusting R_1 to give zero deflection on the meter and in the 'Test' position; the circuit functions exactly as before. The damping capacitor may be switched in or out, at the wish of the operator, by means of S_2.

Potentiometric Manual Polarographs

One means of measuring current, by Kolthoff and Lingane, is to pass it through a standard resistance and measure the potential difference set up. Figure 109 shows a simplified circuit of such a polarograph which is also capable of measuring accurately the potential difference across the cell electrodes. The primary circuit, as in Fig. 107, consists of a battery, switch and an uncalibrated potentiometer.

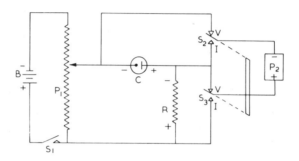

Fig. 109. Circuit diagram of a potentiometric manual polarograph

The resistance R is at least $10\,k\Omega$ and its actual value should be known accurately to within $\pm 0.1\%$. The calibrated potentiometer P_2 is shown schematically. Any type of potentiometer is suitable provided it can measure potentials correctly to ± 1 mV. The galvanometer associated with P_2 should have a period of about 10–20 sec so as to minimise oscillations of the needle. In use, S_2/S_3 is put in position V (voltage) and P_2 is set to the desired polarising voltage. P_1 is adjusted until the galvanometer of the calibrated potentiometer P_2 reads zero. S_2/S_3 is then put in position I (current) and P_2 adjusted until its galvanometer reads zero. The current flowing through the polarographic cell is calculated from Ohm's Law: $i = E/R$ where E is the potential reading of P_2 at the balance point.

Another type of potentiometric manual polarograph, by Jolliffe and Morton, is illustrated in Fig. 110. In this case the calibrated, low resistance potentiometer P_2 is connected in series with the standard resistance R and in a way such that the potential drop across R is opposed. When the potential across R is balanced out by an exactly equal and opposite potential V, then the potential difference across terminals XY will be zero. This is most readily detected by the use of a tuning indicator valve or 'magic eye' of suitable sensitivity.

As in the previous circuit, the polarising potential may be measured by means of suitable switching (not illustrated).

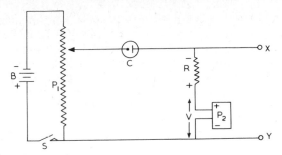

Fig. 110. Circuit diagram of a potentiometric manual polarograph

Recording Polarographs

The potentiometer for supplying the polarising potential and the recording chart are both driven in unison, usually by means of synchronous motors. The records of current flowing *vs* potential are obtained either photographically or by means of a pen recorder.

Photographic recorders. The light beam from a mirror galvanometer is focused on to a sheet of photographic paper which is mechanically driven in step with the polarising potentiometer. As the current varies so does the position of the light spot on the recording paper, which is subsequently developed and fixed in the usual way.

Pen recorders. There are numerous designs of pen recording polarographs on the market. In the Tinsley Polarograph, for example, a sensitive mirror galvanometer is placed in series with the polarographic cell. The polarographic current which passes through the galvanometer deflects the mirror and the light focussed on it is reflected on to a photo-cell. The current produced is then amplified by a DC amplifier so that the output is sufficient to operate a moving coil recorder.

Compensation Circuits

Electrical compensation can be arranged by applying a counter emf or counter current to reduce to zero electrically an interfering wave or residual current.

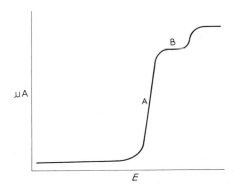

Fig. 111. Polarogram of a mixture of two electroreducible substances

Figure 111 shows the polarogram of two substances, A and B, where the concentration of A is many times greater than that of B. It is not possible to measure accurately the wave-height of B unless wave A is electrically reduced to zero, thus allowing wave B to be examined at a much higher sensitivity.

Figure 112 is a simplified circuit diagram (after Lingane and Kerlinger) to show how a counter current may be applied. This current, the magnitude of which is controlled by the $100\,\Omega$ potentiometer, may be applied to reduce wave A to zero. In practice, the polarogram is obtained for wave A and, when the plateau is reached, the counter current is switched on by operating switch S_2/S_3. The $100\,\Omega$ potentiometer is adjusted until the current reads zero, the meter sensitivity (not shown) is then increased to a suitable value and the polarographic wave for B determined in the usual way.

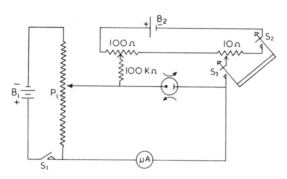

Fig. 112. Polarographic circuit employing counter current to oppose diffusion current of one component

At high sensitivities, it is difficult to measure diffusion currents accurately owing to the large values of residual current (Fig. 113, curve A). By application of a suitable counter current, it is possible to eliminate entirely the effect of the residual current.

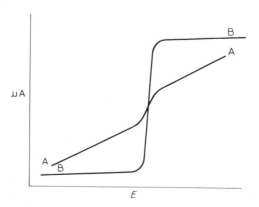

Fig. 113. (A) Uncompensated polarogram at high sensitivity; (B) Compensated polarogram at higher sensitivity

The simplified circuit diagram (Fig. 114, Ilkovic and Semerano) shows one method of producing a counter current to eliminate the residual current. Increased sensitivity may now be applied to the meter so as to obtain curve **B** (Fig. 113) thus enabling the diffusion current to be measured easily and accurately. It is usually necessary to apply damping to the galvanometer when employing electrical compensation since, at the greater sensitivity settings used, the fluctuations due to the dme are magnified.

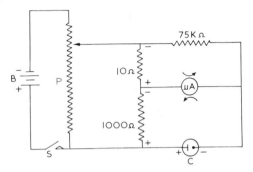

Fig. 114. Polarographic circuit employing counter current to correct for residual current automatically

Derivative or Differential Polarography

If two dropping mercury electrodes of identical characteristics are connected so that one electrode is at a slightly higher potential than the other (Fig. 115,

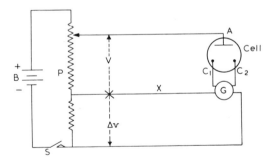

Fig. 115. Circuit diagram of a differential polarograph employing twin dropping mercury electrodes

after Heyrovsky), then it is possible to plot $\Delta i/\Delta E$ vs E (Fig. 116). The potential difference across $C_1A = V$ and that across $C_2A = (V+\Delta v)$. The galvanometer shown is of a specialised design in that it consists of two coils wound in opposite directions. The net result is that the meter measures the *difference* in current flowing for the small fixed difference of voltage Δv between the two electrodes C_1 and C_2. A microammeter inserted at point X will record normal polarograms.

Figure 116 illustrates the type of curve obtained by this method. The height h is proportional to concentration and its peak coincides with the half-wave potential.

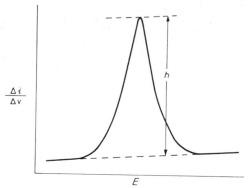

Fig. 116. Differential polarogram

Some electrical compensation has also to be provided (not illustrated) to allow for slight difference in the capillary characteristics.

A derivative curve may also be obtained by using a single dme and placing a large capacitor (2000 μF) in *series* with the galvanometer (Fig. 117, after Leveque and Roth). A short-circuit switch S_2 permits the polarograph to be used normally. It also ensures there is no initial charge on the capacitor which could discharge through the galvanometer and probably damage it.

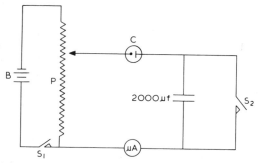

Fig. 117. Differential polarograph employing a capacitor in series with a galvanometer

It is essential that the potential applied to the cell is increased at a uniform rate (e.g. by using a synchronous motor as in a recording polarograph) since the galvanometer will only record a *change* of current in the circuit. The curve Δi vs E will be similar to Fig. 116. It should be noted that this method does not give a true derivative curve due to the time lag in charging the capacitor and, as a result, the value of $E_{\frac{1}{2}}$ recorded is more negative than the true value.

A derivative polarogram may also be obtained using an *AC polarograph*. Here, a small AC voltage is superimposed on the polarising potential, and the AC component of the current produced (Fig. 118) is amplified and measured.

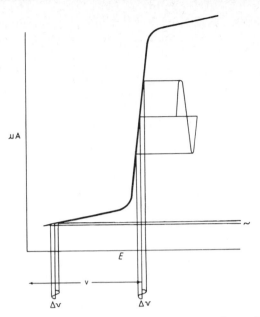

Fig. 118. The effect of a small AC voltage superimposed on the applied DC potential

Maximum AC current is obtained at the half-wave potential, the current being proportional to concentration.

A *square wave polarograph* (Barker and Jenkins) is an extension of the AC polarograph in which a square, rather than a sine, wave is used. The sensitivity of this method is extremely high; a concentration of ions as small as 10^{-8} M and, under certain conditions, 10^{-9} M can be measured.

A *cathode-ray polarograph* is also available. The voltage is kept constant for about 70% of the life of a mercury drop and then increased at a steady rate, over a range of about 0.5 V, for the last 30%. The cv curve during this final period is displayed directly on the screen of a cathode-ray tube and the wave-heights measured. This method is about 100 times more sensitive than the conventional polarograph.

Advantages of derivative differential curves

(1) The position and height of a single point, viz. the peak of the curve, gives both qualitative and quantitative information.

(2) The method can be used to measure concentrations of substances under conditions which would be unfavourable to normal polarography, e.g. when values of half-wave potential are only about 150 mV apart and when reduction waves occur very close to the decomposition potential of the electrolyte.

(3) The method is more rapid and more sensitive than conventional polarography.

Pulse Polarography

In this technique, a voltage pulse is applied to the polarographic cell towards the end of the life of each mercury drop when the drop growth is small. The resulting flow of current comprises a short duration charging current and a longer lived reduction current. The measurement of the current is made about 50 msec after the voltage pulse is applied so that the charging current has decayed significantly. The measured reduction current level is maintained until the next pulse has been applied to provide a typical step-shaped polarogram (similar to Fig. 99). This integral pulse technique may be used for concentrations less than 10^{-7} M.

When pulses of uniform amplitude are superimposed on a sweeping DC voltage, the polarogram has a characteristic derivative shape (as Fig. 116). This differential pulse technique enables the resolution of ions with half-wave potentials less than 40 mV apart to be determined.

POLAROGRAPHIC METHODS OF ANALYSIS

Direct Comparison Method

In this method, the diffusion current obtained for the 'test' solution is compared, under identical conditions, with that of a solution of known concentration. Maximal accuracy is obtained when the diffusion currents of both solutions are about equal.

The most important conditions which must be kept constant in the comparison are temperature, concentration of maximum suppressor (if any), composition of the supporting electrolyte and the characteristics of the dme (i.e. constant m and t values).

For complex substances, such as alloys, it is advisable to keep standard comparison samples of known composition. These comparison samples must approximate closely to the samples being analysed.

Use of Empirical Calibration Curves

The dme is calibrated empirically with various known concentrations of the substance in question and a graph of diffusion current *vs* concentration is plotted. The concentration of the test solution can be read from the graph. It is essential to control the temperature accurately and to check that the capillary characteristics do not vary.

Internal Standard or Pilot Ion Method

This method was suggested by Forche (1938) and is based on the fact that the relative *diffusion current constants*, I, are independent of the particular capillary used, provided the nature and concentration of the supporting electrolyte and the temperature are kept constant.

From the Ilkovic equation:

$$i_d = ICm^{\frac{2}{3}}t^{\frac{1}{6}}$$

where

$$I = 607\,nD^{\frac{1}{2}}$$

For the 'pilot' ion

$$i_{d_1} = I_1 C_1 m^{\frac{2}{3}} t^{\frac{1}{6}}$$

and the 'test' ion

$$i_{d_2} = I_2 C_2 m^{\frac{2}{3}} t^{\frac{1}{6}}$$

hence

$$\frac{i_{d_1}}{i_{d_2}} = \frac{I_1 C_1}{I_2 C_2}$$

The ratio I_1/I_2 is known as the *pilot ion ratio* (symbol R) and is independent of the capillary characteristics.

Thus

$$\frac{i_{d_1}}{i_{d_2}} = R \cdot \frac{C_1}{C_2}$$

Quasi-absolute Method

This method, suggested by Lingane, also makes use of the diffusion current constant. From the Ilkovic equation:

$$C = \frac{i_d}{I m^{\frac{2}{3}} t^{\frac{1}{6}}} \quad (g/l)$$

Expressed in this form rather than in terms of n and D, evaluation of I is simplified because it is very difficult to determine accurate values of the diffusion coefficient which would otherwise be required. I is constant for a given substance in a given supporting electrolyte at a fixed temperature. It is also independent of the solution and the characteristics of the dme provided that the drop time exceeds 1.5 sec.

The absolute method avoids the necessity of time consuming calibration polarograms, the evaluation of $m^{\frac{2}{3}} t^{\frac{1}{6}}$ for a given capillary being much more rapid. However, the method is not as accurate as the more conventional methods of calibration.

Method of Standard Addition

This method was originated by Hohn (1937). The polarogram of the test solution, of known volume, is initially recorded; then an accurately measured quantity of a standard solution of the substance in question is added and a second polarogram is obtained. The original concentration of the substance can be calculated from the increase in the diffusion current from the following formula:

$$C_1 = \frac{C_2 i_1 v_2}{v_1(i_2 - i_1) + i_2 v_2}$$

where C, v and i are the concentration, volume and diffusion current and the subscripts 1 and 2 refer to the test and standard solutions respectively. Maximum precision is obtained when $i_2 \simeq 2i_1$.

EXPERIMENTS IN POLAROGRAPHY

Apparatus

Dropping mercury electrode assembly. A vertical capillary, 6–8 cm long and with a bore diameter of about 50 μm is mounted on a suitable stand. The lower end of the capillary must be cut accurately at right angles to its axis. The upper end of the capillary is connected to a mercury reservoir by means of clean, dry polythene or similar tubing. High grade rubber tubing may be used if traces of surface impurities, such as sulphur, are removed by treating with hot 10% sodium hydroxide (24 hr) and thoroughly washing and drying the tube before use.

The reservoir is then filled with pure triple distilled mercury and recently filtered through sintered glass to remove traces of dust. The tip of the capillary should be immersed in water when not in use and, with reasonable care, should remain serviceable for several months (see Fig. 98, p. 184).

Polarographic cells. The variety of design of polarographic cells is too large to discuss here. The only essential requirements are to provide access to the cell for the capillary, gas supply and the reference electrode connection (Fig. 98).

Reference electrode. A mercury pool anode is often used but, for more accurate work, a saturated calomel electrode, connected to the solution via a suitable salt bridge, is to be preferred.

Deoxygenation apparatus. Either nitrogen or hydrogen gas may be used and it is sometimes necessary to remove traces of oxygen by passing the gas through alkaline pyrogallol before use. In all cases, the gas must be passed through a wash bottle containing the same solvent as in the polarographic cell, before being bubbled through the solution for analysis. This ensures that the gas is saturated with the solvent vapour and thereby prevents undue concentration of the test solution due to removal of solvent.

Thermostat. For accurate work the polarographic cell and gas wash bottles should be at the same temperature. A suitable thermostat, set at $25° \pm 0.1°$, should be used for this purpose.

General Experimental Methods

The following selection of experiments illustrates several of the principles discussed in the theoretical sections of this chapter. In all cases, unless otherwise stated, the following general method should be adopted and full conclusions should be drawn from the results whenever possible.

General Method. Adjust the reservoir height so as to give a drop time of about 3 or 4 sec with the mercury tip immersed in water. Deoxygenate the test solution by passing hydrogen or nitrogen through it for 20–40 min. Rinse the electrode with a little of the solution and immerse the electrode tip in the solution. If a mercury pool anode is to be used, add 1 to 2 ml of mercury and bubble gas through the solution for a few minutes longer. Alternatively, a salt bridge, connected to a saturated calomel electrode, can be inserted into the solution before deoxygenation. If damping of the current fluctuations is required, set the appropriate switch to the desired position. Standardise the potentiometer (except on the simplest of manual polarographs which employ a voltmeter) and set the galvanometer or recording pen to zero.

Connect the electrodes to the polarograph and adjust the galvanometer to the minimum sensitivity, unless the correct setting of the shunt is known. The correct sensitivity is usually found by increasing the polarising voltage to about -1.8 V and adjusting the sensitivity control to give a large, but less than full-scale deflection.

Plot the polarogram, either manually or automatically, by the following general procedure:

(a) *Manual Operation.* Set the potentiometer to zero and observe the meter reading. If the reading is negative, increase the applied potential in 0.1 V increments until a positive value is obtained. [For the majority of experiments it is unnecessary to record the negative values of current. In most instances it is due to the 'tail end' of an anodic wave of the supporting electrolyte anion.] Record further *average* current values for gradually increasing 0.1 V increments of the applied voltage until the current changes start to increase. Now reduce the voltage increments to about 0.02 V until the current changes are small again. Finally, continue with 0.1 V increments until the polarising potential is about −2.0 V. Plot the polarogram graphically and read the diffusion current from the curve.

(b) *Automatic Operation.* After standardising the potentiometer, set the potentiometer to the desired initial value and the recorder pen to the zero lines. Switch on the synchronous motors used for driving the potentiometer slide wire and recorder chart and, assuming the sensitivity setting is correct, the polarogram is plotted automatically. Switch the motors off when the applied potential reaches about −2.0 V or the current readings approach full scale deflection. With photographic recorders the recording chart must be developed and fixed in the usual way before it can be read.

Wash the tip of the dome well with water before rinsing with the next test solution. When all experiments are completed wash the dme thoroughly and then immerse it in water. Lower the mercury reservoir slowly to a point where the mercury *just* ceases to emerge from the capillary.

Quantitative Evaluation of the Polarogram

Of the many methods suggested for the measurement of the diffusion current, the following are the most suitable for general use. It will be assumed that two reduction steps, as in the case of many simple mixtures, are present.

(a) *Exact procedure.* This method is applicable to well-defined waves whose limiting current plateaux are parallel to the residual current curve of the supporting electrolyte. Figure 119 shows how the measurements are made

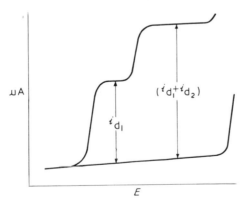

Fig. 119. Evaluation of a polarogram by the exact procedure

where i_{d_1} and i_{d_2} represent the diffusion currents of the first and second reducible ions respectively. By this method, values of diffusion current are automatically corrected for i_r.

(b) *Approximate procedure.* This method is sufficiently accurate for most purposes and is applicable to well-defined waves only. It has the advantage that the residual current polarogram is not necessary. The near horizontal portions of the curve are produced and the values of i_{d_1} and i_{d_2} are obtained as shown (Fig. 120).

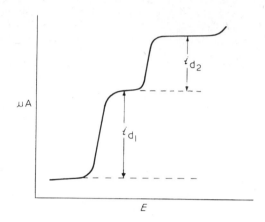

Fig. 120. Evaluation of a polarogram by the approximate procedure

(c) *Modified approximate procedure.* This method is applied when no well-defined plateau separates the two polarographic waves. The value of i_{d_1} is measured by drawing construction lines through the inflection point and parallel to the upper and lower, near horizontal, parts of the polarogram (Fig. 121). The values of i_{d_1} and i_{d_2} are then obtained as shown.

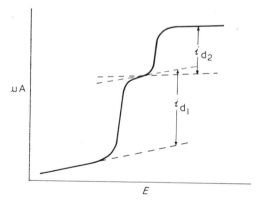

Fig. 121. Evaluation of a polarogram by the modified approximate procedure

Experiment 1. *Oxygen wave, maximum suppression and residual current*

Method. Set up the apparatus and obtain polarograms, by the general method described, as follows:

(a) Electrolyse about 10 ml of potassium chloride (*0.1N*) but do not remove dissolved oxygen at this stage. Note the fluctuations in the current at each applied potential and record the mean value in each case.

(b) Repeat (a) in the presence of one drop of either gelatin (0.2%) or methyl-cellulose (1%).

(c) Repeat (b) in the presence of two, three or more drops of gelatin or methyl-cellulose in successive experiments until the oxygen maximum is completely eliminated (Fig. 101).

(d) Deoxygenate (c) for about thirty minutes and obtain the residual current polarogram.

Note. If this experiment is performed carefully it will be seen that the peak height falls with increased concentration of surfactant. This is the basis of an extremely sensitive method of analysis known as *adsorption analysis*.

It would be to the student's advantage to perform this experiment on the simplest possible (home-made) polarograph and then to repeat on a commercial manual or recording polarograph.

Experiment 2. *Effect of supporting electrolyte on diffusion current*

Method. Obtain polarograms of lead chloride (*0.002M*):

(a) in the absence of supporting electrolyte and measure i_1, the limiting current.

(b) in the presence of potassium chloride (*0.05M*) and measure i_d, the diffusion current.

From the results the migration current, i_m, can be found from the relationship:

$$i_d = i_1 - i_m$$

The transport number of the lead cation, t^+, can also be calculated since:

$$i_m = i_1 t^+$$

Hence,

$$t^+ = 1 - i_d/i_1$$

Experiment 3. *Supporting electrolyte and its effect on the half-wave potential*

Method. Obtain polarograms for a mixture of cadmium chloride (*0.01M*) and lead chloride (*0.008M*) in a supporting electrolyte of:

(a) potassium chloride (*0.1N*)

(b) a mixture of sodium potassium tartrate (*0.5M*) and sodium hydroxide (*0.1N*).

Experiment 4. *Diffusion current and concentration*

Method. Prepare, from a stock solution (*0.1M* $CdSO_4$ in *0.1N* KCl), the following concentrations of cadmium sulphate in potassium chloride (*0.1N*):

(a) $1 \times 10^{-2}M$ (e) $1 \times 10^{-3}M$
(b) $7.5 \times 10^{-3}M$ (f) $1 \times 10^{-4}M$
(c) $5 \times 10^{-3}M$ (g) $1 \times 10^{-5}M$
(d) $2.5 \times 10^{-3}M$ (h) zero, viz. (*0.1N* KCl only)

Obtain a polarogram for solution (a), note the potential range over which the diffusion current plateau is substantially constant and set the potentiometer to an intermediate value, say −1.0 V. Obtain and record diffusion currents for solutions (b) to (g) and the residual current, using solution (h), at the fixed polarising potential. The appropriate sensitivity settings for the galvanometer should be used in each case. Deduct the residual current observed for solution (h) from all the other values recorded and thus obtain the corrected diffusion current. Plot a graph of the logarithms of the diffusion current against the logarithms of the concentration. Also plot a calibration curve, for solutions (a) to (e) inclusive, of current against concentration.

Make at least two suitable dilutions of an 'unknown' cadmium sulphate solution. Electrolyse at -1.0 V and, after deduction of the residual current, use the calibration curve to find the concentration of $CdSO_4$ in the solution.

Experiment 5. *Capillary characteristics*

(a) $i_d/m^{\frac{2}{3}}t^{\frac{1}{6}}$ = constant (see p. 188)
(b) P/m = constant (see p. 188)
(c) i_d^2/P = constant (see p. 188)

Method. Using four different reservoir heights (say, 50, 60, 70 and 80 cm) and solution from Experiment 4(a) above, obtain the corresponding diffusion currents and drop times (average of 10 drops) at an applied potential of -1.0 V. Now transfer the electrodes to a 100 ml beaker containing the same solution, but not deoxygenated. At each reservoir height, and at the same applied potential (-1.0 V), collect 10 drops of mercury in a small glass container. Wash the mercury with water and dry by rinsing with acetone followed by a jet of air. Weigh the mercury on an analytical balance and hence calculate m. Determine P by correcting the reservoir height for back pressure (see p. 188). Conclude whether or not the relationships (a), (b) and (c) above are correct.

Experiment 6. *Diffusion current and temperature*

Method. Use solution from Experiment 4 (a) above and a mercury pool anode; surround the polarographic cell with crushed ice and record the temperature of the solution in the cell by means of a thermometer. Apply a polarising potential of -1.0 V and record the diffusion current and corresponding temperature of the solution. Gradually increase the bath temperature to 40° and record, about every 5°, the corresponding values of diffusion current. Bubble nitrogen through the solution to keep it stirred in between readings. Plot a graph of diffusion current *vs* temperature and calculate the temperature coefficient of the diffusion current at 20°. Conclude how accurately the temperature should be controlled for variations, due to temperature, to be kept within $\pm 1\%$ at 20°.

Experiment 7. *Analysis of a polarographic wave*

Method. (a) Prepare a solution containing cadmium sulphate *(0.01M)* potassium chloride *(0.1N)* and gelatin (0.005%). Place some solution in a polarographic cell which is immersed in a thermostatic bath set at $25° \pm 0.1°$. Deoxygenate and obtain the cv curve from -0.4 to -0.8 *vs* SCE, in 0.01 V increments. It is preferable to use a manual polarograph for this purpose and to choose a suitable galvanometer sensitivity which should remain unchanged throughout the experiment.

(b) Prepare a solution of potassium chloride *(0.1N)* and gelatin (0.005%) and obtain, at 25°, the residual current curve over the same voltage range. Correct the values of current obtained in (a) by subtraction.

(c) Determine the resistance, R, of the polarographic cell by using a conductivity bridge or other suitable instrument. Correct the applied potential recorded in (a) by deducting the corresponding values of iR.

(d) Calculate $\log i/(i_d - i)$ values and plot against the corresponding corrected applied potentials. Determine the number of electrons involved in the reduction by measuring the slope of the graph (see Fig. 102, p. 190).

Experiment 8. *Absolute method for determining concentration*

Method. Prepare a suitable dilution of an 'unknown' cadmium sulphate solution (say, about *0.01M*) so that the final solution also contains potassium chloride *(0.1N)* and gelatin (0.005%). Obtain a cv curve as described in experiment 7(a) and correct for residual current as in 7(b). Finally determine m and t values as in experiment 6, but at an applied potential corresponding with the experimental value of half-wave potential just observed.

Using the formula:

$$C = i_d/(Im^{\frac{2}{3}}t^{\frac{1}{6}})\, g/l$$

discussed on p. 205, calculate the concentration of cadmium in the original solution, given that $I = 3.5$ under the conditions of this experiment.

Experiment 9. *Limit test for trivalent antimony in Sodium Stibogluconate*

Sodium Stibogluconate B.P. contains not more than 0.2 % trivalent antimony when determined by the following method.

Method. To 0.2 g sample, accurately weighed, in 10 ml of water add 2 ml 0.1 % w/v aqueous solution of gelatin, 2 ml concentrated hydrochloric acid and dilute to 20 ml with water. Transfer an aliquot portion to a polarographic cell and bubble nitrogen through the solution for ten minutes. Record a polarogram over the range 0 to $+0.5$ V (viz. an anodic wave) and compare with a standard trivalent antimony calibration curve using, in the above procedure, 0.25 ml of a 0.8 % w/v aqueous solution of a stibophen. The height of the step at a potential of approximately 0.15 V vs SCE is a measure of Sb^{3+}. [A potential of about 0.4 V vs SCE is probably better since $E_{\frac{1}{2}} \simeq +0.15$ V.] The solutions must be examined within 30 minutes of preparation as they are unstable.

The total number of possible experiments and applications of polarography are beyond the scope of this chapter and reference should be made to such books as:

(1) Kolthoff and Lingane, *Polarography*, 2nd edn., Interscience, New York and London, 1952—an extensive treatise in two volumes.

(2) Muller, *The Polarographic Method of Analysis*, 2nd edn. Chemical Education Publishing Co., Easton, Pa., 1951—a small volume, ideal for students, describing many experiments in detail.

AMPEROMETRIC TITRATIONS

Introduction

When a suitable emf is applied between an indicator electrode and an appropriate depolarised reference electrode, the current which flows through the titration cell is measured as a function of the volume of a suitable titrating solution. The end point of the titration is often found at the point of intersection of two straight lines which intersect at the equivalence point. (Compare conductimetric titrations.)

Salomon (1897) first proposed the principle of amperometric titrations. He used two silver electrodes in a dilute solution of potassium chloride, at an applied emf of 0.1 V, and titrated with silver nitrate. The equivalence point was indicated when both electrodes became depolarised and resulted in a large increase in current (Fig. 122). This principle is applied in the electrometric end point method for the titration of sulphonamides (Part I, Chapter 11 and in the determination of water by titration with Karl Fischer Reagent Part I, Chapter 11).

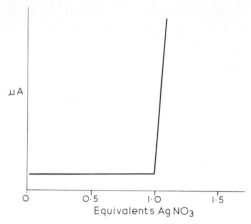

Fig. 122. Amperometric titration of potassium chloride with AgNO$_3$

Amperometric Titrations Using a Dropping Mercury Electrode

Very dilute solutions may be titrated accurately (*ca* $\pm 0.3\%$) and rapidly. Amperometric titration results are independent of the capillary characteristics and temperature, provided that there is no change during a titration; the reaction need not be reversible and substances which are not oxidised or reduced may be titrated if the *reagent* gives a diffusion current. Amperometric titrations involving precipitation may be carried out when the solubility is appreciable and under conditions where potentiometric and indicator methods are inaccurate. Titrations may also be carried out in the presence of large amounts of electrolyte (e.g. potassium chloride) without interference (contrast conductimetric titrations).

Apart from the normal sources of error in volumetric determinations, impurities which give diffusion currents may have to be removed by a preliminary chemical separation, or the conditions so chosen that foreign constituents do not contribute to the current.

Several types of titration curve may be obtained as follows:

(1) Consider a lead solution containing an excess of an indifferent electrolyte. The polarogram (Fig. 123) has a plateau between points A and B, where the current is practically constant. This solution may, therefore, be titrated with a solution of sodium oxalate at any fixed applied emf between the values A and B by measuring the current flowing and the volume of titrant added.

During the titration lead oxalate is precipitated out and this results in a fall of diffusion current until the end point is reached, when only a small residual current flows (Fig. 124).

The slight curvature is due to dilution of the solution with the reagent and is minimised by using a relatively strong titrant (compare conductimetric titrations). In any case, it can be corrected by multiplying the observed values of current by the ratio total volume to initial volume.

(2) It is possible to titrate a non-reducible substance with a reagent which is electroreducible. Lead nitrate, for example, may be titrated with potassium

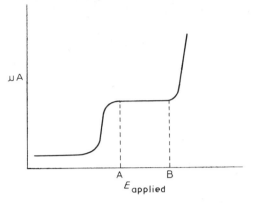

Fig. 123. Polarogram of lead nitrate solution

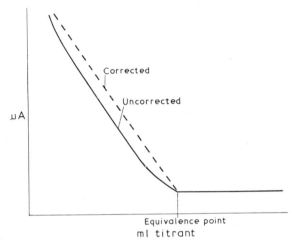

Fig. 124. Amperometric titration of lead with sodium oxalate at 1.0 V vs SCE

dichromate, in acetate buffer, at zero applied volts. (Pb^{2+} is not reduced at this voltage.) When the end point is reached the dichromate ion ($Cr_2O_7^{2-}$) yields a diffusion current (Fig. 125).

(3) The substance to be titrated and the reagent used may both be electro-reducible. Lead nitrate and potassium dichromate, as above, may be titrated at an applied potential of -1.0 V vs SCE. The diffusion current, due to the lead ions, first falls as lead is precipitated out and rises when excess dichromate ions are present (Fig. 126).

(4) Amperometric titrations may also be applied to electro-oxidisable substances. Potassium iodide may, for example, be titrated with mercuric nitrate:

$$Hg^{2+} + 2e \rightleftharpoons Hg$$

$$2I^- + 2Hg \rightleftharpoons Hg_2I_2 + 2e$$

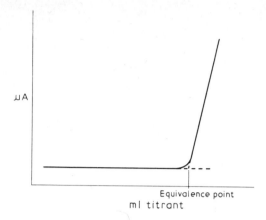

Fig. 125. Amperometric titration of lead with potassium dichromate at zero volts *vs* SCE

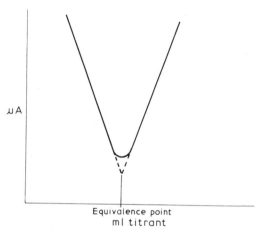

Fig. 126. Amperometric titration of lead with potassium dichromate at −1.0 V *vs* SCE

The titration curve is, more or less, a straight line and the end point is reached when the current is zero (Fig. 127). The negative branch of the curve is the anodic diffusion current due to the iodide and the positive branch is the cathodic diffusion current due to the mercuric ions.

(5) It is occasionally possible to carry out an amperometric titration when neither the substance titrated nor the reagent yields a polarographic wave. A 'polarographic indicator', which has a diffusion current at the applied emf, may be used to react with excess reagent. For example, aluminium salts may be titrated with a fluoride, using a ferric salt as indicator. Both aluminium and iron form complexes with fluoride but the former is more stable. During the titration the diffusion current falls gradually and then very rapidly near the end point which corresponds to zero current.

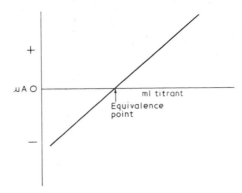

Fig. 127. Amperometric titration of iodide ions with mercuric nitrate

(6) *Amperometric titration involving chelating compounds.* When a metal ion forms a complex or chelate, the reduction potential is displaced to a more negative value. Use can be made of this fact in amperometric titrations using the dropping mercury electrode. Thus, if the voltage is maintained at a value such that *only* the metal ions are discharged, the metal ions may be titrated with a complexing agent until the diffusion current has reached a minimum producing a curve similar to that shown in Fig. 124. Since the stability of the complex is usually very dependent on pH, the pH of the solution must be buffered at a suitable value.

If, however, the complexing or chelating agent also undergoes reduction, its reduction potential being less negative than that of the metal complex, it follows that if the potential is held at a value such that both the metal ion and complexing agent are reduced but not the metal complex, the diffusion current will fall to a minimum, due to the removal of metal ions, and then rise again due to the reduction of the excess chelating agent. The curve obtained is similar to that shown in Fig. 126.

An illustration of these procedures is shown by the work of Stock who titrated copper ions (1 ml samples of 1.75×10^{-3} M) with quinoline-8-carboxylic acid (5×10^{-3}M) in a solution buffered at pH 5. At an applied voltage of -0.4 V *vs* SCE only the copper ions were reduced, producing a curve similar to that shown in Fig. 124. When the applied voltage was -1.35 V *vs* SCE, where the chelating agent but not the metal chelate was reduced, a V-shaped curve like that shown in Fig. 126 was produced, thus enabling a more accurate determination of the end point to be achieved. The advantage of this type of titration is that very low concentrations of metal ions can be determined.

Apparatus

The apparatus required is essentially the same as for ordinary polarography. The titration cell must have adequate capacity, provision being necessary for a burette, electrodes and gas supply. Stirring may be carried out using a magnetic stirrer or with nitrogen bubbling through the solution.

AMPEROMETRIC TITRATION EXPERIMENTS

General Method

Use a conical flask as the titration cell, fitted with a cork with holes for a burette, dropping mercury electrode, agar salt bridge (connected to a SCE) and also for nitrogen inlet and outlet. Pipette a convenient quantity of solution into the flask and add an equal quantity of supporting electrolyte ($0.01M$). Deoxygenate the solution for about twenty minutes.

Apply the required polarising voltage to the cell and measure the diffusion current. Add 1 ml of titrant, stir the solution (either with a magnetic stirrer or with nitrogen) for 20–30 sec and measure the diffusion current. Repeat until well past the equivalence point and plot a graph of diffusion current (corrected for volume change) against the burette reading. The intersection of two straight lines usually marks the position of the equivalence point.

Experiment 1. *Determination of nickel with dimethylglyoxime*

Method. (1) Weigh accurately a sample of nickel salt to give about $0.001M$ nickel solution. Pipette 25 ml into the titration cell and add an equal quantity of supporting electrolyte [a mixture of ammonium hydroxide ($1.0M$) and ammonium chloride ($0.2M$)] and 1 or 2 ml gelatin solution (0.2%).

(2) Deoxygenate the solution and set the applied emf to -1.85 V vs SCE.

(3) Measure the diffusion current.

(4) Titrate with dimethylglyoxime solution ($0.02M$) using the general method. The graph should be V-shaped.

1 ml dimethylglyoxime solution ($0.02M$) $\equiv$ 0.5869 mg Ni

Experiment 2a. *Determination of lead with potassium dichromate solution*

Method. (1) Weigh accurately a sample of lead salt to give about $0.001M$ lead solution. Pipette 25 ml into the titration cell and add an equal quantity of potassium nitrate solution ($0.01M$).

(2) Deoxygenate the solution and set the applied emf to zero.

(3) Measure the diffusion current.

(4) Titrate with potassium dichromate solution ($0.005M$) using the general method. The graph should be ⌐-shaped.

$$1 \text{ ml } 0.01M \text{ K}_2\text{Cr}_2\text{O}_7 \equiv 2.072 \text{ mg Pb}$$

Experiment 2b. As 2a, but at an applied emf of -1.0 V vs SCE
The graph should be V-shaped.

Experiment 3. *Determination of iodide with mercuric nitrate*

Method. (1) Weigh accurately a sample of iodide to give about $0.001N$ iodide solution. Pipette 50 ml of this solution into the titration cell and add an equal quantity of nitric acid ($0.2N$) containing 0.1 % gelatin.

(2) Deoxygenate the solution and set the applied emf to zero V.

(3) Measure the diffusion current.

(4) Titrate with mercuric nitrate solution ($0.01M$). The graph is almost a straight line and the equivalence point is when the current reaches zero.

$$1 \text{ ml } 0.01M \text{ Hg(NO}_3)_2 \equiv 2.538 \text{ mg I}^-$$

COULOMETRIC ANALYSIS

Introduction

Provided no extraneous reaction is involved, Faraday's Law states that one equivalent of chemical change occurs at an electrode during electrolysis with the passage of 96 490 coulombs of electricity. If one can assume that there are no competing side reactions (i.e. 100% current efficiency) then:

$$w = QM/96\,490n$$

where w = weight of substance transformed during electrolysis, Q is the quantity (coulombs) of electricity, M is the molecular (or atomic) weight of the substance being oxidised or reduced and n is the number of faradays involved in the electrolysis. It is clear that there is a direct relationship between the weight of material transformed and the quantity of electricity consumed. Therefore, the process is suitable as a method of chemical analysis providing the variables can be measured accurately.

An inert electrode system and absence of electro-reducible impurities (e.g. oxygen) are essential for quantitative applications; microgram quantities are readily determined, the accuracy depending on the precision of measurement of Q.

Under suitable conditions, a large variety of compounds may be analysed: ascorbic acid, bromoform, carbon tetrachloride, chloroform, cysteine, EDTA, iodoform, methylene blue, oxalic acid, phenol and salicyclic acid, to name but a few. Reference to books on coulometry is recommended for a more detailed study: see, for example, *Coulometry in Analytical Chemistry*, Milner and Phillips, Pergamon, 1967.

Measurement of Parameters

Q may be measured directly by using a silver coulometer and measuring the increase in weight of the cathode. Although this method is theoretically sound it does present certain practical hazards and it is more usual to determine Q by an indirect method.

If a constant current source i (amps) is available and can be measured accurately, then Q can be determined easily by recording the time t (secs) accurately, since

$$Q = it$$

The measurement of current can be direct using a moving coil milliammeter, or indirect by using a potentiometer to measure the potential drop across a standard resistor placed in series with the coulometric cell.

The number of coulombs may be measured by using a calibrated low inertia integrating motor fitted with a suitable revolution counter. The motor is connected across a high precision standard resistor through which known currents are passed for accurately measured times. Since the motor speed is proportional to the voltage (over a small range) across the resistor, the motor may be used to measure the number of coulombs involved in a given reaction even if the current fluctuates.

Coulometric Cell

The cell (Fig. 128) should have two compartments separated by sintered glass or other suitable membrane. Provision for additional indicator electrodes may be necessary since, in some cases, it is convenient to use an electrochemical method to determine the equivalence point of a coulometric reaction rather than an absorptiometric or colorimetric method.

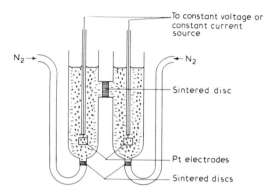

Fig. 128. Coulometric cell (for *in situ* production of titrant)

Constant Current/Constant Potential Coulometry

Either a constant current DC source or a constant potential DC supply is used in coulometric analysis. The former method is rapid and the parameter measured (apart from current) is time which is directly proportional to coulombs. The latter method is much slower since the current continually falls but has the advantage that optimal conditions can be deduced directly from a polarographic curve by selecting a voltage corresponding to a point on the diffusion current plateau.

COULOMETRY EXPERIMENTS

The following experiments are intended to illustrate a fundamental approach to coulometric titrations and in each case the 'titrant' is prepared *in situ*. A simple type of coulometric cell is illustrated in Fig. 128.

Experiment 1. *Coulometric titration of hydrochloric acid*

Method. Fill both electrode compartments of the coulometer cell with *0.01M* sodium sulphate solution and pipette 1.00 ml of *0.1N* hydrochloric acid into the *cathode* compartment together with a few drops of phenolphthalein. Mix and deoxygenate by bubbling nitrogen through the cell. Connect to a suitable constant current source of about 10 mA and note the time (sec) for the indicator to change. Assuming 100% current efficiency, calculate the factor of the hydrochloric acid.

Note. (1) A preliminary trial run may be necessary to ensure suitable conditions.

(2) If a constant current source is not available, then use a voltage source and keep the current as steady as possible by adjustment of a series rheostat (Fig. 129).

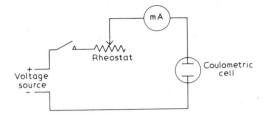

Fig. 129. Circuit for coulometric titration. Direct current measurement

(3) The reactions which occur can be represented by the following equations:

Anode compartment (a) $SO_4^{2-} - 2e = SO_4^*$

(b) $2SO_4^* + 2H_2O + 2H_2SO_4 + O_2$

Cathode compartment (a) $Na^+ + e = Na^*$

(b) $2Na^* + 2H_2O = 2NaOH + H_2$

(4) The anode compartment becomes acidic—this can be checked by addition of a suitable indicator.

(5) A more accurate method of measuring the current is by measuring the potential drop across a series resistor (about $100\,\Omega$) with a potentiometer and calculating the current using Ohm's Law (Fig. 130).

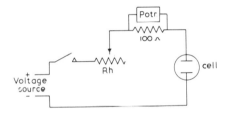

Fig. 130. Circuit for coulometric titration. Indirect current measurement

Experiment 2. *Use of an integrating motor*

An integrating motor (IM) is often used to overcome the difficulty in maintaining a really constant current. The angular velocity of the motor shaft is proportional to the applied voltage within reasonable voltage limits so that the number of revolutions in a given time can be used as a measure of the quantity of electricity which has flowed in that time. The IM has a built-in revolution counter (calibrated in one-hundredth's of a revolution) which must be calibrated before use.

Method. (a) Calibrate the IM by placing it in parallel with a suitable resistor in series with a milliammeter (see Fig. 131).

It is useful to place a voltmeter across the 2 KΩ resistor in the first instance to adjust the voltage to about 20 V (for a 24 V IM). It is essential to take care not to exceed the working voltage of the motor.

Record the counter reading of the IM, switch on the power and start the stop watch simultaneously. Record the current, say, every 10 sec. and switch off at 500 sec. Note the new reading, calculate the total number of coulombs which have flowed and hence calculate the number of coulombs corresponding to 1 revolution of the motor.

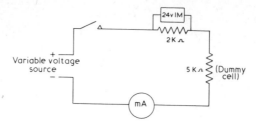

Fig. 131. Circuit for coulometric titration. Calibration of integrating motor

(b) Repeat Experiment 1 by replacing the milliammeter and the 5 KΩ resistor with the coulometric cell. Note the initial and final meter readings and hence calculate the factor of the acid. Compare the results with those obtained in Experiment 1.

(c) Suggest how it would be possible to calibrate the integrating motor so that it reads directly in milli-equivalents.

(d) Devise a method and suitable apparatus to determine the pK_a of a weak acid by coulometric titration. Use it to determine the pK_a values of phosphoric acid.

7 The Basis of Spectrophotometry

W. D. WILLIAMS

Nomenclature and Units

The white light from an incandescent solid such as the filament of an electric lamp is made up of a large number of individual waves of varying wavelength. This is readily shown by passing a beam of the light through a prism when a band of colour, or so-called continuous spectrum, is formed, in which each colour corresponds to waves of a particular wavelength (Fig. 132).

The visible spectrum, however, forms only a small part of the complete spectrum of electromagnetic radiation which extends, as shown in Fig. 132, from the ultra-short wave region of the cosmic rays at one end to that of radio-waves at the other.

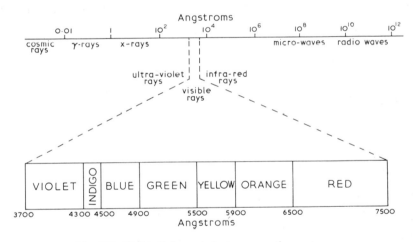

Fig. 132. Visible light and electromagnetic spectrum

Wavelength can be defined as the distance from the crest of one wave to that of the next, as for example A to B, or B to C in Fig. 133.

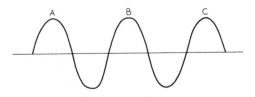

Fig. 133. Wavelength

The wavelength is the distance between any two consecutive parts of the medium whose vibrations are in phase. The units in which wavelength is commonly expressed are recorded in Table 35. By adapting the unit to the appropriate region, the use of cumbersome figures can often be avoided.

Table 35. Wavelength Units

Unit	Symbol	m	Regions where used
Angstrom	Å	10^{-10}	Visible and ultraviolet
Nanometer ⎫ Millimicron ⎭	nm mµ	10^{-9}	Visible and ultraviolet
Micrometer ⎫ Micron ⎭	µm µ	10^{-6}	Infrared

Other units related to wavelength which are often used include wave-number and frequency.

Wave-number is defined as the reciprocal of the wavelength (*in vacuo*) expressed in cm, i.e. the number of waves per cm. It is used particularly in connection with infrared absorption spectra.

Frequency is the number of waves emitted per sec.

The Fresnel is equal to 10^{12} waves per sec.

The inter-relationships of these units can be expressed as follows:

$$\frac{1}{\text{wavelength } in\ vacuo\ (\text{in cm})} = \frac{\text{wave-number}}{} = \frac{\text{frequency}}{\text{speed of light } in\ vacuo\ (\text{cm/sec}^{-1})}$$

$$\frac{1}{\lambda} = \bar{v} = \frac{v}{c}$$

The Absorption of Energy by Molecules

A molecule may absorb energy in three ways,

(a) by raising an electron (or electrons) to a higher energy level,

(b) by increasing the vibration of the constituent nuclei, and

(c) by increasing the rotation of the molecule about its axis.

The relative energies of (a), (b) and (c) are roughly in the order of $10\ 000 : 100 : 1$; and the total energy for any one state is given by the expression:

$$E_{\text{total}} = E_{\text{electronic}} + E_{\text{vibrational}} + E_{\text{rotational}}$$

The potential energy of a diatomic molecule may be illustrated diagrammatically in a potential energy-nuclear distance curve (Fig. 134).

When the nuclei are at infinity, the mutual forces are zero, but as they approach one another, forces of attraction operate and the potential energy decreases. This is readily understood if the heat of formation of substances is borne in mind. As the nuclei get very close to one another, repulsion takes place and the potential energy increases. The atoms can thus vibrate about the minimum position R_e in the vibrational level 0 (Fig. 134). The energy associated with the

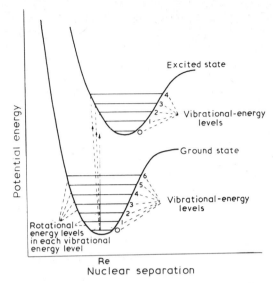

Fig. 134. Diagram of potential energy-nuclear separation curves

molecule is quantised and can only assume the values indicated by the horizontal lines in the diagram. At room temperatures, the molecule is in the lowest vibrational level of the ground state, except in the case of some of the heavier molecules such as iodine whose vibrational quanta are low enough for an appreciable proportion to be in the state which has one quantum of vibrational energy.

In an electronic transition from the ground state to an excited state, the part of the upper curve to which the molecule is transferred is that vertically above the position (internuclear distance) it had in the ground state. This is the most probable and therefore the strongest transition, though transitions to other parts of the potential energy curve involving different amounts of vibrational energy are also possible and occur more weakly.

Changes of rotational energy also accompany the electronic change but they are of smaller magnitude and give rise to a fine structure superimposed on the electronic-vibrational change (Fig. 135).

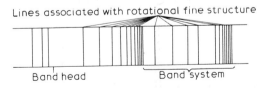

Fig. 135. Diagram of band spectrum

The frequency of the absorption bands associated with the transition is given by

$$hv = E_{\text{excited}} - E_{\text{ground state}}$$

(where h is Planck's constant) and their appearance as a pattern or system of bands arises mainly from transitions to the various vibrational levels of the excited state. This is illustrated in Fig. 136.

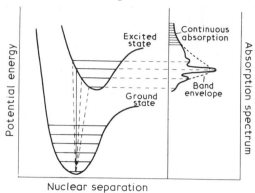

Fig. 136. Diagram of relationship between electronic transitions and absorption curve

With molecules containing more than two atoms, the appearance of the spectrum can be complicated by the possibility of many different types of vibration and by other effects especially in solutions, which blur out the vibration pattern into general continuous absorption. The strong solute-solvent interaction of polar solvents is especially effective in broadening and shifting band systems and, if it is desired to observe as much of the band structure as possible, it is necessary to use non-polar solvents such as cyclohexane.

Excited molecules are extremely short-lived, their lives being of the order of 10^{-8} sec. Therefore energy does not accumulate in the system, but is immediately dissipated in several ways which may be either chemical or physical.

Light Sources

The distribution of energy through a spectrum is mainly a function of temperature; the higher the temperature of the light source the shorter the wavelength of the peak emission. The heating process, however, cannot be carried too far, as changes such as vaporisation would take place with the consequent production of a line spectrum or burning out of a lamp. For example, to obtain ultraviolet light from a tungsten lamp it would have to be run at a gross overvoltage, which would shorten its life if not destroy it.

Common energy sources for the various spectral regions are indicated below.

Infrared radiation. Globar and Nernst glowers are common sources of infrared radiation. The Globar is an electrically heated rod of silicon carbide and the Nernst glower is a small rod of refractory oxides which, when heated, will conduct electricity and thus maintain itself in incandescence. Both these sources operate without a glass envelope which would absorb infrared radiation of wavelength greater than 2 μm.

Visible radiation. The tungsten filament lamp is a satisfactory light source for the region 350 to about 2000 nm. Infrared radiation, which might affect the photocells used in the instrument, can be removed by suitable filters.

Ultraviolet radiation. The most convenient source for spectrophotometers is the hydrogen or deuterium discharge lamp which gives a continuous spectrum of molecular origin in the ultraviolet region. Dissociation of the hydrogen molecule into atoms takes place under the influence of the electric discharge so that the Balmer lines in the spectrum of *atomic* hydrogen can be used to check the wavelength calibration of the instrument. Hydrogen discharge lamps are also made which contain mercury vapour in addition, so that the mercury lines are available for calibration in the ultraviolet region.

When a particularly intense source of ultraviolet radiation is required for the excitation of fluorescence, a mercury vapour lamp in conjunction with suitable filters may be employed or, in spectrofluorimetry (Chapter 10), a xenon high pressure lamp.

Monochromatic Radiation

For absorptiometric measurements, monochromatic radiation (i.e. radiation of one particular wavelength) is necessary, so that absorptiometers and spectrophotometers must have means of producing at least an approximation to such radiation. The methods adopted can be discussed under two headings, filters, and monochromators which may be of the prism or diffraction grating type.

Filters

Glass filters. These are pieces of coloured glass which transmit limited wavelength ranges of the spectrum. The colour is produced by incorporating oxides of such metals as vanadium, chromium, manganese, iron, nickel and copper in the glass. Compounds of non-metallic elements are also used and typical colours produced by inorganic substances are recorded in Table 36.

Table 36. Composition of Glass Filters

Element or Compound	Colour	Remarks
Cobalt	Blue	
Copper	Blue-green	
Manganese	Purple	
Iron	Green	The Fe^{2+} absorbs infrared radiation and is used in the preparation of heat-absorbing filters
Cadmium sulphide	Yellow	
Cadmium sulphide + cadmium selenide	Red	Cadmium sulphoselenide is a solid solution of CdS and CdSe, the colour depending upon the relative proportions
Iron-manganese	Amber	
Rare earths	—	These have sharp absorption bands which are particularly useful for standardising spectrophotometers. A didymium glass filter can be used to absorb sodium radiation when calcium is being determined by flame photometry

The actual colours produced depend upon the ions and their environment and whether the ions enter the lattice structure of the glass or merely modify it.

Glass filters are unaffected by exposure to heat and light so that they maintain their transmission characteristics for long periods. Their disadvantage lies in the broad band-width of the transmitted radiation.

When working in the region 350–400 nm, it is often essential to absorb all visible light to reduce errors from stray radiation. Nickel is used as the colouring agent and the glass appears purple, maximum transmission being about 365 nm. Such filters are therefore also used to isolate radiation of this wavelength in fluorimetry. Conversely, the absorption of ultraviolet radiation but without loss of visible radiation is achieved by the use of cerium oxide.

Glasses coloured with manganese or cadmium sulphoselenide are used to absorb visible light and transmit infrared radiation. The use of ferrous iron for the reverse process is mentioned in Table 36.

Neutral filters show no selective absorption but merely reduce the intensity of the incident light. They are necessary when a feeble light intensity is being balanced in a two-photocell instrument such as the Spekker Fluorimeter. On one side is the weak fluorescence, whereas on the other the direct light of a mercury vapour arc impinges on the photocell. A neutral filter reduces the intensity of the light on the latter photocell and enables a balance to be obtained. The filters owe their properties to the presence of definite proportions of iron, cobalt and nickel oxides or, in certain borosilicate glasses, to a balance between ferrous and ferric iron.

Gelatin filters. The range of coloured organic substances available makes it possible to prepare highly specific filters by the use of suitable mixtures of dyes. The dye mixture is incorporated into gelatin which is then converted into thin sheets and sandwiched between a pair of glass plates. The stability of these filters is very much less than that of glass filters, and a heat-absorbing filter should always be used between the gelatin filter and light source.

Filters are supplied in matched pairs, but since complete matching is impossible, it is important not to substitute one for the other in absorptiometric measurements.

Interferometric filters consist of two parallel glass plates, silvered internally and separated by a thin film of cryolite or other dielectric material. Such filters make use of the interference of light waves rather than absorption to eliminate undesired radiation. They pass light of wavelength λ, dependent upon the thickness (t) of the dielectric, in accordance with the equation:

$$n\lambda = 2t \sin i$$

which for normal incidence ($\sin 90° = 1$) becomes

$$n\lambda = 2t$$

n being an integer. By adopting a suitable thickness for t, of the order of one light wave, it is possible to transmit light of one particular colour depending upon the thickness of t. For example, suppose t is 7500 Å, then the transmitted light would be:

$$\lambda_1 = 2 \times 7500 = 15\,000\,\text{Å} \quad \text{(invisible)}$$

$$2\lambda_2 = 2 \times 7500$$

$$\therefore \quad \lambda_2 = 7500\,\text{Å} \quad \text{(invisible)}$$

Similarly

$$\lambda_3 = 5000\,\text{Å} \quad \text{(visible)}$$

$$\lambda_4 = 3750\,\text{Å} \quad \text{(invisible)}$$

Thus with the postulated filter, the only visible light transmitted is that of 5000 Å. With thicker films, a larger number of specific wavelengths are passed until ultimately white light is transmitted.

With high reflectivities of the silvered surfaces, the band-width of the emergent radiation may be as narrow as 16 nm, but as a consequence, the transmission is usually quite small—about 6%. A recent publication, however, shows a marked improvement in transmission values to about 25%.

Such filters serve as relatively inexpensive monochromators for a specific purpose, for example the isolation of calcium radiation from that of sodium in the flame photometry method for Na^+ and Ca^{2+} in the same solution. Ordinary filters are incapable of this and cause large errors in the calcium content.

A guide to the characteristics of each type of filter is given in Table 37.

Table 37. Filter Characteristics

Type	Bandwidth (nm)	Per cent Transmission
Glass	150+	25–90
Gelatin	25–50	5–30
Interferometric	10–20	5–30

Under certain conditions, monochromatic radiation can be obtained with filters. A line source of radiation is required such that a particular radiation has no very bright competitor within about ± 30 nm. It may then be effectively isolated by means of a filter of the correct nominal wavelength and with say 50 nm bandwidth (Fig. 137). The bandwidth of the resultant radiation may be 1 nm or less, but the method is of limited application because of lack of suitable line sources.

Monochromators

Dispersion by a prism. When a beam of monochromatic light passes through a prism, it is bent or refracted, the amount of the deviation being dependent on the wavelength, blue light being bent more than red. If white light is substituted for monochromatic radiation, a separation of the different radiations leads to the formation of a spectrum (Fig. 138), from which suitable portions can be selected for use as light sources of different wavelength.

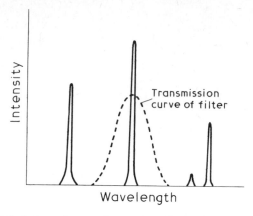

Fig. 137. Isolation of monochromatic radiation from a line source

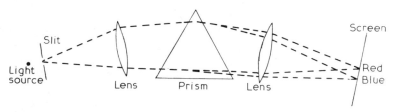

Fig. 138. Formation of spectrum

For resolution of two such dispersed radiations of wavelength λ and $(\lambda + \Delta\lambda)$ respectively, Rayleigh's criterion is usually adopted. This requires that the maximum intensity of the diffraction pattern of one radiation should occur at the first minimum of the other (Fig. 139). If $\Delta\lambda$ is the closest wavelength

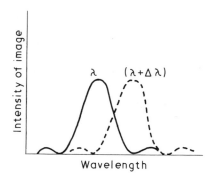

Fig. 139. Resolution (Rayleigh's criterion) of rays of wavelength λ and $\Delta\lambda$ respectively

spacing which can be resolved at a particular wavelength λ, then the resolving power of the instrument is given by the term $\lambda/\Delta\lambda$. The resolution is dependent upon the refractive dispersion (μ) of the material from which the prism is made and upon the size of the prism

$$\frac{\lambda}{\Delta\lambda} = \text{base of prism} \times \frac{d\mu}{d\lambda}$$

In most spectrophotometers, the simple arrangement of producing a spectrum shown in Fig. 138 is replaced by other arrangements involving mirrors rather than lenses. In all arrangements, however, the spectrum can be scanned across the exit slit by appropriate rotation of the prism or a mirror, and different parts of the spectrum can be selected at will by a 'wavelength' drive.

Prisms are made of quartz for use in the ultraviolet region of the spectrum since glass absorbs wavelengths shorter than about 330 nm. Glass prisms are preferable for the visible region of the spectrum, as the dispersion is much greater than that obtained with quartz. For the infrared regions, the transparent substances usually used for prisms are sodium chloride ($2-15\ \mu m$), potassium bromide ($12-25\ \mu m$), lithium fluoride ($0.2-6\ \mu m$), and caesium bromide ($15-38\ \mu m$).

Dispersion by gratings. A diffraction grating consists of a very large number of equispaced lines (200 to 2000 per mm) ruled on a glass blank coated with a thin film of aluminium. Plastic replicas made from rulings on metal blanks can be made much more cheaply than the original rulings and are often employed as the dispersing element in monochromators. They can be used either as transmission gratings or, when aluminised, as reflection gratings. Rotation of the grating permits appropriate parts of the spectrum to emerge from the exit-slit of the monochromator.

The theory of the plane transmission gratings is given below. Figure 140 shows part of a diffraction grating in which the gaps represent the transparent spaces. The distance d between consecutive corresponding elements is called the grating space. A parallel beam of monochromatic light falls upon the

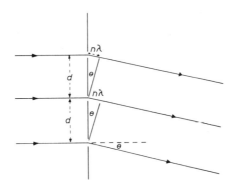

Fig. 140. Formation of nth order spectrum: $d \sin \theta = n\lambda$

grating at normal incidence. If the path difference for light diffracted through an angle θ from consecutive elements is $n\lambda$, then the various rays will reinforce each other. The values of θ in the equation $d \sin \theta = n\lambda$ ($n = 0, 1, 2, 3$) correspond to the angles of diffraction of different orders. When white light is used instead of monochromatic light, first and successive order images give rise to spectra (Fig. 141) due to the variation of θ with λ for a given n but the zero order is undispersed. The rulings of a grating can be shaped to concentrate the light in certain orders to give greater efficiency than when the light is spread

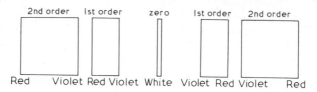

Fig. 141. Diagrammatic appearance of spectra from a grating

over many spectra, and this is especially important in the infrared. Gratings can be plane or concave, the latter being capable of focusing its own spectrum without the use of lenses or mirrors. However, for reasons of convenience in scanning, plane gratings are most frequently used in monochromators which can be made to cover all spectral regions from 180 nm to 15 μm. To eliminate the effect of overlapping orders, filters have sometimes to be used.

Measurement of the Intensity of Radiation

For the accurate determination of substances by their light absorption, precise determinations of the ratio of the intensity of the incident to the transmitted beams are necessary. Photoelectric detectors are most frequently used for this purpose. They must be employed in such a way that they give a response linearly proportional to the light input, and they must not suffer from drift or fatigue. One of the simplest detectors is the barrier-layer photocell which has the advantage that it requires no power supplies but gives a current, which, under suitable conditions, is directly proportional to the light falling on it.

Barrier-layer cells. These consist of a metallic plate, usually copper or iron, upon which is deposited a layer of selenium or sometimes cuprous oxide. An extremely thin transparent layer of a good conducting metal, for example, silver, platinum or copper, is formed over the selenium to act as one electrode, the metallic plate acting as the other (Fig. 142).

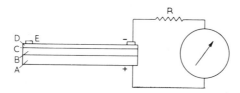

Fig. 142. Barrier-layer cell and circuit. A, metal base-plate; B, selenium layer; C, theoretical barrier-layer; D, transparent metal layer; E, collecting ring; R, external resistance

Ordinarily, selenium and metallic oxides and sulphides have extremely small electrical conductivities, the electrons being in energy levels where they are not mobile. Light of suitable frequency, however, imparts sufficient energy to the electrons so that they leave the selenium and enter the transparent metal layer. If the two electrodes are now connected through a galvanometer, a current will flow as shown by the deflection of the galvanometer needle.

If the resistance in the external circuit is small, the current produced by such a cell is very nearly proportional to the intensity of the illumination (Fig. 143). With higher resistances leakage back into the selenium layer and

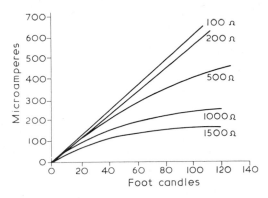

Fig. 143. Current output of barrier-layer cell with various resistances in the external circuit

hence loss of linearity occurs. The current output also depends upon the wavelength of the incident light, the sensitivity being similar to that of the human eye when an appropriate correction filter is used (Fig. 144).

The photocell is more responsive than the human eye to blue and red light so that the correction filter must absorb light in these regions and hence reduce the response of the cell.

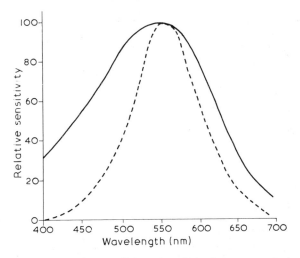

Fig. 144. Sensitivity of barrier layer cell (——) and the human eye (—————) to visible radiation

The cells must be used in combination with a sensitive galvanometer because their low internal resistance precludes amplification of the current. Fatigue effects may also be observed, i.e. the current rapidly reaches a maximum which is not maintained. A gradual decrease in the reading takes place, indicating loss of sensitivity, but this is readily recovered by keeping the cell in the dark. The time required for the current to reach a steady value in normal work varies from cell to cell but is usually less than 10–15 sec. The popularity of these cells for measuring light intensities is due in large measure to their cheapness and hard-wearing qualities.

Photo-emissive cells. Electrons are liberated when light falls on a metal surface, and if this is enclosed in an evacuated envelope and kept at a negative voltage, a current which is proportional to the incident light can be drawn from it. The essential feature for a sensitive surface in these cells is that the electrons should be liberated easily from the metal. Elements of high atomic volume, for example, potassium or caesium, are commonly used, and in order to increase the sensitivity, composite coatings such as caesium/caesium oxide/silver oxide have proved of more value than the metal alone. The sensitive surface is enclosed in a high vacuum and forms the cathode of the cell (Fig. 145).

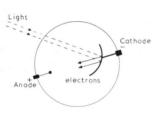

Fig. 145. Diagram of photo-emissive cell

Application of a sufficiently high potential between cathode and anode ensures that all the electrons liberated by the action of light reach the anode. A saturation photocurrent which exhibits a linear relationship with intensity of illumination is then obtained. By using different metals, the cells can be made to respond to different regions of the spectrum. In spectrophotometers two cells are usually used, one being responsive to ultraviolet and visible radiation of wavelengths up to about 620 nm, and the other to radiation of wavelengths 620 to 1000 nm.

A modification of the above type of cell involves the inclusion of an inert gas at a low pressure to give gas-filled cells. The underlying principle is that the liberated electrons cause ionisation of the gas with consequent increase in the photocurrent, but owing to lack of linear response, the cells cannot be used in instruments designed for accurate intensity measurement.

In order to obtain greater sensitivity to very weak light intensities, multiplication of the initial photoelectrons by secondary emission is employed. Several anodes at a gradually increasing potential are contained in the one bulb (Fig. 146). Electrons from the photocathode are attracted to anode 1 and liberate more electrons which travel to anode 2 because of its higher potential relative to anode 1. This process is continued to the last anode, and the result is a final photocurrent many times greater than the primary current which still shows a linear response with increase in the intensity of illumination. Photomultiplier tubes are ideal for measuring weak light intensities such as occur in fluorescence and in the determination of trace elements by their emission spectra.

Photoconductive detectors. Lead sulphide, and many other similar materials, show decreased electrical resistance on exposure to radiation. They form

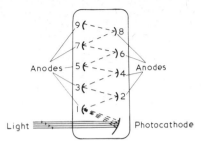

Fig. 146. Diagram of photomultiplier tube

photoconductive cells of particular use in the near infrared region, and in many cases their sensitivity extends in a lesser degree to the visible and ultra-violet regions.

Thermocouples. A thermocouple consists of elements of two different metals joined together at one end, the other ends being attached to a sensitive galvanometer. When radiant energy impinges on the junction of the metals, a thermo-electromotive force is set up which causes a current to flow. Thermocouples are used in the infrared region, and to assist in the complete absorption of the available energy the 'hot' junction or receiver is usually blackened.

Bolometers. These make use of the increase in resistance of a metal with increase in temperature, for example, if two platinum foils are suitably incorporated into a Wheatstone bridge, and radiation is allowed to fall on one foil, a change in resistance is produced. This results in an out-of-balance current which is proportional to the incident radiation. Like thermocouples, they are used in the infrared region.

Thermistors. The principle of operation is similar to that described under bolometers but thermistors are constructed of semi-conducting material which has a high negative coefficient of resistivity, i.e. the resistance decreases with increase in temperature.

Golay detector. In this detector, the absorption of infrared radiation causes expansion of an inert gas in a cell chamber. One wall of the chamber consists of a flexible mirror and the resulting distortion varies the intensity of illumination falling on a photocell from a reflected beam of light. The current from the photocell is proportional to the incident radiation.

Spectrophotometers

The various parts of a spectrophotometer, viz. light source, monochromator and detector, are shown in a typical arrangement in Fig. 147 for a double-beam (in time) recording instrument operating on the optical null principle. The principle is that there should be no difference between the intensities of the sample and reference beams at all times.

Light from the appropriate source, which is selected automatically, passes through the Littrow type prism. The beam is switched alternately through the sample and reference cells by a balanced sector mirror M7. Any difference in the intensities of the sample and reference beams produces an electrical signal from the photomultiplier. The signal, after amplification, is applied to

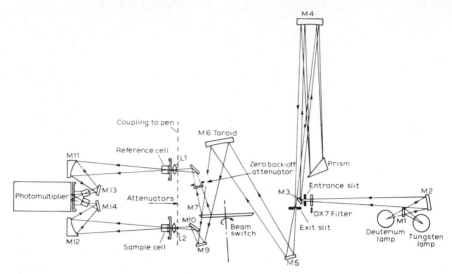

Fig. 147. Double beam spectrophotometer

a servo motor-generator which drives the beam attenuators (combs) in the direction which reduces the difference in intensities of the two beams. The attenuators are mounted on a single trolley which is connected to the pen of the recorder. Therefore, as the wavelength range is scanned by turning the prism a record of absorbance *vs* wavelength is obtained.

In spectrophotometers for use in the ultraviolet and visible regions of the spectrum the sample compartment is placed so that radiation emerging from the monochromator passes through solutions. The intensity of such radiation at any particular wavelength represents a fraction of that from the source itself and the risk of photochemical decomposition is thereby reduced. As the compartment is close to the photomultiplier a light-tight cover is required with a switch or shutter so that the detector is protected on opening the compartment.

In the infrared region, risk of decomposition is negligible and a different arrangement is possible with an open sample-handling area adjacent to the light source (Fig. 178), Chapter 11.

In recording instruments there is little to choose between prism and grating instruments except that in the latter, filters are required to absorb unwanted orders of spectra, e.g. with an absorption band at about 600 nm an orange filter would be necessary to transmit radiation of 600 nm but absorb radiation of 300 nm which would be present in the second order spectrum. The latter radiation would otherwise constitute stray radiation. Gratings have the advantage of constant dispersion over the spectral region whereas, to maintain a constant bandwidth, prisms require a decrease in slit width with increase in wavelength.

8 Analytical Applications of Absorption Spectra

W. D. WILLIAMS

INTRODUCTION

When monochromatic radiation passes through a cell which contains a solution of an absorbing substance, the effects occurring are reflection at cell surfaces, absorption by the solute and transmission of unabsorbed radiation. The intensity of the total incident light is then given by

$$I = I_{\text{reflected}} + I_{\text{absorbed}} + I_{\text{transmitted}}$$

$I_{\text{reflected}}$ can be compensated for by means of a control cell containing solvent only, so that under the experimental conditions the intensity of the incident light can be represented by I_0, where

$$I_0 = I_{\text{absorbed}} + I_{\text{transmitted}}$$

Lambert's Law

Lambert investigated the relationship between the incident light (I_0) and the transmitted light (I_T) for various thicknesses of substance, and found that the rate of decrease in intensity of the light with the thickness of the medium was proportional to the intensity of the incident light. Expressed mathematically

$$-\frac{dI}{dt} \propto I \quad \text{or} \quad -\frac{dI}{dt} = \mu I$$

where

$$t = \text{thickness of substance}$$
$$\mu = \text{proportionality constant}$$

$$\therefore \quad -\frac{dt}{dI} = \frac{1}{\mu I}$$

Integrating,

$$-t = \frac{1}{\mu} \ln I_T + C$$

where I_T = intensity of light transmitted at thickness t. When $t = 0$

$$C = -\frac{1}{\mu} \ln I_0$$

$$\therefore \quad -t = \frac{1}{\mu} \ln I_T - \frac{1}{\mu} \ln I_0$$

$$\therefore \quad \ln \frac{I_0}{I_T} = \mu t$$

μ is known as the *absorption coefficient*, but on conversion to common logarithms, the expression becomes

$$\log_{10} \frac{I_0}{I_T} = \frac{\mu}{2.3026} t = Kt$$

where K is the *extinction coefficient*. It is generally defined as the reciprocal of the thickness (t in cm) required to reduce the intensity of the incident light to $\frac{1}{10}$ its original intensity.

Beer's Law

Beer studied the effect of concentration of a coloured substance in solution on the absorption and arrived at a similar equation

$$\log_{10} \frac{I_0}{I_T} = kc \qquad (c = \text{concentration})$$

where k is the proportionality constant corresponding to μ (p 235). Beer's Law is usually stated in the form: absorption is proportional to the number of absorbent molecules in the light path.

To combine both laws, consider a cell of unit thickness containing a solution of an absorbing substance. Then

$$\log_{10} I_0 - \log_{10} I_{T_1} = kc$$

Now consider an identical cell placed after the first cell. The incident light on the second cell is I_{T_1}. Hence

$$\log_{10} I_{T_1} - \log_{10} I_{T_2} = kc$$

Continuing in this way

$$\log_{10} I_{T_2} - \log_{10} I_{T_3} = kc$$
$$\cdot \;\; \cdot \;\; \cdot \;\; \cdot \;\; \cdot \;\; \cdot \;\; \cdot \;\; \cdot \;\; \cdot \;\; \cdot$$
$$\log_{10} I_{T_{t-1}} - \log_{10} I_{T_t} = kc$$

Adding

$$\log_{10} I_0 - \log_{10} I_{T_t} = kc \times t$$

and in general

$$\log_{10} \frac{I_0}{I_T} = kct$$

This is a combination of the Beer and Lambert laws and is the fundamental relationship used in spectrophotometry.

The value of the constant k depends upon how c is expressed. In this connection the following proportionality factors are often used:

$$\log_{10} \frac{I_0}{I_T} = E = \text{Optical Density, Extinction or Absorbance}$$

$$E = Kt \qquad\qquad K = \text{extinction coefficient}$$
$$E = 0.4342\, \mu t \qquad\qquad \mu = \text{absorption coefficient}$$
$$E = E\,(1\,\%, 1\,\text{cm}), \quad \text{for convenience expressed as } E_1^1,$$

where

$$c = 1\% \text{ w/v}$$
$$t = 1 \text{ cm}$$

$$E = \epsilon c t$$

where

$$\epsilon = \text{molecular extinction coefficient}$$
$$c = \text{g-mol per litre}$$
$$t = 1 \text{ cm}$$

$$E = -\log_{10} T$$

where

$$T = \text{Transmission } \frac{I_T}{I_0}$$

$$\epsilon = E_1^1 \times \frac{\text{Molecular Weight}}{10}$$

Of these, E_1^1 is the usual method of expressing absorption in the British Pharmacopoeia and is useful when the molecular weight of a compound is unknown or doubtful.

Instead of *extinction* the term *absorbance* (A) is now in use but as the British Pharmacopoeia has retained the former term, except for determinations involving infrared absorption, it is also used here in connection with the assays.

The apparent deviations from Lambert's Law with a stable solution arise from instrumental defects such as stray radiation and the small amount of energy available at the extremes of the range of the instrument. Deviations from Beer's Law, on the other hand, may be caused by such defects, but physico-chemical changes may also play a prominent part. Molecules of methylene blue, for example, associate to form dimers with increase in concentration, and the absorption curve is modified. Similarly, changes in pH with change in concentration of solute may also cause undesirable effects. The validity of Beer's Law must, therefore, be confirmed for the actual substance under examination. This is easily done by preparing a series of dilutions and measuring the extinction for each in cells of standard length. On plotting E against concentration a straight line which passes through the origin should be obtained (curve A Fig. 148). Faulty colour development would give a curve as B in Fig. 148.

Emphasis was laid at the beginning of this section on monochromatic radiation for which the Beer–Lambert Law strictly applies. When general radiation is used, the absorption will vary with the wavelength and the resulting absorption curve can be plotted in many ways. One of the more usual forms is shown in Fig. 149.

For publication purposes, the ordinate is calibrated in terms of E_1^1 if the subject matter is analytical, but where structural features of a molecule are

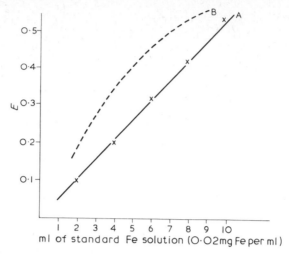

Fig. 148. Calibration curve for the determination of Fe by means of thioglycollic acid

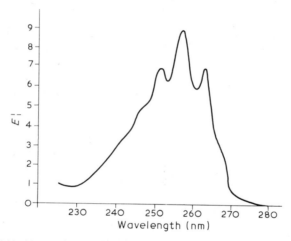

Fig. 149. Absorption curve of ephedrine hydrochloride in *0.1N* HCl

correlated with absorption data then molecular extinction coefficients are more convenient. These are sometimes expressed in terms of $\log_{10} \epsilon$ particularly when absorption curves which are markedly different in intensity are compared.

ABSORPTIOMETRIC ASSAYS OF THE BRITISH PHARMACOPOEIA

Adrenaline in Lignocaine and Adrenaline Injection. Characteristic colours, which vary in hue with change in pH, are given by the reaction of ferrous ions with phenols which contain two adjacent hydroxyl groups. Thus, when a buffer solution is added to a solution of adrenaline and a ferrous salt, a blue colour develops at pH 6.5 and gradually changes to a characteristic red-blue colour (λ_{max} 540 nm) which reaches a maximum intensity at pH 8–8.5.

A reference curve prepared by treating suitable aliquots of a solution of adrenaline acid tartrate by the same process is used to calculate the content of adrenaline.

Cognate Determinations

Adrenaline in Lignocaine and Adrenaline Injection. This assay is of interest because sodium metabisulphite is added to avoid obtaining low results. Further addition of the buffer solution precipitates lignocaine so that an extraction with ether is included to remove the base and give a clear solution.

Adrenaline in Procaine and Adrenaline Injection. This is determined as described under Lignocaine and Adrenaline Injection except for the omission of the ether extraction.

Isoprenaline Tablets

Methyldopa Tablets. An E_1^1 of 89 at 530 nm is given for this compound.

Beclomethasone Dipropionate. Most of the official steroids which contain either an α-ketol group or a potential α-ketol group are now determined by making use of the reducing action of the group on triphenyltetrazolium chloride. In the presence of alkali (tetramethylammonium hydroxide) triphenylformazan is formed and the extinction is measured against a blank of reagents. Precautions should be taken throughout the assay against the effect of light and air.

$$\begin{array}{c}
C_6H_5 \quad C_6H_5 \\
| \qquad | \\
N\text{---}N^+ \quad Cl^- \\
| \qquad \| \\
N \diagdown_C \diagup N \\
| \\
C_6H_5
\end{array} + H_2O + 2e \rightarrow \begin{array}{c}
NH\text{---}N\text{=}C\text{---}N\text{=}N \\
| \qquad | \qquad | \\
C_6H_5 \quad C_6H_5 \quad C_6H_5
\end{array} + Cl^- + OH^-$$

Triphenyltetrazolium   Triphenylformazan
chloride

Cognate Determinations

In the assays for the following compounds, the extinction is measured at 485 nm and the E_1^1 values are for the compounds treated as described in the assay process. Preparations of these substances must be extracted and the steroid obtained finally in solution in aldehyde-free dehydrated ethanol ready for reaction with the reagent.

Substance	E_1^1	Substance	E_1^1
Cortisone Acetate	392	*Hydrocortisone Acetate*	390
Deoxycortone Acetate	405	*Methylprednisolone*	421
Dexamethasone	402	*Prednisolone*	438
Dexamethasone Acetate	363	*Prednisolone Acetate*	392
Fludrocortisone	374	*Prednisone*	440
Hydrocortisone	435	*Prednisone Acetate*	394

These values are taken from the 1968 British Pharmacopoeia but it should be noted that the 1973 edition specifies that the colour produced should be compared with that from the appropriate BCRS or AS standard.

Bendrofluazide Tablets. The assay is based upon the decomposition of the bendrofluazide to yield a fragment containing a primary aromatic amine.

A colour is developed by diazotisation and coupling with N-(1-naphthyl)-ethylenediamine hydrochloride in acid solution, the excess nitrous acid being removed during the assay by sulphamic acid. The colour is compared at 518 nm with that from a known amount of bendrofluazide, treated in the same way.

Cognate Determination

Hydroflumethazide Tablets

Benzhexol Tablets. Tertiary amines and quaternary ammonium compounds form chloroform-soluble complexes with indicators such as bromothymol blue and bromocresol purple. Special conditions may be necessary for complex formation, and for benzhexol a slightly acid, buffered solution of bromocresol purple is added to a solution of the sample. Extraction with chloroform gives a solution of the complex, the extinction of which is measured at 408 nm and compared with that obtained from benzhexol hydrochloride treated in the same way. Note that the buffered indicator solution is extracted once with chloroform, before adding to the sample, in order to remove traces of coloured impurities.

Calciferol. One of the most sensitive reagents for califerol is antimony trichloride, particularly when a small proportion of acetyl chloride is present. Although the reagent is corrosive and unpleasant to handle it posseses the advantage of not reacting with any decomposition products of the vitamin.

The Pharmacopoeia formulates two preparations of calciferol—tablets and a solution in a vegetable oil. For tablets, the assay involves continuous extraction of the vitamin by means of light petroleum. Evaporation of the light petroleum extract, with precautions against oxidation of calciferol, leaves a residue of the vitamin together with traces of fatty matter from the tablet bases. These impurities do not interfere with the colorimetric estimation so that solution of the residue in ethanol-free chloroform and addition of the antimony trichloride solution to duplicate aliquot portions is satisfactory. Duplicates are essential as the development and fading of the brown-orange colour is rapid. The extinctions are measured at two wavelengths [500 nm (max.) and 550 nm] and the extinction at 550 nm must be subtracted from that at 500 nm in order to apply a partial correction for any general absorption in the sample. The content of calciferol is then calculated by comparison with extinctions obtained by treating a known amount of calciferol in ethanol-free chloroform in the same way.

The assay of Calciferol Solution is complicated by the presence of the vegetable oil which contains phytosterols that interfere in the colour reaction. The unsaponifiable matter which contains the calciferol is therefore extracted in the normal manner with additional precautions to avoid oxidation of the vitamin. Chromatography of the residue on an alumina column from 15 to 20% anaesthetic ether in hexane separates the calciferol from its decomposition products and most of the phytosterols. The position of the calciferol in the effluent fractions can be found by testing aliquot portions with the antimony trichloride reagent. Once found, its position in future assays remains the same *provided the same batches of alumina and solvents are used.* Evaporation of the appropriate fractions and completion of the assay follow the line indicated under Calciferol Tablets.

Camphorated Opium Tincture. The morphine which has been isolated (see Part I) is dissolved in acid solution and an aliquot portion is treated with sodium nitrite and ammonia as in the usual nitroso-phenol reaction. A similar aliquot portion is now made alkaline with ammonia and a solution of standard colour (morphine and nitrite reagent) is added until the colour of the test solution is matched after adjusting both solutions to the same volume. The volume of reagent added is a measure of the morphine in the aliquot portion taken and hence of that in the original sample.

Cognate Determination
Morphine Sulphate Injection

Carbachol Injection. Like most quaternary ammonium compounds, carbachol gives an insoluble reineckate with solution of ammonium reineckate, but the Pharmacopoeia directs that a preliminary hydrolysis be carried out. The precipitate, choline reineckate, is soluble in acetone to give a deeply coloured solution with maximum absorption at 526 nm due to the reineckate portion of the molecule. By reference to a calibration curve prepared by treating known amounts of carbachol in the same way, the concentration of carbachol in the injection may be determined.

Chlorproguanil Tablets. Several oxidation procedures are available for biguanide compounds, and for Chlorproguanil, sodium hypobromite is used in the presence of cetrimide and isopropanol. The colour is measured at 480 nm and a calibration curve is required using Chlorproguanil Hydrochloride AS.

Cognate Determinations
Metformin Tablets
Phenformin Tablets

Both these biguanides are treated with an alkaline solution of sodium nitroprusside and potassium ferrocyanide in the presence of strong hydrogen peroxide solution.

Diamorphine Injection. The diamorphine is hydrolysed to morphine by boiling with dilute hydrochloric acid. A colour is developed by treatment with sodium nitrite followed by addition of ammonia. The standard colour is obtained from anhydrous morphine.

Dienoestrol Tablets. The assay is based upon the typical reducing action of phenols on sodium molybdophosphotungstate in alkaline solution to give a coloured reduced compound of molybdenum, molybdenum blue. The development of the colour is slow, and although one hour is allowed, an increase in intensity does take place on further standing. It is essential, therefore, to measure the extinction of the test and standard solutions after the same time interval. Maximum absorption occurs at 780 nm, and in the absence of a spectrophotometer, Kodak filter No. 8 (red), is satisfactory. Occasionally, a white precipitate forms in the solution whilst the colour is developing, but by shaking with ether, coagulation occurs and centrifuging gives a clear aqueous solution.

Digoxin Tablets. Chemical assays of digitalis glycosides are usually based upon colour reactions involving either the sugar moiety of the molecule or the α,β-unsaturated lactone ring. The former (Keller–Kiliani) reaction is used in this instance.

The tablets are extracted with a suitable solvent and insoluble matter is removed by filtration. An aliquot portion of the filtrate is treated directly with a reagent consisting of glacial acetic acid which contains a trace of ferric chloride and 2% v/v of sulphuric acid. After allowing to stand for 1.5 hr the colour is compared at 590 nm with that obtained from a standard amount of digoxin BCRS.

Cognate Determinations

Digoxin Injection. The ethanol is removed by evaporation and the residue treated as described for Digoxin Tablets except that filtration is not necessary. The same reaction is used to determine the Uniformity of Content of the tablets.

Digitoxin. A preliminary separation from other glycosides is carried out by chromatography on a short column of kieselguhr using aqueous dimethylformamide as stationary phase and chloroform-benzene (1:3) as eluant. An aliquot part of the effluent is evaporated to dryness, dissolved in ethanol 95%, and a colour developed with Baljet's reagent (freshly prepared alkaline picrate). The extinction is measured at 495 nm using the appropriate blank, and is compared with digitoxin As. The reagent reacts with the α,β-unsaturated lactone group of the glycoside (cf. digoxin).

Ergometrine Maleate. A reaction characteristic of ergot alkaloids is the formation of a deep blue colour with a sulphuric acid solution of *p*-dimethylaminobenzaldehyde. When first used in the 1932 British Pharmacopoeia, exposure to bright light was necessary for maximum development of colour, but by the inclusion of a trace of ferric chloride in the reagent, the maximum intensity is obtained in five minutes.

Cognate Determinations

Ergometrine Injection
Ergometrine Tablets

Ergotamine Injection
Ergotamine Tartrate
Ergotamine Tablets
Methylergometrine Maleate
Methylergometrine Injection

Folic Acid; Folic Acid Tablets. Folic acid is reductively cleaved by zinc amalgam in acid solution to *p*-aminobenzoylglutamic acid. Zinc amalgam is used for this reduction as strong reducing agents such as Zn and hydrochloric acid, or $TiCl_3$ may carry the reduction too far. The primary aromatic amino

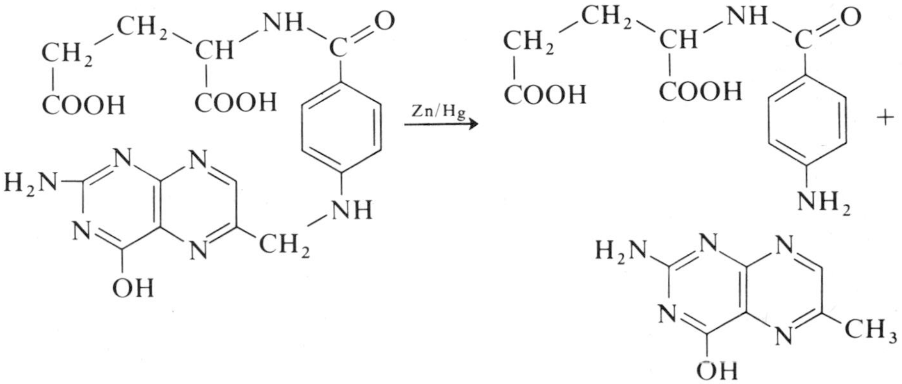

group so produced is then diazotised in the normal manner, with precautions against the action of light, and coupled in acid solution with *N*-(1-naphthyl)-ethylenediamine hydrochloride. The colour has maximum absorption at 550 nm and the extinction is compared with a calibration curve obtained from *p*-aminobenzoic acid which has been diazotised and coupled in the same way as the *p*-aminobenzoylglutamic acid.

To ensure that the measurements obtained refer solely to folic acid, and that they do not include a contribution from a free primary aromatic amino group present in a decomposition product, a blank determination is carried out on unreduced solution and an appropriate correction is applied. The colour then corresponds to a definite quantity of *p*-aminobenzoic acid which in turn corresponds to a definite quantity of $C_{19}H_{19}O_6N_7$, the relationship being 1 g *p*-aminobenzoic acid is equivalent to 3.22 g of $C_{19}H_{19}O_6N_7$.

Glyceryl Trinitrate Tablets. The glyceryl trinitrate is extracted from the chocolate basis of the tablets by shaking with glacial acetic acid. A colour is then developed in an aliquot portion of the acetic acid solution by means of phenoldisulphonic acid, and intensified by the addition of excess of ammonia, according to the reactions set out below. The standard substance in this assay is potassium nitrate, which conforms to the nitric acid released by acidolysis in the test solution.

$$CH_2.O.NO_2$$
$$CH.O.NO_2 \rightarrow 3HNO_3$$
$$CH_2.O.NO_2$$

$$\text{HNO}_3 + \underset{\underset{\text{SO}_3\text{H}}{\bigcirc}}{\overset{\text{OH}}{\bigcirc}}\text{SO}_3\text{H} \longrightarrow \underset{\underset{\text{SO}_3\text{H}}{\bigcirc}}{\overset{\text{OH}}{O_2N{-}\bigcirc}}\text{SO}_3\text{H} \xrightarrow{\text{NH}_4\text{OH}} \underset{\underset{\text{SO}_3^-}{\bigcirc}}{\overset{O}{{}^-O{-}\overset{+}{N}{=}\bigcirc}}\text{SO}_3^-$$

Cognate Determination

Pentaerythritol Tablets

Isocarboxazid Tablets. An orange colour is produced when isocarboxazid is treated in acetone solution with ammonium molybdate in dilute hydrochloric acid. Maximum absorption occurs at about 420 nm.

Cognate Determination

Nialamide Tablets. Maximum adsorption occurs at about 441 nm.

Lymecycline and Procaine Injection. The procaine hydrochloride is determined by the reaction with *p*-dimethylaminobenzaldehyde in buffered acid solution. It is a reaction which is tending to increase in use for the determination of primary aromatic amines. Maximum absorption of the Schiff base appears at about 454 nm.

Methyltestosterone Tablets. Methyltestosterone is an α,β-unsaturated ketone with an intense absorption band in the ultraviolet region of the spectrum at 241 nm. A direct measurement at this wavelength however would be of doubtful value in extracts from tablets because of the possibility of irrelevant absorption. The pharmacopoeial method involves isolation of methyltestosterone dinitrophenylhydrazone and measurement of the extinction at 390 nm in chloroform solution. The absorption band in this region is due to the dinitrophenylhydrazone portion of the molecule. For calculation purposes, the extinction is compared with that obtained by treating methyl testosterone in the same way.

Cognate Determination

Methandienone Tablets

Nandrolone Decanoate. See Part 1, Chapter 14.

Oestradiol Monobenzoate Injection. The injection is an oil; and partition between solvents is adopted to remove the oestradiol monobenzoate into a more convenient solvent, viz., ethanol (70%). Evaporation of the ethanolic extract in the presence of sodium carbonate causes hydrolysis of the benzoate. After the addition of sodium hydroxide solution, further extraction of the aqueous phase with trimethylpentane removes traces of oily matter which would interfere later in the assay. Sodium carbonate alone would not retain the oestradiol in the aqueous phase, hence the addition of sodium hydroxide. The oestradiol is recovered by acidification of the alkaline liquid and extraction into benzene. Traces of fatty acid in the extract must be removed by washing with sodium carbonate solution.

The colour reaction is carried out on an evaporated aliquot portion of the benzene solution and is based upon the brown-red colour given by oestrogens with phenol-sulphuric acid mixture. The reaction is normally a two-stage process, the development of a yellow colour which is then converted to a red colour. Traces of iron have a marked effect on colour formation, and with the *iron-phenol* reagent of the Pharmacopoeia, complete conversion of the yellow phase of α- and β-oestradiols to stable red colours is obtained in a single-stage heating operation. Ferrous iron is not as efficient as ferric iron in this process.

The comparison colour is developed by treating a solution of 1.00 mg of oestradiol benzoate in 5 ml of benzene in exactly the same way as described for the injection. The extinctions for comparison are obtained by deducting half the extinction at 420 nm from that at about 520 nm and under these conditions the amount of oestradiol benzoate in the volume of injection taken is given by A/B where A is the test value and B the standard value. By adopting this method, a correction for non-oestrogenic impurities is applied since under the experimental conditions the following relationship holds for non-oestrogenic substances:

$$\frac{\text{Extinction at 520 nm}}{\text{Extinction at 420 nm}} = 0.5$$

Reserpine and Reserpine Tablets. Treatment with nitrous acid (from sodium nitrite) gives rise to a yellow colour, the extinction of which is measured at 390 nm. Reserpine AS is treated in the same way for comparison purposes.

Stilboestrol and Stilboestrol Tablets. The determination of stilboestrol depends upon a photochemical reaction as follows.

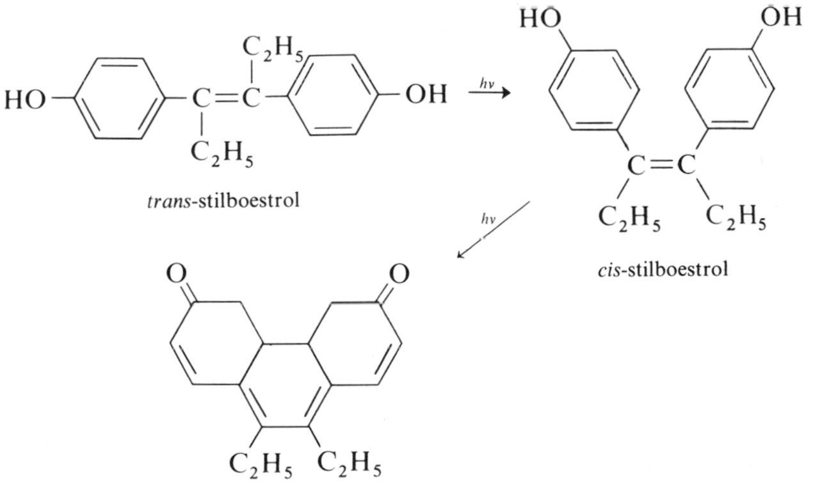

trans-stilboestrol

cis-stilboestrol

The highly conjugated di-keto system is produced by irradiation of the stilboestrol solution in a closed spectrophotometer cell for 10 min with a 15-watt short wave ultraviolet lamp. The extinction is measured at 418 nm and compared with stilboestrol BCRS treated in the same way.

Thiambutosine Tablets. The assay depends upon the formation of a copper complex, the extinction of which is measured at 400 nm. Methyl alcohol is used as the solvent.

Thymol in Tetrachloroethylene. Many phenols in the presence of sulphuric acid give yellow colours with titanium dioxide. The pharmacopoeial test is designed to ensure that the content of thymol falls between the two limits 0.008 and 0.012% w/w by using solutions of those strengths as standards. The colour, obtained by shaking 0.5 ml of sample with 5 ml of carbon tetrachloride and 5 ml of a 0.1% w/v solution of titanium dioxide in sulphuric acid, should then lie between those developed in the same manner from 0.5 ml of the standard solutions. The colour is in the lower, i.e. the sulphuric acid layer, and adequate time must be allowed for separation of the layers.

Cognate Determinations

Thymol in Trichloroethylene. A 0.175% w/v solution of thymol in carbon tetrachloride is used to prepare the standard concentrations whereas a 0.193% w/v solution is used in the tetrachloroethylene example. This difference is to allow for the difference in the specific gravities of tetrachloroethylene and trichloroethylene.

Halothane. A 0.225% w/v solution of thymol in carbon tetrachloride is used to prepare the standard concentrations.

Tyrosine in Hyaluronidase. Hydrolysis of Hyaluronidase gives tyrosine as one of the constituent amino acids, and the red-brown colour which is produced with 'mercuric nitrite' is due to the phenolic group in the molecule. The intensity and shade of colour are affected by the concentration of mercuric sulphate and sulphuric acid, so that the same reagents must be used to develop the test and standard colours. Under the conditions used in the Pharmacopoeia, maximum absorption occurs at 540 nm.

SPECTROPHOTOMETRY

Spectrophotometric assays make use of the light absorption characteristics of the various medicinal compounds which are listed in Table 44. These characteristics may also serve as identity tests. The compounds in Table 45 (p. 283) are listed in order of their principal absorption maxima, but also with the wavelength of secondary maxima to assist in the identification of unknown compounds.

Qualitative Control of Purity

The identification of a substance by the absorption characteristics cannot be accepted as conclusive evidence of identity and other tests, chemical and physical, must also be used. The absorption, however, does offer an easy method for the routine control of organic substances and its use can be extended to improve the value of the test. Thus for Cyanocobalamin, Procyclidine Hydrochloride and Sodium Aminosalicylate the extinctions at several wavelengths are measured, and the calculated ratios are limited to certain values. For example, the extinction of the solution used in the assay of Cyanocobalamin is measured at 278, 361 and 550 nm. The ratios of the extinctions at 278 and 550 nm to that at 361 nm should be about 0.57 and about 0.3 respectively. This method is particularly useful in the detection of irrelevant absorption, i.e. absorption caused by impurities. When the medicinal compound is transparent, impurities which absorb light are readily detected. This method is used in the Pharmacopoeia to detect or limit the amount of noradrenalone in Noradrenaline Acid Tartrate. The latter, although absorbing strongly at 279 nm, shows no absorption at 310 nm at which wavelength noradrenalone does absorb.

Determination of Medicinal Substances

Once the light absorption characteristics of the substance are known, and the validity of Beer's Law is confirmed, then solution of the substance in a suitable solvent and measurement of the extinction at a suitable wavelength is all that is necessary for assay purposes.

Example 1. Ergometrine Maleate

The characteristics of Ergometrine Maleate are, E_1^1 at 312 nm = 183.

The extinction of a 0.002783% solution in water at 312 nm was found to be 0.505 (1 cm cells).

$$\therefore \quad \% \text{ Ergometrine Maleate} = \frac{0.505 \times 100}{0.002783 \times 183}$$

$$= 99.2\%$$

Example 2. Riboflavine

The complete absorption spectrum of riboflavine exhibits four peaks at 223, 267, 375 and 444 nm respectively, each of which shows variations with change in pH. This is readily understood if the formula is examined because the molecule can become a cation in acid solution and an anion in alkaline solution. The peaks at 267, 375 and 444 nm, however, are stable between pH 2 and pH 7 so that solutions of the substance itself and from the tablets are made up in the presence of a small amount of sodium acetate–acetic acid buffer. In the measurement of the extinctions, the intense peak at 267 nm is used in the assay of riboflavine itself, but that at 444 nm is adopted for the tablets. This illustrates the general principle that where irrelevant absorption is likely to be present (in this instance arising from the tablet basis), it is better to adopt an absorption peak in the near ultraviolet or visible region for measurements. When no such peak exists, it may be possible to introduce one by preparing a derivative, as for example in the assay of Methyl Testosterone Tablets.

Determination of Mixtures of Medicinal Substances

The spectrophotometric method is also applied to preparations which contain only one ingredient, or to the determination of a single substance in a mixture provided the absorption of the other components can be neglected. Procaine and Adrenaline Injection is a typical example of the latter type in which procaine hydrochloride can be determined by simple dilution with water and measurement of the extinction at 290 nm. The absorption caused by adrenaline and chlorocresol is negligible at the high dilution used. The constants for procaine hydrochloride are, E_1^1 at 290 nm $= 680$. In an actual determination, the injection was diluted by a factor of 2000. The extinction found was 0.684 in 1 cm cells,

$$\therefore \quad \% \text{ Procaine Hydrochloride} = \frac{0.684}{680} \times 2000$$

$$= 2.01 \% \text{ w/v}$$

(a) *Two components with dissimilar spectra*

(i) If the absorption characteristics of a mixture of two substances are very different, it is usually possible to determine both substances by measuring the extinction at two wavelengths, e.g. Injection of Pethidine Hydrochloride:

> In *0.1N* HCl E_1^1 for pethidine hydrochloride at 279 nm $= 0$
> E_1^1 for pethidine hydrochloride at 257 nm $= 7.3$
> E_1^1 for chlorocresol at 279 nm $= 105$
> E_1^1 for chlorocresol at 257 nm $= 20$.

Let the observed extinction at 257 nm $= A$ at dilution factor x, and at 279 nm $= B$ at dilution factor y. The percentage of chlorocresol is calculated as for a simple solution because the pethidine has no absorption at 279 nm.
  Hence

$$\% \text{ Chlorocresol} = \frac{B}{105} \times y$$

For pethidine hydrochloride, however, account must be taken of the contribution of the chlorocresol to the absorption at 257 nm. This is easily calculated, from the percentage of chlorocresol found, in the following way.
  For a 1% solution of chlorocresol the extinction would be 20, hence for a $B/105 \times y\%$ solution the extinction would be $B/105 \times y \times 20$. But the solution

has been diluted by a factor of x,

$\therefore$ Contribution of chlorocresol to the absorption at 257 nm

$$= \frac{B}{105} \times y \times \frac{20}{x} = C$$

$\therefore$ % Pethidine Hydrochloride $= \dfrac{(A-C)}{7.3} \times x$

(ii) When both substances contribute to the absorption at each wavelength, the calculation is as follows:

> let component a have E_1^1 values of a_1 at λ_1 and a_2 at λ_2
> let component b have E_1^1 values of b_1 at λ_1 and b_2 at λ_2

The extinction of a solution of the mixture is determined at λ_1 and λ_2, and the E_1^1 values are calculated—S_1 and S_2 at λ_1 and λ_2 respectively.
  Then

$$100S_1 = a_1 x + b_1 y$$

and

$$100S_2 = a_2 x + b_2 y$$

where

$$x = \text{concentration (as \% w/w) of component } a$$
$$y = \text{concentration (as \% w/w) of component } b$$

Solving the simultaneous equations for x and y,

$$x = 100 \left(\frac{b_1 S_2 - b_2 S_1}{b_1 a_2 - b_2 a_1} \right)$$

$$y = 100 \left(\frac{a_1 S_2 - a_2 S_1}{a_1 b_2 - a_2 b_1} \right)$$

It should be confirmed that the extinctions of the two substances are additive at the wavelengths selected.

(b) Mixtures of three or more components

For mixtures of three components, the calculation is more involved and the analysis calls for an extremely high degree of accuracy if precise results are to be obtained. It is often better to separate the components chemically and apply the method for single substances to each. This procedure is essential when the components of a mixture have similar absorption curves. Typical examples to which such a method has been applied are Morphine Sulphate Injection, Apomorphine Hydrochloride Injection and Strychnine Hydrochloride Injection. Both the active ingredient and chlorocresol can then be determined spectrophotometrically.

(c) *Natural products*

Irrelevant absorption by impurities in the substances so far mentioned is likely to be very small or absent though the possibility of its presence can never be ignored. When, however, crude drugs are examined, such absorption is almost certain to be present so that careful purification procedures must be adopted. For some products it is possible to correct for the irrelevant adsorption, and a good example is the Morton–Stubbs correction as applied to the much investigated determination of Vitamin A in fish liver oils and concentrates. The assumption is made that the irrelevant absorption is linear over the appropriate range of the absorption curve of Vitamin A. From a consideration of the absorption spectrum of the pure vitamin, three wavelengths are selected one of which is the wavelength of the peak of the absorption band and the other two, one on each side of the peak, are chosen so that

$$\frac{E_{\lambda_{max}}}{E_{\lambda_1}} = \frac{E_{\lambda_{max}}}{E_{\lambda_2}} = \frac{7}{6}$$

Let Curve 1 in Fig. 150 be the observed absorption for the sample in cyclohexane and Curve 2 be the irrelevant absorption. The subtraction Curve 3 is then the absorption due to the vitamin alone.

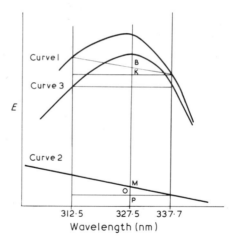

Fig. 150. Geometrical construction for obtaining correction formula for vitamin A

The total irrelevant absorption is given by $MO + OP$. Of these, MO is easily obtained as

$$MO = BK = (E_{312.5} - E_{337.7}) \times \frac{337.7 - 327.5}{337.7 - 312.5}$$

$$= (E_{312.5} - E_{337.7}) \times \frac{10.2}{25.2}$$

Further

$$\frac{E_{327.5} - MO - OP}{E_{337.7} - OP} = \frac{7}{6} \text{ (conditions)}$$

$$\therefore \quad 6E_{327.5} - 6MO - 6OP = 7E_{337.7} - 7OP$$

$$\therefore \quad OP = 7E_{337.7} - 6_{327.5} + 6MO$$

$$\therefore \quad \text{total correction} = MO + OP = 7E_{337.7} - 6E_{327.5} + 7MO$$

$$= 7E_{337.7} - 6E_{327.5} + \frac{71.4}{25.2}E_{312.5} - \frac{71.4}{25.2}E_{337.7}$$

$$= 4.167E_{337.7} - 6E_{327.5} + 2.833E_{312.5}$$

$$\therefore \quad E_{327.5} \text{ (corrected)} = E_{327.5} - (4.167E_{337.7} - 6E_{327.5} + 2.833E_{312.5})$$

$$= 7(E_{327.5} - 0.595E_{337.7} - 0.405E_{312.5})$$

The correction procedure can be applied when the observed absorption curve is not grossly distorted, otherwise the sample or fraction thereof must be purified, for example, by chromatography on alumina. The wavelength of maximum absorption differs slightly when the vitamin is present as an ester from that when it is present as the free alcohol, so that for extracts which have been saponified, the figures for Vitamin A alcohol are used. The constants for Vitamin A acetate are satisfactory as standards for untreated oils. A different correction formula is therefore necessary for halibut-liver oil and concentrated solutions of Vitamin A, which are examined directly in solution in cyclohexane, from that for cod-liver oil which undergoes a saponification procedure. The pharmacopoeial formulae apply to solutions in cyclohexane; ethanolic solutions require different formulae because of solvent effect.

This correction procedure is modified by selecting wavelengths of 316, 328 (maximum) and 340 nm for solutions of Vitamin A in ester form in cyclohexane. If the absorption curve corresponds closely to that of pure Vitamin A ester as shown by relative extinction values, the potency in units per g is calculated directly from the expression

$$E_1^1, 328 \text{ nm} \times 1900$$

When irrelevant absorption is present, a corrected extinction is used for calculation under carefully specified conditions. The formula

$$E_{328} \text{ (corr)} = 3.52 (2E_{328} - E_{316} - E_{340})$$

may be obtained from the figures given in the Pharmacopoeia by applying a geometrical procedure similar to the above. The proof is left to the student as an exercise.

For 'other Vitamin A', a saponification procedure is specified and isopropanol is used as solvent. A corrected extinction is calculated from the expression

$$E_{325} \text{ (corr)} = 6.815E_{325} - 2.555E_{310} - 4.260E_{334}$$

which is used to obtain the potency in units per g by means of the formula

$$E_1^1, 325 \text{ nm (corr)} \times 1830$$

Conditions, under which the corrected extinction may be used, are imposed so that an undue amount of irrelevant absorption will be detected. If much is present, a chromatographic procedure must be adopted for purifying the unsaponifiable matter.

The Morton–Stubbs correction procedure uses only three points on the absorption curve and, moreover, assumes that the irrelevant absorption is linear over the range of wavelengths selected. It is natural, therefore, that attention should be directed to mathematical methods of characterising absorption curves in such a way as to eliminate the effects of all irrelevant absorption.

Ashton and Tootill (1956) developed a formula based on Legendre polynomials and seven points on the absorption curve of griseofulvin in order to determine the antibiotic in fermentation liquors. The method, however, is not readily amenable to routine use. By the application of orthogonal functions to absorption curves, Glenn (1963) has shown that the effect of linear irrelevant absorption can be eliminated completely. Providing the irrelevant absorption is not too similar to the absorption curve of the substance being examined, Glenn's method provides a correction procedure even where absorption is not strictly linear. Calculations involving up to twelve points on the absorption curve can be handled conveniently with the aid of a desk calculator. A general discussion is given by Wahbi and Abdine (1973) including derivation of the formula used by Ashton and Tootill (1956). The method is further discussed by Wahbi and Ebel (1974).

(d) Body Fluids

The determination of drugs and metabolites in body fluids forms part of several major fields of investigation—bioavailability, biochemistry, drug metabolism and toxicology. Pharmaceutical analysts are necessarily concerned in all these aspects because information of this type is required in submissions of data on new drugs to the Committee on Safety of Medicines (Part 1, Chapter 2).

The examination of body fluids is more difficult than that of pharmaceutical preparations in several ways:

(a) a small quantity of drug or metabolite is usually present in a large volume of blood, urine or tissue;

(b) solvent extraction of body fluids gives rise to an extract which may contain, in addition to the drug, endogenous pigments or compounds which make optical methods of analysis subject to error unless great care is taken in the analytical conditions, e.g. choice of solvent, pH of extraction and subsequent purification methods. Even so, it is often necessary to use chemical methods as an intermediate step in the spectrophotometric method of analysis to introduce some degree of specificity or to increase absorbance;

(c) the drug may occur both free and combined as conjugates, e.g. glucuronide or ethereal sulphates, both the latter being polar and water-soluble;

(d) protein binding of the drug may occur and this leads to poor recoveries unless the protein is denatured during the extraction procedure;

(e) the use of several extractions for quantitative recovery of drug may lead to difficulties, e.g. emulsion formation with plasma samples and large volumes of solvent for evaporation. These difficulties can be mitigated by using a single extraction with a large solvent : sample ratio, and by carrying out control analyses with normal body fluids to which the drug to be determined has been added.

Structural Analysis of some Official Substances

A detailed account of the application of absorption spectra to the determination of the structure of organic compounds is beyond the scope of this book, but elementary principles can be cited to help in understanding the appearance and position of the absorption bands of many pharmacopoeial substances.

Absorption of light in the visible and ultraviolet regions of the spectrum is due to the presence of 'chromophores' in the absorbing molecule. The term 'chromophore' was originally used for unsaturated groups of atoms which were thought to be essential for colour. With the extension of light absorption studies into the ultraviolet region, the term includes such multiple bonds as those of ethylenic, acetylenic and carbonyl groups in which the electrons are more loosely bound than those in fully saturated compounds. Unfortunately these simple chromophores absorb in a region inaccessible to ordinary spectro-photometers, because special instruments and light sources are needed to obtain reliable results at wavelengths less than 210 nm. In Table 38 are recorded representative examples of chromophores and their constants.

Table 38. Typical Chromophores and their Constants

Substance	Chromophore	λ_{max} (nm)	ϵ_{max}
Methane	—	125	—
Oct-3-ene	$>$C$=$C$<$	185	8000
Acetone	$>$C$=$O	188	900
		279	15
Acetoxime	$>$C$=$N$-$	190	5000
Acetic acid	$-$C$=$O $\mid$ OH	204	40

The electrons involved in the absorption of energy may be considered under three headings:

σ *electrons* occur in fully saturated systems such as the bonds of alkanes and they require so much energy for excitation that the absorption bands lie in the vacuum ultraviolet region of the spectrum. They do not concern us further except to say that interaction with π electron systems gives rise to hyperconjugation and to the small effects noted on pp. 254–6.

n electrons are non-bonding, such as those of the lone pairs in O, N and S and the transition n $\rightarrow \pi^*$ i.e. to an excited bonding orbital accounts for the low intensity absorption of the carbonyl group at about 280 nm. Interaction with electrons of conjugated systems gives rise to the effects noted under 'auxochromes' (p. 254).

π *electrons* are those of multiple bonds, the so-called 'mobile' electrons. The basic absorption, due to the transition $\pi \rightarrow \pi^*$, lies in the 180–200 nm region.

The absorptions of two or more chromophores which are separated by more than one bond are usually additive, but when such chromophores are conjugated, i.e. separated by a single bond, pronounced effects are produced. The maximum absorption is shifted to longer wavelengths thus bringing it into the working range of spectrophotometers. Such an effect is called a *batho-chromic* shift and the increase in ϵ_{max} which often accompanies such a shift is known as a *hyperchromic* effect. The reverse changes are known as *hypsochromic shifts* and *hypochromic* effect, respectively, and occur quite often when a chromophoric system is changed, for example, by alteration of pH. In Table 39 are recorded typical examples of conjugated chromophores and auxochromes.

Table 39. Typical Conjugated Chromophores and Auxochromes

Compound	Chromophore	λ_{max} (nm)	ϵ_{max}
Butadiene	$>C=C-C=C<$	217	21 000
Crotonaldehyde	$>C=C-C=O$	217	16 000
Sulphanilamide in NaOH solution	$NaHNO_2S-\!\!\langle\rangle\!\!-NH_2$	251	16 300
Sulphanilamide in HCl solution	$H_2NO_2S-\!\!\langle\rangle\!\!-\overset{+}{N}H_3$	218 265	12 700 1080

Changes in absorption spectra can also be produced by fully saturated groups attached to a chromophoric system and these groups are called auxochromes. Unlike chromophores, which are covalently unsaturated (for example, double bonds), auxochromes are covalently saturated. They are of two types, (*a*) co-ordinatively unsaturated for example $-NH_2$, $-S-$, which contain lone pairs of electrons and (*b*) co-ordinatively saturated, for example $-\overset{+}{N}H_3$. Auxochromes of type (*a*) are generally more effective in modifying absorption spectra. The influence of alkyl groups is usually very small and merely affects the position of the absorption peak (Table 40).

The correlation of absorption with structure is still largely empirical as different types of compound may absorb in the same region. Hence for complete identification of a compound, absorptiometric measurements must be supplemented by chemical tests and other physical measurements.

Before proceeding to more complicated chromophoric systems, the absorption characteristics of pharmacopoeial substances which contain simple combinations of double bonds can now be discussed by reference to the above effects. Vitamin A_1 is a particularly good example of the effect of conjugation

$$[CH=CH-C(Me)=CH]_2CH_2OH$$

Vitamin A_1 λ_{max} 326 nm
ϵ_{max} 51 000

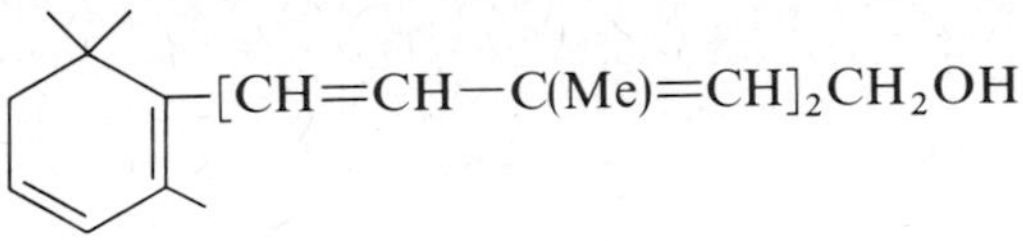

$$-[CH=CH-C(Me)=CH]_2CH_2OH$$

Vitamin A$_2$ λ_{max} 287 nm
ϵ_{max} 22 000
λ_{max} 351 nm
ϵ_{max} 41 000

on the position and intensity of the absorption maximum. The additional double bond in vitamin A$_2$ has changed the absorption maximum to 351 nm. This compound is also of interest because, in addition to the absorption caused by the six ethylenic bonds, there is evidence of absorption by the partial chromophoric system $-[CH=CH-CMe=CH]_2$ at 287 nm, an effect which is enhanced by the steric effects of the methyl groups in the ring.

Comparison of the absorption of Calciferol with that of its stereo-isomers iso-vitamin D$_2$ (*all-trans*) and precalciferol illustrate the effects produced by *cis-trans* isomerism and also by steric factors. The shift of maximum from 265 nm in Calciferol to 287 nm in iso-vitamin D$_2$ is typical of the increase of

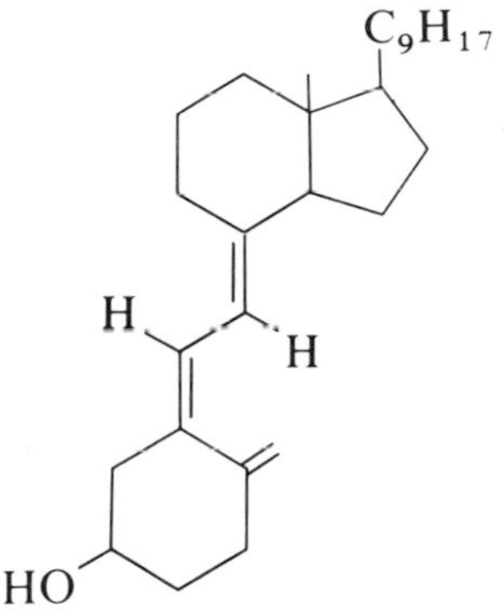

Calciferol
λ_{max} 265 nm
ϵ_{max} 18 200

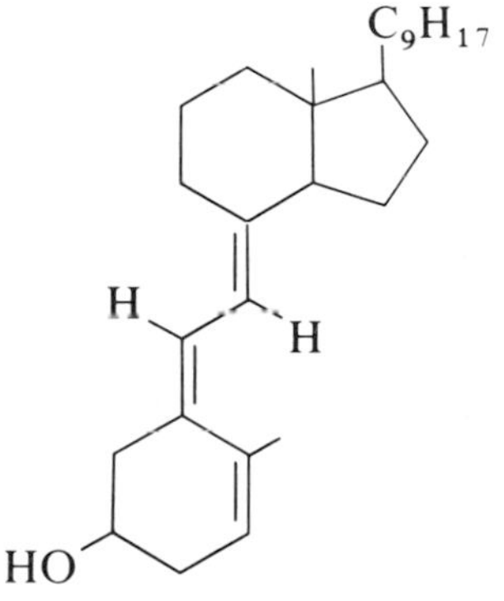

iso-Vitamin D$_2$(*all-trans*)
λ_{max} 287 nm
ϵ_{max} 44 100

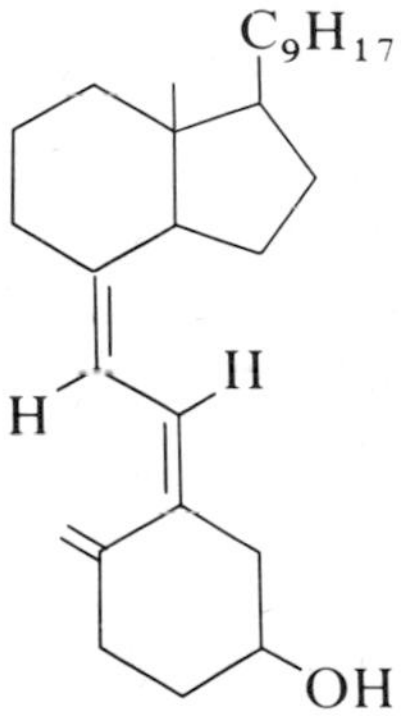

Precalciferol
λ_{max} 265 nm
ϵ_{max} 9800

chromophore length in passing from a *cisoid* to a *transoid* diene system. The chromophoric systems in Calciferol and precalciferol are identical but, owing to steric hindrance between the hydrogen atoms on $C_{(7)}$ and $C_{(15)}$ in precalciferol, the intensity of the absorption is decreased.

The intense ultraviolet absorption of many keto-steroids is due entirely to the conjugated system 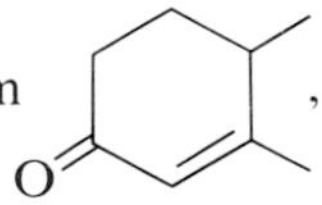 , and even when a third double bond is present in conjugation as in Prednisolone and Prednisone, very little change is observed in the position of maximum absorption. These are examples of

Progesterone
λ_{max} 241 nm

Prednisone
λ_{max} 241 nm

Prednisolone
λ_{max} 241 nm

crossed conjugation, and under these conditions the absorption is always characteristic of the main chromophoric system.

In order to produce a marked change in wavelength, conjugation must extend the chromophoric system as with vitamins A_1 and A_2. The identity of the auxochromes present in the compounds discussed may not be clear at first, but small modifications in the basic absorption of the chromophoric systems are caused by substituents. Increasing alkyl substitution causes bathochromic displacement of the absorption band, and the compounds in Table 40 illustrate

Table 40. Effect of Alkyl Substitution on Absorption Maximum

Compound	λ_{max} (nm)	Substituents
Crotonaldehyde	217	1
α-Ionone	228	2
Progesterone	241	3

this point although other factors such as the position of double bonds relative to ring systems (see R. B. Woodward, 1942, 1941) also operate in fixing the position of the absorption bands. The effect is best observed with an isolated ethylenic bond. Bladon, Henbest and Wood, (1952) selected 210 nm (not a maximum) as a convenient wavelength for observation because of the limitation of the instrument used. The bathochromic shift of the end-absorption associated with increasing alkyl substitution caused significant increases in ϵ values sufficient to identify the degree of substitution, e.g. cholest-2-ene (disubstituted, $\epsilon_{210} = 200$); cholest-8-ene, (tetrasubstituted, $\epsilon_{210} = 4400$).

Certain arrangements of chromophores appear to act as a unit, as for example in benzene, the ultraviolet absorption spectrum of which is capable of modification by other chromophores and by auxochromes. Benzene itself shows a high intensity absorption band at 200 nm, and a second band at 255 nm of low intensity, which exhibits very fine vibrational structure. Substitution of one of the hydrogens by an alkyl group reduces the fine structure but it is still evident in all of the following pharmacopoeial compounds which exhibit typical benzenoid absorption with a maximum at 257 nm:

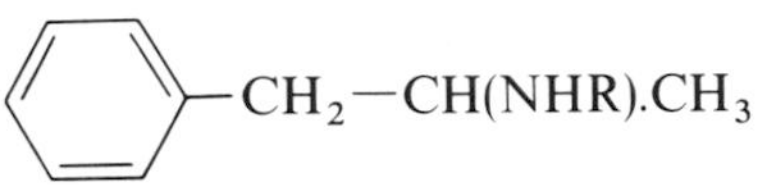

Amphetamine   R = H
Methylamphetamine   R = $-CH_3$

Ephedrine

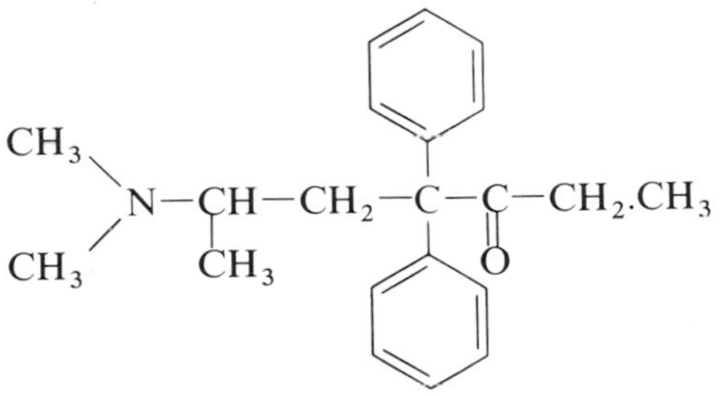

Methadone

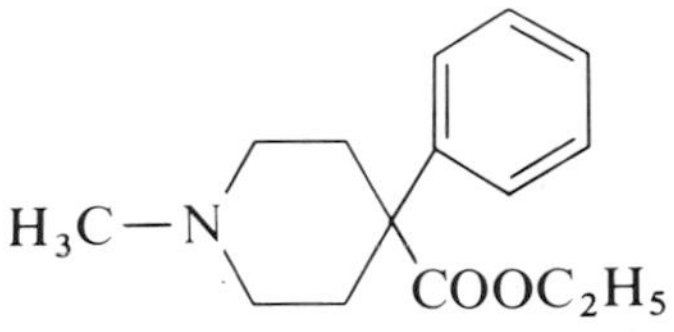

Pethidine

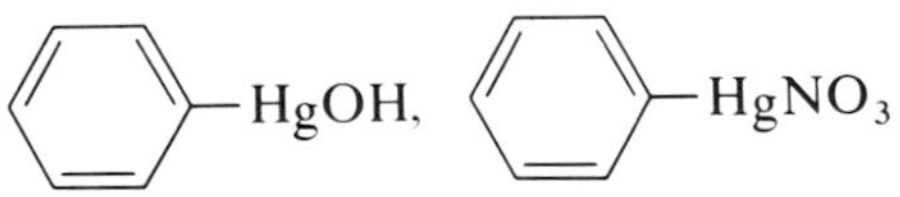

Phenylmercuric Nitrate

When phenolic hydroxyl groups are present, a marked bathochromic displacement of the absorption to about 280 nm occurs, and the effect of solvent on the absorption is considerable because, whereas in neutral or acid media the auxochrome is $-OH$, in alkaline media it becomes $-O^-$. For example, the absorption maximum for chlorocresol in acid solution occurs at 279 nm, but in alkaline media the peak is found at 296 nm. Typical examples of phenols are Morphine Hydrochloride, Nalorphine Hydrobromide and Ethinyl Oestradiol, whilst Adrenaline and Isoprenaline are examples of o-dihydroxyphenols.

The amino group is a powerful auxochrome when attached directly to a benzene system, and aniline exhibits high intensity absorption at about 230 nm and typical low intensity benzenoid absorption at 280 nm. Unfortunately there are no examples in the Pharmacopoeia of such a system; additional chromophores are always present so that a cumulative effect is obtained. Thus in Procainamide the 200 nm *high intensity absorption* of benzene suffers a bathochromic displacement to 280 nm.

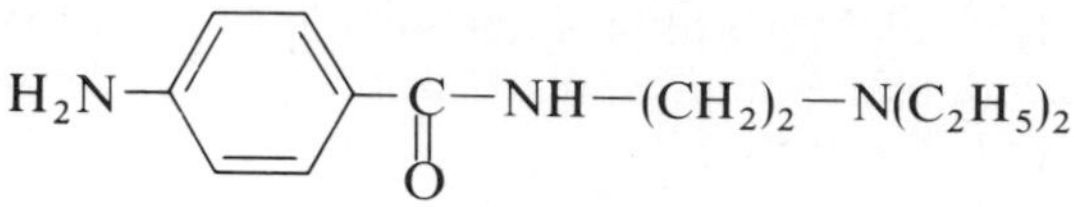

$$H_2N-\langle\ \rangle-\overset{\overset{\displaystyle}{\|}}{\underset{\displaystyle O}{C}}-NH-(CH_2)_2-N(C_2H_5)_2$$

Procainamide

Similarly, the sulphonamide drugs of general formula

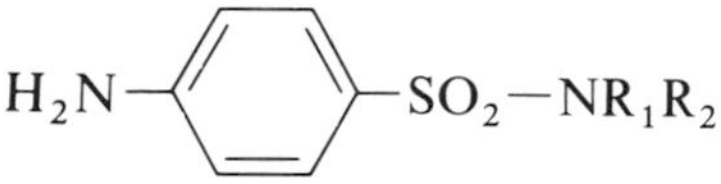

$$H_2N-\langle\ \rangle-SO_2-NR_1R_2$$

show high intensity absorption at about 251 nm. The effect of change in pH in the solvent used for the ultraviolet measurements is very striking (Fig. 151).* In alkaline solution, the absorbing system is as given above, but in acid media the amino $-NH_2$ group is replaced by $-\overset{+}{N}H_3$ which is considerably less efficient as an auxochrome.

p-Aminobenzoic acid is similar in absorption characteristics to Procainamide and Procaine in neutral or alkaline media, but in acid solution the absorption curves of all three approach that of benzoic acid.

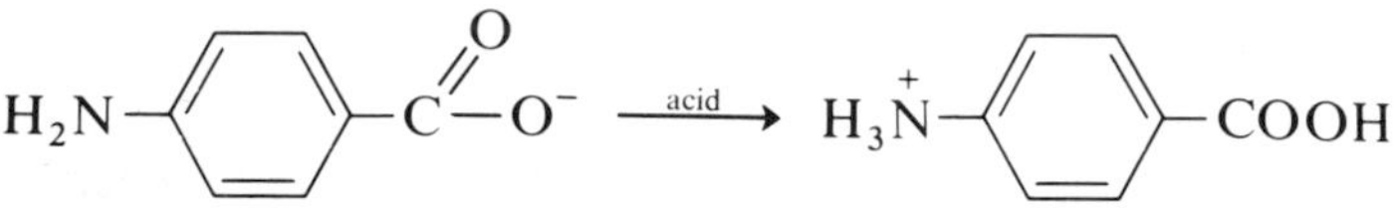

$$H_2N-\langle\ \rangle-\overset{\overset{\displaystyle O}{\|}}{C}-O^- \xrightarrow{\text{acid}} H_3\overset{+}{N}-\langle\ \rangle-COOH$$

Sodium p-aminosalicylate in alkaline solution has an additional auxochrome, the phenate ion, so that the high intensity absorption noted in the p-amino

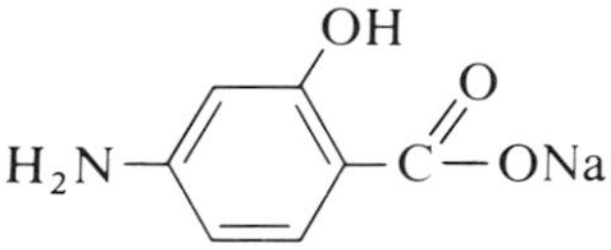

$$H_2N-\langle\ \overset{OH}{}\ \rangle-\overset{\overset{\displaystyle O}{\|}}{C}-ONa$$

Sodium p-aminosalicylate

benzoic acid (max. 280 nm) now appears at 265 nm with the typical phenate absorption of lower intensity at 299 nm. The absorption curves for p-amino-benzoic acid, benzoic acid and p-aminosalicylic acid are shown in Figs. 152 and 153.

---

* Students should note the various ways in which extinctions may be expressed by comparing this and subsequent figures.

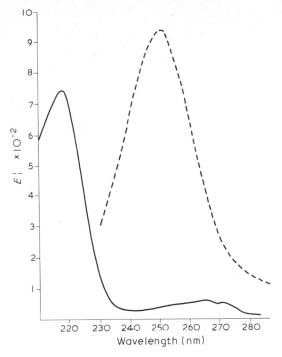

Fig. 151. Absorption curve for sulphanilamide in N hydrochloric acid (——) and in N sodium hydroxide (————)

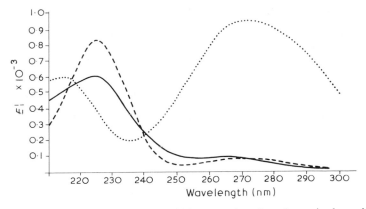

Fig. 152. Absorption curves for benzoic acid in water (——), and p-aminobenzoic acid in water (· · · ·) and in $0.1N$ hydrochloric acid (————)

The fusion of two or more benzene rings causes bathochromic displacement of the 200 nm and 255 nm bands, and, in general, three regions, distinguished by the intensity of the absorption, can be discerned (Table 41). Auxochromes and conjugated chromophores cause modification in the basic absorption in

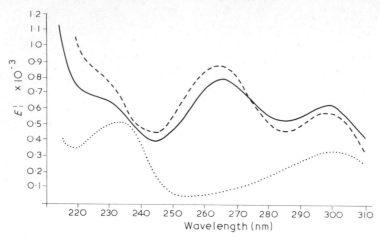

Fig. 153. Absorption curves for *p*-aminosalicylic acid in water (———) in *0.1N* hydrochloric
acid (· · · ·), and in *0.1N* sodium hydroxide (– – – –)

Table 41. Absorption Characteristics of Fused-ring Polycyclic Aromatics

Compound	Region					
	I		II		Benzenoid	
	λ_{max} (nm)	$\log \epsilon_{max}$	λ_{max} (nm)	$\log \epsilon_{max}$	λ_{max} (nm)	$\log \epsilon_{max}$
Benzene	—	—	200	3.65	255	2.35
Naphthalene	220	5.05	275	3.75	314	2.50
Anthracene	250	5.20	380	3.80	—	—
Phenanthrene	250	4.50	295	4.10	330	2.90

the manner already described for benzene so that quite complex absorption
spectra result.

The absorption spectra of pyridine and its derivatives are completely analogous to those of benzene and its derivatives, as shown by the following comparison of benzene and benzoic acid with pyridine, nicotinic acid and its
derivatives (Table 42). Changes in pH will cause changes in the absorption
spectra of pyridine derivatives

$$\begin{array}{c} \diagdown \diagup \\ N \end{array} \diagup \xrightarrow{H^+} \begin{array}{c} \diagdown \diagup \\ \overset{+}{N} \diagup \\ | \\ H \end{array}$$

so that the solvent used should always be specified. The absorption band at
266 nm (in *0.01N* hydrochloric acid) is used as an identity test for Isoniazid.
The increase in absorption at about 263 nm in acid conditions as compared
with that in alkaline conditions is a valuable property of pyridine compounds
as it enables them to be determined in the presence of those compounds for
which no such ΔE value is observed, e.g. triprolidine in the presence of codeine.

Table 42. Absorption Characteristics of Pyridine Derivatives

Substance	Formula	$\lambda_{1\,max}$ (nm)	$\lambda_{2\,max}$ (nm)	Solvent
Benzene		198	255	Ethanol
Pyridine		195	250	Hexane
Benzoic acid	COOH	230	270	Ethanol
Nicotinic acid	COOH	212	263	0.1N NaOH
Nicotinamide	$CONH_2$	212	263	0.1N NaOH
Nikethamide	$CON(C_2H_5)_2$	212	263	0.1N NaOH
Isoniazid	$CO.NH.NH_2$	215	266	0.01N HCl

Comparisons may be drawn in the same way between the quinolines and naphthalene, and between acridine and anthracene.

When a compound exists in tautomeric forms it may be possible, by careful selection of pH values for the solution of the substance, to obtain selective

pH 2   pH 10   pH 13

absorption in the ultraviolet region of the spectrum. For example Phenobarbitone shows strong absorption under alkaline conditions (pH 13) at about 255 nm (Fig. 154), due to the $-C=N-C=O$ chromophoric system. At pH 10 maximum absorption occurs at about 240–245 nm.

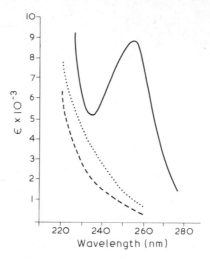

Fig. 154. Absorption spectra of phenobarbitone in *0.1N* hydrochloric acid (– – – –), in water (· · · ·) and in *0.1N* sodium hydroxide (——)

The purines, Caffeine and Theophylline, similarly show characteristic absorption because of the chromophoric system shown in heavy type.

Caffeine

Because of the tautomeric nature of nitrosophenols a pronounced increase in colour can be obtained by making solutions of such compounds alkaline.

This is actually done in the colorimetric assay for morphine in Camphorated Opium Tincture. The sensitivity of the method is increased by measuring the absorption of the nitroso-morphine in alkaline solution.

The subject could be expanded greatly but enough has been said to indicate that inspection of the formula of a substance will reveal whether or not interesting features of absorption in the ultraviolet region can be expected.

EXPERIMENTS IN ABSORPTIOMETRY AND SPECTROPHOTOMETRY

The variations in design of absorptiometers and spectrophotometers make it inadvisable to relate experiments to any one particular instrument. The following practical exercises are therefore designed to illustrate principles of method and theory rather than to repeat official processes of which explanations have already been given.

Instrumental

Experiment 1 (a). *To illustrate the principles underlying the choice of filters in absorptiometry*

Prepare solutions of copper sulphate containing 1, 2, 4, 6, 8 and 10% w/v of $CuSO_4,5H_2O$ and plot a calibration curve using in turn a blue, yellow, orange and red filter.

The calibration curves obtained with a Spekker absorptiometer and Kodak filters are shown in Fig. 155.

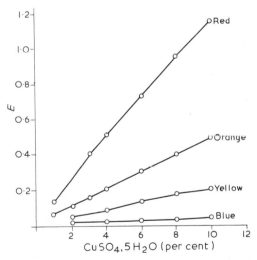

Fig. 155. Calibration curves for solutions of copper sulphate with various filters

An instructive extension to the exercise is to determine the actual absorption curve of the 2% copper sulphate solution using a Unicam SP600 or Hilger

Uvispek spectrophotometer. The absorption curve is recorded in Fig. 156 and a calibration curve using the wavelength of maximum absorption is shown in Fig. 157.

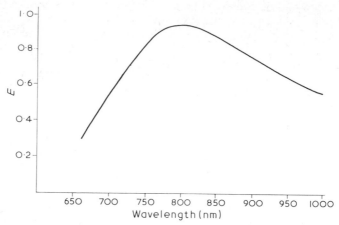

Fig. 156. Absorption curve of 2% copper sulphate solution

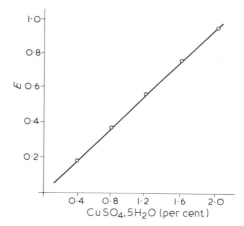

Fig. 157. Calibration curve for solutions of copper sulphate at 810 nm

For reasonable compliance with the Beer–Lambert Law, the incident radiation should be as near monochromatic as possible, the wavelength being such that the solution shows maximum absorption. Monochromatic radiation is an unrealisable ideal in practice, and in simple absorptiometers, filters are used to limit the wavelength range of the light passing through the solution.

The transmission properties of filters are characterised by a nominal wavelength and bandwidth, for example Kodak filter No. 568 has maximum transmission at 580 nm but also transmits all wavelengths between about 560 and 610 nm in varying degrees. The bandwidth is therefore 50 nm and this is typical of many filters.

In order to obtain the best sensitivity when determining concentration, it is necessary to use a filter which gives as steep a calibration curve as possible. In the exercise the filter of choice would be Kodak 570 (red), though even here deviations from Beer's Law are evident (Fig. 155).

The ideal way of choosing a filter is to record first the shape of the absorption curve of the solution using a spectrophotometer. This is then compared with the transmission curves of a set of filters. Thus, referring to Fig. 156, the ideal filter for copper sulphate solutions would be one which shows maximum transmission at 810 nm which is in the near infrared.

This explains the deviations observed in Fig. 155, and by plotting a calibration curve using light of about 810 nm, complete agreement with Beer's Law is found (Fig. 157).

In the absence of a spectrophotometer, an empirical approach to the selection of a filter can be adopted by choosing a filter whose colour is complementary to that of the solution. A colour wheel (Fig. 158), where complementary

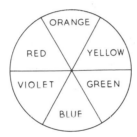

Fig. 158. A colour wheel

colours are opposite one another, may be used. Thus, for a blue, green or yellow solution, an orange, red or violet filter, respectively, could be used.

Experiment 1 (b)

Prepare solutions of potassium chromate in *0.05N* potassium hydroxide to contain 0.0005, 0.001, 0.002, 0.003, 0.004, 0.005 and 0.006% w/v K_2CrO_4. Plot the extinctions of the solutions with *0.05N* potassium hydroxide as reference solution in 1 cm cells using a mercury vapour arc in conjunction with Wood's glass filters. Repeat the experiment using 0.025, 0.05, 0.06, 0.07 and 0.08% w/v solutions of K_2CrO_4 but with a tungsten filament lamp and blue filters instead of the mercury vapour arc and Wood's glass filters respectively.

The absorption peak for potassium chromate in alkaline solution is at 372 nm. In the first part of the experiment, radiation of 365 nm wavelength is isolated because the Wood's glass filters absorb all visible radiation. The conditions are therefore comparable to those used in the extension to Experiment 1a, and a linear graph which passes through the origin is obtained. In the second part, however, the filter transmits radiation in the 400–450 nm region and the calibration curve shows deviations from Beer's Law. Note the loss in sensitivity under these conditions as shown by the much higher concentrations of potassium chromate required.

Experiment 2. *To check the wavelength and extinction scales on a photoelectric spectrophotometer.*

Potassium nitrate, potassium chromate and potassium dichromate have known absorption characteristics and are in frequent use as standards.

Method. Prepare an 0.006% w/v solution of potassium dichromate (AnalaR) in distilled water which contains dilute sulphuric acid (5 ml) per litre. Plot the absorption curve of the solution over the range 225–390 nm using 1 cm cells and water as reference solution. Pay particular attention to the regions of maxima (two) and minima (two) and calculate the E_1^1 values at these points. Compare the results with the standard values (Table 43).

Table 43. Absorption Characteristics of Potassium Dichromate

λ_{max} (nm)	E_1^1	λ_{min} (nm)	E_1^1
257	145.2	235	124.8
350	106.7	313	48.7

Prepare a new solution of potassium dichromate and measure the extinctions at the maxima and minima. Calculate the E_1^1 values and compare the results with those of the first experiment.

Preparation of the Solutions

The solutions used in spectrophotometry are generally extremely dilute, and in order to conserve expensive material or specially purified solvents, small quantities of material are weighed. In the absence of a micro-balance, the following method of weighing small quantities (10–15 mg) on an ordinary analytical balance has been found to give consistent results.

Method. Weigh accurately a clean dry weighing bottle, note the weight and allow to stand for 5 min on the balance pan. Check the weight, and if different from the first weight, allow to stand for a further 5 min and check again. Repeat if necessary until the weight is constant. This check is to ensure that the bottle is in equilibrium with the surrounding air in the balance case and that static electricity is not affecting the weight. Add the approximate amount of substance carefully from a spatula making sure *you do not touch the bottle with the fingers.* Re-stopper the bottle using forceps or tongs and weigh again accurately. Do not attempt to adjust the weight to exactly that required by the addition or removal of small quantities of substance. As the bottle is in equilibrium with the surroundings, there is no need to check the weight after 5 min at this stage. Dissolve the substance in an appropriate solvent and make up to volume in the normal manner.

Check on Cell Matching

Before using the solution, a check should be carried out on the cells in use. Normally these are matched, but it is of interest to note how often the following will indicate differences in the cells.

Method. Fill both cells with the solvent being used. Dry the outside of the cells with paper tissues. *Do not touch the polished faces with the fingers but handle the cells by the ground glass sides only.* Check the transmission of one cell against the other as blank at a selected wavelength. Note the reading, reject the solvent in one cell and fill again with solvent.

Check the reading and repeat the emptying and filling until consistent readings are obtained. Now reject the solvent in the other cell and repeat the above procedure. Any residual difference at this stage can be used to correct extinctions. When the difference becomes large (0.005–0.010) then cleaning of cells must be undertaken.

The solution under examination may now replace the solvent in one cell and the absorption curve may be determined.

Check on the Wavelength Scales

Convenient checks on the wavelength scale are given by line spectra, for example the hydrogen lines at 6563 Å (red) and 4861 Å (green) may be used with a hydrogen discharge lamp as light source. A visual check is easily made by observing the intensity of the light emerging from the exit slit by means of a mirror in the cell compartment. A sudden increase and decrease in the intensity as the wavelength drum is slowly rotated indicates the position of the line. The wavelength on the drum at the maximum intensity should then correspond to the known wavelength. Other convenient sources are mercury and sodium discharge lamps; for example mercury shows satisfactory radiation of the following wavelengths, 5461 (green) 3650 and 2537 Å, and sodium gives rise to the bright yellow lines at 5890 and 5896 Å. If the lines occur in the ultraviolet region of the spectrum, the instrument itself can be used to indicate the position of maximum intensity.

Provided the precautions indicated above have been taken, the results obtained by any one person on a particular instrument should agree to well within 1%. It is of interest to note that the differences between the results of different laboratories are considerably more than this value.

Experiment 3. *To show the effect of slit width*

Prepare a 0.002% solution of phenol in cyclohexane and examine the absorption peak at 278 nm using narrow and wide slit widths. The typical effect using a Hilger Uvispek spectrophotometer is shown in Fig. 159.

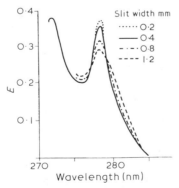

Fig. 159. Effect of slit width on extinction

The effect of an increase in the slit width is an increase in the intensity of the emergent radiation until a point is reached at which any further increase is accompanied by an increase in band-width. When this occurs, the proportion of the radiation absorbed clearly decreases if the absorption peak is very sharp, and the extinction therefore falls.

It might be thought that by decreasing the slit width the emergent radiation would tend more and more to monochromaticity. This however is not so and a limiting value (optimum slit width) is reached beyond which any reduction merely reduces the intensity of the emergent radiation. It should be noted that in some instruments, e.g. the Unicam SP600, conditions of maximum sensitivity are always being used and separate adjustment of slit widths as described above is not possible. Loss of efficiency as a consequence of filming of mirrors or prism leads to the use of wider slit widths and hence wider bandwidths.

Experiment 4. *To show the effect of stray radiation*

Method. Potassium dichromate solution 0.003 % w/v is satisfactory for this purpose. Measure the extinction over the range 290–320 nm using 1 cm cells and (a) a tungsten filament lamp as light source, (b) a hydrogen discharge lamp as light source. The typical effect is shown in Fig. 160.

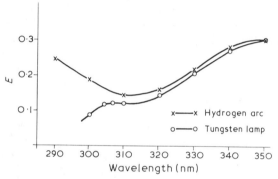

Fig. 160. Effect of stray radiation. Absorption curves of a 0.0026 % w/v solution of $K_2Cr_2O_7$ in slightly acidified water

Stray radiation is any radiation of wavelengths other than those which are absorbed by the solution being examined, and includes any radiation which reaches the photocell without having passed through the absorbing solution. It may arise in many ways. The effects noted in Experiments 1 and 3 are caused by stray radiation induced by incorrect choice of filters and wide bandwidth respectively. Fluorescence of the absorbing layer will give rise to stray radiation and the Pharmacopoeia requires that solvents used in absorptiometric work must be free from fluorescence at the wavelength of measurement.

Owing to the imperfections in optical systems, a small amount of stray radiation may be present and this increases with ageing of mirrors and light sources. Even a small proportion of such radiation may exert a pronounced effect in the 200–210 nm region of the spectrum where the available energy from the hydrogen lamp rapidly decreases with decrease in wavelength. The exercise is designed to illustrate this point but at a longer wavelength. Here, the tungsten filament lamp emits a small proportion of radiation at 300 nm and the monochromator is required to filter out a large amount of visible radiation. The stray radiation error is increased to such an extent that a spurious maximum is obtained. These spurious maxima are common at wavelengths below about 210 nm but can be detected by the marked deviations from the Beer–Lambert Law when solutions of different concentrations are examined.

Spurious maxima are not confined to the 210 nm region but can also occur at longer wavelengths if the solvent absorbs strongly at the wavelength being used. The conditions used in the exercise are then reproduced—the energy available for absorption is very feeble and stray radiation effects are consequently increased.

The presence of stray light may be demonstrated visibly as described below:

Visible demonstration of stray light in Spectrophotometer. Switch on the tungsten lamp and adjust the lamp mirror, if necessary. Open to maximum slit-width and set the wavelength scale to about 540 nm. Place a mirror at an angle of about 45° near the exit slit and position oneself so that the light emerging is clearly visible. Rotate the wavelength dial to the ultraviolet or near infrared (i.e. invisible) end of the spectrum and observe the white light emerging.

The Pharmacopoeia requires that the extinction of the solvent cell and its contents should not exceed 0.4 and shall in general be less than 0.2 when measured with reference to air at the same wavelength.

Experiment 5. *To show the effect of a constant galvanometer error at various values of extinction*

Method. Prepare solutions of bromocresol green in water which has been made slightly alkaline by the addition of a suitable buffer, for example one of pH 8. The concentrations should be such that extinctions ranging from 0.05 to over 2.0 will be obtained. Treat each solution in the following way. Measure the extinction at about 620 nm using 1 cm cells and water as reference solution. Note the reading (E_1) and adjust the extinction scale until a convenient deflection of the galvanometer is obtained. Note the reading' (E_2) on the extinction scale for this particular position of the galvanometer. Record the results in the following form.

Conc. (ml of solution)	E_1	E_2	$E_1 - E_2$	$\dfrac{(E_1 - E_2)}{E_1} \times 100$
0.1	0.056	0.037	0.019	33.9
0.2	0.108	0.087	0.021	19.45

Plot the ratio $(E_1 - E_2)/E_1 \times 100$ against E_1 to obtain a curve similar to that obtained in an actual experiment and shown in Fig. 161.

The exercise illustrates in a practical manner that which can be derived from theoretical considerations. Suppose that a constant error occurs in the measure of the intensity of the transmitted light. Then if I_{T_1} is the true and I_{T_2} is the apparent transmitted light the error in the measurement dE is given by

$$dE = \log \frac{I_0}{I_{T_1}} - \log \frac{I_0}{I_{T_2}} = \log \frac{I_{T_2}}{I_{T_1}}$$

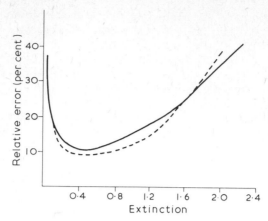

Fig. 161. Effect of constant galvanometer error on relative error. Observed (– – – –). Calculated for a 4% error in transmission (——)

But

$$I_{T_2} = I_{T_1} + K \quad \text{(where } K = \text{constant error)}$$

$$\therefore \quad dE = \log \frac{I_{T_1} + K}{I_{T_1}} = \log\left(1 + \frac{K}{I_{T_1}}\right)$$

The relative error

$$\frac{dE}{E} = \frac{\log(1 + K/I_{T_1})}{E}$$

may be evaluated by giving suitable values to K and I_{T_1}. E the extinction will depend upon the value selected for I_{T_1}. The curve obtained should be similar to that obtained in the exercise, and suggests that an optimum region exists for the determination of extinctions. The relative error is shown to be at a minimum in the range 0.3–0.7 and the solutions are usually adjusted so that the extinctions fall within this range.

It must be emphasised, however, that this is rather an arbitrary method of determining the optimum working range of an instrument. Moreover it implies that all instruments have similar optimum ranges. For an accurate assessment of the optimum working range, a large number of solutions must be examined and the extinction coefficients calculated from observed extinctions, which as before should cover a wide range. Under these conditions, the optimum working range is given by the precision of the extinction coefficients, and may well extend from about 0.3 to well over 1.0, the actual range varying with the instrument and its condition.

Solvent Effects

Experiment 6. *To show the effect of solvent upon the absorption curve of phenol*

Method. Prepare a 0.002% w/v solution of phenol in (a) water and (b) cyclohexane. Plot the absorption curve of each solution over the range 230–300 nm using 1 cm cells and the appropriate solvent as a blank. Typical absorption curves are shown in Fig. 162.

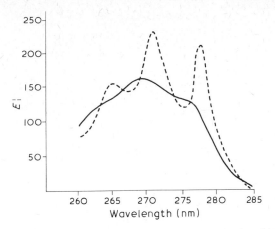

Fig. 162. Absorption curves of phenol in cyclohexane (– – – –) and in water (——)

In cyclohexane, little interaction between solvent and solute occurs and vibrational fine structure is observed. In water however, solvation of the solute and hydrogen bonding are possible so that the fine structure is almost eliminated, only the band envelope being obtained. Similarly, if the two solvents are aqueous and differ only in their ionic character, for example different strengths of the same buffer (salt) solution, it is possible for a 'salt effect' to cause slight differences in the absorption band of a compound.

Experiment 7. *To show the effect of* pH *upon the absorption curve of sulphanilamide*

Method. Prepare 100 ml of a 0.1% w/v aqueous solution of sulphanilamide. Dilute 1 ml of this solution to 100 ml using N hydrochloric acid and also dilute 1 ml to 100 ml using N sodium hydroxide. Plot the absorption curves of each solution over the range 210–300 nm for the acid solution and 230–300 nm for the alkaline solution, using 1 cm cells and appropriate reference solutions. Typical absorption curves are shown in Fig. 151.

In alkaline solution the primary amino group is retained as the auxochrome, but in acid solution quaternisation occurs to a co-ordinatively saturated auxochrome.

$$H_2N-\langle\bigcirc\rangle-\overset{\overset{O}{\uparrow}}{\underset{\underset{O}{\downarrow}}{S}}-NH_2 \underset{\text{alkali}}{\overset{\text{acid}}{\rightleftharpoons}} H_3\overset{+}{N}-\langle\bigcirc\rangle-\overset{\overset{O}{\uparrow}}{\underset{\underset{O}{\downarrow}}{S}}-NH_2$$

This is much less effective in modifying absorption and the characteristic benzenoid absorption is obtained at 265 nm.

Note that solutions of sodium hydroxide absorb radiation below about 230 nm, so that unless the concentration of alkali is very small, the readings will be unreliable.

Structural

Experiment 8. *To show typical benzenoid absorption*

Method. Prepare a 0.05% solution in *0.1N* hydrochloric acid of any of the following chemicals:

> Pethidine hydrochloride
> Ephedrine hydrochloride
> Methylamphetamine hydrochloride
> Atropine sulphate
> Methadone hydrochloride
> Phenylmercuric nitrate

Plot the absorption curve over the range 220–280 nm using 1 cm cells and *0.1N* hydrochloric acid as reference solution.

The absorption curve of ephedrine hydrochloride is shown in Fig. 149 and is typical of all the above compounds because all possess the general formula

⬡—R. Strong auxochromes such as $-NHR$, $-OH$, are not directly attached to the benzene ring and chromophores such as $>C=O$ are not in conjugation so that the absorption corresponds closely to that of toluene.

Experiment 9. *To show typical pyridine absorption*

Method. Prepare a 0.002% w/v solution of nicotinic acid (or nicotinamide) in water and plot the absorption curve over the range 230–290 nm.

The absorption curve for nicotinamide is shown in Fig. 163.

The chromophoric system is (pyridine ring with $\overset{O}{\overset{\|}{C}}-R$) but the absorption caused by the pyridine nucleus occurs at 265 nm. Compare the curve with those due to the benzene system, in particular that of benzoic acid (see below).

Cognate Experiments

Determine the absorption curves of
Quinoline
Naphthalene
Acridine

Experiment 10. *To show the influence of conjugated chromophores and auxochromes on chromophores*

Method 1. The effect of conjugation on the carbonyl chromophore. Prepare a 0.01% solution of cyclohexanone in ethanol, a 0.001% solution of testosterone in ethanol and a 0.001% solution of *p*-dimethylaminobenzaldehyde in ethanol and plot the absorption curves over the range 210–320 nm.

The carbonyl chromophore absorbs strongly at about 185 nm and weakly at about 280 nm, but when conjugated with an ethylenic bond as in testo-

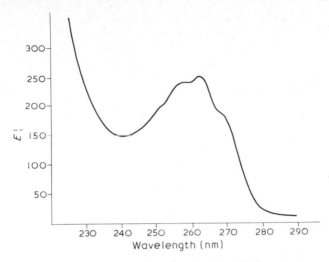

Fig. 163. Absorption curve for nicotinamide in water

sterone, the absorptions are shifted to about 240 nm and 300 nm respectively. Note that in the absorption curve of *p*-dimethylaminobenzaldehyde, the benzenoid absorption is shifted to a longer wavelength (about 280 nm). A molecule shows, not only the absorption corresponding to the interacting chromophores, but also that corresponding to an individual chromophore. Where the latter is weak, it is often masked by the strong absorption of the conjugated chromophoric system.

Method 2. The effect of substituents on the benzene chromophore. For this exercise the following solutions are required:

Aniline	0.005 % w/v in water
Benzoic Acid	0.001 % w/v in water
p-Aminobenzoic Acid	0.0005 % w/v in water
Cinnamic Acid	0.0005 % w/v in water

Plot the absorption curves over the range 210–320 nm using 1 cm cells and water as reference solution.

The $-NH_2$ auxochrome is comparable with the $-COOH$ chromophore in its effect on the absorption, cf. Figs. 152 and 153. They can readily be distinguished, however, by determining the absorption curves in acid solution. The absorption curves of benzoic and cinnamic acids will remain unaffected whereas those of aniline and *p*-aminobenzoic acid will alter appreciably because of the change in the chromophoric system. Compare the curves with those of pethidine etc. (Experiment 8).

Experiment 11. *To determine the isobestic point of an indicator and to show its use in analysis*

Method. Pipette 2.0 ml of bromocresol green solution (0.02%) into each of 8×50 ml graduated flasks. Adjust to volume with buffer solutions of pH 3.0, 3.4, 3.8, 4.2, 4.6, 5.0, 5.4

and 6.0 respectively. Mix well and determine the absorption curve of each solution over the range 470–700 nm using 1 cm cells. Plot each curve on the same graph and note the isobestic point, i.e. the point of equal extinction regardless of pH, and the position of the absorption maximum.

Draw a graph of the ratio

$$\frac{Extinction \ at \ Wavelength \ of \ Maximum \ Absorption \ at \ pH \ 3.0}{Extinction \ at \ Isobestic \ point}$$

and of the ratios determined with the other buffers, as ordinate against pH as abscissa.

After the completion of the graph, any pH between about 3.1 and 5.2 can be determined by adding a few drops of bromocresol green indicator to the test solution and measuring the extinction at two wavelengths—one at the isobestic point and the other at the maximum absorption. Calculation of the ratio and reference to the graph gives the required pH. In this method the quantity of indicator need not be known, because the ratio of the extinctions would be the same regardless of that quantity.

This principle has been applied to the determination of Thiamine in pharmaceutical products. The vitamin shows marked changes in absorption at about 245 nm with change in pH, but no change in extinction at about 273 nm—the isobestic point. If a solution of the vitamin is examined at 245 nm using pH values of 1 and 7 respectively the difference in extinctions can be related directly to the vitamin content of the solution *provided that irrelevant absorption is absent*. This proviso can be confirmed by measuring the extinction of both solutions at 273 nm when no change should be observed. Any change which does occur must be due to irrelevant absorption or change in solvent, and a correction can be applied to the extinction at 245 nm. The difference in the extinctions at 245 nm then becomes directly proportional to the concentration of the vitamin.

Mixtures

A number of mixtures have been considered in Part 1, Chapter 14.

Body Fluids

In the following examples it must be remembered that an important preliminary part of the investigation has been omitted. It is essential that the compound to be determined is identified in the sample by appropriate techniques, e.g. TLC or GLC. Thus in Experiment 16 bases other than Diphenhydramine will react in a manner similar to the latter.

Experiment 12. *Determination of atropine in urine*

Poisoning by solanaceous alkaloids may arise by ingestion of the attractive belladonna berries or in suicidal attempts with atropine sulphate eye-drops. Treatment by forced diuresis has been used in one case described by Groden and Williams (1964), and in assessing the effectiveness of the treatment, atropine in samples of the patient's urine was determined by the well-known Vitali reaction.

Method. Transfer the urine (50 ml) to a separator, make slightly acid with dilute HCl and extract with ether (50 ml) (Note 1). Run off the lower aqueous layer into a second separator, wash the ether layer with water (10 ml) and add the washing to the acid urine. Reject the ether layer. Make alkaline with ammonia and extract with ether (3 × 30 ml). Bulk the ether extracts, wash with water (10 ml) and reject the washing.

Evaporate the ether to dryness, dissolve the residue in ethanol (2 ml) and transfer quantitatively to a 5 ml volumetric flask with ethanol (Note 2). Pipette the solution (2 ml) into a porcelain dish, evaporate to dryness, add fuming nitric acid (0.3 ml) and carefully tilt the dish to ensure complete coverage of the residue by the acid. Evaporate to dryness on a boiling water-bath in a fume cupboard.

Cool, add dimethylformamide (5 ml) (Note 3) and 0.5N ethanolic KOH (0.5 ml). Mix and immediately measure the extinction in a 1 cm cell at 560 nm using water as reference.

Calculate the amount of atropine present in the urine by reference to a calibration curve constructed from atropine sulphate (0, 0.2, 0.4 and 0.6 mg) treated in the same way.

Note 1. This extraction removes some colouring matter but not all.

Note 2. In poisoning cases the concentration of atropine in the urine is initially quite high and by making up to volume a suitable aliquot portion may be used if the first test gives too intense a colour. Later samples may be examined directly but the effect of pigments when large volumes of urine are used in attempts to increase detection limits makes the results unreliable.

Note 3. Various other solvents, such as acetone and ethanol have been used.

Experiment 13. *Determination of a barbiturate in plasma or urine*

The determination depends upon the formation of a conjugated system in the molecule by change in pH and subsequent measurement of a ΔE value to overcome the effect of irrelevant absorption. The relevant formulae are given earlier in this chapter (p. 261).

The complete absorption curves can be used to differentiate barbiturates if ΔE values are determined at various wavelengths [Goldbaum (1952) and Broughton (1956)] but with the advent of TLC and GLC this aspect becomes of lesser importance, particularly if a mixture of barbiturates is present. The ΔE values at various wavelengths, however, are important in that the presence of a barbiturate can be confirmed and some confidence in the result be obtained. The difficulties associated with the determination have been discussed by Bogan and Smith (1967) and by Stone and Henwood (1967).

Reagents. Borate Buffer. M solution of boric acid and potassium chloride.

Approximately $0.45N$ Sodium Hydroxide. Adjust the normality so that a pH of 10.5 is obtained on mixing 2 volumes of alkali with 1 volume of borate buffer.

Solvents. Wash chloroform (10 volume) with approximately N NaOH (1 volume) and water (2 × 1 volume).

Method. Transfer the plasma (5 ml) (Note 1) to a separator and add chloroform (75 ml). Shake thoroughly and allow to separate. Filter the chloroform extract and transfer an aliquot portion of the filtrate (50–60 ml accurately measured) to a dry separator. Add approximately $0.45N$ NaOH (5 ml) and shake thoroughly. Run off the lower chloroform layer, transfer the alkaline layer to a centrifuge tube and centrifuge until clear. Dilute the clear supernatant (2 ml) with (a) $0.45N$ NaOH (1 ml) and (b) borate buffer (1 ml) in 1 cm spectrophotometric cells (Note 2). Record the absorption curves of each solution against the appropriate blank of water (5 ml) carried through the procedure, over the range 225 to 305 nm in a 1 cm cell.

Examine the curves at 260 and 240 nm for a ΔE effect. The differences should be positive (260 nm) and negative (240 nm) if a barbiturate is present. If this is confirmed calculate the concentration of barbiturate by reference to the ΔE value obtained for the known barbiturate (previously identified by TLC or GLC) carried through the same procedures (Note 3).

Note 1. For urine samples considerable amounts of interfering substances are to be expected in the chloroform extract. The extract should be purified by shaking with a phosphate buffer pH 7.4 (5 ml) before continuing as described.

Note 2. A melting point tube, sealed at both ends, is a convenient stirrer for this purpose.

Note 3. In simulating a case of poisoning the body fluid should contain about 20 μg of barbiturate/ml. Providing the differences are of the stated values, confirming the presence of a barbiturate, a large ΔE value for calculation purposes may be obtained by adding N H_2SO_4 (1 ml) instead of borate buffer to the alkaline extract (2 ml).

Experiment 14. *Determination of Diphenylhydantoin* (*Phenytoin*) *in blood or urine*

The determination is based upon the conversion of a weak ultraviolet absorbing system to one which is more highly absorbing by oxidation as described by Wallace (1966, 1968).

Phenytoin

Benzophenone
λ_{max} 257 nm

Method. Adjust the pH of the sample to about 6–7 by dropwise addition of *0.5N* HCl and place 10 ml in a separator. Add chloroform (100 ml) and shake vigorously (3 min). Filter through a phase separating paper into a glass-stoppered graduated measuring cylinder and record the volume. Add N NaOH (5 ml) and shake thoroughly (3 min). Allow to separate and transfer the alkaline layer (4 ml) to a 250 ml round-bottomed flask and evaporate to approx. 1 ml by means of a rotary evaporator at 50–60° under reduced pressure (Note 1). Cool, add a solution of potassium permanganate in $7N$ NaOH (1%; 20 ml), (Note 2) heptane (5 ml) and reflux the contents, with stirring for 30 min. Cool and record the absorption curve of the heptane layer over the region 220–340 nm using a blank of heptane.

Calculate the amount of diphenylhydantoin by reference to a calibration curve (0–20 μg/ml) prepared by treating aqueous solutions of the drug to identical procedures.

Note 1. It is essential that thoroughly clean glassware is used and a solvent-free residue is obtained otherwise loss of permanganate and poor oxidation occurs.

Note 2. It is essential that the permanganate solution is purple in colour. The solution turns dark-green on storage and should not then be used. Caddy, Mullen, Tranter and Fish (1973) recommend using separate solutions and mixing when required. An alternative simple method is to weigh approx. 6 g of NaOH pellets into a sample tube, mark the volume produced, and use as a measure for the quantity of NaOH in future assays. Potassium permanganate (0.2 g) is weighed for each determination. Water (15 ml) should be added to the reaction flask before addition of reagents.

Experiment 15. *Determination of Glutethimide* (*Doriden*) *in body fluids*

The determination is based upon the formation of a conjugated chromophore as described for barbiturates.

$$\lambda_{max}\ 235\ nm$$

Unfortunately, glutethimide is unstable in alkaline solution and the extinction at 235 nm decreases rapidly in aqueous solution but more slowly in mainly ethanolic solution (Goldbaum, Williams and Koppanyi, 1960). Under the latter conditions it is possible to determine the extinction at intervals over a period of 30–40 min and to extrapolate the curve log E vs time to zero time and hence obtain the extinction corresponding to the original amount of glutethimide present. The method involves a rate study and if several solutions of glutethimide are treated in the same way several straight lines are obtained all parallel to one another. If now, the result from the sample gives a straight line parallel to those of the standards it can be stated with some confidence that glutethimide is present in a relatively pure state and the extrapolated extinction is proportional to its concentration. If, however, irrelevant absorption is present the straight line is not parallel to those of the standards and sufficient time must be allowed to enable E_∞ to be determined as a correction. As this may take up to 3 hr a quicker method for E_∞ is to extrapolate the curve to $t = 180$ min as the half-life period is about 20 min.

Reagents. Ethanol (for spectroscopy). Before use, heat the ethanol on a boiling water-bath to remove CO_2, and cool.

0.2N KOH. This is conveniently prepared by dilution of N KOH in ethanol with CO_2-free water.

Standards. 0.001, 0.002, 0.003 % of glutethimide in ethanol.

Method. Standards. Transfer the standard solution (3 ml) to a 1 cm cell and add *0.2N* KOH (1 ml). Mix, stopper (Note 1) and record the absorption curve using a blank of ethanol (3 ml) and *0.2N* KOH (1 ml). Note the time of mixing. Record the absorption curve at 5, 10, 20, 30 and 40 minutes. Plot log E_{235} against time and extrapolate to zero time. Repeat for the remaining standard solutions and plot $E_{(t=0)}$ against concentration for a calibration curve. Sample. Transfer the body fluid (2 ml) to a separating funnel and add chloroform (25 ml). Shake thoroughly, allow to separate and run off the chloroform layer into a second separator. Wash the chloroform extract with *0.5N* NaOH (2 × 5 ml) (Note 2) and *0.5N* HCl (5 ml) and reject the washings. Evaporate a 20 ml aliquot portion of the chloroform to low volume and complete the evaporation to dryness with a stream of air. Dissolve the residue in ethanol (4 ml) and treat the solution (3 ml) as described for the standard solutions. Proceed according to the type of line obtained.

Calculate the percentage of glutethimide in the sample.

Note 1. It is essential to prevent access of CO_2 otherwise a turbidity develops. Such turbidity may also occur if much fatty material is present in sample extracts (Goldbaum, Williams and Johnston, 1962).

Note 2. Glutethimide is such a weak acid that no loss occurs in this wash.

Extension. Determine the effect of (*a*) water and (*b*) ethanol as solvent on the rate of decomposition of Bemigride.

Experiment 16. *Determination of Methaqualone and Diphenhydramine in blood or tissue*

The two drugs occur together in Mandrax tablets and in cases of poisoning by the tablets both appear in body fluids.

The determination of Methaqualone depends upon its intense ultraviolet absorption at 234 nm. Diphenhydramine is weakly absorbing in this region because of its isolated benzene ring systems and because it occurs at a much lower concentration (25 mg) in the tablets than does Methaqualone (250 mg). Its determination therefore depends upon increasing the sensitivity of detection. Ion-pair formation is possible with methyl orange and, as excess methyl orange can be removed, a measurement of the remaining indicator's absorption is also a measure of the concentration of diphenhydramine. References to the method are Dill and Glazko (1949) and Akagi, Oketani and Takada (1963).

Reagents. Hexane A.R.

Chloroform containing 1% v/v of iso-amyl alcohol. The presence of the alcohol is essential to avoid adsorption of the diphenhydramine-indicator complex on glass surfaces; otherwise extremely variable results are obtained.

Methyl Orange Solution. *0.5M* boric acid saturated with methyl orange indicator. Store over a layer of the chloroform iso-amyl alcohol solvent.

Acid-alcohol. 2% w/v H_2SO_4 in ethanol.

Method. To blood (5 ml) or macerated tissue (5 g) add *0.1N* NaOH (5 ml) and extract with hexane (25 ml). Reject the lower aqueous layer and extract the hexane layer with *2N* HCl (6 ml). Measure the extinction of the acid layer in a 1 cm cell using *2N* HCl as blank.

Transfer the acid extract to a small separator (Note 1) and add *2N* NaOH (10 ml). Extract with the special chloroform solvent (10 ml) and reject the aqueous phase. Shake the organic layer with methyl orange solution (0.5 ml) and centrifuge the organic layer to remove excess reagent (Note 2). To the clear chloroform extract (5 ml) add acid-alcohol (0.5 ml) and measure the extinction at 535 nm (Note 3).

Calculate the concentration of methaqualone and diphenhydramine by treating solutions of the two compounds in the same way.

Note 1. All the acid extract must be used and washing with small volumes of water will assist in the recovery.

Note 2. The volume of methyl orange reagent is kept small as the presence of more water would cause dissociation of the complex. It is for this reason that centrifugation is employed rather than washing with water.

Note 3. The addition of acid-ethanol changes the yellow colour of the indicator complex to the red colour of the indicator itself.

Table 44. Ultraviolet Absorption Characteristics of Drugs in Common Use

Substance	Solvent	λ (nm)	E_1^1
Acetazolamide	*0.1N* Hydrochloric acid	265	474
Alprenolol			
Injection	*0.1N* Hydrochloric acid	271	59.5
Tablets	Ethanol (95%)	271	65
Amitriptyline Hydrochloride			
Injection	*0.1N* Hydrochloric acid in methanol	239	462

Table 44 (*continued*)

Substance	Solvent	λ (nm)	E_1^1
Tablets	*0.1N* Hydrochloric acid	239	445
Apomorphine Hydrochloride	*0.1N* Hydrochloric acid	273	550
Azathioprine Tablets	*0.1N* Hydrochloric acid	280	600
Benztropine Tablets	*2N* Hydrochloric acid	288	11.5
Betamethasone Sodium	Water	241	297
Phosphate Injection	*0.002N* Sodium chloride	241	391
			(as Betamethasone)
Bisacodyl Tablets	Chloroform	264	148
Carbamazepine Tablets	Ethanol (95%)	285	490
Carbenoxolone Tablets	*0.02N* Sodium Carbonate in Methanol (50%)	256	199
Carbimazole	*0.1N* Hydrochloric acid	291	557
Carbimazole Tablets	Water	291	557
Chloramphenicol Capsules	Water	278	288
Chlorcyclizine Tablets	*0.1N* Sulphuric acid	231	512
Chlordiazepoxide Tablets	*0.1N* Hydrochloric acid	308	292
Chlorothiazide Tablets	*0.1N* Sodium hydroxide	292	430
Chlorotrianisine Tablets	Chloroform	310	412
Chlorpheniramine Maleate			
Injection	Water	262	147
Tablets	*0.5N* Sulphuric acid	265	212
Chlorpromazine Hydrochloride			
Injection	*0.1N* Hydrochloric acid	254	915
Tablets	*0.1N* Hydrochloric acid	254	915
Chlorthalidone	*0.04N* Hydrochloric acid in methanol	275	57.4
Choline Theophyllinate Tablets	*0.01N* Sodium hydroxide	275	415
Clomiphene Citrate Tablets	*0.1N* Hydrochloric acid	292	175
Colchicine Tablets	Dehydrated ethanol	350	425
Co-trimoxazole Tablets for Trimethoprim	Acetic acid (0.6%)	271	204
Cyanocobalamin	Water	361	207
Cyclizine Tablets	*0.1N* Sulphuric acid	225	390
Cyclomethycaine Sulphate	*0.01N* Hydrochloric acid	261	400
Cycloserine	*0.1N* Hydrochloric acid	219	342
Cyproheptadine Hydrochloride	Ethanol (95%)	286	355
Deoxycortone Trimethylacetate	Dehydrated ethanol	240	405
Desipramine Tablets	*0.1N* Hydrochloric acid	250	270
Diazepam Capsules Tablets	Sulphuric acid (0.5%) in Methanol	284	446
Dihydrocodeine Tartrate	Water	284	35.7
Dichlorphenamide Tablets	*0.1N* Sodium hydroxide	285	42.5
Dimethisterone	Dehydrated ethanol	240	467

Table 44 (*continued*)

Substance	Solvent	λ (nm)	E_1^1
Dimethisterone Tablets	Hexane	232	443
Diphenhydramine Injection	0.01N Hydrochloric acid	258	16.0
Dydrogesterone Tablets	Ethanol	286	838
Endrophonium Chloride Injection	0.1N Hydrochloric acid	273	110
Ethinyloestradiol	Dehydrated ethanol	280	71
Ethisterone	Dehydrated ethanol	240	520
Ethopropazine Hydrochloride Tablets	Ethanol (95%)	252	845
Ethosuximide Capsules	Ethanol (95%)	248	8.5
Fluocortolone Hexanoate	Methanol	242	340
Fluocortolone Pivalate	Ethanol	242	350
Fluphenazine Tablets	0.1N Hydrochloric acid in ethanol (80%)	258	620
Fluoxymesterone	Dehydrated ethanol	240	495
Frusemide Tablets	0.1N Sodium hydroxide	271	595
Gallamine Triethiodide Injection	0.01N Hydrochloric acid	225	525
Griseofulvin	Dehydrated ethanol	291	686
Haloperidol Tablets	0.01N Hydrochloric acid in methanol	245	341
Hydrochlorothiazide	0.002N Sodium hydroxide	273.5	507
Hydrocortisone Hydrogen Succinate	Dehydrated ethanol	240	341
Hydrocortisone Sodium Succinate	Dehydrated ethanol	240	327
Injection	Dehydrated ethanol	240	435 (Calculated for hydrocortisone)
Hydroxocobalamin	Acetate buffer	351	190 (chloride) 188 (sulphate)
Hydroxocobalamin Injection	Acetate buffer	351	196 (anhydrous)
Hydroxyprogesterone	Ethanol	240	395
Imipramine Hydrochloride Tablets	0.1N Hydrochloric acid	250	264
Indomethacin Capsules Suppositories	Methanol/Buffer pH 7.2	318	193
Levallorphan Tartrate Injection	0.016N Sulphuric acid	279	46.5
Tablets	0.016N Sulphuric acid	279	46.5
Levorphanol Tartrate Injection	0.1N Sulphuric acid	279	46.0
Lucanthone Hydrochloride Tablets	0.1N Hydrochloric acid	256	1092
Megestrol Acetate	Ethanol	287	630

Table 44 (*continued*)

Substance	Solvent	λ (nm)	E_1^1
Mepyramine Maleate Injection	*0.01N* Hydrochloric acid	316	206
Mercaptopurine Tablets	*0.1N* Hydrochloric acid	325	1165
Metaraminol Tartrate Injection	Water	272	111
Methandienone	Dehydrated ethanol	245	516
Methyltestosterone	Dehydrated ethanol	240	535
Metyrapone Capsules	*0.1N* Hydrochloric acid	260	500
Nandrolone Decanoate	Ethanol	240	407
Nandrolone Phenylpropionate	Dehydrated ethanol	240	430
Nialamide	*0.01N* Hydrochloric acid	265	197
Nicoumalone	*0.1N* Hydrochloric acid in methanol (90%)	306	521
Nitrazepam Tablets	Hydrochloric acid (0.5%) in methanol	280	910
Nitrofurantoin	Dimethylformamide and acetate buffer	367	765
Norethandrolone	Methanol	240	565
Norethisterone	Dehydrated ethanol	240	571
Nortriptyline Hydrochloride Tablets	Water	239	517
Orciprenaline Sulphate			
Injection	*0.01N* Hydrochloric acid	276	72.3
Tablets	*0.01N* Hydrochloric acid	276	72.3
Orphenadrine Citrate Tablets (Slow)	*0.1N* Hydrochloric acid	264	13.65
Paracetamol Tablets	*0.01N* Sodium hydroxide	257	715
Pentagastrin	*0.01N* Ammonia	280	70
Perphenazine Tablets	Ethanol	258	921
Phenazocine			
Injection	*0.1N* Hydrochloric acid	278	50.3
Tablets			
Phenindione Tablets	*0.1N* Sodium hydroxide	278	1310
Phenoxybenzamine Hydrochloride	Chloroform (ethanol-free)	272	56.3
Phenoxymethylpenicillin	Sodium bicarbonate solution (0.04%)	268	34.8
Phenoxymethylpenicillin Calcium Phenoxymethylpenicillin Potassium	Water	268	34.8 $(C_{16}H_{18}N_2O_5S)$
Phenylephrine Hydrochloride Injection	*N* Sulphuric acid	273	95
Phytomenadione	Trimethylpentane	249	420
Practolol			
Injection	Methanol	248	620
Tablets			

Table 44 (*continued*)

Substance	Solvent	λ (nm)	E_1^1
Prednisolone Pivalate	Ethanol	240	337
Prednisolone Sodium Phosphate	Water	247	312
Probenecid	Ethanol (95%) containing a trace of hydrochloric acid	248	332
Prochlorperazine Tablets	Dehydrated ethanol containing strong ammonia solution (0.01%)	258	615
Prochlorperazine Methane-Sulphonate Injection	Dehydrated ethanol containing strong ammonia solution (0.01%)	258	635
Progesterone	Dehydrated ethanol	240	540
Promazine Hydrochloride Injection and Tablets	Water acidified with dilute hydrochloric acid	251	912
Promethazine Hydrochloride Injection	0.01N Hydrochloric acid	249	910
Promethazine Hydrochloride Tablets	Water containing a trace of of hydrochloric acid	249	910
Promethazine Theoclate Tablets	Ethanol	255	755
Propranolol Hydrochloride			
Injection	Methanol	290	210
Tablets	Methanol	290	210
Protriptyline Tablets	0.1N Hydrochloric acid	292	465
Pyridostigmine Bromide Injection and Tablets	Water	269–270	186
Riboflavine	Acetate buffer	444	323
Salbutamol Tablets	Water	276	71.1
Sodium Anoxynaphthonate	Phosphate buffer	570	555
Sodium Cromoglycate Cartridges	Buffer pH 7.4	326	164
Spironolactone Tablets	Methanol	238	470
Testosterone	Dehydrated ethanol	240	560
Testosterone Phenyl-propionate	Dehydrated ethanol	240	560
Testosterone Propionate	Dehydrated ethanol	240	490
Theophylline in Aminophylline Aminophylline Tablets	0.01N Sodium hydroxide	275	650
Thioridazine Tablets	Chloroform	316	121
Tranylcypromine Sulphate	0.2N Sulphuric acid	271	10.7 $(E_{271}^1 - E_{282}^1)$
Triamcinolone Acetonide	Ethanol	240	354
Trifluoperazine Hydrochloride	Diluted hydrochloric acid	256	630
Trimipramine Tablets	0.1N Hydrochloric acid	250	300
Tripelennamine Hydrochloride Tablets	Methanol	245	550

Table 44 (*continued*)

Substance	Solvent	λ (nm)	E_1^1
Triprolidine Hydrochloride Tablets	0.1N Hydrochloric acid	290	290
Tubocurarine Chloride	Water	280	118 (calculated as anhydrous)
Tubocurarine Chloride Injection	Water	280	105
Viprynium Embonate	Methoxyethanol	508	785
Vitamin A Ester form	Cyclohexane	} Special conditions	
Vitamin A	Isopropanol	} and formulae	
Warfarin	0.01N Sodium hydroxide	308	431

Table 45. Ultraviolet Absorption Characteristics of Drugs in Common Use Classified by Principal and Subsidiary Absorption Maxima

Wavelength of principal absorption maximum (nm)	Compound	Concentration %	E_1	Solvent
219	Cycloserine	0.00125	0.43	0.1N Hydrochloric acid
225	Gallamine Triethiodide	0.001	0.55	0.1N Hydrochloric acid
	Probenecid	0.002		Ethanol (95%)
226	Clofibrate	0.001	0.46	Ethanol
228	Frusemide (271)	0.0005	0.53	0.1N Sodium hydroxide
229	Tolbutamide	0.001	0.49	Methanol
230	Chlorcyclizine Hydrochloride	0.001	0.44	Ethanol (95%)
	Iopanoic Acid	0.001	0.68	0.01N Sodium hydroxide
	Meclosine Hydrochloride	0.001	0.33	Ethanol
	Triprolidine Hydrochloride (276)	0.001	0.47	Water
232	Chlorpropamide	0.006	0.36	0.01N Hydrochloric acid
233	Cocaine	0.001	0.43	0.01N Hydrochloric acid
	Cocaine Hydrochloride	0.001	0.39	0.01N Hydrochloric acid
234	Methaqualone	0.0005	0.65	0.1N Hydrochloric, acid

Table 45 (*continued*)

Wavelength of principal absorption maximum (nm)	Compound	Concentration %	E_1	Solvent
	Phenformin hydrochloride	0.0005	0.3	Water
237	Cephalothin Sodium	0.002	0.68	Water
238	Beclomethasone Dipropionate	0.002	0.59	Ethanol
	Iodipamide Methyl-glucamine Injection (acid component)	0.001	.0.61	Methanol
	Methyl Salicylate (306)	0.001	0.57	Ethanol (95%)
239	Amitriptyline Hydrochloride	0.0012	0.55	Methanol
	Mepyramine Maleate	0.002	0.90	*0.01N* Hydrochloric acid
	Nortriptyline Hydrochloride	0.001	0.48	Methanol
	Propyliodone (281)	0.001	0.26	Ethanol
	Sodium Diatrizoate	0.001	0.52	Water
240	Betamethasone	0.001	0.37–0.40	Ethanol
	Betamethasone Valerate	0.001	0.32–0.34	Ethanol
	Cephaloridine	0.002	0.72–0.79	Water
	Cortisone Acetate	0.001	0.39	Ethanol
	Deoxycortone Acetate	0.001	0.44	Ethanol
	Dequalinium Acetate (326, 335)	0.0008	0.65	Water
	Dequalinium Chloride (326, 335)	0.0008	0.65	Water
	Dexamethasone	0.001	0.4	Methanol
	Dexamethasone Acetate	0.001	0.35	Ethanol
	Edrophonium Chloride (294)	0.002	1.1	*0.1N* Sodium hydroxide
	Ethisterone	0.001	0.52	Ethanol
	Fludrocortisone Acetate	0.001	0.4	Ethanol
	Fluocinolone Acetonide	0.0015	0.54	Ethanol
	Hydrallazine Hydro-chloride (260, 303, 315)	0.001	0.58	Water
	Hydrocortisone	0.001	0.43	Ethanol
	Hydrocortisone Acetate	0.001	0.39	Ethanol
	Levallorphan Tartrate (299)	0.005	1.02	Sodium hydroxide
	Levorphanol Tartrate (299)	0.005	0.99	Sodium hydroxide
	Methylprednisolone	0.001	0.4	Ethanol
	Norethisterone	0.001	0.57	Ethanol

Table 45 (*continued*)

Wavelength of principal absorption maximum (nm)	Compound	Concentration %	E_1	Solvent
	Norethisterone Acetate	0.001	0.51	Ethanol (95%)
	Prednisolone	0.001	0.4	Ethanol
	Prednisolone Acetate	0.001	0.37	Ethanol
	Prednisolone Trimethylacetate	0.001	0.34	Ethanol
	Prednisone	0.001	0.41	Ethanol
	Prednisone Acetate	0.001	0.38	Ethanol
241	Antazoline Hydrochloride (291)	0.001	0.5	0.1N Hydrochloric acid
	Stilboestrol	0.001	0.6	Ethanol
242	Crotamiton	0.001	0.32	Cyclohexane
	Dihydrotachysterol (251, 261)	0.0005	0.44	Methanol
	Iothalamic Acid	0.001	0.55	0.1N Hydrochloric acid
	Meglumine Iothalamate Injection	0.001	0.55	Ethanol (95%)
243	Colchicine (350)	0.001	0.73	Ethanol
	Phytomenadione (249, 261, 270)	0.001	0.4	Trimethylphentane
245	Dichlorophen (304)	0.001	0.65	0.1N Sodium hydroxide
	Tripelennamine Hydrochloride (305)	0.001	0.49	Water
246	Propantheline Bromide (282)	0.006	0.7	Methanol
	Sulthiame	0.001	0.4	Methanol
247	Chlorotrianisene (307)	0.001	0.32	Ethanol (95%)
	Cinchocaine Hydrochloride (319)	0.001	0.65	N Hydrochloric acid
248	Bisacodyl	0.001	0.65	0.1N Potassium hydroxide in methanol
	Ethosuximide	0.05	0.43	Ethanol (95%)
	Probenecid (25)	0.002	0.66	Ethanol
249	Paracetamol	0.001	0.9	Methanol
	Phytomenadione (243, 261, 270)	0.001	0.42	Trimethylpentane
	Promethazine Hydrochloride (300)	0.001	0.9	0.01N Hydrochloric acid
250	Desipramine Hydrochloride	0.0015	0.41	0.1N Hydrochloric acid
	Imipramine Hydrochloride	0.002	0.53	0.01N Hydrochloric acid
	Papaverine Hydrochloride (284, 310)	0.0005	0.8	N Hydrochloric acid

Table 45 (*continued*)

Wavelength of principal absorption maximum (nm)	Compound	Concentration %	E_1	Solvent
	Phenmetrazine Hydro-chloride (256, 261, 267)	0.05	0.39	*0.01N* Hydrochloric acid
251	Dihydrotachysterol (242, 261)	0.0005	0.5	Methanol
	Ephedrine Hydrochloride (257, 263)	0.05		Water
	Pethidine Hydrochloride (257, 263)	0.05		Water
	Phenylpropanolamine Hydrochloride (257, 262)	0.05	0.37	*0.1N* Hydrochloric acid
	Procyclidine Hydro-chloride (257, 263)	0.08	0.42	Water
	Promazine Hydrochloride (301)	0.001	0.91	Water
	Tricyclanol Chloride (257, 263)	0.1	0.45	Water
252	Diphenoxylate Hydro-chloride (258, 264)	0.05	0.55	Methanol
	Ethopropazine Hydro-chloride (303)	0.0005	0.42	Ethanol (95%)
	Glutethimide (258, 264)	0.025	0.43	Ethanol
	Poldine Methylsulphate (258)	0.05	0.48	Water
	Trimetaphan Camphor-sulphonate (258, 264)	0.05	0.38	Water
253	Diphenyhydramine Hydrochloride (258)	0.025	0.34	*0.01N* Hydrochloric acid
	Perphenazine (305)	0.0008	0.65	*0.1N* Hydrochloric acid
	Pyridoxine Hydrochloride (324)	0.001	0.18	pH 7.0
	Sulphafurazole	0.0005	0.39	*0.01N* Sodium hydroxide
254	Chlorpromazine Hydro-chloride (300)	0.0005	0.46	*0.1N* Hydrochloric acid
	Dextromoramide Tartrate (259, 264)	0.075	0.53	*N* Hydrochloric acid
	Oxyphenbutazone	0.001	0.75	*0.01N* Sodium Hydroxide
256	Benzalkonium Chloride Solution (262, 268)	0.1	0.49	Water
	Benztropine Mesylate	0.05	0.55	Water
	Folic Acid (283, 365)	0.0015	0.82	*0.1N* Sodium hydroxide
	Lucanthone Hydrochloride (330)	0.0005	0.55	*0.1N* Hydrochloric acid

Table 45 (*continued*)

Wavelength of principal absorption maximum (nm)	Compound	Concentration %	E_1	Solvent
	Phenmetrazine Hydro-chloride (250, 261, 267)	0.05	0.55	*0.01N* Hydrochloric acid
	Trifluoperazine Hydro-chloride	0.001	0.63	*0.01N* Hydrochloric acid
257	Chloroquine Phosphate (329, 343)	0.001	0.29	*0.01N* Hydrochloric acid
	Chloroquine Sulphate (329, 343)	0.001	0.39	*0.01N* Hydrochloric acid
	Ephedrine Hydrochloride (251, 263)	0.05		Water
	Hydroxychloroquine Sulphate (329, 343)	0.001	0.39	*0.01N* Hydrochloric acid
	Pethidine Hydrochloride (251, 257)	0.05		Water
	Phenylpropanolamine Hydrochloride (251, 262)	0.05	0.47	*0.1N* Hydrochloric acid
	Procyclidine Hydrochloride (251, 263)	0.08	0.5	Water
	Tolnaftate	0.001	0.36	Methanol
	Tricyclamol Chloride (251, 263)	0.1		Water
258	Benztropine Methane-sulphonate	0.05	0.56	Water
	Cetylpyridinium Chloride	0.005	0.6	Water
	Di-iodohydroxyquinoline (335)	0.0005	0.53	Ethanol/dioxan
	Diloxanide Furoate	0.0007	0.49	Ethanol (95%)
	Diphenhydramine Hydrochloride (253)	0.025	0.40	*0.01N* Hydrochloric acid
	Fluphenazine Hydro-chloride	0.001	0.6	*0.01N* Hydrochloric acid in Methanol
	Glutethimide (252, 264)	0.025	0.46	Ethanol
	Methotrexate (303, 365)	0.001	0.6	*0.1N* Sodium hydroxide
	Orphenadrine Citrate (264, 271)	0.03	0.34	Ethanol (95%)
	Orphenadrine Hydro-chloride (264, 271)	0.03	0.54	Ethanol (95%)
	Poldine Methylsulphate (252)	0.05	0.55	Water
	Prochlorperazine Methanesulphonate	0.0007	0.44	Ethanol/NH$_3$
	Tranylcypromine Sulphate (264, 271)	0.025	0.36	*0.2N* Sulphuric acid
	Trimetaphan Camphor-sulphonate (252, 264)	0.05	0.39	Water

Table 45 (*continued*)

Wavelength of principal absorption maximum (nm)	Compound	Concentration %	E_1	Solvent
259	Dextromoramide Tartrate (254, 264)	0.075	0.58	N Hydrochloric acid
	Ethotoin (265)	0.05	0.55	Ethanol
	Phenindamine tartrate	0.002	0.44	Water
260	Cephalexin	0.002	0.47	Water
	Dapsone	0.0005	0.36	Methanol
	Hydrallazine Hydrochloride (240, 303, 315)	0.001	0.54	Water
	Melphalan	0.0005		Methanol
	Sulphinpyrazone	0.001	0.55	0.01N Sodium Hydroxide
261	Dihydrotachysterol (242, 251)	0.0005	0.38	Methanol
	Phenmetrazine Hydro-chloride (250, 256, 267)	0.05	0.39	0.01N Hydrochloric acid
	Phytomenadione (243, 249, 270)	0.001	0.385	Trimethylpentane
262	Benzalkonium Chloride (256, 268)	0.1	0.59	Water
	Chlorpheniramine Maleate	0.004	0.58	Water
	Pentamidine Isethionate	0.001	0.46	0.01N Hydrochloric acid
	Phenylpropanolamine Hydrochloride (251, 257)	0.05	0.36	0.1N Hydrochloric acid
263	Benzylpenicillin	0.18		Water
	Ephedrine Hydrochloride (251, 257)	0.05		Water
	Pethidine Hydrochloride (251, 257)	0.05		Water
	Procyclidine Hydro-chloride (251, 257)	0.08	0.36	Water
	Tricyclamol Chloride	0.1	0.42	Water
264	Dextromoramide Tartrate (254, 259)	0.075	0.50	N Hydrochloric acid
	Orphenadrine Citrate (258, 271)	0.03	0.36	Ethanol (95%)
	Orphenadrine Hydro-chloride (258, 271)	0.03	0.57	Ethanol (95%)
	Glutethimide (252, 258)	0.025	0.36	Ethanol
	Phenylbutazone	0.0005	0.33	0.01N Sodium hydroxide
	Procyclidine Hydrochloride (257)	0.08	0.37	Water
	Tranylcypromine Sulphate (258, 271)	0.025	0.41	0.2N Sulphuric acid

Table 45 (*continued*)

Wavelength of principal absorption maximum (nm)	Compound	Concentration %	E_1	Solvent
	Trimetaphan Camphor-sulphonate (252, 258)	0.05	0.29	Water
265	Calciferol	0.001	0.46	Cyclohexane
	Chlorpheniramine Maleate	0.002	0.43	*0.1N* Sulphuric acid
	Ethotoin (259)	0.05	0.51	Ethanol
	Isoniazid	0.001	0.42	*0.01N* Hydrochloric acid
	Phanquone (272, 292)	0.001	0.50	*N* Sulphuric acid
	Sodium Aminosalicylate (299)	0.001	0.86	*0.1N* Sodium hydroxide
	Primaquine Phosphate (282)	0.0015	0.5	*0.01N* Hydrochloric acid
266	Nitrofurantoin (367)			Acetate buffer and dimethylformamide
267	Phenmetrazine Hydro-chloride (250, 256, 261)	0.05	0.29	*0.01N* Hydrochloric acid
	Riboflavine			Assay solution sodium bicarbonate
268	Benzalkonium Chloride	0.1	0.46	Water
	Capreomycin Sulphate	0.002	0.6	*0.1N* Hydrochloric acid
	Clioquinol	0.0005	0.51	Hydrochloric acid (dilute)
	Domiphen Bromide (274)	0.02	0.60	Water
	Phenoxymethylpenicillin (274)	0.02		Sodium bicarbonate solution (0.05%)
	Reserpine	0.002	0.54	Ethanol (95%)
	Pyrazinamide (310)	0.001	0.66	Water
	Sulphafurazole (253)	0.001	0.48	*0.01N* Hydrochloric acid
269	Methaqualone	0.0005	0.16	*0.1N* Hydrochloric acid
269–270	Pyridostigmine Bromide	0.0025	0.46	Water
270	Ethacrynic Acid	0.005	0.58	*0.01N* Hydrochloric acid in methanol
	Theophylline	0.001	0.53	*0.1N* Hydrochloric acid
	Thiambutosine	0.0005	0.35	Ethanol (95%)
271	Alprenolol Hydrochloride (277)	0.01	0.65	Ethanol (95%)
	Bupivacaine Hydrochloride	0.04	0.45	*0.01N* Hydrochloric acid
	Frusemide	0.0005	0.30	*0.1N* Sodium hydroxide

Table 45 (*continued*)

Wavelength of principal absorption maximum (nm)	Compound	Concentration %	E_1	Solvent
	Hydrochlorothiazide (318, 274, 320)	0.001	0.66	Ethanol (95%)
	Naphazoline Hydro-chloride (281, 288, 291)	0.002	0.46	*0.01N* Hydrochloric acid
	Naphazoline Nitrate (281, 288, 291)	0.002	0.43	*0.01N* Hydrochloric acid
	Orphenadrine Citrate (258, 264)	0.03	0.24	Ethanol (95%)
	Orphenadrine Hydro-chloride (258, 264)	0.03	0.37	Ethanol (95%)
	Tranylcypromine Sulphate	0.025	0.29	*0.2N* Sulphuric acid
272	Caffeine	0.001	0.47	*0.1N* Hydrochloric acid
	Metaraminol Tartrate	0.01	0.60	Water
	Phanquone (265, 292)	0.001	0.50	*1N* Sulphuric acid
	Phenoxybenzamine Hydrochloride (279)	assay solution	0.68	Chloroform (ethanol free)
	Pyrimethamine	0.001	0.32	*0.005N* Hydro-chloric acid
273	Bendrofluazide	0.0015	0.62	*0.1N* Sodium hydroxide
	Cyclopenthiazide (320)	0.001	0.44	*0.01N* Sodium hydroxide
	Edrophonium Chloride (240, 294)	0.005	0.55	*0.1N* Hydrochloric acid
	Methoserpidine (295)	0.001	0.32	Ethanol
	Phenylephrine Hydrochloride	0.005	0.48	*N* Sulphuric acid
274	Chlortetracycline Hydrochloride	0.001	0.75	*N* Sulphuric acid
	Domiphen Bromide (268)	0.02	0.50	Water
	Hydrochlorothiazide (271, 318, 320)	0.001		*0.002N* Sodium hydroxide
	Hydroflumethiazide (333)	0.001	0.46	*0.01N* Sodium hydroxide
	Hydroxocobalamin (351, 528)			Acetate buffer
	Phenoxymethylpenicillin (268)	0.02		Sodium bicarbonate solution (0.05%)
	Piperocaine Hydrochloride	0.001	0.32	*0.01N* Hydrochloric acid
275	Chlorthalidone (284)	0.01	0.6	Ethanol (95%)
	Choline Theophyllinate	0.001	0.42	*0.01N* Sodium hydroxide
	Dodecyl Gallate	0.001	0.30	Methanol
	Octyl Gallate	0.001	0.38	Methanol

Table 45 (*continued*)

Wavelength of principal absorption maximum (nm)	Compound	Concentration %	E_1	Solvent
	Procainamide Hydrochloride	0.001	0.60	*0.02N* Sodium hydroxide
	Propyl Gallate	0.001	0.49	Methanol
	Theophylline	0.001	0.65	*0.01N* Sodium hydroxide
276	Salbutamol	0.004	0.28	*0.1N* Hydrochloric acid
	Triprolidine Hydrochloride (230)	0.001	0.47	Water
	Alprenolol Hydrochloride (271)	0.01	0.6	Ethanol (95%)
277	Metronidazole	0.001	0.38	*0.1N* Hydrochloric acid
278	Butylated Hydroxytoluene	0.005	0.43	Ethanol
	Chloramphenicol	0.002	0.595	Water
	Dextromethorphan Hydrobromide	0.01	0.54	Water
	Orciprenaline Sulphate	0.0075	0.5	*0.01N* Hydrochloric acid
	Phenindione (330)	0.0004	0.54	*0.1N* Sodium hydroxide
	Phentolamine Mesylate	0.002	0.5	Water
279	Diamorphine Hydrochloride	0.01	0.40	Water
	Iodoxuridine	0.004	0.65	*0.01N* Sodium hydroxide
	Levallorphan Tartrate (240, 299)	0.01	0.47	Water
	Levorphanol Tartrate (240, 299)	0.01	0.46	Water
	Mefenamic Acid	0.002	0.72	*0.01N* Hydrochloric acid in methanol
	Noradrenaline Acid Tartrate	0.005	0.40	*0.1N* Hydrochloric acid
	Phenoxybenzamine Hydrochloride (272)	assay solution	0.54	Chloroform (ethanol free)
280	Adrenaline	0.003	0.45	*0.1N* Hydrochloric acid
	Adrenaline Acid Tartrate	0.006	0.48	*0.1N* Hydrochloric acid
	Benzylpenicillin (263)	0.18	0.18	Water
	Chlorphenesin	0.01	0.67	Water
	Danthron	0.001	0.51	*0.01N* Ethanolic sodium hydroxide
	Ethinyloestradiol	0.01	0.71	Ethanol
	Isoprenaline Sulphate	0.005	0.50	Water

Table 45 (*continued*)

Wavelength of principal absorption maximum (nm)	Compound	Concentration %	E_1	Solvent
	Methicillin Sodium	0.01	0.55	Water
	Methyldopa	0.004	0.46	*0.1N* Hydrochloric acid
	Oestrone	0.01	0.85	Ethanol
	Procainamide Hydro-chloride Injection	0.0005	0.30	Water
	Propanidid	0.005	0.41	Ethanol (95%)
281	Naphazoline Hydro-chloride (271, 288, 291)	0.002	0.53	*0.01N* Hydrochloric acid
	Naphazoline Nitrate (271, 288, 291)	0.002	0.51	*0.01N* Hydrochloric acid
	Propyliodone (239)	0.001	0.26	Ethanol
282	Primaquine Phosphate (265)	0.0015	0.5	*0.1N* Hydrochloric acid
	Propantheline Bromide (246)	0.006	0.37	Methanol
283	Folic Acid (256, 365)	0.0015	0.80	*0.1N* Sodium hydroxide
	Nicoumalone (306)	0.001	0.65	*0.1N* Hydrochloric acid in methanol (90%)
	Pholcodine	0.01	0.38	*0.1N* Sodium hydroxide
	Pholcodine	0.01	0.40	*0.1N* Hydrochloric acid
284	Chlorthalidone (275)	0.01	0.45	Ethanol (95%)
	Codeine Phosphate	0.01	0.39	Water
	Papaverine Hydrochloride	0.0005	0.087	*N* Hydrochloric acid
285	Acetomenaphthone	0.003	0.74	Ethanol
	Morphine Hydrochloride	0.01		Water
	Morphine Sulphate	0.01	0.40	Water
	Nalorphine Hydrobromide	0.01	0.39	Water
286	Cyproheptadine Hydro-chloride	0.0016	0.5	Ethanol (95%)
	Dichlorphenamide (296)	0.005	0.30	*0.1N* Hydrochloric acid
287	Trimethoprim	0.002	0.49	*0.1N* Sodium hydroxide
288	Naphazoline Hydrochloride (271, 281, 291)	0.002	0.37	*0.01N* Hydrochloric acid
	Naphazoline Nitrate (271, 281, 291)	0.002	0.35	*0.01N* Hydrochloric acid
289	Camphor	0.25	0.53	Ethanol (95%)
290	Ethionamide	0.001	0.42	Ethanol (95%)
	Methyprylone	0.25	0.50	Ethanol (50%)

Table 45 (*continued*)

Wavelength of principal absorption maximum (nm)	Compound	Concentration %	E_1	Solvent
	Propranolol Hydrochloride (306, 319)	0.002	0.42	Methanol
	Triprolidine Hydrochloride	0.002	0.6	*0.1N* Sulphuric acid
291	Naphazoline Hydrochloride (271, 281, 288)	0.002	0.37	*0.01N* Hydrochloric acid
	Naphazoline Nitrate (271, 281, 288)	0.002	0.34	*0.01N* Hydrochloric acid
	Noscapine (310)	0.006	0.60	Ethanol (95%)
	Nystatin (305, 319)	0.001		Methanol containing glacial acetic acid (0.05%)
	Pyridoxine Hydrochloride (254, 324)	0.001	0.43	*0.1N* Hydrochloric acid
292	Chlorothiazide	0.001	0.43	*0.01N* Sodium hydroxide
	Phanquone (265, 272)	0.001	0.48	*N* Sulphuric acid
294	Edrophonium Chloride (273, 240)	0.002	0.34	*0.1N* Sodium hydroxide
295	Methoserpidine (273)	0.001	0.20	Ethanol
296	Dichlorphenamide (286)	0.005	0.30	*0.1N* Hydrochloric acid
298	Morphine Sulphate	0.01	0.70	*0.1N* Sodium hydroxide
	Morphine Hydrobromide	0.01	0.6	*0.1N* Sodium
	Nalorphine Hydrobromide	0.01	0.6	*0.1N* Sodium hydroxide
299	Levallorphan Tartrate (240, 279)	0.005	0.35	*0.1N* Sodium hydroxide
	Sodium Aminosalicylate (265)	0.001		*0.1N* Sodium hydroxide
300	Chlorpromazine Hydrochloride (254)	0.0005		*0.1N* Hydrochloric acid
	Promethazine Hydrochloride (249)	0.001		*0.01N* Hydrochloric acid
301	Promazine Hydrochloride (251)	0.001		Water
303	Ethopropazine Hydrochloride	0.0005		Ethanol (95%)
	Hydrallazine Hydrochloride (240, 260, 315)	0.001	0.27	Water
	Methotrexate (258, 365)	0.001	0.55	*0.1N* Sodium hydroxide
304	Dichlorophen (245)	0.001	0.27	*0.1N* Sodium hydroxide

Table 45 (*continued*)

Wavelength of principal absorption maximum (nm)	Compound	Concentration %	E_1	Solvent
305	Nystatin (291, 319)	0.001		Methanol containing glacial acetic acid (0.05%)
	Perphenazine (254)	0.0008		*0.1N* Hydrochloric acid
	Tripelennamine Hydrochloride (245)	0.001		Water
306	Methyl Salicylate (238)	0.001	0.28	Ethanol (95%)
	Nicoumalone (283)	0.001	0.5	*0.1N* Hydrochloric acid in methanol (90%)
	Propranolol Hydrochloride (290, 319)	0.002	0.25	Methanol
	Solapsone	0.001	0.35	Water
307	Chlorotrianisene (247)	0.001	0.21	Ethanol (95%)
	Novobiocin Calcium Novobiocin Sodium	0.001		Methanol containing potassium hydroxide
308	Chlordiazepoxide Hydrochloride	0.0015	0.44	*0.1N* Hydrochloric acid
310	Noscapine (291)	0.006	0.72	Ethanol (95%)
	Papaverine Hydrochloride (250, 284)	0.0005	0.114	*N* Hydrochloric acid
	Pyrazinamide (268)	0.001	0.055	Water
312	Ergometrine Maleate	0.0035	0.64	Water
315	Hydrallazine Hydrochloride (240, 260, 303)	0.001	0.21	Water
319	Cinchocaine Hydrochloride (247)	0.001	0.24	*N* Hydrochloric acid
	Nystatin (291, 305)	0.001		Methanol containing glacial acetic acid (0.05%)
	Liothyronine Sodium	0.01	0.65	*0.1N* Sodium hydroxide
	Propranolol Hydrochloride (290, 306)	0.002	0.15	Methanol
320	Cyclopenthiazide (273)	0.001	0.06	*0.01N* Sodium hydroxide
	Hydrochlorothiazide (273, 323)	0.001	0.1	*0.01N* Sodium hydroxide
324	Pyridoxine Hydrochloride (291, 254)	0.001	0.35	pH 7.0
325	Thyroxine Sodium	0.01	0.76	*0.1N* Sodium hydroxide
	All-*trans* Vitamin A Acetate (328, 326)	0.0003	0.4575	Isopropanol

Table 45 (*continued*)

Wavelength of principal absorption maximum (nm)	Compound	Concentration %	E_1	Solvent
326	All-*trans* Vitamin A Acetate (328, 325)	0.0003	0.4635	Ethanol
327	Dequalinium Acetate (240)	0.001	0.48	Water
	Dequalinium Chloride (240)	0.001	0.51	Water
	Phytomenadione (243, 249, 261, 270)	0.01	0.69	Trimethylpentane
328	All-*trans* Vitamin A Acetate (325, 326)	0.0003	0.4545	Cyclohexane
329	Bendrofluazide (273)	0.0015	0.12	*0.01N* Sodium hydroxide
	Chloroquine Phosphate (257, 343)	0.001	0.32	*0.01N* Hydrochloric acid
	Chloroquine Sulphate (257, 343)	0.001	0.44	*0.01N* Hydrochloric acid
	Hydroxychloroquine Sulphate (257, 343)		0.43	*0.01N* Hydrochloric acid
330	Lucanthone Hydrochloride (256)	0.0005	0.11	*0.1N* Hydrochloric acid
	Phenindione (278)	0.0004	0.16	*0.1N* Sodium hydroxide
333	Hydroflumethiazide (274)	0.001	0.095	*0.01N* Sodium hydroxide
335	Di-iodohydroxy-quinoline (259)	0.0005		Ethanol + 5% dioxan
343	Amodiaquine Hydrochloride	0.001	0.37	*0.1N* Hydrochloric acid
	Chloroquine Phosphate (257, 329)	0.001	0.37	*0.01N* Hydrochloric acid
	Chloroquine Sulphate (257, 329)	0.001	0.46	*0.01N* Hydrochloric acid
	Hydroxychloroquine Sulphate (257, 329)		0.47	*0.01N* Hydrochloric acid
349	Doxycycline Hydrochloride	0.001	0.30	*0.01 N* Hydrochloric acid in methanol
350	Colchicine	0.001	0.42	Ethanol
	Mefenamic Acid (279)	0.002	0.58	*0.01N* Hydrochloric acid in methanol
351	Hydroxocobalamin (274, 528)			Acetate buffer
353	Oxytetracycline Dihydrate	0.002	0.56	Solution pH 2.0
	Oxytetracycline Hydrochloride	0.002	0.56	Solution pH 2.0
354	Dithranol	0.001	0.44	Chloroform
358	Vibrynium Embonate (508)	0.001	0.38	Methoxy-ethanol
360	Triamterene	0.0005	0.42	Acetic acid 10%

Table 45 (*continued*)

Wavelength of principal absorption maximum (nm)	Compound	Concentration %	E_1	Solvent
365	Folic Acid (256, 283)	0.0015	0.28	*0.1N* Sodium hydroxide
	Methotrexate (258, 303)	0.001	0.2	*0.1N* Sodium hydroxide
367	Nitrofurantoin (266)		0.76	Acetate buffer and dimethylformamide
380	Tetracycline Hydrochloride	0.001	0.376	
385	Demethylchlortetracycline Hydrochloride	0.001	0.357	
508	Vibrynium Embonate (358)	0.001	0.75	Methoxyethanol

References

Akagi, M., Oketani, Y. and Takada, M., *Chem. Pharm. Bull. Japan.* (1963) **11**, 62.
Ashton, G. C. and Tootill, J. P. R., *Analyst* (1956) **81**, 225, 232.
Bladon, P., Henbest, H. B. and Wood, G. W., *J. chem. Soc.* (1952), 2737.
Bogan, J. and Smith, H., *J. Forens. Sci. Soc.* (1967) **7**, 37.
Broughton, P. M. G., *Biochem. J.* (1956) **63**, 207.
Caddy, B., Fish, F., Mullen, P. W. and Tranter, J. *J. Forens. Sci. Soc.* (1973) **13**, 127.
Dill, W. A. and Glazko, A. J., *J. biol. Chem.* (1949) **179**, 395.
Glenn, A. L., *J. Pharm. Pharmac.* (1963) **15**, *Suppl.* 123T.
Goldbaum, L. R., *Anal. Chem.* (1952) **24**, 1604.
Goldbaum, L. R., Williams, M. A. and Koppanyi, T., *Anal. Chem.* (1960) **32**, 81. *See also:* Goldbaum, L. R., Williams, M. A. and Johnston, E. H., *J. Forens. Sci.* (1962) **7**, 499.
Groden, B. M. and Williams, W. D., *Postgrad. Med. Journal* (1964) **40**, 28.
Stone, H. M. and Henwood, C. R., *J. Forens. Sci. Soc.* (1967) **7**, 51.
Wallace, J. E., *J. Forens. Sci.* (1966) **11**, 552; *Anal. Chem.* (1968) **40**, 978.
Wahbi, A. M. and Abdine, H., *J. Pharm. Pharmac.* (1973) **25**, 69.
Wahbi, A. M. and Ebel, S., *J. Pharm. Pharmac.* (1974) **26**, 317.
Woodward, R. B., *J. Am. chem. Soc.* (1941) **63**, 1123; *J. Am. chem. Soc.* (1942) **64**, 72, 76.

9 Flame Photometry and Atomic Absorption Spectrophotometry

W. D. WILLIAMS

FLAME PHOTOMETRY

Emission Spectra

When a solid body is heated to incandescence, radiation is emitted as a continuous spectrum, i.e. one which is uniform and without the appearance of bright or dark lines. On the other hand, excitation of gaseous or vaporised material (i.e. in the atomic or molecular form) causes the emission of band or line spectra at specific wavelengths, characteristic of the material. The energy necessary to excite the emission of spectral lines varies from element to element, and although, theoretically, it is possible to detect all elements in this way, the method is usually limited in practice to the detection and determination of metals, metalloids and certain non-metals, such as silicon.

The energy available from a coal gas-air flame, which reaches a temperature of the order of 1800 to 2000°K, is able to excite only a very limited number of elements to a level where emission will occur. At higher temperatures, in the oxy-acetylene flame ($T = 2800°K$) and in the direct current arc ($T = 4000$ to $8000°K$), the number of elements which can be detected increases rapidly with temperature, so that whereas only 30 to 40 emit in the oxy-acetylene flame more than 70 can be detected in the direct current arc. Sensitivity, however, is not related solely to temperature; thus the condensed spark which can reach extremely high effective temperatures for very short times is considerably less sensitive than the arc in detecting small amounts of various elements.

For quantitative work with the direct current arc, the intensity of the emitted radiation may be measured by means of photomultiplier units attached directly to the spectrograph, or it may be determined by photographic photometry. Variation in the position and intensity of the arc itself, and its relatively short duration, make it unsuitable for use in combination with normal photoelectric spectrophotometers. Similar objections do not apply to the flame method which uses a steady source. Despite the fact that it cannot detect such small amounts as the arc method, it is easy to use since flame temperatures are lower and more readily controlled. The high sensitivity of spectrophotometric instruments permits a satisfactory standard of accuracy in the determination of those elements for which the flame method is suitable.

The characteristic radiation of the element being determined is isolated by the monochromator of the spectrophotometer. However, when the spectrum consists of a small number of lines then it may be possible to isolate the required light by means of filters alone. Sodium light may be effectively isolated from that of potassium in this way and vice versa. For calcium light in the presence of sodium light, however, a monochromator is preferable although an efficient filter based on interference principles may be used. For such elements as Li, Na, K and Ca, therefore, it is possible to use less elaborate apparatus and in a commercial instrument, the EEL Photoelectric Flame Photometer. Filters are used together with a barrier layer photocell to measure the intensity of the emitted light.

To excite the radiation, a solution of a salt of the element is subjected to the action of an air-acetylene, or air-coal gas flame in a burner which incorporates a nebuliser unit. The solution, in the form of a fine spray, is thus injected continually into the flame at a rate of about 1 to 2 ml per minute. The effect of other ions, or neutral substances in the solution, on the emission spectrum may be considerable; ethanol increases the intensity whereas dextrose, citrates, sulphates and phosphates decrease it. These effects are minimised by preparing standard solutions which are similar in composition to that undergoing test, and also by dilution.

A number of examples of the application of flame photometry are given in official compendia and include limit tests for sodium (Na^+) in certain potassium salts and Magnesium Chloride, in which K^+ and Ca^{2+} are also subject to control. It is also used to determine K^+ and Na^+ in Haemodialysis Solutions and Na^+ in Intraperitoneal Solutions.

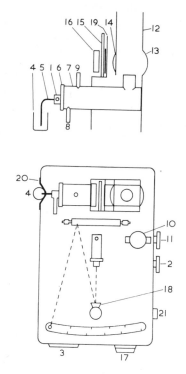

Fig. 164. Diagram of the components of the EEL flame photometer 1, atomiser; 2, control valve; 3, pressure gauge; 4, beaker; 5, stainless steel capillary tube; 6, ebonite plug; 7, mixing chamber; 8, draining tube; 9, gas inlet tube; 10, pressure stabilizer; 11, control valve; 12, chimney; 13, reflector; 14, lens; 15, optical filters; 16, photocell; 17, potentiometer; 18, taut-suspension galvanometer unit; 19, window; 20, plate; 21, zero control

The EEL Flame Photometer

The simple exercises described below may be carried out on the EEL Flame Photometer, a description of this instrument is given here because it illustrates the theoretical principles which have been given earlier. A diagrammatic view of the components is shown in Fig. 164.

Compressed air is supplied to a small annular-type atomiser (1) through a control valve (2) at a pressure of 12 lb per sq inch as indicated on the gauge (3) mounted on the front of the instrument. The flow of air through the atomiser draws the sample from the beaker (4) up the stainless steel capillary tube (5) and sprays it as a fine mist through the ebonite plug (6) into the mixing chamber (7). Here the large droplets fall out and flow to waste through the drain tube (8). Gas is introduced into the mixing chamber through the inlet tube (9) from the internal gas pressure stabiliser (10) and control valve (11). The gas-air mixture passes to a multi-jet burner mounted above the mixing chamber where it burns as a broad flat flame, and the hot gases pass up the well ventilated chimney (12). The light emitted by the flame is collected by a reflector (13) and focused by a lens (14) through the interchangeable optical filters (15) onto an 'EEL' barrier-layer photocell (16). The current generated by this cell is taken through a potentiometer (17) to a Tinsley taut-suspension galvanometer unit (18). A glass window (19) is interposed between the lens and filter for cooling purposes.

Operation. With the flame burning at constant air and gas pressures, and with the appropriate filter in position, the strongest standard solution is sprayed by moving the beaker containing the solution up the recessed plate (20) on the side of the instrument. This automatically places it in position relative to the capillary tube up which the liquid is then drawn. The sensitivity control is adjusted to give a full scale reading. Water is now sprayed and the zero control (21) used to zero the instrument. The spraying of standard and water and the adjustments are repeated until the full scale deflection corresponds to the strongest standard solution, for example to 10 ppm of K^+. The intermediate dilutions are now sprayed in turn and the readings are noted to prepare a calibration curve. The use of this instrument in the determination of concentrations is described in Experiment 1.

Practical Exercises

Experiment 1. *Determination of the concentration of* K^+ *in a dilute aqueous solution of potassium chloride*

Calibration Curve

Method. Prepare a stock solution of potassium chloride (AnalaR) by dissolving the salt (0.477 g) in sufficient water to produce 500 ml. The concentration of K^+ is 500 ppm. Dilute aliquot portions of this stock solution with water to obtain standard solutions containing 0.2, 0.4, 0.6. 0.8 and 1.0 mg of K^+ per 100 ml. This series of solutions should be used to prepare the calibration curve.

Solution to be examined. If the concentration of K^+ is completely unknown, it is possible to avoid the preparation of a large number of dilutions by making use of the potentiometer sensitivity control. This is calibrated with a relative concentration scale which may be used in the following way to determine the approximate dilution required.

Method. Spray a standard solution, say the 6 ppm standard, with the potentiometer set at full sensitivity, i.e. 1 on the dial. Note the scale reading. Reduce the sensitivity to zero and readjust while spraying the solution to be examined until the same scale reading is obtained. The reading of the potentiometer will then give the *approximate* dilution required.

Prepare and spray the dilution after standardising the instrument for full scale deflection with the strongest standard solution in the normal way. Note the reading and calculate the concentration of K^+ by means of the calibration curve.

Cognate Determinations

Haemodialysis Solutions. Determination of Na^+ and K^+.
Intraperitoneal Dialysis Solutions. Determination of Na^+.

Experiment 2. *To show the effect of (a) ethanol and (b) dextrose on the intensity of the light emitted.*

Method. (*a*) Dilute an aliquot portion of the stock solution of potassium chloride with ethanol (10 % v/v) to give a solution containing 6 ppm of K^+, i.e. 0.6 mg K^+ per cent. Compare the reading obtained when spraying this solution with that given by the standard solution of the same strength. When setting the instrument to zero, spray ethanol (10 % v/v) to overcome the objection that the increase in the reading may be due to traces of K^+ in the ethanol. (*b*) Repeat the above, but use a solution of dextrose (10 %) in water in place of the ethanol (10 % v/v).

The concentrations of ethanol and dextrose in the above solutions are grossly in excess of those which would be encountered in practice, but the exercise illustrates the importance of including in the standard solutions the same substances that are present in the solutions under examination. The concentration of these other substances should be of the same order in standard and test solutions.

Interference effects may be roughly classified as arising from the influence of other cations, anions or neutral substances. Cationic interference such as that of Na^+ in Ca^{2+} determinations may be largely eliminated by efficient monochromators. Anionic interference results in a reduction of the intensity of the radiation from the element being determined, the reduction being greater for some elements than for others, e.g. the emission from potassium is influenced to a greater extent than that for sodium. For each anion, there is a limiting concentration below which little interference occurs and very dilute solutions are, therefore, preferable in flame photometry.

Neutral substances may increase or decrease the emission and it is difficult to forecast what the effect of a particular substance will be.

Experiment 3. *Determination of Ca^{2+} in Magnesium Chloride*
The limit for Ca^{2+} is 0.01 % whereas that for Na^+ is 0.5 %. There is, therefore, considerable risk of error by interference from sodium emission unless selection of the Ca^{2+} emission line at 423 nm is possible. This requires a suitable monochromator and a more elaborate instrument than the EEL flame photometer. It is instructive, however, to use the emission at 650 nm isolated by means of an interference filter, and to investigate the effect of Na^+ on the apparent Ca^{2+} content when added to standard solutions of calcium. The method suggested in the British Pharmacopoeia is Method II, which allows for the presence

of the sample by adding standard Ca^{2+} solutions to the sample. It does not, however, allow for interference from sodium which is the reason for selecting 423 nm if possible.

Method. Add the sample (5 g) to each of four 50 ml volumetric flasks A, B, C, D and add water to volume to flask A. To flasks B, C and D add respectively 0.5, 1.5 and 2.5 ml of calcium solution FP (400 ppm) and dilute to volume with water. The concentration of added Ca^{2+} in flasks B, C and D is therefore 4, 12 and 20 ppm. Use the 20 ppm solution to set the instrument and spray each solution 3 times, with a water wash between each and plot the average of each 3 readings against concentration of added standard. Extrapolate the line joining the points to cut the concentration axis. Read off the concentration of the sample solution and calculate the amount of Ca^{2+} in the sample.

ATOMIC ABSORPTION SPECTROPHOTOMETRY

Flame photometry has met with considerable success in analysis, but, even so, there are limitations, e.g. in the number of elements detectable and also in the number of atoms of an element excited by the flame. In any population of atoms at a temperature T, the ratio:

$$\frac{\text{Number of atoms in first excited state}}{\text{Number of atoms in ground state}} = \frac{N_2}{N_1} = \frac{P_2}{P_1} e^{-(E_2 - E_1)/kT}$$

where P_2 and P_1 are the statistical weights of the respective states and are related to the total quantum number J of the atom by $P = 2(J + 1)$, $(E_2 - E_1)$ is the excitation energy, k is the Boltzmann constant. The ratio for zinc in a flame at $3100°K$ is 1.13×10^{-9} and most of the atoms in the flame are, therefore, in the ground state.

Atomic absorption spectroscopy uses the flame, into which the solution is sprayed, as if it were a cell containing an absorbing solution, the light source being a hollow cathode lamp of the *element being determined*. The intensity of a selected emission line is measured before and during the spraying of the solution in the flame to give an extinction as in spectrophotometry (p. 236). The flame will, of course, emit some light of the same wavelength as that of the source, but even if this interference is quite large, it is eliminated by modulation of the light source and incorporation of a suitably-tuned amplifier to respond only to the modulation.

The absorption technique is useful when one element is being determined in the presence of a large amount of another which interferes in flame photometry. If interference is due to actual chemical combination, however, the absorption technique will be affected also. Examples of elements which are difficult to determine by flame photometry, but readily so by atomic absorption spectroscopy are magnesium and zinc (David, 1960; Menzies, 1960).

Apparatus

Figure 165 shows the basic design of the EEL 140 atomic absorption spectrophotometer.

The sequence of events in the determination of an element is as follows. Sample solution is drawn through the atomiser (1) into the mixing chamber (2)

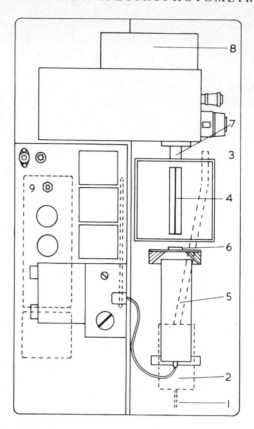

Fig. 165. EEL Atomic Absorption Spectrophotometer

where it mixes with the fuel and the support gas, either air or nitrous oxide. Large droplets run to waste through the drain (3) while the rest of the mixture passes to the burner (4), where the sample's chemical bonds are dissociated.

Emission from the lamp (5) which has a cathode made wholly or partly of the element to be determined, is modulated at 50 Hz from the line supply. The beam is focused by the lens (6), on the centre of the flame. Absorption occurs. The transmitted remnant of the beam is focused on to the inlet slit of the Czerny–Turner monochromator (7). The isolated resonance line emerges from the exit slit to energise the photomultiplier (8), the output of which is then amplified.

The amplifier (9) electronically switched in synchronisation with the lamp modulation, acts as a phase-sensitive detector discriminating against all signals originating outside the lamp. The amplified signal therefore represents only the intensity of the resonance line transmission, and nothing else.

Light Sources. Conventional sources such as tungsten lamps and deuterium or xenon discharge lamps are not satisfactory for atomic absorption because the band-width of the radiation emerging from the monochromator is very broad by comparison with the absorption band of the ground state atom.

The absorption line width for the latter may be from 0.001 to 0.01 nm by comparison with a band-width of 1 nm from the monochromator of a conventional spectrophotometer. The light source must supply radiation of a line width less than that of the absorption line width of the element being determined. This is achieved by vapour discharge lamps for certain easily excited elements, for example, sodium and potassium. Normally, however, hollow cathode lamps of the appropriate element are used. When a current flows between the anode and cathode in these lamps, metal atoms are sputtered from the cathode cup, and collisions occur with the filler gas. A number of the metal atoms become excited and give off their characteristic radiation. The choice of the filler gas depends on its emitted spectrum compared with the spectrum lines of the element of interest and very often either argon or neon is used.

Most commercial types of hollow cathode lamps are low power devices. This is necessary to prevent the heat developed by the discharge from melting the cathode filling, or the filling being lost due to evaporation. Examples are lead, zinc and bismuth, which are highly volatile, and can be run only at low excitation energies, although alloys of elements with low melting points can be run at considerably higher currents than those lamps with pure cathodes. Another reason for running hollow cathode devices at low currents is their tendency to self-absorb the radiation emitted, thereby giving lower sensitivity than could otherwise be obtained without self-absorption. This effect is clearly demonstrated on lamps for the alkali metals and zinc.

It is possible to obtain multi-element lamps but care is needed in their use as interference from adjacent lines of the element not being determined could give rise to non-linear calibration curves or spectral interference if the elements are also present in solution. Two-element lamps that are useful, however, are Ca/Mg, Cu/Zn and Na/K, and a 3-element lamp is obtainable for Ca/Mg/Zn.

Present day lamps are more stable and of longer life than those used previously so that although a large initial capital outlay is required for a selection of lamps a useful life of several years may be expected.

The Flame. The optimum conditions for the flame depends upon the physical dimensions of the burner, its position relative to the beam from the light source, temperature and fuel/air supplies. A long path length is desirable for adequate sensitivity and typical dimensions of the burner slot are 10 cm × 0.4 mm.

An air/acetylene mixture gives a temperature of about 2300° which is satisfactory for most elements. The temperature of the flame using nitrous oxide/acetylene is about 3000° and this is required for refractory elements, e.g. aluminium. Special precautions are, however, required in using this last mixture.

The ultimate sensitivity attained with the flame also depends upon the efficiency of the nebuliser unit for aspiration of the sample.

Flameless units. The flame consumes up to 1 ml or more per minute and sensitivity may be materially increased if a solid sample can be ignited. Developments in this field include sample boats for evaporation of solution and insertion into the flame; graphite furnaces with temperatures up to 3000° into which a sample is injected and a tantalum ribbon onto which a sample is deposited for ignition. The volumes of solution required for the last two systems are about 100 and 50 μl respectively and a considerable increase in sensitivity is obtained in spite of the small volumes.

Cold vapour systems for particular elements are possible and that for Hg is described fully in the practical section.

The Monochromator. The radiation from the source consists of the line spectrum of the element forming the cathode, along with emission from the gas in the lamp. Isolation of the required radiation by grating monochromators is, therefore, fairly readily accomplished. However, in view of the presence of other elements in sample solutions, monochromators giving suitable dispersion are required. Dispersion will vary according to the quality of the monochromator and ranges from 1 nm/mm to 10 nm/mm. Clearly, for resolution of closely spaced lines a good monochromator is essential but for many of the routine determinations less expensive systems operate quite well.

The wavelength scale associated with the monochromator need not be calibrated in subdivisions of nm units as it is always necessary to adjust the setting manually in conjunction with the response of the galvanometer as described below.

Situation. It is unlikely that toxic levels of metals will occur in the atmosphere on spraying solutions but it is a useful precaution to have some means of drawing off possibly high local concentrations of for example lead and mercury in the vapour above the flame.

Interferences

Atomic absorption is less subject to the physical factors which affect flame photometry. These include decrease in intensity of emission by energy transfer in the excited state, and interference by radiation from elements present in solution, as in the determination of calcium ions in the presence of sodium ions. Scatter of radiation in the 220 nm region of the spectrum may, however, cause serious errors in absorbance under certain circumstances as described for the determination of lead in Calcium Carbonate.

Chemical Effects. These effects are operative in atomic absorption as well as in flame photometry and may be such as to decrease or increase the absorbance value. Ionisation is one such cause as the ion formed does not absorb radiation of the same wavelength as that for the ground state atom. Barium has two useful absorbing wavelengths, viz. 554 nm for the resonance radiation and 455 nm for the ion line. It is essential for reliable results using the 554 nm radiation that the ionisation effect in the flame be reduced by the addition of a salt of an easily ionised metal, in this case, potassium chloride at 2000 ppm. The response is then due to the ground state barium atom. A similar effect, i.e. reduced response, is observed with calcium if a nitrous oxide/acetylene flame is used. The addition of sodium or potassium chloride is required in order to overcome it.

Chemical combination with anions in solution may give rise to stable compounds which are less dissociated in the flame than, for example, a halide salt. The absorbance of a calcium chloride solution, for example, is reduced in the presence of SO_4^{2-}, PO_4^{3-} and SiO_3^{2-}. Such interference can often be overcome by the addition of a complexing agent, e.g. disodium edetate, or a releasing agent, i.e. a cation which forms a more stable salt with the interfering anion. In the determination of calcium, strontium or lanthanum chloride may be used.

Interference by cations may also occur, e.g. iron interferes in the determination of chromium. The addition of lanthanum chloride is useful in this example also.

Setting-up procedure

The operation of the spectrophotometer will vary from one instrument to another but there are common features that occur in operation.

Fuel and air supplies. The pressures are critical for optimum performance and once the air pressure is fixed the pressure of the fuel supply can be adjusted to give maximum absorbance when spraying a standard solution into the flame.

Selection of emission line. The light source will have several emission lines but one will usually show a much greater sensitivity to the element than the others even although the intensity of emission may be less. For example, the lead line at 217 nm is more sensitive than that at 283.3 nm although the latter is about 10 times more intense. On the other hand the most useful line for copper at 324.8 nm is more intense than the other possible ones in the spectrum. Generally, it is the most sensitive line that is chosen except where other factors must be considered, e.g. in the determination of lead excessive scatter at 217 nm could well make the selection of 283.3 nm radiation preferable if reasonable levels of lead are present (10–75 ppm).

Difficulty sometimes occurs, particularly with relatively simple monochromators, where the wavelength indicator may be one or two nanometers different from the true value. The monochromator should be carefully adjusted to give maximum transmission in the region of the known wavelength of the emission line. Immediately check that absorbance occurs when a solution of the particular element is sprayed. In a particular example the wavelength indicator read 1.5 nm low and it was quite easy to confuse the lead ion line at 220 nm with the lead line at 217 nm with the result that no absorption occurred when standard lead solutions were sprayed.

Spraying technique. A standard method should be adopted to ensure reproducibility of results and reduce contamination and corrosion of the nebuliser and burner. It is usual to take three readings for each solution, with a spray of water between each, and to average the readings for plotting or calculation purposes. Immediately the determination is completed 20 to 30 ml of water should be sprayed through the nebuliser and burner unit to sweep out all traces of previous solutions. Normally, solutions are very dilute but in the determination of trace impurities they may contain large amounts of corrosive salts.

Standard Solutions and Calibration Curves. Method 1. The concentrations for standard solutions vary according to the instrument itself and must, therefore, be in accord with the instrument's capabilities. For example, in the determination of zinc using the EEL 140, the concentration range is 0–5 ppm whereas the EEL 240 requires 0–2 ppm and the Perkin-Elmer 290 B 0–1 ppm.

The calibration curve may be obtained in the same way as for spectrophotometric assays, viz. by setting the instrument to zero and infinite absorbance and reading the absorbance for standard solutions (and samples) directly. Alternatively, the strongest standard solution may be sprayed and the instrument set to a convenient absorbance value, e.g. 1.0 if possible. The remaining solutions are sprayed and the readings noted, much in the same way as described under flame photometry. The advantage of this method is that future assays may be done using one standard solution only which is set to the known reading for that standard. All readings then fall on the previous calibration curve.

Method 2, also described in the Pharmacopoeia, is one which avoids the effects of the sample itself on the emission or absorbance of radiation (matrix effect). Known amounts of standard solution are added to constant aliquots of the sample solution and diluted to volume. The readings obtained are plotted in the normal way. Extrapolation of the curve cuts the extrapolated concentration axis at the concentration of the sample solution. This method does not solve all the problems in that any slight curvature of the line joining the points makes extrapolation less reliable. Further, any factor which gives rise to irrelevant absorption, e.g. molecular absorption or light scattering, still operates and high results are obtained (see determination of Pb in Calcium Carbonate below).

Practical Experiments

Experiment 1. *Determination of* Zn^{2+} *in Globin Zinc Insulin Injection (40 units per ml)*

The determination of zinc in solution forms part of the general quality control of the injection. The content of Zn ions is stated to be 0.3–0.35 mg per 100 units of insulin.

Stock Solution of Zinc (5000 ppm). Dissolve zinc metal (AnalaR) in $5N$ hydrochloric acid (20 ml) and dilute to 500 ml with water.

Method. Dilute the injection (5 ml, 200 units) to 200 ml with water. Spray the solution using the standard procedure and read off the concentration of Zn^{2+} from a calibration curve prepared with solutions containing 0.5, 1, 2 and 3 ppm of Zn^{2+}.

Experiment 2. *Determination of Lead in Calcium Carbonate*

There are many examples in the British Pharmacopoeia of chemicals which require long, tedious methods for the determination of traces of lead. For such compounds atomic absorption may offer a considerable advantage in time if they can be brought into solution quickly and easily.

Formerly Calcium Carbonate was in this class of compound and although the method described in the European Pharmacopoeia is now very simple the application of atomic absorption to the determination of traces of Pb is described to illustrate several useful points. These are (*a*) effect of strong solutions on absorbance, (*b*) scale expansion and (*c*) a correction procedure sometimes applicable to solutions which show marked scattering of radiation.

The limit for Pb in Calcium Carbonate was formerly 10 ppm and the solution of sample must necessarily contain a fairly high concentration of salt. This leads to difficulties such as matrix effects, weakening of radiation and low intensity of absorption by the trace element being determined. The last can be overcome to a certain extent by a scale expansion accessory which is a valuable addition to an atomic absorption spectrophotometer. The effect of the presence of calcium chloride is to cause scatter and molecular absorption, both of which can be allowed for by making use of the lead ion line at 220 nm. The principle underlying the determination is therefore to obtain a reading at 217 nm (lead + other effects) and at 220 nm (other effects only). The reading at 220 nm is used to correct the absorption at 217 nm and the difference is related to the lead content of the sample. This convenient correction method for Pb is based on the reasonable assumption that scatter and molecular absorption are similar at the two *adjacent* wavelengths. It would be quite wrong to use radiation, if any were present, situated 10–20 nm from the line of interest because scatter of radiation decreases with increase in wavelength.

A correction can, of course, be applied if a sample of Calcium Carbonate of known lead content is available for comparison. More recently Clay (1973) has suggested the use of a deuterium hollow cathode lamp as a continuum source for background correction.

Stock Solution of Lead. Dissolve lead nitrate (3.995 g) in water to produce 500 ml of solution.

Method. To Calcium Carbonate (10 g) add water (20 ml) and hydrochloric acid (22 ml) *slowly* (Note 1). Boil to remove carbon dioxide, cool and dilute to 50 ml with water. Spray the solution adopting the standard procedure and record the response at 217 nm and 220 nm using scale expansion (Note 2). Measure the difference between the readings and calculate the amount of lead in the sample by reference to a calibration curve made at the same time with standard lead solutions containing 0.5, 1.0, 1.5 and 2.0 ppm Pb (Note 3).

Note 1. Care should be taken to avoid loss of sample by the vigorous effervescence which occurs.

Note 2. The scale expansion should be such that the response for the sample solution should be almost full scale. It follows that the response at 220 nm will be less and, therefore, on scale. It is also important, to illustrate underlying theory, to add some lead solution to the sample solution after completion of the exercise and re-determine the absorption at both wavelengths. The difference between the readings should be greater than that found in the exercise but the response at 220 nm should be much the same, indicating that the ground state atom does not absorb the radiation characteristic of the lead ion.

Note 3. These values were suitable for a relatively simple instrument and should be well within the capabilities of most atomic absorption spectrophotometers. The volume of hydrochloric acid used in this determination is quite large and allowance should be made in the final calculation for any Pb that may be present in the acid.

Cognate Determination

Dextrose. Determination of lead. Use 20% solutions and Method 2, i.e. the method of standard addition. Note that irrelevant absorption is absent when spraying this solution, in marked contrast to that shown by the calcium solution.

Experiment 3. *Determination of the active components in a soothing skin cream* The formula is stated to be:

Chlorcyclizine Hydrochloride	2%
Calamine	8%
Water-miscible base to	100%

Probably the simplest method for Calamine is ashing and conversion of the percentage of zinc oxide obtained to terms of Calamine. However, any other metal salts in the water-miscible base would give rise to error and only the analyst in the manufacturer's laboratory would know of this. A method specific for zinc is, therefore, required.

Moody and Taylor (1972) have examined a large number of official and proprietary preparations of zinc, including oils and creams, by atomic absorption. The skin cream, however, is more complicated than the oily preparations and a method is also required for the determination of the chlorcyclizine hydrochloride. It is convenient to determine both components on a single weighing.

Method. Chlorcyclizine Hydrochloride. Accurately weigh the sample (about 1 g) into a small beaker, add N NaOH (15 ml), warm to disperse the cream and to dissolve the calamine. Transfer the turbid mixture to a small separator with the aid of N NaOH (5 ml) and extract the chlorcyclizine base with chloroform (30, 20, 20, 20 ml) (Note 1) washing each chloroform extract with the same volume of water (10 ml). Bulk the chloroform extracts and reserve the alkaline aqueous layers for the determination of zinc.

Extract the chloroform layer with $0.1N$ H_2SO_4 (20, 15 10 ml), bulk the acid extracts and warm to remove traces of chloroform. Cool and make up to 100 ml with $0.1N$ H_2SO_4 in a volumetric flask. Record the absorption curve over the range 200–250 nm in 2 cm cells using $0.1N$ H_2SO_4 as blank and a standard solution of chlorcyclizine hydrochloride (0.02%) for comparison *on the same chart.* Calculate the concentration of chlorcyclizine hydrochloride in the cream.

Calamine. Bulk the reserved aqueous liquors and wash through the separators with hydrochloric acid (10 ml) and water (10 ml), adding the washings to the bulked aqueous liquid. Warm to remove chloroform globules, cool and dilute to 100 ml with water. Dilute the solution (10 ml) to 1000 ml with water and determine the zinc by comparison with standard zinc solutions containing 1, 2, 3, 4 and 5 ppm Zn^{2+} at 214 nm. (Note 2).

Calculate the concentration of Zn^{2+} in the sample and convert to % calamine by use of the factor 1.75 (Note 3).

Note 1. Ether gives low results in the determination.

Note 2. It is convenient to adjust the reading obtained with the 5 ppm solution to 0.5.

Note 3. The factor is empirical in that some ferric oxide is present in Calamine and is based on a residue on ignition of 71%. Alternatively Calamine may be used as standard but this offers no real advantage to an analyst outside the manufacturer's laboratory.

Experiment 4. *Determination of Thiomersal (0.002%) as mercury in a solution for contact lenses*

The direct examination of the solution by atomic absorption is complicated by (*a*) the relatively low sensitivity of the method for mercury when using a flame, (*b*) the different response given by Hg^+ and Hg^{2+} and (*c*) the organic combination of mercury in the bacteriostat (Thiomersal). A considerable increase in sensitivity is obtained by converting organically combined mercury to the mercuric salt followed by reduction to $Hg°$ and sweeping the elemental mercury as vapour through a gas cell with end-windows of quartz. Absorption of 254 nm radiation gives adequate response for less than 0.1 μg of mercury. The method is the basis for the determination of the very low levels of mercury encountered in pollution studies.

Apparatus. For the Pye-Unicam, EEL 240 and Perkin-Elmer instruments the gas cell may be purchased to fit into the space occupied by the burner, but no such accessory is available for the EEL 140. It may, however, be easily prepared from glass tubing and two end-windows of quartz or silica. Figure 166 illustrates the cell and general arrangement. The path length of the cell is about 11 cm with end-windows 22 mm diameter fixed with any convenient resin which also serves to hold the glass tube for locating the cell in position in the place normally occupied by the burner.

The coarse sinter-glass filter should be fairly large but capable of being covered by 5 ml of solution. A 50 ml Quickfit boiling tube is convenient for

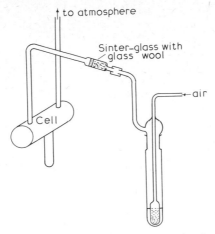

Fig. 166. Diagram of apparatus for the determination of mercury

holding the sample. The delivery of air should be about 1000 ml per min as supplied by a small pump. The system as described is open-ended and the vapour from the sample is vented to the atmosphere. A closed circuit system is also possible but a more elaborate pump is required and all leaks must be eliminated: this sometimes proves difficult.

Reagent. Stannous chloride solution made by dissolving stannous chloride A.R., (20 g) in hydrochloric acid (20 ml) and water (20 ml), boiling in the presence of tin granules and diluting to 100 ml with water.

Method. Evaporate the sample (5 ml) to dryness on a boiling water-bath, add sulphuric acid (1 ml) and continue heating with the occasional addition of a few drops of hydrogen peroxide (100 vol) until a clear solution is obtained (Note 1). Cool and dilute the solution to 200 ml with water.

Transfer the solution (1 ml) to the boiling tube, add water (4 ml) (Note 2) and stannous chloride solution (4 drops). Immediately connect up the apparatus, switch on the pump and record the absorbance as the vapour is swept through the cell (Note 3). Repeat the determination on aliquot portions of a solution of thiomersal treated in the same way.

Note 1. This simple treatment is sufficient for material low in organic matter otherwise more drastic conditions are necessary, e.g. the normal wet-oxidation with sulphuric and nitric acids as prescribed for Penicillamine.

Note 2. It is essential for reproducible results that conditions are constant for each determination. The volume (5 ml) should be maintained and when aliquot portions of a standard solution are used, e.g. 0, 1, 2, 3 and 4 ml, water (5, 4, 3, 2 and 1 ml) should be added.

Note 3. For the conditions and apparatus described 0.1 μg of mercury in the boiling tube should easily give a response almost full scale on the recorder. Scale expansion should be adjusted to achieve this. As the mercury vapour is swept out of solution the recorder pen reaches a maximum and then returns to the base line.

Cognate Determinations

Investigate the application of the method to the following using 0.1 g and oxidation with nitric and sulphuric acids. Standard solutions of mercury should be prepared with mercuric chloride.

Danthron
Penicillamine
Penicillamine Capsules
Penicillamine Tablets
Penicillamine Hydrochloride

References

Clay, A. F., *Scan*, Pye Unicam (1973) **3**, 20.
David, D. J., *Analyst* (1960) **85**, 779.
Menzies, A. C., *Anal. Chem.* (1960) **32**, 898.
Moody, R. R. and Taylor, R. B., *J. Pharm. Pharmac.* (1972) **24**, 848.

10 Spectrofluorimetry

W. D. WILLIAMS

Fluorescence

Re-emission of energy as fluorescence occurs in molecules in which the electron system is shielded in some way from normal de-activation processes, so that complete de-excitation by collision or chemical reaction is discouraged. Such a molecule, on excitation, may possess higher vibrational energy in the excited state than it possessed in the ground state. This vibrational energy is lost by collisional processes in the higher electronic state after which the molecule fluoresces, i.e. it drops back to the ground state with the emission of energy as radiation. The potential energy diagram (Fig. 167) shows that the

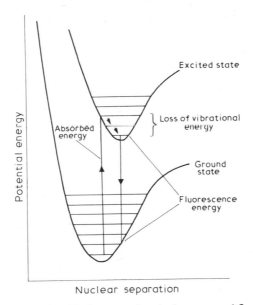

Fig. 167. Diagram of relationship between absorbed energy and fluorescence energy

fluorescence energy is less than the incident energy, i.e. it is of longer wavelength; for example excitation by ultraviolet light may give rise to visible fluorescence. The fluorescence ceases immediately the exciting radiation is removed and this fact distinguishes it from phosphorescence which continues for some time after cessation of excitation.

With increase in concentration of the fluorescent molecules, a limiting intensity of fluorescence corresponding to complete absorption of the exciting light is not obtained. Generally, at a certain concentration, the intensity goes through a flat maximum and decreases with further increase in concentration. Thus the same value of fluorescence could correspond to two different values of concentration. This phenomenon is known as self-quenching or concentration-quenching.

Fluorescence is also sensitive to other substances that may be present in solution, and quenching, from either cause, must always be considered if large errors in quantitative work are to be avoided (see p. 322 for the usual method of detecting quenching). These effects may occur by internal processes within each molecule (intramolecular quenching), or may require a collision between two molecules either of the same or a different species (intermolecular quenching). Both types are examples of true quenching, i.e. the light energy is absorbed by the molecule but is not re-radiated because of the process described. It should be distinguished from other processes which may also reduce the fluorescence efficiency, i.e. the ratio

$$\text{Fluorescence efficiency} = \frac{\text{fluorescence quanta emitted}}{\text{light quanta absorbed}}$$

Re-absorption of the fluorescence by light-absorbing non-fluorescent impurities in the sample or solvent, chemical changes, the type of cell used to hold the sample and the angle at which it is irradiated, all contribute to a reduction in efficiency which seldom reaches the theoretical unity.

A prerequisite for fluorescence is an absorbing system but, unfortunately, not all such systems fluoresce. The relationship between structure and fluorescence is being rapidly explored and certain limited predictions concerning chromophores, auxochromes and fluorescence can be made. Thus, aromatic compounds containing hydroxyl, methoxyl, amino and alkylamino groups are capable of fluorescence, whereas those containing acetylamino, carboxyl, halogeno and nitro groups tend to be non-fluorescent.

Spectrofluorimetry

Spectrofluorimetry bears the same relationship to fluorimetry as spectrophotometry does to absorptiometry, i.e. the intensity of fluorescence can be measured at different wavelengths to give a *fluorescence emission spectrum*. The technique, however, provides additional analytical data in that it is possible to reproduce the absorption curve of a fluorescent substance as an *excitation spectrum* at concentrations very much less than those required in spectrophotometry. In the latter method, for weak concentrations of sample, the detector must distinguish and measure a small difference between two relatively intense beams of light (blank and test). In spectrofluorimetry, on the other hand, a distinction is required between a very weak intensity of fluorescence and that of the still weaker fluorescence of a blank; this difference is very much smaller than that measurable in spectrophotometry.

As spectrofluorimetry may be used in two ways, the limitations of light sources and measuring devices assume real significance. These limitations are variations in the intensity of available energy with wavelength, and variation in the response of the detector to light of different wavelength (Experiment 6).

In spectrophotometry both these factors are not immediately evident, because comparison of blank and test solution is carried out under identical conditions and the absorption curve recorded is true (within the limitations of the instrument). The excitation and fluorescence spectra may, however, be grossly distorted versions of the true spectra if the instrument is not specially adapted.

Excitation Spectra. Before a compound can fluoresce energy must be absorbed, and with an ideal light source of constant intensity at different wavelengths the most intense fluorescence is produced by radiation corresponding in wavelength to that of the absorption peak of the substance. Therefore, if the intensity of the fluorescence is plotted as a function of the wavelength of the radiation used to excite the fluorescence, an activation or excitation spectrum will result. This will be identical with the absorption spectrum when corrected for instrumental effects because the fluorescence efficiency is generally independent of wavelength. As the intensity of fluorescence is measured at a particular wavelength the disadvantage of variation in sensitivity of the detector with wavelength does not appear. However, in practice, the light source is not ideal and the output from the monochromator used to supply exciting radiation will vary according to wavelength. The detector will therefore respond to variations in intensity of fluorescence caused by more (or less) absorption of energy by the sample, and also by more (or less) excitation energy available from the light source. A curve of intensity of exciting light as a function of wavelength can be prepared for the light source and may be used to correct the apparent excitation curve obtained (Experiment 6).

Emission Spectra (Fluorescence). When a monochromatic source of constant light intensity is used to irradiate a sample the fluorescence may be analysed in a monochromator at constant slit width to give an apparent emission spectrum. The true spectrum is obtained by applying a correction for change in detector sensitivity with wavelength and for changes due to the fluorescence monochromator, viz. half-band width of emergent light and light losses.

Fluorescence emission spectra arise because of transitions from the first excited state (Fig. 167) and their shapes are, therefore, independent of the light used to excite fluorescence. If the substance has one absorption band, the emission spectrum often bears a mirror-image relationship to it when plotted on a *frequency* scale but, if several bands occur this relationship may be highly distorted because of overlapping of absorption and fluorescence bands.

Instrumentation. When both excitation and emission spectra are to be recorded two monochromators are essential, one for the light source and one for the fluorescence. The light source must provide a high level of ultraviolet and visible radiation and a compact high-pressure xenon-arc lamp is commonly used. As many experiments will almost certainly entail the measurement of very weak fluorescence the detector must be a highly sensitive photomultiplier tube of low dark current.

If the main interest lies in fluorescence emission spectra one monochromator may be dispensed with and a suitable light source and filter used instead. The rather poor luminosity associated with the monochromator, even with a xenon-arc lamp, is replaced by the much more intense light from a source such as a mercury vapour lamp from which a suitable activation beam is isolated by means of the filter. This arrangement partially overcomes one of the difficulties inherent in spectrofluorimetry, viz. that so much of the available light is lost.

The arrangement of typical instruments is shown in Figs. 168 and 169. Figure 168 is of the Aminco–Bowman spectrophotofluorimeter which re-

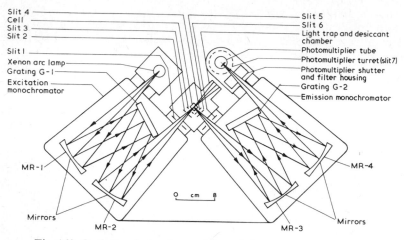

Fig. 168. Optical unit showing position of slits of Aminco–Bowman spectrophotofluorimeter

quires an additional unit if true spectra are to be recorded directly. Figure 169 illustrates the principle involved in instruments capable of recording a true excitation spectrum. A portion of the excitation energy is directed to a solution of a fluorescent standard. The intensity of fluorescence is (ideally) proportional to the intensity of the incident light [eq. (3), p. 316], and a reference signal with which the sample signal is compared is obtained.

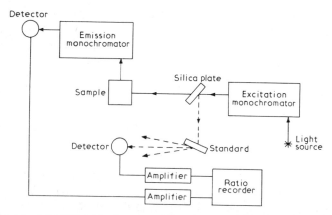

Fig. 169. Diagram (after Parker) of a spectrofluorimeter to show method of obtaining a reference signal for ratio recording of excitation spectra

Analytical Factors

In addition to the instrumental factors already discussed, the following require consideration in the development of a spectrofluorimetric assay.

Absorption by sample or other solute. The fundamental equation for measurement of fluorescence may be obtained from that of the Beer–Lambert law. In terms of quanta

$$\frac{I_0}{I_T} = 10^{Kct} \quad \text{(Chapter 8, p. 237)}$$

$$\therefore \quad I_T = I_0 \times 10^{-Kct}$$

But fluorescence $(F) = (I_0 - I_T) \times Q$ where Q = fluorescence efficiency

$$\therefore \quad F = (I_0 - I_0 \times 10^{-Kct})Q$$
$$= I_0(1 - 10^{-Kct})Q$$
$$= I_0(1 - e^{-2.3Kct})Q$$

Now

$$e^x = 1 + x + \frac{x^2}{2!} + \frac{x^3}{3!} + \cdots$$

$\therefore$ let

$$-2.3\,Kct = x$$

Then

$$F = I_0\left\{\left(1 - 1 - x - \frac{x^2}{2!} - \frac{x^3}{3!}\right) + \cdots\right\}Q$$

$$\therefore \quad F = I_0\left\{\left(-x - \frac{x^2}{2!} - \frac{x^3}{3!}\right) + \cdots\right\}Q$$

$$\therefore \quad F = I_0\left\{2.3Kct - \frac{(-2.3Kct)^2}{2!} + \cdots\right\}Q \qquad (1)$$

Now I_0 and Q are constant for a particular instrument and sample respectively, and eq. (1) may be reduced to the useful relationship

$$F = I_0(2.3\,Kct)Q \qquad (2)$$
$$= Ac \text{ where } A = 2.3I_0Ktq$$
$$= \text{a constant}$$

That is, the fluorescence is proportional to concentration, but as all the terms except the first have been neglected in eq. (2) it must be emphasised that the equation applies only to small values of fluorescence and hence low concentrations of sample. Thus it is true to about 5% when the extinction of a solution of the sample is about 0.05.

Under these conditions, the amount of light absorbed by the test solution is very small, and for a linear calibration curve the extinction should not be greater than about 0.02. With stronger solutions, the fluorescence tends to concentrate more and more at the irradiated face of the cell, an effect which is clearly visible in the simple experiment with quinine (p. 322). The appearance in the cells is shown in Fig. 170.

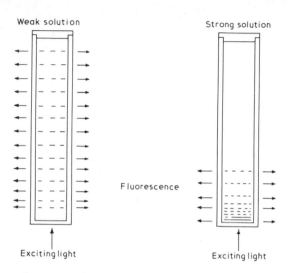

Fig. 170. Effect of concentration of solute on emission of fluorescence from cell

The absorption of the exciting radiation in this way or by a non-fluorescent absorbing solute, is an example of an inner-filter effect which distorts both the excitation and the emission spectrum. A similar effect is self-absorption of the fluorescence by the solution; this may be caused by too high a concentration of solute or by any non-fluorescent absorbing solute or impurity that may be present. When the absorbing impurity is present in *small* amount, it will not interfere, providing the fraction of light absorbed is also small. When the concentration of impurity is high, so much incident radiation may be absorbed that very little is left to excite fluorescence, i.e. the fluorescence is extinguished.

If the concentration of fluorescent substance is so great that all incident radiation is absorbed, equation (1) reduces to:

$$F = I_0 Q \tag{3}$$

That is, fluorescence is independent of concentration, and proportional to the intensity of the incident radiation, a feature which may be used to determine the approximate emission curve of a light source (Experiment 6).

Adsorption. The extreme sensitiveness of the method requires very dilute solutions, 10–100 times weaker than those employed in spectrophotometry. Adsorption of the fluorescent substance on the container walls may therefore present a serious problem, and strong stock solutions must be kept and diluted as required. Quinine is a typical example of a substance which is adsorbed on to cell walls.

Light. Monochromatic light is essentially for the excitation of fluorescence in quantitative work because the intensity will vary with wavelength. The purity of the irradiating beam especially if obtained *via* filters should, therefore, be checked by examination of the light scattered by a slightly turbid solution. The trace obtained should show one peak only, corresponding in wavelength to that expected.

Even when strictly monochromatic light is thus ensured, a further possible source of error is the Raman emission from the solvent. This emission, as with fluorescence, occurs at a longer wavelength than that of the exciting beam and its effect becomes greater with decrease in concentration of solute, because, for an adequate response to the fluorescence the sensitivity of the detector must be increased. It may be possible to use radiation of a wavelength far removed from that of the fluorescence peak, and therefore, interference from Raman scatter will not occur.

If Raman emission is suspected, a small change in frequency of exciting radiation will produce a corresponding change in Raman emission frequency. For aqueous solutions, the difference in frequency between excitation and Raman emission is very nearly 3.4×10^3 cm^{-1}. In terms of wavelength some quoted figures for the Raman peak of common solvents with excitation by the mercury line at 366 nm are: water, 416 nm; ethanol, 409 nm; chloroform, 410 nm. For comparison, the peak fluorescence emission of quinine and thiochrome occur at 450 and 430 nm respectively.

Methods of illumination. The method of illumination described for the Spekker Fluorimeter (Fig. 172 p. 321) is known as the right-angle method and is the one commonly adopted in analysis. It has the advantage of giving a smaller blank value for scattered light and fluorescence from container walls than that given by the alternative method of frontal illumination (Fig. 171).

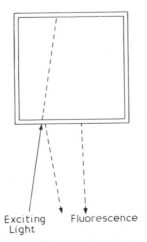

Exciting   Fluorescence
Light

Fig. 171. Frontal illumination

The frontal method must be adopted for opaque solutions and for solids. It can be used for both dilute and concentrated solutions, but inner-filter effects with consequent distortion of spectra also occur with this method.

Oxygen. The presence of oxygen may interfere in two ways: by direct oxidation of the fluorescent substance to non-fluorescent products, or by quenching of fluorescence. It is a useful precaution, therefore, to check the de-aerated solution and compare the result obtained with that from the oxygen-containing solution. Anthracene is well known to be susceptible to the presence of oxygen.

pH. It is to be expected that alteration of the pH of a solution will have a significant effect on fluorescence if the absorption curve of the solute is changed. Many phenols, for example, are fluorescent in both dissociated and undissociated forms. Consequently, the fluorescence from a solution of the phenol will show two peaks, one being due to the ionic form; strongly acidic solutions may be necessary to suppress the peak due to the ionic form.

Photodecomposition. In spectrophotometry, the intensity of the radiation passing through solutions is weak by photochemical standards although adequate for measurements; decomposition of solute is, therefore, not very likely. Spectrofluorimetry, on the other hand, requires high intensity illumination for irradiation and the risk of photochemical change is thereby increased. Parker and Barnes (1957) have calculated for a hypothetical case involving quantities to be expected in normal practice that an error of 20% could quite easily arise. It may be possible in unfavourable cases to select radiation of a wavelength which is not strongly absorbed so that the extent of photochemical change is reduced, whilst at the same time adequate sensitivity is retained.

Temperature and Viscosity. Variations in temperature and viscosity will cause variations in the frequency of collision between molecules. Thus an increase in temperature or decrease in viscosity is likely to decrease fluorescence by de-activation of the excited molecules by collision. Similarly, many substances not normally fluorescent at room temperature are capable of emitting light when excited at a low temperature or when in a viscous solvent or glassy matrix.

Applications

Fluorimetry. Compounds which are intrinsically fluorescent are readily determined with simple instruments as the solution for examination is normally obtained by solution of the sample in a suitable solvent. Table 46 lists some typical examples.

Single substances which are, in themselves, non-fluorescent may be determined as a result of chemical change. This method is useful for both inorganic and organic compounds, and many inorganic compounds form highly fluorescent complexes by combination with organic reagents. The determination of selenium illustrates the increase in sensitivity which can be obtained with fluorimetry as compared with that for absorptiometry. Thus 0.3 μg of Se may be determined by measurement of the extinction of its complex with 3,3'-diaminobenzidine, but by using the fluorescence of the latter, 0.04 μg of Se can be measured. The sensitivity is further increased to 0.002 μg of Se with 2,3-diaminonaphthalene as reagent.

Table 46. Some Fluorescent Compounds

Compound	Solvent	Wavelength (nm) Excitation	Fluorescence
Aminacrine	*0.1N* Hydrochloric acid	245, 380, 408	450
Ergometrine	1% Tartaric acid	325	465
Proflavine	*0.1N* Hydrochloric acid	440	510
Quinine	*0.1N* Sulphuric acid	350	450
Riboflavine	Aqueous buffer pH 6	270, 370, 445	520

Thiochrome

Aneurine hydrochloride in pharmaceutical preparations such as tablets and elixirs and in foodstuffs such as flour is relatively easily determined by oxidation to the highly fluorescent thiochrome. The product is soluble in isobutyl alcohol and hence is easily extracted from the reaction mixture for measurement.

Many examples of this type exist and a selection is recorded in Table 47.

Table 47. Compounds Readily Converted to Fluorescent Products

Compound or Element	Reagent	Wavelength (nm) Excitation	Fluorescence
Hydrocortisone	75% v/v Sulphuric acid in ethanol	460	520
Nicotinamide	Cyanogen chloride	250	430
Chlordiazepoxide	Photo-oxidation after hydrolysis	380	480
Cyanide ion	Quinone monoxime benzenesulphonate ester	440	500
Aluminium	Pontachrome blue black R	470, 580	630
Boron	Benzoin	370	580

Spectrofluorimetry. With the advent of commercial spectrofluorimeters the analyst is no longer confined to work in the visible region of the spectrum. Many more single substances may be determined directly as given in Table 48, which lists but a few of the fluorescent compounds of pharmaceutical interest.

Table 48. Fluorescent Compounds

| Compound | pH | Wavelength (nm) | | Minimum concentration required (μg/ml) |
		Excitation	Fluorescence	
Adrenaline	1	295	335	0.1
Allylmorphine	1	285	355	0.1
Amytal	14	265	410	0.1
p-Aminosalicylic acid	11	300	405	0.004
Aureomycin	11	355	445	0.02
Chloroquin	11	335	400	0.05
Chlorpromazine	11	350	480	0.1
Chlorpromazine sulphoxide	7	335	400	0.02
Cinchonidine	1	315	445	0.01
Cinchonine	1	320	420	0.01
Cyanocobalamin	7	275	305	0.003
Epinephrine	1	295	335	0.07
Folic Acid	7	365	450	0.01
Lysergic acid diethylamide	7	325	465	0.002
Menadione		280	320	0.07
Noradrenaline	1	285	325	0.006
Norepinephrine	1	295	335	0.06
Oxychloroquin	11	335	380	0.08
Pamaquin	11	300, 370	530	0.06
Pentobarbitone	13	265	440	0.1
Pentothal	13	315	530	0.1
Phenobarbitone	13	265	440	0.5
Procaine	11	275	345	0.01
Procainamide	11	295	385	0.01
Physostigmine	1	300	360	0.04
Quinine	1	250, 350	450	0.002
Rescinnamine	1	310	400	0.008
Reserpine	1	300	375	0.008
Salicylic acid	11	310	435	0.01
Terramycin	11	390	520	0.05
Thymol	7	265	300	0.1
Vitamin A	—	325	470	0.01

For mixtures of two components it may be possible to select exciting radiation of appropriate wavelengths, such that one compound only fluoresces at any one time. Even if this is not possible, measurement of fluorescence at two wavelengths may be sufficient to determine the composition of the mixture. Techniques which have proved their worth in spectrophotometry are also of considerable value in spectrofluorimetry, e.g.

(a) application of separation methods and determination of each component separately,

(b) changes in pH with concomitant changes in fluorescence, e.g. both morphine and codeine show fluorescence at 355 nm but only that of morphine is eliminated on making the solution alkaline.

(c) preparation of derivatives, e.g. atropine forms a fluorescent chloroform-soluble complex with eosin (compare indicator extraction methods, Part 1). The technique is therefore useful in solving the many difficult analytical problems that arise when traces of compounds are to be determined in biological materials.

PRACTICAL EXPERIMENTS

Fluorimetry

Because fluorescence is a re-emission of most of the radiation which has been absorbed, the optical system in the instruments for its measurement must be such that the exciting radiation does not impinge upon the photocell. The arrangement in the Spekker Fluorimeter is a modification of that in the Spekker Absorptiometer in which two very sensitive barrier layer photocells are used to measure the intensity of the fluorescence. A photomultiplier detector is used in place of the barrier layer cells in modern versions of the instrument. (Fig. 172). The neutral glass filter (p. 226) is inserted at C (Fig. 172) and when the

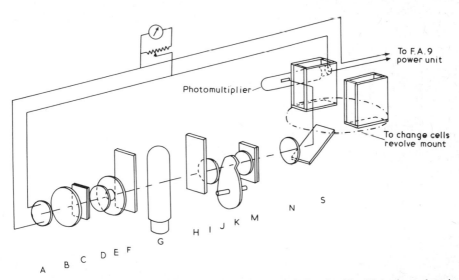

Fig. 172. Simplified diagram of optical system and photocell circuit of Spekker photoelectric fluorimeter. (A) Photocell; (B) Window; (C) Filter; (D) Lens; (E) Iris diaphragm; (F) Heat absorbing filter; (G) Mercury lamp; (H) Heat absorbing filter; (I) Window; (J) Measuring disc; (K) Lens; (M) Filter (u.v.); (N) Lens; (S) Reflector

fluorescence is very weak, it may be necessary to insert two such filters before the instrument can be balanced.

Experiment 1. *To show quenching of fluorescence*

Method. (*a*) *Barrier layer cell instrument.* Dissolve 12.5 mg of quinine sulphate in sufficient *0.1N* sulphuric acid to produce 250 ml of solution. By dilution with the *0.1N* sulphuric acid, prepare two series of solutions so that the percentage of quinine sulphate is (a) 0.0001, 0.0002, 0.0003, 0.0004, 0.0005, and (b) 0.001, 0.002, 0.003, 0.004 and 0.005 respectively.

Measure the fluorescence of the solutions in the first series using the 0.0005 % solution as reference solution. Plot the measurements against concentration in the normal manner. Repeat the above using the stronger solutions and the 0.005 % solution as reference solution. Typical curves are shown in Fig. 173 and Fig. 174.

(*b*) *Photomultiplier instrument.* Prepare dilutions with *0.2N* sulphuric acid as above, so that the percentages of quinine sulphate are 0.00002, 0.00004, 0.00006, 0.00008 and 0.0001 respectively. Measure the fluorescence of the solutions using the 0.0001 % solution as reference. Plot the measurements against concentration in the normal manner, when a straight line should be obtained.

(*c*) *Photomultiplier instrument.* Using the 0.0001 % quinine sulphate prepared above and a solution of potassium iodide (*0.1M*), prepare solutions of 0.00008 % quinine sulphate containing *0.005, 0.001, 0.0015, 0.002, 0.0025, 0.005, 0.0075, 0.01* and *0.015M* potassium iodide. Plot the drum reading against the concentration of potassium iodide.

The fluorescence of the weak solutions shows a linear relationship with concentration. The stronger solutions show marked quenching effects due in this instance to most of the fluorescence being concentrated at one end of the cell (Fig. 170). It is thus possible for a particular intensity of fluorescence to correspond to two concentrations of the substance. In fluorimetric assays, therefore, the usual check to avoid or detect this source of error is the following. Determine the apparent content of the fluorescing substance in the tablet, solution or preparation and then repeat the determination after adding a

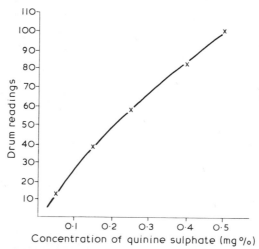

Fig. 173. Relationship between fluorescence and concentration of weak solutions of quinine sulphate

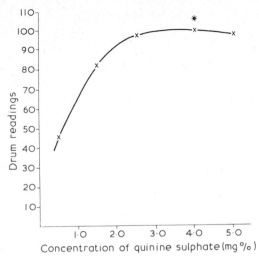

Fig. 174. Relationship between fluorescence and concentration of strong solution of quinine sulphate.* 0.004% Solution used as reference and point obtained by calculation

known amount of the substance being determined. The increase in fluorescence should correspond to the quantity of the substance added. If this is not the case, then quenching effects are present caused either by the substance itself (concentration quenching) or by other substances in solution. The latter effect is observed in the presence of potassium iodide [Experiment 1(c)] when the drum readings are seen to decrease with increasing iodide concentration.

Experiment 2. *To determine the concentration of quinine sulphate in Ferrous Phosphate Syrup with Quinine and Strychnine*

Method. Dilute the syrup (about 2 g accurately weighed) with *0.1N* sulphuric acid to 100 ml. Make a further dilution of this solution with *0.1N* sulphuric acid so as to bring the concentration of quinine into the range 0.0002–0.0004%. Compare the fluorescence of this solution with that of the 0.0005% solution of quinine sulphate used in Experiment 1(*a*) and read off the concentration from the graph.

Determine the weight per ml of the syrup and calculate the concentration of anhydrous quinine (as % w/v) in the sample.

Experiment 3. *To determine the concentration of Proflavine Hemisulphate in Proflavine Cream*

Method. Thoroughly mix the sample and transfer about 1 g, accurately weighed, to a separator. Dilute with ether (50 ml) and extract the mixture with *0.1N* hydrochloric acid (4 × 20 ml). Wash each extract with ether (20 ml), transfer to a 500 ml volumetric flask and make up to volume with *0.1N* hydrochloric acid. Dilute 10 ml of this solution to 100 ml with *0.1N* hydrochloric acid. Compare the fluorescence of the solution with that of a standard solution of proflavine hemisulphate (0.00005% in *0.1N* hydrochloric acid) using filters of maximum transmission at about 440 and 510 nm for excitation and fluorescence respectively.

Read off the concentration of flavine from a calibration curve prepared from solutions containing 0.00001, 0.00002, 0.00003 and 0.00004% proflavine hemisulphate in *0.1N* hydrochloric acid.

Spectrofluorimetry

Experiment 4. *To determine the excitation and emission spectra of a substance for which no data are available*

Method. Prepare a 0.0001 % solution of the substance in water, ethanol or cyclohexane, as appropriate, and transfer to the fluorimeter cell. Set the excitation monochromator to an arbitrary wavelength in the ultraviolet region (say 300 nm), and using high sensitivity, examine for fluorescence by turning the fluorescence monochromator manually. Note the wavelength for which a maximum response is obtained and set the monochromator to this wavelength (λ_1).

Obtain the excitation spectrum of the substance by recording the intensity of fluorescence at the wavelength λ_1 as the excitation monochromator scans the spectrum from 200 nm. Note the excitation wavelength for which the fluorescence is a maximum and set the excitation monochromator to this wavelength.

Obtain the fluorescence emission spectrum by recording the intensity of fluorescence as a function of wavelength as the fluorescence monochromator scans the spectrum from a wavelength shorter than λ_1.

The spectra are obtained in a reasonably short space of time but it must be emphasised that with so many factors involved, considerable manipulation of sensitivity controls may be required before the optimum conditions are found.

Experiment 5. *To determine the fluorescence emission spectrum of quinine sulphate solution (0.00001 %) at two different wavelengths of excitation*

Method. Examine the excitation spectrum of quinine and note the two maxima at about 250 and 350 nm respectively. Record the fluorescence emission spectrum under the same conditions of sensitivity using first one and then the other wavelength for excitation.

Fluorescence occurs because of transitions from the first excited state to the ground state. Even when the second and higher excited states are present, reversion to the first excited state occurs before fluorescence is emitted. Therefore the position of the fluorescence emission spectrum is the same in both tracings as would be expected. The difference between the tracings is one of intensity caused by the incident intensity of the radiation at about 350 nm being so much greater than that available at 250 nm. This difference is shown in the following experiment.

Experiment 6. *To determine the approximate emission characteristics of the light source and its monochromator using quinine sulphate as standard*

Method 1. Prepare a 0.001 % solution of quinine sulphate in *0.1N* sulphuric acid and determine the absorption curve of the solution in the normal way (curve 1).

Dilute the solution with *0.1N* sulphuric acid to give a 0.00001 % solution of quinine sulphate and determine the apparent excitation curve over the range 200–400 nm with the fluorescence monochromator set at 450 nm (curve 2). Determine the blank values for *0.1N* sulphuric acid and subtract from curve 2 when comparing the two curves. Calculate a factor at each 10 nm interval by dividing the observed fluorescence (corrected curve 2) by the extinction found in curve 1 at each wavelength. Plot the values against wavelength to give a graph, representing, in relative figures, the intensity of emission from the monochromator at each wavelength.

Method 2. Prepare a 1% solution of quinine sulphate in *N* sulphuric acid and determine the apparent excitation curve of the solution over the range 200–400 nm with the fluorescence monochromator set at 450 nm. The sensitivity controls must be adjusted to maintain the response of the photomultiplier at a reasonable level. Plot the readings (after allowing for changes in sensitivity) against wavelength to give a graph similar to that obtained in method 1.

The excitation spectrum is a function of the absorption of radiation by the sample, the fluorescence efficiency of the sample, and the intensity of the light source. Of these factors the fluorescence efficiency is normally constant over a wide range of wavelengths; therefore, a comparison of the absorption and excitation curves of a well characterised standard substance should show the variation of the intensity of the light source with wavelength. This is the principle of method 1 and among the standard substances available are:

> quinine sulphate
> 3-aminophthalimide
> *m*-nitromethylaniline
> 4-dimethyl-4'-nitrostilbene
> aluminium chelate of 2,2'-dihydroxy-1,1'-azonaphthalene-4-
>   sulphonic acid.

Argauer and White (1964) discuss the method and substances in detail but note that for accurate results the wavelength scales of spectrofluorimeter and spectrophotometer must correspond very closely. As the intensity of the light source falls off rapidly at the shorter wavelengths the possible error is greater at these wavelengths.

The response curve obtained by method 2 is possible because all the light is absorbed and equation (3) is now relevant. Frontal illumination of the cell is preferable but this cannot be done in the Aminco–Bowman instrument unless the cell holder is removed completely and the cell is carefully adjusted manually to reflect the light into the fluorescence monochromator.

Both methods give figures which are proportional to the *number of quanta*. If, however, a thermopile is used to obtain the relative intensity of the emission of the light source with wavelength the result is in terms proportional to *energy units*.

To convert from one form to the other use the relationship

$$\text{No. of quanta} = \frac{\text{No. of energy units}}{h\nu} = \frac{\text{No. of energy units}}{hc} \times \lambda$$

in which the symbols have their usual significance. Therefore, multiply the figures obtained for the energy curve by the corresponding wavelengths to obtain the curve representing intensity in terms proportional to quanta.

Experiment 7. *To determine the composition of a mixture of α- and β-naphthol*

Method. Prepare stock solutions of α- and β-naphthol (about 0.001–0.003%) in water, *making both solutions identical in concentration*. Dilute each solution (5 ml) with water to 500 ml, including whilst doing so a buffer solution (pH 7.5, 5 ml). Determine the excitation and fluorescence emission curves for each solution, and select one excitation wavelength such that a reasonable response to the fluorescence is obtained with each phenol.

Prepare a solution of the mixed naphthols (about 0.003 %) and dilute 5 ml with water and buffer as described above. Determine for each solution (standards and sample), the fluorescence emission curve, recording each on the same chart. Record also, the blank for the diluted buffer solution. Calculate the percentage composition of the mixture as described below.

The calculation is similar to that for a two-component mixture in spectrophotometry (p. 249) and is best illustrated by reference to an actual determination. From a consideration of the excitation and fluorescence emission curves, 300 nm was selected as the excitation wavelength. With the instrument set at this value the traces shown in Fig. 175 were obtained.

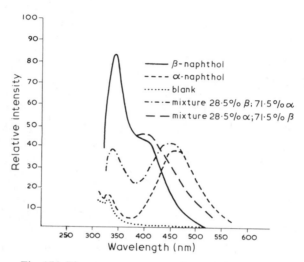

Fig. 175. Fluorescence spectra of α- and β-naphthols

Data:	Concentration of standards:	$1 \times 10^{-4} \%$
	λ_{max} for β-naphthol:	350 nm
	λ_{max} for α-naphthol:	465 nm
	Reading at 350 nm standard: β-naphthol	76.5 (net)
	Reading at 465 nm standard α-naphthol	36.7 (net)
	Reading at 465 nm standard β-naphthol	10.0 (net)
	Reading at 350 nm standard α-naphthol	2.5 (net)
	Reading at 350 nm (mixture)	30.2 (net)
	Reading at 465 nm (mixture)	40.5 (net)

Calculation:

$$\text{Reading at 350 nm} = 76.5 \times \beta + 2.5\alpha$$

$$\text{Reading at 465 nm} = 10\beta + 36.7\alpha$$

where α and β are the proportions of α- and β-naphthol respectively.

$$\therefore \quad 30.2 = 76.5\beta + 2.5\alpha$$
$$40.5 = 10\beta + 36.7\alpha$$

Solving for α and β

$$\alpha = 1.003$$
$$\beta = 0.367$$

$$\therefore \quad \text{Percentage } \alpha\text{-naphthol} = 26.8$$
$$\text{Percentage } \beta\text{-naphthol} = 73.2$$

Note that with a mixture in the solid state the actual concentration need not be known, but in a crude product and also when the naphthols are present in solution, the concentration must be obtained before the percentage can be determined. This information is available from the curves as both naphthols give the same intensity of fluorescence at about 435 nm. By making the concentrations the same for the standards, this point is clearly seen on the recording. Note this reading (26.6 net) and also that for the mixture (36.8, net).

Then,

$$\text{concentration of mixture} = \frac{36.8}{26.6} \times \text{concentrations of standards}$$

$$= 1.38 \times 10^{-4}\%$$

The setting of the excitation monochromator is critical, as the scale is far shorter than those encountered on spectrophotometers. Therefore, it is difficult to reproduce the setting with the desirable degree of accuracy and it is better to examine the solutions in the way described rather than to use constants for application to assays carried out after an interval.

Experiment 8. *To determine the content of* $C_{20}H_{24}O_2$ *in an Ethinyloestradiol Tablet of average weight*

Method. Weigh and powder twenty tablets. Place an accurately weighed quantity of the powder equivalent to about 0.2 mg of Ethinyloestradiol in a 50 ml graduated flask previously rinsed with methanol (Note). Add water (2 ml), warm gently and add methanol (20 ml). Shake for 5 min and dilute to volume with methanol. Allow to stand for tablet base to settle, and irradiate 2 ml of the clear supernatant liquid with light of about 290 nm wavelength. Measure the fluorescence at about 320 nm and add N sodium hydroxide (0.2 ml). Mix carefully and again measure the fluorescence. Compare the difference between the two values with that obtained by measuring under the same conditions, the fluorescence of a standard solution of Ethinyloestradiol (Authentic Specimen, $4\,\mu g$ per ml, 2 ml) in methanol, before and after the addition of N sodium hydroxide (0.2 ml).

Note. Purify the methanol before use by boiling under reflux over sodium hydroxide, and distillation.

The determination is based upon the fluorescence of the phenolic portion of the molecule, and the effect of pH is to give increased specificity to the assay.

Experiment 9. *Determination of morphine and codeine in admixture*

The determination depends upon the fact that codeine fluoresces in both acid and alkaline media whereas morphine fluoresces in acid media only. The method was used by Chalmers and Wadds (1970) for the determination of both compounds in opium. They extended their work to the determination of papaverine and noscapine which were differentiated by the effect of protonation in chloroform solution on the fluorescence spectra.

Standard Solutions. 0.0005 % solutions of codeine and morphine in *0.1N* HCl and *0.1N* NaOH.

Method. Determine and record the fluorescence emission curves for both solutions using appropriate blanks. Make a trial dilution of the solution of morphine and codeine to obtain almost full-scale reading in acid solution. When this is achieved make appropriate dilutions in *0.1N* HCl and *0.1N* NaOH and record the fluorescence emission curves as for the standard solutions. Calculate the concentration of morphine and codeine.

Experiment 10. *Determination of total alkaloids (calculated as emetine) in Ipecacuanha Tincture*

The determination depends upon the fluorescence of ipecacuanha alkaloids in acid solution and the large dilution of the tincture such that the effect of the colour of the tincture may be neglected.

Method. Dilute the tincture (1 ml) to 200 ml with water in a volumetric flask. To the dilution (10 ml), add *0.1N* H_2SO_4 (10 ml), and dilute to 100 ml with water. Compare the fluorescence at about 320 nm (excitation at about 280 nm) with that from a 0.0001 % solution of emetine hydrochloride in *0.01N* H_2SO_4 (Note). Calculate the percentage of total alkaloids (calculated as emetine) from the relationship.

$$\% \text{ alkaloids} = \frac{F_{\text{sample}}}{F_{\text{std}}} \times 0.0001 \times 2000 \times \frac{480.7}{552.7}$$

Note. Emetine Hydrochloride BP contains 15–19 % H_2O and that of the USP contains 8–14 %. It is therefore advisable to dry the standard material *in vacuo* at 100° before use to arrive at material of definite composition.

Extension. (*a*) Determine the effect of pH on the fluorescence characteristics of emetine and cephaeline.

(*b*) Determine the effect of light on the fluorescence characteristics of emetine and cephaeline.

Cognate Determination

Ipecacuanha Powder. Weigh accurately the well mixed sample in fine powder (0.1 g), add dimethylsulphoxide (5 ml) and warm in a water bath (15 min). Cool, dilute to 200 ml with *0.01N* H_2SO_4, mix well, and filter a portion. Dilute 10 ml to 100 ml with *0.01N* H_2SO_4 and proceed as described for Ipecacuanha Tincture.

Experiment 11. *Determination of dissolution rate for Digoxin Tablets*

The tablets contain either 0.25 mg or 0.0625 mg of Digoxin and a sensitive method of detection is required. Although digoxin contains an α, β-unsaturated lactone ring the intensity of absorption is not sufficient at the concentrations obtained in practice for ultraviolet absorption to be used. Wells, Katzung and Meyers (1961) developed a spectrofluorimetric method based on the conversion of digoxin to a mono-anhydro compound with the double bond situated at $\Delta^{8(14)}$.

Reagents. Ascorbic acid (0.1%) in methanol (AR)

Hydrogen Peroxide (*0.009M*) prepared from strong hydrogen peroxide solution previously standardised with potassium permanganate

  Hydrochloric Acid AR

  Water, freshly distilled

  Digoxin BCRC (100 mg) in ethanol (80%, 100 ml)

stored at 4° and diluted immediately before use as follows: 10 ml to 100 ml with water, and this solution is further diluted (10 ml) to 100 ml with water.

  Apparatus. As described in Part 1, Chapter 14 using a stirrer speed of 120 ± 5 rpm and the basket so positioned that the bottom of the basket is 2 cm from the base of the beaker.

  Method. Place 6 tablets in the basket and lower into the beaker containing water (600 ml) at 36.5 to 37.5°. Set the stirrer in motion and at intervals of 0.5, 1.0, 2.0 and 4.0 hours withdraw a sample (5 ml) from a point midway between the basket and the wall of the beaker and level with the middle of the basket. Replace the sample with water (5 ml).

  Filter through a Millipore filter (0.8 μm) and reject the first 1 ml of filtrate. To the filtrate (1 ml) in a 10 ml volumetric flask, add, with thorough mixing, ascorbic acid solution (3 ml), hydrogen peroxide solution (0.2 ml), and dilute to volume with hydrochloric acid. Measure the fluorescence after exactly 2 hours, using an excitation wavelength of 360 nm and a fluorescence emission wavelength of 490 nm.

  The instrument should be set to zero with water and to 100 with the diluted standard digoxin solution (1.0 ml for 0.25 mg tablets, and 0.25 ml + 0.75 ml water for 0.0625 mg tablets) treated in the same way. Calculate the percentage of available digoxin in solution and plot the results.

  Extension. Examine and compare the results for tablets obtained from different manufacturers.

Experiment 12. *Determination of 11-hydroxysteroids in plasma*

The determination is based upon the development of a fluorescence when such steroids are treated with sulphuric acid. It is not a specific reaction and Stenlake, Davidson, Williams and Downie (1970) have shown that cholesterol and cholesterol esters are some of the major interfering fluorogens. Nevertheless, the determination is routinely done in clinical laboratories by a method essentially that of Mattingly (1962) and studied collaboratively (1971).

  Reagents. Methylene Dichloride. Purify by extraction with sulphuric acid ($\frac{1}{10}$th volume, 5 extractions), washing with water ($\frac{1}{10}$th volume) until the washings are neutral. Add anhydrous sodium sulphate, shake well and decant into a thoroughly clean distillation apparatus. Recover the methylene dichloride by distillation (Note 1).

  Water. Glass-distilled water should be used.

  Fluorescence Reagent. Add ice-cold sulphuric acid AR (7.5 volumes) to ethanol (for spectroscopy, 2.5 volumes) *slowly, with cooling in ice.* The mixture should be colourless and used within a few days.

  Stock Standard. Dissolve hydrocortisone (50 mg) in ethanol (20 ml), and dilute 1 ml of the solution to 100 ml with water (concentration = 25 μg/ml). Store at 4°.

  Working Standard. Dilute the stock standard (1 ml) to 100 ml with water.

Apparatus. Scrupulous cleanliness is essential and all glass apparatus should be cleaned with chromic acid, followed by washing with sodium metabisulphite solution, tap water and finally with purified water.

Extraction of Plasma. Pipette the plasma (2 ml) into a glass-stoppered centrifuge tube (20 ml) and add methylene dichloride (15 ml). Moisten the stopper with water to ensure a good fit and avoid loss of solvent, and shake gently, horizontally, for 2 min. Allow the phases to separate and remove the plasma by aspiration. Duplicate blanks (water, 2 ml) and standard (2 ml) are treated in the same way.

Method. To the methylene dichloride extract (10 ml) in a 20 ml glass-stoppered tube, add fluorescence reagent (5 ml) and shake vigorously for 20 sec. Repeat the procedure with blanks and standards at 1 min intervals (Note 2). Transfer the acid layer to a spectrofluorimeter cell and measure the intensity of fluorescence at 530 nm using an excitation wavelength of 470 nm exactly 12 min after mixing the extracts with fluorescence reagent. Calculate the concentration of plasma corticosteroids using the formula

$$\mu g/100\ ml = \frac{F_T - F_B}{F_S - F_B} \times 25$$

where

F_T = Intensity of Fluorescence of sample

F_S = Intensity of Fluorescence of standard

F_B = Intensity of Fluorescence of blank

Note 1. Methylene dichloride vapour is toxic and care should be exercised in its use. If further purification is required, the solvent may be refluxed with and distilled from phosphorus pentoxide as described by Spencer-Peet, Daly and Smith (1965).

Note 2. Strict timing is essential and no more than six tubes should be handled in one experiment (2 samples, 2 blanks, 2 standards). The fluorescence intensity for corticosteroids decreases with time, whereas that for cholesterol and its esters increases, an effect which is readily shown if the various compounds in fluorescence reagent are incubated at 60° over a period of time.

References

Argauer, R. J. and White, C. E., *Anal. Chem.* (1964) **36**, 368 (and correction, 1022).
Chalmers, R. A. and Wadds, G. A., *Analyst* (1970) **95**, 234.
Collaborative Report, *Brit. Med. J.* (1971) **ii**, 310.
Mattingly, D., *J. clin. Path.* (1962) **15**, 374.
Spencer-Peet, J., Daly, J. R. and Smith, V., *J. Endocr.* (1965) **31**, 235.
Stenlake, J. B., Williams, W. D., Davidson, A. G. and Downie, W. W., *J. Endocr.* (1970) **46**, 209.

Reading

Argauer, R. J. and White, C. E., *Fluorescence Analysis*, Marcel Dekker, New York, 1970.
Hercules, D. M. (ed.), *Fluorescence and Phosphorescence Analysis*, Interscience, New York, 1966.
Parker, C. A., *Photoluminescence of Solutions*, Elsevier, Amsterdam, 1968.
Udenfriend, S., *Fluorescence Assay in Biology and Medicine*, Academic Press, New York, 1965 and Vol. II, 1969.

11 Infrared Spectrophotometry

W. D. WILLIAMS

Introduction

The production of relatively inexpensive double-beam recording infrared spectrophotometers has made a powerful analytical tool available to most laboratories. Provided that reasonable care is taken, an air-conditioned room for the instrument is no longer necessary and it may take its place alongside the familiar spectrophotometers for the ultraviolet and visible regions of the spectrum. In this context, except for Experiment 11, infrared refers to that part of the spectrum between 2.5 and 15 μm, wherein absorption of energy causes changes in vibrational energy in the ground state of the molecule. The transition from vibrational level 0 to vibrational level 1 (Fig. 134) gives rise to the fundamental absorption of the molecule and overtones or harmonics are caused by the transitions 0–2, 0–3 and so on, though the intensity of absorption for these overtones is very much less than that for the fundamental frequencies. For energy to be transferred from the light source to the molecule the frequency of vibration of each must coincide and, moreover, must be accompanied by a change in the dipole moment of the molecule.

Just as ultraviolet absorption spectra are strictly caused by changes in electronic, vibrational and rotational energy, so infrared absorption spectra are due to changes in vibrational energy accompanied by changes in rotational energy. For simple molecules in the gas phase this leads to results from which information may be obtained on force constants, bond lengths, moments of inertia and, where adequate resolution is available, on isotopes in an easily accessible region of the spectrum. Experiment 1 for hydrochloric acid illustrates some of these points. For non-linear polyatomic molecules the number of normal modes of vibration is given by the expression $3n-6$ ($3n-5$ for a linear molecule) where n = number of atoms. All these vibrations occur at the same time and an infrared spectrophotometer has been likened to a stroboscope in that it enables the particular frequencies of vibration to be recorded. Particular groups in the molecule, e.g. hydroxyl, carbonyl and amines also have characteristic absorption frequencies known as group frequencies which are almost independent of the nature of the rest of the molecule and which therefore, occur in particular regions of the absorption spectrum. Neighbouring groups and interaction between atoms, e.g. hydrogen bonding, will, however, have some influence on the frequency and so changes of considerable value for use in structural analysis are observed. The terms used in describing various modes of vibration are shown in Figs 176 and 177.

The regions of the spectrum where these vibrations occur are shown under 'Assignments' on the Spectra-Structure Correlation chart on p. 335.

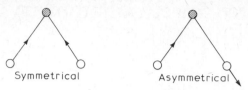

Fig. 176. Stretching modes

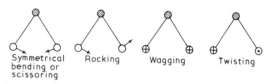

Fig. 177. Deformation modes. (+) below plane of paper; (·) above plane of paper

Instrumentation

The arrangement of a typical infrared spectrophotometer is shown in diagrammatic form in Fig. 178.

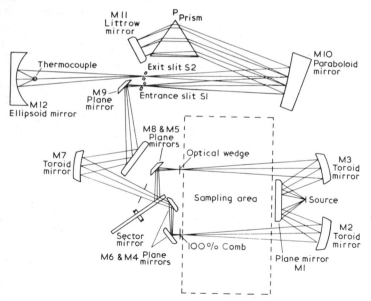

Fig. 178. Optics of infrared spectrophotometer

The various components of a spectrophotometer, viz. light source, monochromator, and detector are easily recognised and have been discussed in general terms in Chapter 7. The rotating sector mirror (Fig. 178) reflects the reference beam from M3 into the apparatus during one half of its rotation and allows the sample beam from M2 through during the other half. Any difference between the two beams causes an out-of-balance signal; at balance the two beams have the same intensity and produce a DC signal but when unbalanced an AC signal results which causes movement of the optical wedge into or out of the reference beam until the two beams are balanced. The optical wedge is coupled to a pen and an absorption curve is drawn as the spectrum is scanned. This system is known as the optical null method. The wavelength of the radiation emerging from the exit slit depends upon the position of the Littrow mirror M11 relative to the prism. This mirror, rotating at a definite speed, moves the spectrum across the exit slit so that an absorption curve covering the region 2.5–15 μm can be obtained. The 100% comb in the sample beam is manually operated to adjust the pen to 100% transmission in the initial setting-up stage.

Prisms of sodium chloride are of use for the whole of the region from 4000–650 cm^{-1} but suffer from the disadvantage of low resolution at 4000–2500 cm^{-1}. Gratings are much better in this respect; typical rulings are 240 lines per mm for the 4000–1500 cm^{-1} region and 120 lines per mm for the 1500–650 cm^{-1} region. The presence of various orders of spectra passed or reflected by gratings makes it necessary to use either filters or a fore-prism to ensure that unwanted radiation is eliminated; construction of the instrument is simplified if filters are used. The sample and reference beams are in close proximity to one another so that absorption by atmospheric moisture and carbon dioxide can be neglected unless so much is present that the available energy in certain parts of the region is reduced. Flushing of the instrument with dry nitrogen must then be carried out.

Qualitative Uses

The number of absorption bands in an infrared spectrum may be considerable, and therein lies its value as a means of identification. Two compounds are one and the same if their spectra agree in all respects, i.e. in the position and relative intensity of the bands. If the spectra are not identical and have been obtained by examination of the compounds in the solid state, further tests must be carried out. The pharmacopoeia describes the following additional tests:

(a) recrystallise both compounds from the same solvent and repeat the determination.
(b) dissolve the sample in a suitable solvent and measure the absorption spectrum against the solvent as blank. Compare with the authentic substance under the same conditions.

Different absorption spectra, after taking these precautions, indicate non-identity of the compounds. This procedure is necessary because various crystalline forms of the same substance may give rise to different spectra, e.g. Cortisone Acetate has been obtained in five different crystalline forms.

Comparison of spectra in this way is probably the simplest use of infrared absorption. Very often, however, it is necessary to identify a completely unknown substance or substances in, perhaps, a single tablet. All the resources of the analyst must be brought to bear on the problem including chromatography and ultraviolet absorption. As a first step separation of the active principle from inert tablet base will be necessary and partition between

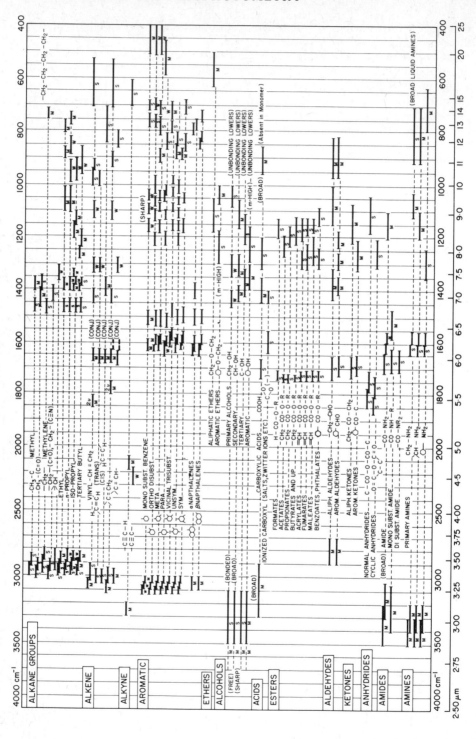

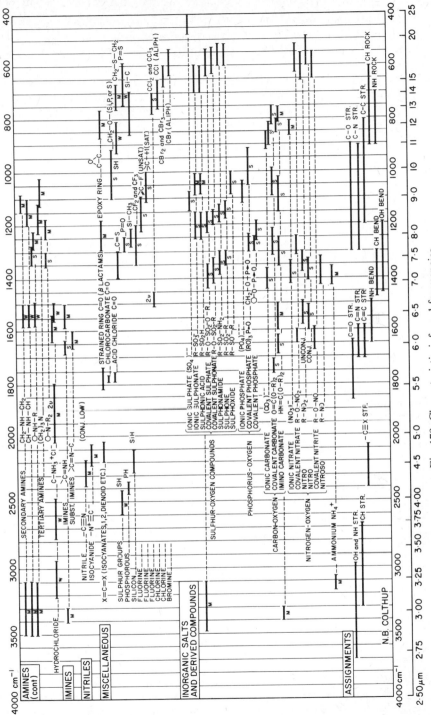

Fig. 179. Characteristic infrared frequencies

immiscible solvents under controlled conditions of pH is very useful for this purpose. It may be possible to use the residue from an appropriate extract directly for infrared analysis from the results, of which the presence or absence of functional groups may be inferred. An indication may thus be obtained of the type of compound for which a search should be made in the reference files. The residue may, however, be a mixture and require further treatment. In this connection gas chromatography combined with infrared analysis of fractions will undoubtedly become of more importance for the isolation and identification of pure components.

Interpretation of Infrared Spectra

Interpretation of absorption spectra is considerably simplified by charts or tables which correlate the frequency at which a band occurs with molecular structure, and Fig. 179 illustrates such a comprehensive chart. The region between 4000 and $1500 \, \text{cm}^{-1}$ is probably easier to interpret than that between 1500 and $650 \, \text{cm}^{-1}$, because the latter includes many skeletal vibrations which are typical of the molecule as a whole. It is outside the scope of this book to give detailed analyses of the assignments and recourse must be made to the references at the end of this chapter. The following spectra are therefore presented as an introduction to the interpretation of bands in various regions of the spectrum.

In the discussion of the spectra, note that not all the bands are assigned to particular groups. Such assignment would be impracticable because many skeletal vibrations occur, and the first step therefore is to interpret the region between 3500 and $1500 \, \text{cm}^{-1}$. Strong absorption bands in the region 850–700 cm^{-1} are often indicative of aromatic systems but no indication is given here of the wealth of information afforded by the pattern of the bands on the degree and position of substitution of the benzene nucleus. A further aid to interpretation is to examine the spectra of a series of compounds derived from a simple parent compound (see Table 49).

Table 49. Chemically Related Compounds for Comparison of Infrared Spectra

Parent Compound	Derived Compounds
Heptane	Heptenes, methylhexanes, heptanol, heptanal, heptanoic acid and methyl heptanoate, methylcyclohexane
Benzene	Phenol, benzaldehyde, benzoic acid, methyl benzoate, nitrobenzene
Toluene	Benzyl alcohol, phenylacetic acid and methyl phenylacetate

A careful comparison of the individual spectra with those in the same and different series enables the effect of cyclisation and aromatisation to be observed as well as the appearance of bands typical of group frequencies.

Examples

(*a*) The absorption spectrum (Fig. 180) consists of few bands and the sub-stance is, therefore, relatively simple. The presence of O—H is indicated by the broad absorption at 3300 (stretching) and at 1020 cm^{-1} (C—OH stretching/deformation). Aromatic absorption is absent in the diagnostic regions (1600, 1500 and 850–700 cm^{-1}) but medium intensity absorption at 1650 may be due to unsaturation (C=C). This is confirmed by the much stronger adsorption at 920 and 990 cm^{-1} typical of the system H_2C=CH—.

(*b*) The presence of the hydroxyl group is indicated immediately by the absorption at 3300 and 1020 cm^{-1} (Fig. 181). This region also shows absorp-

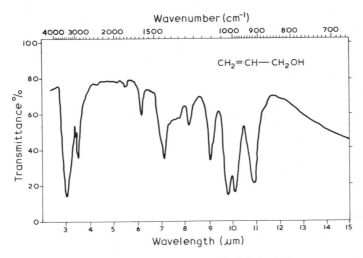

Fig. 180. Infrared spectrum of allyl alcohol

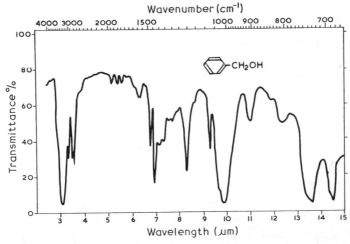

Fig. 181. Infrared spectrum of benzyl alcohol

tion due to C—H in HC=C and CH$_2$ at 3010 and 2800 cm^{-1} respectively. That the unsaturation is of an aromatic nature is evident from the characteristic absorption in the regions 1600, 1500 and 850–700 cm^{-1}. Aromatic patterns are also shown between 2000 and 1700 cm^{-1}. The hydroxyl group cannot be attached directly to the aromatic system otherwise enhancement of the absorption at 1600 cm^{-1} would be noticeable. Aromatic absorption at 695 and 735 cm^{-1} indicates monosubstitution of the benzene ring.

(c) The broad absorption band at 3000–2500 cm^{-1} (Fig. 182) indicates hydrogen-bonded O—H of —COOH. Hydrochlorides of organic bases give somewhat similar absorption but the carbonyl group is more likely in view of the strong carbonyl absorption at 1690 cm^{-1}. Confirmation is given by the bands at 1280 and 1310 cm^{-1}.(—C—O—) and by the characteristic broad band at 940 cm^{-1} due to bonded O—H deformation in dimeric acids.

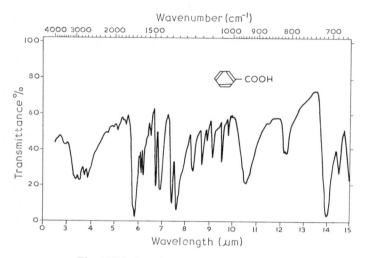

Fig. 182. Infrared spectrum of benzoic acid

An aromatic system is indicated by the absorption at 1600, 1500, 720 and 695 cm^{-1}, the latter bands being characteristic of monosubstitution in the benzene ring.

(d) The appearance of the absorption in the 3000–2000 cm^{-1} (Fig. 183) is typical of O—H stretching in a carboxylic acid dimer. Strong bands at 1690 and 1300 cm^{-1} are also indicative of this latter group. Other carbonyl absorption is also present at 1750 cm^{-1}, confirmed by the characteristic ester absorption at 1180 and 1220 cm^{-1}.

The compound is aromatic as shown by the bands at 1600, 1490 and 850–700 cm^{-1}. The absorption at 1600 cm^{-1} is very strong for an aromatic compound but it can be explained as being enhanced by the presence of polar substituents.

(e) Aliphatic C—H is present only in small amount because of the very weak absorption at about 3000 cm^{-1} (Fig. 184). The band at 3200 cm^{-1} may be due to hydroxyl but is probably (from the intensity and appearance), due

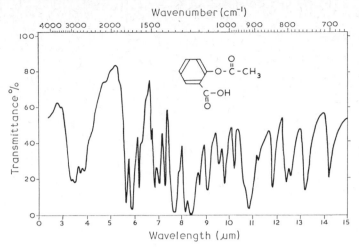

Fig. 183. Infrared spectrum of acetylsalicylic acid

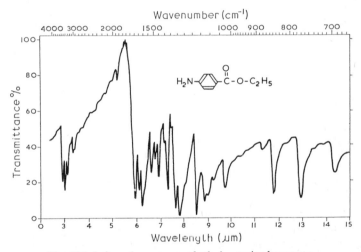

Fig. 184. Infrared spectrum of ethyl p-aminobenzoate

to NH stretching. An aromatic system is confirmed by the bands at 1580, 1500 cm^{-1} and by those in the 850–700 cm^{-1} region.

Strong carbonyl absorption at 1680 cm^{-1} might easily be interpreted as amide —C=O with the band at 1580 cm^{-1} as the second peak for amide absorption. This is in agreement with the N—H stretching frequency at 3200 cm^{-1}. However, typical ester absorption (C—O) appears at 1260 and 1280 cm^{-1} so that a better interpretation would be the presence of ester and amino groups.

(*f*) The absorption at about 3200 cm^{-1} (Fig. 185) may be due to the NH stretching vibration rather than to —OH; compare the normal appearance of

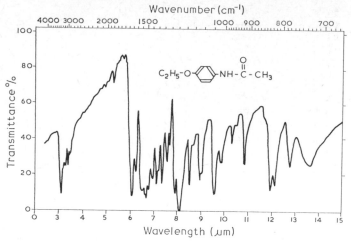

Fig. 185. Infrared spectrum of phenacetin

the latter absorption (Figs. 180 and 181) with that in this curve. An aromatic system is probably present because of the low intensity of CH_2 vibrations. This is confirmed by the absorption at 1600, 1500 and 850–700 cm^{-1}.

Carbonyl absorption is most likely to account for the band at 1650 cm^{-1}, as double bond systems give rise to much less intense absorption. It is, however, rather low in frequency for an aldehyde, ketone or ester. The band at 1550 cm^{-1} might be caused by NH deformation in an amide, by an amine or by a nitro group. When considered with the band at 1650 cm^{-1}, the most likely explanation is the presence of an amide group.

The absorption at 1250 cm^{-1} is certainly in the C—O stretching region of an ester group, but aromatic ethers also absorb strongly in this region. Some confirmation for the group is given by the band at 1040 cm^{-1}.

Quantitative Analysis

The principles underlying quantitative ultraviolet spectrophotometry apply also to infrared work. Moreover, infrared absorption spectra possess an advantage over those in the ultraviolet region in the greater number of bands present. It may often be possible to select a fairly strong band for each component in a mixture such that little or no interference occurs one with another. A calibration curve of extinction against concentration may be constructed or, if Beer's Law is shown to be obeyed, a direct comparison of sample and standard absorption may be used. Strictly, the integrated areas of the absorption bands should be compared but with sharp bands, peak heights can be used in calculations.

Infrared and ultraviolet spectrophotometry differ, however, in the concentrations used. Ten per cent solutions are common in infrared work and such a concentration is necessary because all solvents have some absorption in one part or another of the infrared region. This means that very short path lengths of 0.025 to 0.1 mm must be used in many assays. The high concentration of solute makes the accurate cancellation of solvent absorption very difficult but errors may be reduced by applying a base-line technique. The assumption is made that absorption due to solvent (or a second component) is constant or varies linearly with frequency over the region of the absorption band. The method of calculation is illustrated by reference to Fig. 186. *abc* is the recorded absorption of component A and *def* is the absorption caused by solvent and other components. Draw the line *agc* connecting the two minima *a* and *c* or between two suitable wavelengths on each side of the band. The point *g* is obtained by dropping a line perpendicular to the zero transmittance line to meet *ac* at *g*. The extinction $\log I_0/I_T$ is calculated from the distances shown in Fig 186.

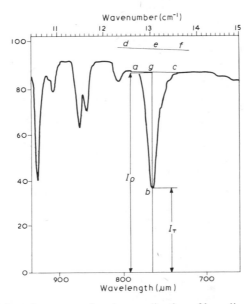

Fig. 186. Infrared spectrum showing application of base-line technique

Even when bands do interfere, simultaneous equations may be set up as described for ultraviolet work. As many more components may be determined from an infrared spectrum than is possible with an ultraviolet absorption spectrum the work involved in solving a large number of simultaneous equations is considerable. Computers and matrix methods for solving simultaneous equations are, therefore, being used increasingly in analytical work. Infrared absorption assays also differ from ultraviolet absorption assays in the initial period of investigation necessary to obtain reproducible conditions. It may take some time to select the right solvent, cell thickness, slit width and absorption band.

For the determination of small quantities of substances, present either as impurity or as solute in a preparation, compensation for absorption by the major component or solvent in the preparation may be achieved by introducing that component into the reference beam. A spectrum of the minor component only is thereby obtained. This method has been applied to the determination of β-picoline, 2,6-lutidine and 2-ethylpyridine in γ-picoline and also to solutions of testosterone propionate in arachis oil. Sufficient energy must be available for reliable results and the validity of Beer's Law for appropriate standard solutions will normally confirm this.

Practical Experiments

Experiment 1. *Determination of the vibration-rotation spectrum of hydrogen chloride, and calculation of the force constant, moment of inertia and bond length for* HCl

Method. Place a thin layer of sulphuric acid in a 4 cm quartz or fused silica cell such as is used in ultraviolet spectrophotometry. Add carefully a few crystals of sodium chloride and cover the cell immediately with a polythene cap. Record the absorption curve over the region 3500–2500 cm⁻¹ by means of a grating instrument using an empty quartz cell in the reference beam, and

- (a) determine the position (in cm⁻¹) of the Q branch, i.e. the 'gap' in the spectrum (Fig. 187) and hence calculate k the force constant.
- (b) count the number of bands in the P and R branches and the range in cm⁻¹. Calculate the average separation in wave numbers and hence obtain I, the moment of inertia.
- (c) calculate the bond distance from the relationship $I = \mu r^2$ (see below).

Adequate resolution for the complete experiment is obtained with a simple grating instrument or a more expensive large prism machine. If a simple prism instrument is available, the Q branch may still be observed and part (a) of the calculation can be done. The 4 cm quartz cell enables a gas cell to be

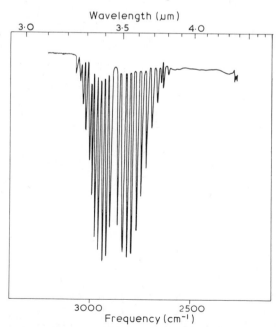

Fig. 187. Infrared spectrum of hydrogen chloride

dispensed with although the latter is essential if HI or HBr are to be examined.

The change in vibrational energy observed in this experiment is due to the transition $V_0 \rightarrow V_1$ (Fig. 134). Changes in rotational energy accompany this transition so that the net result is given by

$$\Delta E = \Delta E_{\text{vibrational}} \pm \Delta E_{\text{rotational}} \tag{1}$$

The rotational energy of the molecule is given by

$$E_{\text{rotational}} = \frac{h^2}{8\pi^2 I} J(J+1)$$

where

$$J = \text{rotational quantum number}$$

$$I = \text{moment of inertia}$$

When changes occur ΔJ is restricted to ± 1 so that

$$\Delta E_{\text{rotational}} = \frac{h^2}{8\pi^2 I} 2(J+1)$$

or

$$v = \frac{h}{8\pi^2 I c} \times 2(J+1)$$

$$= 2B(J+1)$$

where

$$c = \text{velocity of light}$$

B is a constant and therefore according to expression (1) above, the spectrum should consist of equally spaced peaks on each side of a central position corresponding to $\Delta E_{\text{vibrational}}$. As a result of the selection rule $\Delta J = \pm 1$, $(J+1)$ may be ± 1, ± 2, $\pm 3 \ldots$ and therefore absorption corresponding to $\Delta E_{\text{vibrational}}$ alone will not be observed. This 'gap' in the spectrum is known as the Q branch and shows clearly in the absorption spectrum obtained in an actual experiment and recorded in Fig. 187. When $(J+1)$ is equal to $+1, +2, +3 \ldots$ the R branch is observed, and when it is equal to $-1, -2, -3 \ldots$ the P branch is recorded.

The frequency of the radiation associated with the transition $V_0 \rightarrow V_1$ is loosely referred to as the vibrational frequency of the molecule. By treating the two atoms H and Cl as two masses connected by a spring and applying Hooke's law the frequency is shown to be, as a first approximation

$$\bar{v} = \frac{1}{2\pi c} \sqrt{\left(\frac{k}{\mu}\right)}$$

where

$$c = \text{velocity of light}$$

$$k = \text{force constant}$$

$$\mu = \text{reduced mass of system}$$

$$= \frac{m_1 m_2}{m_1 + m_2} \times \frac{1}{N}$$

where

$$m_1, m_2 = \text{respective } \textit{atomic weights of the atoms}$$

$$N = \text{Avogadro number.}$$

$\bar{v}$ is obtained from the position of the Q branch, hence k may be determined. From the average separation of the peaks, say x cm^{-1}

$$x = \frac{h}{8\pi^2 Ic} \times 2 \quad (J+1 = 1)$$

$$\therefore \quad I = \frac{h}{4\pi^2 xc}$$

Now

$$I = \mu r^2$$

where

$$r = \text{internuclear distance}$$

$$\therefore \quad r^2 = \frac{I}{\mu}$$

$$r = \sqrt{\left(\frac{I}{\mu}\right)}$$

$$= \sqrt{\left\{(I \times N) \middle/ \frac{1.008 \times 35.46}{36.47}\right\}}$$

From Fig. 187, it may be seen that the separation between the peaks in the P and R branches is not strictly uniform and therefore not in accord with the simplified discussion presented above. The discrepancy is due to coupling between the rotation and vibration of the molecule. Also the peaks vary in intensity due to the fact that the population of the various rotational energy levels varies in the lower state ($V = 0$). For details of these and other factors which complicate the spectra, see Banwell (1966) or Barrow (1962).

Application of the formula for the frequency of vibration (in cm^{-1}) enables *approximate* figures to be obtained from some of the group frequencies, for example for C = O, CH, N—H, and so on. The value of k used in the calculation is 5×10^5, 10×10^5 and 15×10^5 dynes/cm for single, double and triple bonds respectively.

Experiment 2. *Determination of the absorption spectrum of a liquid—cyclohexanol.*

Method. Check the instrument for 100% and 0% transmission and select the slit width programme (minimum slit width). Place a drop of the liquid on a sodium chloride plate of a demountable cell and cover with another sodium chloride plate thus forming a capillary film of liquid. Insert the cell into the sample beam. If the film is too thick, carefully tighten the screws holding the cell until a transmission of about 5% is obtained for the strongest absorption band. Do not overtighten or the plates may crack.

After use, dismantle the cell, rinse the plates with chloroform and allow to dry under a lamp to avoid condensation of water vapour as the chloroform evaporates.

The thickness of the film can be varied by means of lead spacers but the method adopted above proves satisfactory in most cases. The liquid chosen here illustrates, in conjunction with the results of Experiment 5, the effect of hydrogen bonding on the absorption spectrum. Phenols, alcohols and carboxylic acids show this effect, and for further discussion see Experiment 5.

Experiment 3. *Determination of the absorption spectrum of a solid using the mull technique*

Method. Powder thoroughly about 15–20 mg of sample in a small agate mortar and add 2 drops of purified liquid paraffin. Continue the trituration until a smooth paste is obtained and transfer it to one of the sodium chloride plates of a demountable cell. Assemble the cell and continue as described for a liquid in Experiment 2.

It is essential to reduce the particle size of the sample to below 3 μm otherwise scattering of radiation will give rise to a poor absorption spectrum. With care, as little as 3 mg of material may be used with a small amount of liquid paraffin. In this method unless the material dissolves in the paraffin, the spectrum obtained is that of the *solid* sample. Crystal forces and hydrogen bonding may, therefore, influence the trace obtained. Strong bands due to paraffin itself appear at 2920–2850 and 1460–1380 cm^{-1}.

Experiment 4. *To determine the path length of a sealed cell*

Method. Record the spectrum of the *empty* cell versus air and note the wavelengths between which lie an integral number of fringes. Calculate the thickness of the cell by means of the equations:

$$\text{(i)} \ t \text{ (in cm)} = \frac{n}{2}\left(\frac{1}{\bar{v}_1 - \bar{v}_2}\right) \ \text{where } n = \text{number of fringes}$$

$$\bar{v}_1, \bar{v}_2 - \text{wavenumbers selected}$$

$$\text{or (ii)} \ t \text{ (in } \mu m) = \frac{n}{2}\left(\frac{\lambda_1 \times \lambda_2}{\lambda_2 - \lambda_1}\right) \ \text{where } \lambda_1, \lambda_2 = \text{wavelengths selected (}\mu m)$$

The 'spectrum' obtained in this experiment is that of a typical interference fringe pattern and may be obtained for cells up to about 0.5 mm in path length. To illustrate the calculation refer to the fringe pattern in Fig. 188.

$$t = \frac{6}{2}\left(\frac{5 \times 10.1}{10.1 - 5}\right)$$

$$= 29.7 \ \mu m$$

$$= 0.030 \text{ mm}$$

The appearance of the pattern itself often indicates the quality of the cell: a poor fringe pattern is given by cells in which the plates are not exactly parallel or have become worn after use. Compare the patterns for *a* and *b* in Fig. 189.

For cells longer than about 0.5 mm the path length may be determined in several ways:

(i) by examination of a substance whose extinction coefficient is accurately known

(ii) by use of a travelling microscope
(iii) by comparison of a compound in the cell with the same compound in a previously calibrated variable path length cell.

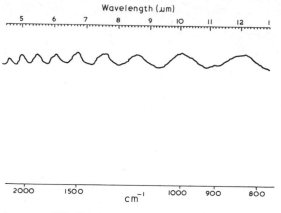

Fig. 188. Interference fringe pattern

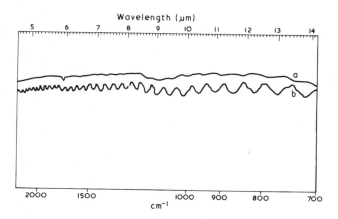

Fig. 189. Interference fringe pattern

Experiment 5. *Determination of the absorption spectra of solutions of cyclohexanol to show the effect of dilution on the* O—H *absorption*

Method. Prepare two solutions (10% w/v and 1% w/v) of cyclohexanol in carbon tetrachloride. Obtain the absorption spectrum for each in a 0.1 mm cell (10% solution) and a 1 mm cell (1% solution) over the range 4000–2500 cm^{-1}.

Compare the curves with that obtained in Experiment 2. In the liquid state, a broad absorption band is recorded and this is retained in the strong (10%) solution. Note how this is reduced in the 1% solution; note also the increase in absorption at shorter wavelengths. This effect is due to a change from polymeric to dimeric and finally to monomeric hydroxyl absorption:

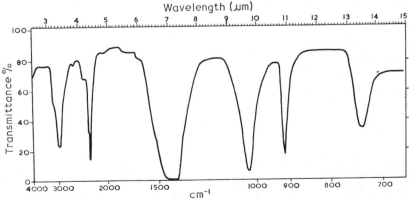

max ≈ 3300 cm⁻¹   max ≈ 3610 cm⁻¹
(broad)   (sharp)

The formation of hydrogen bonds will tend to weaken the O—H bond, hence k, the force constant, will decrease, and the observed vibrational frequency will be less than that for the monomeric form. Carboxylic acids, however, remain dimeric even on dilution. If the hydrogen bond is intramolecular, no change in the O—H absorption will occur on dilution.

Solution spectra are useful in that difficulties inherent in solid state spectra are absent. However, solvent effects may occur as a shift of the absorption frequency of a substance when one solvent is substituted for another. All solvents absorb in some part of the infrared region and, therefore, several solvents are necessary to cover the whole spectrum. Figures 190–199 show the absorption of some common solvents. In this experiment carbon tetrachloride transmits most of the incident light and no reference cell is required.

The concentration needed for solutions will vary according to the substance and cell thickness, but a useful concentration for preliminary investigation is 10 per cent w/v in a 0.1 mm cell.

Very thin films of aqueous solutions can be examined in cells of barium fluoride, Irtran-2 or silver chloride but these solutions and, indeed, almost any kind of sample can be handled by a technique called attenuated total reflectance,

Fig. 190. Infrared spectra of common solvents:
Acetonitrile (0.1 mm cell)

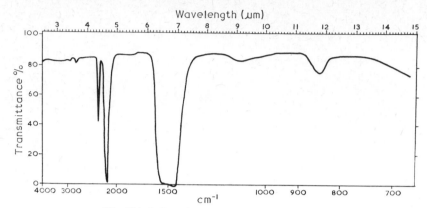

Fig. 191. Infrared spectra of common solvents:
Carbon disulphide (0.1 mm cell)

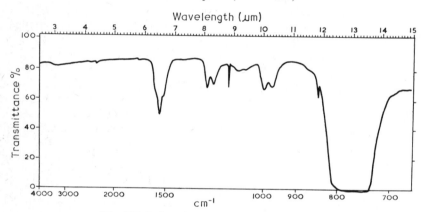

Fig. 192. Infrared spectra of common solvents:
Carbon tetrachloride (0.1 mm cell)

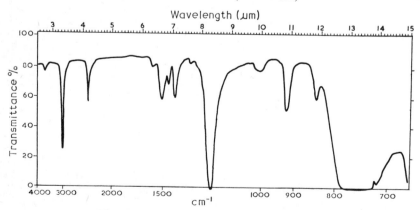

Fig. 193. Infrared spectra of common solvents:
Chloroform (ethanol free) (0.1 mm cell)

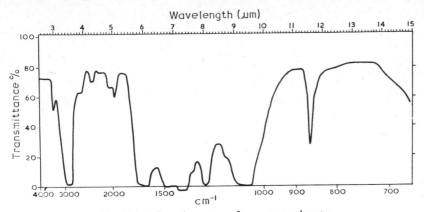

Fig. 194. Infrared spectra of common solvents:
Dimethylformamide (0.1 mm cell)

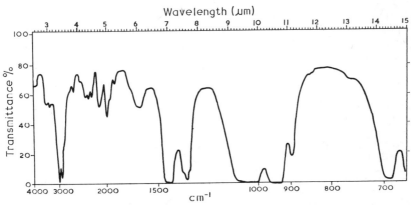

Fig. 195. Infrared spectra of common solvents:
Dimethyl sulphoxide (0.1 mm cell)

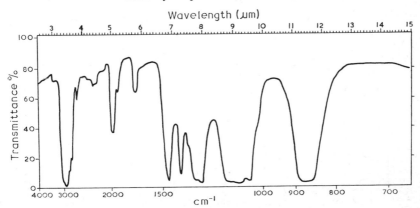

Fig. 196. Infrared spectra of common solvents:
Dioxan (0.1 mm cell)

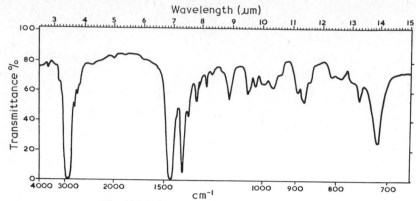

Fig. 197. Infrared spectra of common solvents:
n-Hexane (0.1 mm cell)

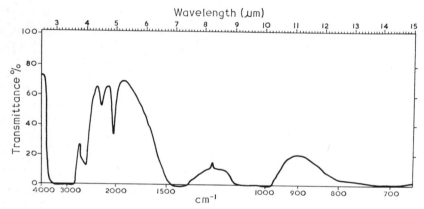

Fig. 198. Infrared spectra of common solvents:
Methanol (0.1 mm cell)

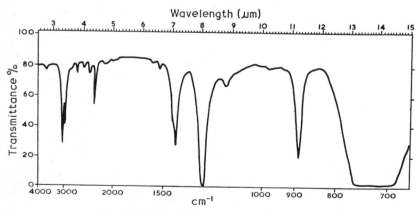

Fig. 199. Infrared spectra of common solvents:
Methylene chloride (0.1 mm cell)

which is based upon total internal reflectance. The efficiency of the interface between two media as a mirror is dependent on the difference in refractive indices of the media. As the refractive index of a material changes rapidly in the region of an absorption band these changes result in a decreased amount of energy reflected at the interface. The resulting spectra are, strictly, plots of reflected energy against wavelength and appear very similar to conventional absorption bands; they provide essentially the same information.

The technique requires an accessory for the infrared spectrophotometer and the basic design for the unit handling micro-samples is shown in Fig. 200.

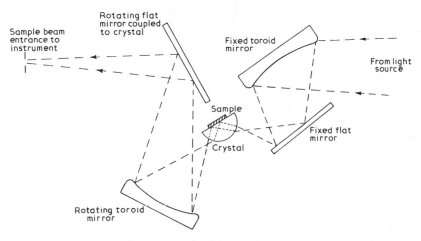

Fig. 200. Unit for handling micro samples

Samples are held in contact with the flat face of the crystal hemicylinder and different crystals are required to provide a satisfactory range of refractive indices to cover all samples. Typical crystal materials are KRS-5, germanium and silver chloride; giving a high, medium and low refractive index respectively. The rotating mirrors are necessary to allow various angles of incidence on the crystal so that optimum attenuated total reflectance spectra are obtained.

Cognate Experiments

Benzoic acid
Catechol
o,o' and p,p'-Dihydroxydiphenyl

Experiment 6. *Determination of the absorption curve of a solid using the potassium chloride disc technique*

Method. Finely powder the sample (1 mg) and mix with dry finely powdered potassium chloride (200 mg). Triturate the mixture thoroughly for about 2 min and transfer to a suitable die and press. Attach the die to a vacuum pump, allow to stand for 1 min under vacuum and compress the powder at 10–12 tons/sq. inch for 2 min. Release the pressure and vacuum, and extract the disc for insertion in the sample beam of the spectrophotometer.

For potassium chloride alone the disc should be completely transparent, but it may become translucent when the sample is present. The disc is about 13 mm diameter and the above quantities will, therefore, give a weight of sample between 10 and 15 $\mu g/mm^2$.

Attention has already been drawn to the care necessary when using this technique for comparison purposes. It has the advantage, however, of giving spectra free from solvent peaks, and is, therefore, extremely useful as a routine method.

The disc method has not entirely superseded the mull technique for the spectra of solid samples because differences have been observed in the results obtained by using the two techniques; the differences are related to the treatment the sample received during the preparation of the disc. In finely ground material, for example, interaction may occur between vibrations of the sample and the potassium halide lattice. For a discussion of the infrared analysis of solid substances; see Duyckaerts (1959).

The method is also used in the examination of the very small quantities of material eluted from columns in gas chromatography. The fraction is trapped in about 300 mg of potassium chloride in the form of a short column placed immediately after the detector. The solid is powdered, pressed into a disc in the normal way and the absorption spectrum of the trapped fraction is determined.

Experiment 7. *Determination of the amount of phenobarbitone in Phenobarbitone Tablets*

Method. Prepare a series of solutions of phenobarbitone in chloroform containing 20, 40, 60, 80, and 100 mg in each 10 ml. Transfer each solution in turn to a 0.1 mm cell and measure the absorption curve over the region 4.4–6.5 μm. Calculate the absorbance using the base-line technique for each solution at the peak absorption (about 1730 cm^{-1}) and plot a graph of absorbance against concentration in the normal manner.

Weigh and powder 20 tablets. Extract with chloroform an accurately weighed quantity of powdered tablets equivalent to about 80 mg of phenobarbitone. Evaporate the extract to dryness and dissolve in sufficient chloroform to produce 10 ml of solution. Determine the absorption curve as described above and calculate the absorbance at the selected peak.

Read off the number of mg of phenobarbitone from the graph. Hence calculate the number of mg of phenobarbitone in a tablet of average weight.

Careful standardisation of instrumental conditions is essential in quantitative analysis. Of these the 100% and 0% transition and the slit width settings are very important. If the calibration curve obeys Beer's Law, a direct comparison of the absorbance of test and standard solutions at the peak absorption may be used. This direct comparison was adopted in the United States Pharmacopoeia XVI for the assay of Acetazolamide and Diethyltoluamide.

Failure to obey Beer's Law means that reference must be made to a calibration curve for the determination of concentration.

Experiment 8. *Determination of the amount of acetylsalicylic acid and phenacetin in Acetylsalicylic Acid and Phenacetin Tablets*

Method. (i) *Calibration curves.* Carefully check the 100% and 0% transmission over the 8–10 μm range and adjust, if necessary. Prepare the following solutions using chloroform as solvent:

Acetylsalicylic acid   1, 2, 3 and 4% w/v
Phenacetin   1, 1.5, 2 and 2.5% w/v

Record in a 0.1 mm cell the absorption curve of each solution against a blank of chloroform in the reference beam. With chloroform alone the 100% line is obtained. Calculate absorbances for each solution at two wavelengths—about 8.95 μm at which phenacetin absorbs and at about 9.22 μm at which acetylsalicylic acid absorbs. Plot the calibration curves as shown in Fig. 200 which were obtained in an actual experiment.

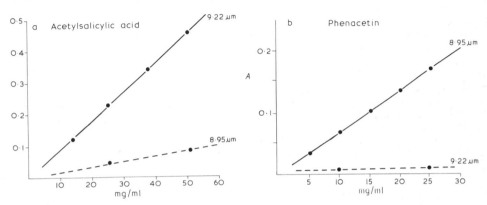

Fig. 201. Calibration curves for (a) acetylsalicylic acid and (b) phenacetin

(ii) *Tablets.* Weigh and powder 20 tablets. Transfer an accurately weighed quantity of the powder equivalent to about 0.3 g of acetylsalicylic acid to a sintered-glass crucible and wash with small portions of chloroform until the acetylsalicylic acid and phenacetin are completely extracted. Collect the washings and evaporate to dryness. Dissolve the residue in chloroform and transfer quantitatively to a 10 ml flask. Make up to volume with chloroform, mix, and record the absorption curve over the 8–10 μm region using a blank of chloroform. Calculate the absorbances at about 8.95 and 9.22 μm and, by reference to the calibration curves, work out the concentrations of acetylsalicylic acid and phenacetin in the solution, and hence in the tablets.

(iii) *Calculation.* From the absorbance at 9.22 μm calculate the concentration of acetylsalicylic acid by reference to the calibration curve. Read off the correction for phenacetin corresponding to this apparent concentration and apply it to the observed absorbance at 8.95 μm. Read off the apparent concentration of phenacetin and also the correction to be applied to the absorbance for acetylsalicylic acid. Apply this correction and repeat the above procedure.

Although the base-line technique described in Experiment 7 may be applied to acetylsalicylic acid and to phenacetin when present alone it is not possible to apply it in this instance: Fig. 202 shows the effect. The absorption curve of each is not linear over the region selected in this assay so that the reference points for calculation are the carefully checked 100% and 0% transmission lines. Equations may be developed for calculating the concentrations but the method outlined above is quick and simple in practice as the following numerical example shows.

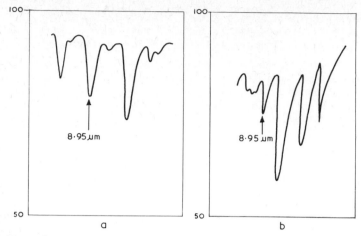

Fig. 202. Absorption curve of (a) phenacetin (10.1 mg/ml) in chloroform, and (b) phenacetin (10.1 mg/ml) in chloroform containing acetylsalicylic acid (25.4 mg/ml)

| Test Solution: | Acetylsalicylic Acid | 25.4 mg/ml |
| | Phenacetin | 10.12 mg/ml |

A (8.95 μm) = 0.109

A (9.22 μm) = 0.236

First calculation using E (9.22 μm) gave acetylsalicyclic acid as 28.0 mg/ml. The correction to the absorbance at 8.95 μm was 0.045 (from graph)

$$\therefore \quad \text{Phenacetin} = 9.4 \text{ mg/ml.}$$

Second calculation. The correction to the absorbance for acetylsalicylic acid because of the phenacetin was 0.0175.

$$\therefore \quad \text{Acetylsalicylic acid} = 25.4 \text{ mg/ml.}$$

Using this concentration the correction to be applied to the absorbance for phenacetin was 0.040.

$$\therefore \quad \text{Phenacetin} = 10.1 \text{ mg/ml.}$$

Third calculation.

Correction for acetylsalicylic acid = 0.018.

$$\therefore \quad \text{Acetylsalicylic acid} = 25.4 \text{ mg/ml.}$$

Correction for phenacetin = 0.40

$$\therefore \quad \text{Phenacetin} = 10.1 \text{ mg/ml.}$$

The results of the second and third calculations are identical and, therefore, no further calculations need be performed.

Experiment 9. *To determine the percentage v/v of chloroform in Chloroform Spirit*

Method. Prepare a series of solutions of ethanol-free chloroform in ethylene glycol to contain 0.2, 0.5, 1.0, 1.5 and 2.0% w/v of chloroform. Treat each solution in turn in the following way. Transfer the solution (0.1 ml) to a flask approximately 1 ml in volume and attach to an evacuated gas cell (Fig. 203). Open tap A and carefully heat the flask for 2 min at 150° in an oil bath. With the flask still in the bath close tap A. Insert the gas cell into the sample beam and record the absorption spectrum over the range 820–700 cm^{-1}. Using the base-line technique calculate the absorbance of each solution and plot the values against concentration of chloroform to construct a calibration curve.

Dilute the sample (5 ml) to 25 ml with ethylene glycol and mix well. Treat 0.1 ml of solution as described above and determine the percentage w/v of chloroform in the sample by reference to the calibration curve. Convert the result to percentage v/v by dividing by the weight per ml of chloroform (1.48).

The apparatus (Fig. 203) is a normal gas cell of about 150 ml volume and reproducible results are obtained by using the same cell, flask and pipette for each determination. An internal standard is not necessary providing extreme care is taken in using the pipette.

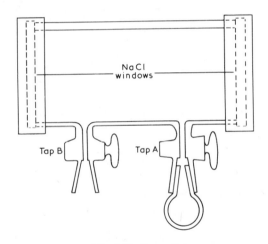

NaCl windows

Tap B

Tap A

Fig. 203. Gas cell

Tap B (Fig. 203) remains closed in this determination but when aqueous solutions are examined a small flask containing phosphorus pentoxide may be used to maintain a low level of water vapour during the determination.

Cognate Determinations

Compound Chloroform and Morphine Tincture
Chloroform Liniment

Experiment 10. *To determine the percentage of methanol in Orciprenaline Sulphate*
The determination of solvent residues in solid chemicals is a normal part of quality control and is included in such terms as loss on drying or volatile

matter. Ordinary drying processes do not, however, guarantee complete elimination of water or organic solvent and a number of examples exist where specific methods of determination are applied—water by titration with Karl–Fischer reagent and acetone, dichloromethane, isopropanol and methanol by GLC (Chapter 3).

Some of these are readily determined by vapour phase infrared absorption with sample weights considerably less than those required for GLC. The infrared spectrum also identifies the solvent.

Method. Accurately weigh the sample (about 0.03 g) into the small bulb (Fig. 203), add diethylene glycol (0.2 ml) and attach to an evacuated gas cell. Complete the determination as described for chloroform (above) but using aliquot portions of methanol in diethylene glycol for calibration purposes.

Note. Methanol (1 mg) gives a suitable absorbance.

Cognate Determinations

Colchicine (0.1 g) for determination of ethyl acetate or chloroform.
Novobiocin Calcium (0.15 g) for determination of acetone.
Novobiocin Sodium (0.15 g) for determination of acetone.
Commercial batches of Novobiocin salts contained acetone as solvent but this is not stated in the Pharmacopoeia which presumably allows other solvents.
Streptomycin Sulphate (0.15 g) for determination of methanol.
Warfarin Sodium (Clathrate) (0.1 g) for determination of isopropanol.

Experiment 11. *To determine the percentage of menthol in Peppermint Oil using the near infrared region of the spectrum*

Method. Prepare standard solutions of menthol in carbon tetrachloride to contain 0.5, 1.0, 1.5, 2.0 and 2.5% w/v of menthol. Measure the absorption curve of each solution in 4 cm glass or quartz cells over the region 8000–7000 cm^1. Calculate the absorbance at the peak absorption (about 7200 cm^{-1}) using the base-line technique and construct a calibration curve.

Accurately dilute the peppermint oil with carbon tetrachloride to give a concentration of menthol of about 1.5% and determine the absorbance at about 7200 cm^{-1} as described above. By reference to the calibration curve, calculate the percentage w/v of menthol in the oil and finally convert to percentage w/w by using the weight per ml of the oil.

The near infrared region of the spectrum is becoming of increasing interest with the advent of sensitive detectors. The spectra consist mainly of overtones of the fundamental bands and are less intense than those of the latter. Nevertheless, because solvents can be used in cells of long path length, the region has potential value for determination of those groups which show reasonable absorption. Figure 204 illustrates the groups which may be determined and the regions of absorption.

The assay for menthol depends upon the determination of the hydroxyl group, and is therefore less specific than an assay which involves absorption due perhaps to a skeletal vibration characteristic of menthol. The result of the assay should be similar to that obtained by the Pharmacopoeial method.

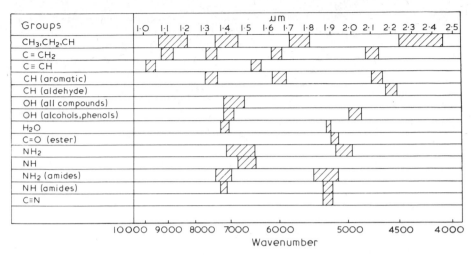

Fig. 204. Characteristic absorptions in near infrared

Experiment 12. *To determine the cis-trans isomer ratio in Clomiphene Citrate*

The shape of molecules often plays a significant role in the medicinal action of many compounds and if isomers are formed in the synthesis it may be important to include in the specification a test for the relative proportions of one to the other. This control ensures uniformity of composition from batch to batch and one method is described for the *cis*-isomer of Tranylcypromine under Gas Chromatography (Chapter 3). It is well known that *cis*- and *trans*-substituted double bonds have slightly different infrared absorption bands in the 13 μm region and this feature is the basis of the present determination.

Method. Dissolve *trans*-clomiphene citrate AS (22.5 mg) and *cis*-clomiphene citrate AS (52.5 mg) in water (10 ml) in a separator and add sodium hydroxide solution (5%, 1 ml). Extract the mixture of bases with ether (3 × 10 ml) and wash the combined ether extracts with water (2 × 10 ml). Dry the ether with Na_2SO_4 (anhydrous), filter, evaporate to dryness and dissolve the residue in carbon disulphide (1 ml). Record the absorption curve in a 0.2 mm cell over the range 12.50 to 14.00 μm and calculate the absorbance for the peaks at about 13.16 and 13.51 μm using the base-line technique between the minima at about 12.66 and 13.89 μm.

Repeat the determination on a 1:1 mixture (75 mg) of *cis* and *trans*-clomiphene citrates AS and on Clomiphene Citrate (75 mg). Calculate the ratio

$$\frac{\text{Absorbance at } 13.16 \ \mu\text{m}}{\text{Absorbance at } 13.51 \ \mu\text{m}}$$

for each determination and confirm that the ratio for the sample lies between the ratios for the standards indicating that the sample contains 50–70% *cis*-clomiphene citrate.

Experiment 13. *To determine the percentage of thiotepa* $C_6H_{12}N_3PS$ *in Thiotepa Injection*

The preparation consists of a sterile powder of Thiotepa, Sodium Bicarbonate and Sodium Chloride in a sealed container to which Water for Injection is added before use.

Method. Carry out the determination of Uniformity of Weight (Method B, Part I, Chapter 14) and from the mixed contents weigh a quantity containing the equivalent of 75 mg of Thiotepa into a sinter-glass crucible. Extract with carbon disulphide (3×5 ml) (Note) and allow to filter through into a small flask. Evaporate the filtrates to low volume, transfer quantitatively to a 10 ml volumetric flask and dilute to volume with carbon disulphide. Measure the absorbance in a 0.1 mm cell over the region 9–12 μm and calculate the value at the maximum at about 10.75 μm using the base-line technique. Use Thiotepa AS treated in the same way for comparison and calculate the content of $C_6H_{12}N_3PS$ in the average contents of one ampoule.

Note. The utmost care should be exercised in the handling of carbon disulphide because of its toxicity and inflammability.

Experiment 14. *Determination of Cyclophosphamide in Cyclophosphamide Tablets*

The purpose of this experiment is to illustrate the use of the potassium chloride disc in quantitative work by means of an internal standard. It has a great degree of specificity and, although rather long, is still more rapid than the pharmacopoeial method.

The internal standard must necessarily have few absorption bands and ferric thiocyanate meets this requirement. The method is described in the National Formulary XII (1965) but is incorrect in the USP XVIII in that the equivalent of 4 ml of internal standard solution is used for the tablets and reference is made to a standard curve prepared from 5 ml of internal standard solution.

Internal Standard Solution. Extract a mixture of ferric ammonium sulphate (2.63 g) and ammonium thiocyanate (26.3 g) in water (250 ml) with ethyl acetate (6×100 ml). Dry the extracts ($MgSO_4$, dried) and filter. Make up to 1000 ml with ethyl acetate.

Standard Curve. Use three quantities of Cyclophosphamide (50, 100 and 150 mg) accurately weighed and treat each as follows.

Dissolve in water (50 ml), extract with chloroform (3×10 ml) collecting the extracts in a 50 ml volumetric flask. Add the internal standard solution (4 ml) and dilute to volume with chloroform. Transfer the solution (1.00 ml) to a mortar containing potassium chloride (500 mg) and evaporate the solvent with the aid of a current of air and finally a vacuum (2 min). Prepare a disc from the dried powder and record the infrared absorption spectrum. Calculate the absorbance (Note) of the internal standard band at 4.9 μm and the Cyclophosphamide band at 9.5 μm by the base-line technique. Calculate the ratio $A_{9.5}/A_{4.9}$ and construct a calibration curve.

Method. Treat the finely powdered tablets, accurately weighed (equivalent to about 350 mg Cyclophosphamide), in a sinter-glass crucible with chloroform as described for Phenobarbitone Tablets (p. 352). Transfer the filtrate to a 200 ml volumetric flask, add the internal standard solution (16.0 ml) and dilute to volume with chloroform. Proceed as described for the standard curve from the words 'Transfer 1.00 ml..' Calculate the ratio $A_{9.5}/A_{4.9}$ and read off the concentration of Cyclophosphamide from the curve. Calculate the content of Cyclophosphamide in the tablets.

Note. The term 'absorbance' instead of 'extinction' is used throughout this Chapter in conformity with current practice.

Routine Maintenance

Demountable Cells

The cells for use in infrared spectrophotometry require considerable care in storage and handling, as the sodium chloride or potassium bromide windows are fogged by traces of moisture and are easily scratched. A small cabinet fitted with a simple heater, e.g. a 60 watt lamp and a dish of silica gel as drying agent should be used for storing all cells, windows or plates, and relevant apparatus such as the die for preparing potassium chloride discs. The cabinet and contents are thereby maintained at about 10–20° above room temperature and there is little risk of fogging from atmospheric moisture.

In spite of all precautions, however, the windows eventually become opaque to visible radiation, and it is necessary to repolish them if the transmission of infrared radiation is much reduced by scatter. A polishing kit can be purchased which normally consists of a glass plate with a coarse and finely ground surface, a polishing cloth stretched on a firm base, polishing powder, e.g. jeweller's rouge, and an optical flat. They repay their cost in a short time as all types of defect, e.g. fogged or scratched surfaces in the windows, may be remedied.

The following system, however, has proved satisfactory in obtaining service-able windows from sodium chloride 'blanks' within 6 to 7 min. Blanks offer a considerable saving in cost over the price of polished windows and they normally occur with surfaces similar to those obtained from the glass roughing and smoothing laps of a polishing kit.

Materials. Sodium chloride blanks
    Two polishing cloths
    Two thick glass or wooden bases on which the polishing cloths are firmly
      spread
    Polishing powder
    Ethanol containing a few crystals of sodium chloride.

Method. To one of the polishing cloths add a small amount of powder and saturate with ethanol. Place the blank over the powder and move it with firm pressure, rapidly, in a linear motion (Note). At short intervals rotate the blank and continue the polishing using more powder and ethanol. Examine the blank after one or two minutes by lightly drawing it over a clean portion of the cloth and continue the treatment if the surface is not smooth and transparent. Complete the polishing of the one side of the blank with rapid movements on the second clean and dry polishing cloth. Repeat for the other side of the disc.

Note. Some polishers prefer circular motion but this is a matter of personal choice.

This method will not remove deep scratches for which ground glass surfaces or very fine sand paper are necessary. Gloves are not essential for handling sodium chloride blanks.

Sealed Cells

These consist of two sodium chloride or potassium bromide windows separated by a lead spacer, the whole being held together in a metal frame along with protective washers. Unless the transmission of the cell has deteriorated appreciably and the cell is in obvious need of renovation it is unwise to attempt to dismantle it. However, teflon spacers instead of lead spacers are now being used, and this simplifies considerably the dismantling and cleaning of cells.

Method. Note the construction of the cell, isolate the window assembly and carefully separate the two windows and lead spacer. These are normally firmly stuck but the insertion of a razor blade into the join and *gentle* leverage at each corner will generally separate them. Peel off the lead spacer and polish the two windows in the normal way.

To re-assemble the cell, prepare a new lead spacer of the required thickness by lightly rubbing it all over with mercury (Note 1). Place it on the polished window and add the second polished window. Press lightly together and incorporate into the metal frame using the washers (Note 2). Tighten the screws in the frame carefully to avoid cracking the windows and set aside for 24 hours (Note 3). Test the cell for leakage, using chloroform, by filling and allowing to stand for an hour; the solvent should be retained.

Note 1. Mercury is poisonous by absorption as vapour or through the skin. Gloves should be worn in this preparation and care taken to avoid undue delay in completing the assembly. All residues should be transferred *immediately* to a closed container.

Note 2. Take care to line up the holes in the plate with those in the washer.

Note 3. During this time the lead spacer absorbs mercury and it swells, thus ensuring a good seal.

Potassium Chloride

The method of Hales and Kynaston (1954) is satisfactory in yielding a uniform powder which compresses readily into discs.

Method. Prepare a saturated solution of potassium chloride (AnalaR, 500 ml) and filter (sintered glass) from undissolved salt. Divide the filtrate into two (300 ml and remainder of filtrate) and add the 300 ml portion to ice-cold hydrochloric acid (300 ml) with stirring. Filter off the precipitate on a sintered-glass filter using a Buchner flask. Drain well and wash the precipitate with the reserved saturated solution in portions, draining well in between each wash. Transfer to filter paper and press gently to remove as much liquid as possible and dry in a large porcelain dish at 120° for two hours. Break up the powder with a glass rod and complete the drying at 400° overnight in a muffle furnace (Note 1). Allow to cool and distribute the contents among weighing bottles each holding about 10 g. Store in a desiccator and use the contents of each bottle in turn (Note 2).

Note 1. Heating at 400° removes the last traces of water, and hydrochloric acid, which is difficult to remove completely, is reduced to negligible amounts.

Note 2. Distribution among several containers avoids constantly exposing all the material to the air during the period of its use.

References

Banwell, C. N., *Fundamentals of Molecular Spectroscopy*, McGraw-Hill, London, 1972.

Barrow, G. M., *Introduction to Molecular Spectroscopy*, McGraw-Hill, London, 1962.

Bellamy, L. J., *Infrared Spectra of Complex Molecules*, 2nd edn., Methuen, London, 1958.

Conley, R. T., *Infrared Spectroscopy*, Allyn and Bacon, Boston, 1966.

Cross, A. D., *Practical Infrared Spectroscopy*, Butterworths, London, 1964.

Duyckaerts, G., *Analyst* (1959) **84**, 201.

Hales, J. L. and Kynaston, W., *Analyst* (1954) **79**, 702.

12 Nuclear Magnetic Resonance Spectroscopy

R. T. PARFITT

INTRODUCTION

Nuclear magnetic resonance (NMR) spectroscopy is a technique that permits the exploration of a molecule at the level of the individual atom (usually hydrogen), and affords information concerning the environment of that atom. Only about one half of known element isotopes, when placed in a magnetic field, absorb energy from the radiofrequency region of the electromagnetic spectrum. Of these isotopes ^{1}H, and ^{13}C are the most important from the viewpoint of the organic and pharmaceutical chemist. The precise frequency from which energy is absorbed gives an indication of how an atom is bound to, or located spatially with respect to, other atoms. Thus NMR offers an excellent physical means of investigating molecular structure and molecular interactions. NMR spectra may also be used for compound identification, by a fingerprint technique similar to that employed in infrared spectroscopy, and sometimes as a specific method of assay for the individual components of a mixture.

Theory

Many atomic nuclei have an angular momentum, that is they spin. A charged spinning particle generates a magnetic dipole along its spin axis and therefore an isolated nucleus stripped of its electrons may be thought of as a small bar magnet. From quantum mechanics the magnitude of the nuclear angular momentum (P) is given by the expression:

$$P = h/2\pi\sqrt{\{I(I+1)\}}$$
$$= \hbar\sqrt{\{I(I+1)\}} \tag{1}$$

Where

h = Planck's constant

$\hbar = h/2\pi$ = modified Planck's constant

I = spin quantum number

The value of the spin quantum number (I) depends upon the particular nucleus or isotope and can be $0, \frac{1}{2}, \frac{2}{2}, \frac{3}{2}, \frac{4}{2}, \frac{5}{2}$ etc; i.e. integral or half integral multiples.

Furthermore from quantum mechanical considerations there are ($2I+1$) possible orientations, and thus corresponding energy levels, for a nucleus with a magnetic moment under the influence of an external magnetic field. A comprehensive list of spin quantum values and other nuclear properties is provided by Pople, Schneider and Bernstein (1959). To establish the value of I empirically the following rules may be applied:

(i) $I = 0$ when both the mass and atomic number are even numbers. The nuclei of such isotopes have a spherically symmetrical charge distribution. They do not possess angular momentum, and do not give nuclear magnetic resonance spectra. They include ^{12}C, ^{16}O, and ^{32}S.

(ii) $I = \frac{1}{2}, \frac{3}{2}, \frac{5}{2}$ etc. (half integral multiples) when the isotope mass number is odd and atomic number is odd or even.

(iii) $I = 1, 2, 3$ etc. when the mass number is even and the atomic number is odd.

By far the most important group of nuclei from the standpoint of organic chemistry and drug chemistry are those with $I = \frac{1}{2}$. Examples are ^{1}H, ^{3}H, ^{19}F, ^{31}P, ^{13}C, ^{15}N, ^{29}S. These nuclei possess spin and have a spherically symmetrical charge distribution. Examples where $I = 1$ are ^{2}H and ^{14}N and where $I > 1$ are ^{10}B, ^{11}B, ^{35}Cl, ^{17}O and ^{27}Al. Isotopes with a spin value equal to, or greater than unity have an ellipsoidal charge distribution and possess spin. They have a *nuclear electric quadrupole moment* (Q).

This chapter deals mainly with nuclei of spin $\frac{1}{2}$ and most examples and applications will be from proton magnetic resonance (PMR) spectroscopy.

In a homogeneous magnetic field an atomic nucleus ($I > 0$) will assume one of $(2I + 1)$ orientations. Thus nuclei of spin $\frac{1}{2}$ will have two modes of alignment with respect to the applied field. They can align themselves either 'with' the field or 'against' the field, thus corresponding to the spin quantum values of $I = -\frac{1}{2}$ and $I = +\frac{1}{2}$. The energetically favourable, or ground, state is that in which the nuclei are aligned 'with' the field, rather like a compass needle aligned naturally in the earth's magnetic field. The unfavourable ($I = +\frac{1}{2}$) orientation may be considered as the excited state.

The energy difference between the ground and excited states may be expressed by the equation

$$\delta E = \gamma \hbar H_0. \tag{2}$$

where

$$H_0 = \text{applied magnetic field}$$

and

$$\gamma = \text{magnetogyric ratio}$$

The *magnetogyric ratio*, a proportionality constant unique for each isotope, is given by the expression

$$\gamma = \frac{\mu}{I\hbar}$$

where μ is the maximum observable component of the magnetic moment.

A spinning nucleus in a magnetic field (H_0) experiences H_0 as a torque about an axis perpendicular to the axis of rotation. This causes the spin axis to precess about the direction of the field (Fig. 205). An increase in the strength

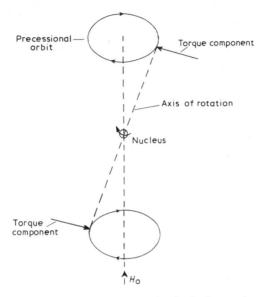

Fig. 205. Precession of spin axis of spinning nucleus

of H_0 does not produce an energy transition of the nucleus, but simply increases the precessional frequency. For a given value of H_0 the precessional or *Larmor frequency* is given by the Larmor equation:

$$v_0 = \frac{\gamma}{2\pi} \cdot H_0 \tag{3}$$

Introducing the Planck constant (h) to both sides, eq. (3) becomes:

$$hv_0 = \frac{\gamma h}{2\pi} \cdot H_0 \tag{4}$$

$$= \gamma \hbar H_0$$

which from eq. (2) $= \delta E$.

Thus, if the nucleus is irradiated with radiofrequency energy where $hv_0 = \gamma \hbar H_0$, then when it is in the ground state it can absorb radiation energy. Radiofrequency energy equal to the precessional frequency of the nucleus will initiate resonance in the precessing spin axis and cause the nucleus to 'flip' into a higher energy state. The radiation energy required for such a transition falls within the radiofrequency region of the electromagnetic spectrum and, in an NMR experiment, is applied to the sample at right angles to the magnetic field H_0.

The NMR Signal

When nuclei in an homogeneous magnetic field are scanned with a radio-frequency signal, then as the frequency approaches the precessional frequency, becomes equal to it, and finally exceeds it, resonance develops, attains a maximum, and subsides (Fig. 206). Maximum resonance is observed when the radiofrequency (v) is equal to the precessional frequency of the nuclei. Likewise a resonance signal results if v is kept constant and the magnetic field (H_0) is varied, since $v \propto H_0$. The magnetic field sweep is usually employed in practice.

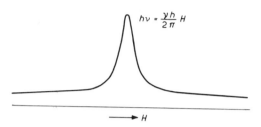

Fig. 206. Appearance of an NMR signal. Ordinate: absorption energy; abscissa: strength H of the external magnetic field or frequency of the r/f generator

All nuclei of a single isotope subjected to the above conditions do not automatically assume the most favourable alignment, but adopt a *Boltzmann distribution* with respect to the two adjacent energy levels. This distribution has a small excess (a few nuclei per million) of nuclei in the lower energy state. Only at absolute zero do all nuclei adopt the lower energy orientation, at higher temperatures thermal motion causes randomisation. It is the absorption of energy by this small excess of nuclei which is responsible for nuclear magnetic resonance spectra and this accounts for the low sensitivity of the technique. An equilibrium state is established by the applied radiofrequency working to reduce the lower spin state excess to zero, and relaxation processes acting to maintain the Boltzmann distribution. A state of *saturation*, when the number of nuclei in each energy state is equal, may be attained by irradiation of the sample with a powerful radiofrequency signal (see nuclear double resonance, p. 394).

Relaxation Processes

Mechanisms must exist whereby nuclei in an excited state can revert to the ground state. Otherwise, the energy levels would equalise rapidly and no further energy absorption would occur. The mechanisms responsible for the maintenance of an NMR signal may be classified under two general headings.

(*a*) *Radiation emission.* Radiofrequency energy emitted by those nuclei moving from an excited to the ground state under the influence of an external electromagnetic field is exceeded by the opposite transition, and this leads to the net absorption of energy. Spontaneous emission of electromagnetic energy in the radiofrequency region is negligible.

(*b*) *Radiationless transitions,* usually referred to as *relaxation processes,* not only maintain a state of energy absorption but govern resonance line width.

They are of prime importance in NMR spectroscopy. Relaxation processes are classified as either *spin-lattice relaxation* or *spin-spin relaxation.*

Spin-lattice or *longitudinal relaxation* involves the transfer of energy from a nucleus in an excited state to the molecular lattice. Here 'lattice' is used to describe the molecular structural environment in which a nucleus is situated irrespective of whether it is in a solid, a liquid or a gas. During this phenomenon, also known as spin-cooling, the nucleus returns to the ground state and the lattice gains thermal energy. The efficiency of spin-lattice relaxation (T_1) is expressed as the half-life required for the system to establish an equilibrium state. T_1 is a function of Brownian motion in fluids and lattice vibration in solids. Short relaxation times (T_1) imply efficient relaxation and lead to broad resonance bands. Organic liquids, for example, of T_1 approximately 1 sec, give optimum line widths for practical purposes. Pure solids have large values of T_1, often many hours, and in the absence of other factors would result in very narrow absorption bands. Although for liquids T_1 is usually greater than 10^{-2} sec and less than 10^2 sec, the presence of paramagnetic impurities, such as oxygen, gives rise to very rapid spin-lattice relaxation and thus to broader resonance bands. Similarly nuclei with $I > \frac{1}{2}$ possess a *nuclear electric quadrupole moment* which causes local magnetic field fluctuations leading to short relaxation times and consequently line broadening.

Spin-spin or transverse relaxation occurs mainly in solids. It involves a mutual exchange of spins between an excited nucleus and a neighbour, without altering the overall spin state of the system. The time T_2 is a measure of the efficiency of spin-spin relaxation. The total magnetic field felt at any particular nucleus is H_0 plus or minus a factor for the small magnetic fields generated by the nuclei of its immediate environment. Thus, the greater the number of spin-state exchanges the greater are the local field fluctuations, leading to a greater range over which radiofrequency energy may be absorbed, and therefore to line broadening. Large values of T_2 result in narrow resonance lines, small values give broad lines. In liquids and gases local magnetic field variations average to zero, whereas in solids such interactions are finite and lead to very broad absorption bands.

Instrumentation

Several high resolution nuclear magnetic resonance spectrometers are available commercially; all may be considered as consisting of five major units:

(a) a magnet producing a strong homogeneous field,
(b) a means of varying the magnetic field over a narrow sweep range (milligauss),
(c) a radiofrequency oscillator,
(d) a radiofrequency receiver,
(e) recorder and integrator.

(a) The *magnet*. Either a permanent or an electromagnet may be used to supply a field of high homogeneity. Permanent magnets provide an ever-present field of good homogeneity and stability, but field variation is not possible. Electromagnets, on the other hand, require expensive field stabilisers but the field strength may be varied. Ideally, all points within the pole gap of the magnet should experience an identical value of H_0. Because the strength of a resonance signal depends upon the magnetic field strength, the latter should be as great as possible.

(b) *Variation of the magnetic field* or *magnetic field sweep*, the method employed in most instruments to produce NMR spectra, is accomplished by passing a direct current through coils wound around the magnet pole pieces or through a pair of Helmholz coils flanking the sample. The rate of sweep is important, too slow a sweep leads to saturation effects, whereas a fast sweep results in 'ringing' (Fig. 207). Provided ringing is not excessive and does not distort the resonance signal it is indicative of good field homogeneity.

(c) *A radiofrequency oscillator* supplies the signal required to induce transitions in the nuclei of the sample from the ground to the excited state. The source is often a highly stable crystal-controlled oscillator, the output of which is multiplied to the desired frequency. The signal arises from a coil situated in the pole gap of the magnet and the sample rests within the coil.

(d) The resonance signal is detected by one of two methods. In *single coil* instruments a radiofrequency bridge, rather like a Wheatstone bridge, is employed. The applied signal is balanced against the received signal and the absorption or resonance signal is recorded as an out of balance emf, which may be amplified and recorded mechanically. In the *double coil* or *nuclear induction* method transmitter and receiver coils are set at right angles to each other about the sample. Figure 208 illustrates a double coil NMR spectrometer.

Fig. 207
Ringing

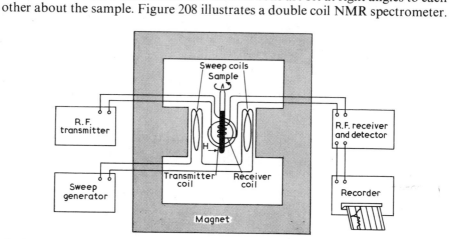

Fig. 208. Schematic diagram of an NMR spectrometer (Courtesy of Varian Associates Palo Alto, California)

(e) *Recorder and Integrator.* Spectra from modern stable high resolution instruments are recorded mechanically on pre-calibrated charts zeroed with respect to a reference compound (p. 383). Resonance line intensity in an NMR spectrum is proportional to the number of nuclei responsible for the signal.

The area under the signal is a direct measure of the intensity and is determined on a modern instrument in a cumulative manner by an automatic electronic integrator (Fig. 209).

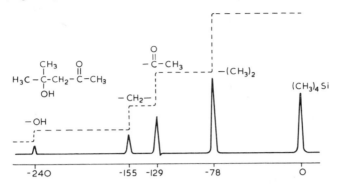

Fig. 209. Proton NMR spectrum of 2-hydroxy-2-methylpentan-4-one at 60 MHz and 14×10^3 gauss, with tetramethylsilane as internal standard. (– – –): Integral absorption intensity. Abscissa: strength of the magnetic field referred to the position of the $(CH_3)_4Si$ resonance signal (Hz)

Most instruments currently in use for proton studies operate at 60 megahertz (MHz) and 14 092 gauss (G) (or more correctly 1.4092 Tesla), 90 MHz and 21 140 G or 100 MHz and 23 490 G. Higher resolution and sensitivity result from the use of fields produced by liquid helium-cooled superconducting solenoids and operating at either 220 MHz or 300 MHz. This improved resolution is illustrated by the spectra of *N-sec*-butylaniline in Fig. 210 (p. 368).

PRACTICAL CONSIDERATIONS

Sample Spinning

A simple method of increasing the effective homogeneity of the applied magnetic field is to spin the sample at approximately 30 rps about its longitudinal axis. Variation of the position of a nucleus in the pole gap effectively averages the field it experiences. Spinning at too high a rate may cause turbulences which seriously affect resolution. If the spin rate is too slow field averaging is incomplete and resonance lines tend to broaden. Also, under certain circumstances strong resonance bands are flanked by side-bands, one to each side of the parent. *Spinning side-bands* may be distinguished from true resonance signals by varying the spin rate, an increased rate causes the side-bands to move away from the parent signal.

Solvents

Samples normally encountered in NMR studies are either liquids or solids; gases occur rarely and because of their low signal to noise ratio are examined under pressure. Liquids usually give excellent spectra 'neat', but samples of high viscosity benefit from dilution with a mobile solvent. Solids require a solvent in which they have at least 10% solubility and because a sample

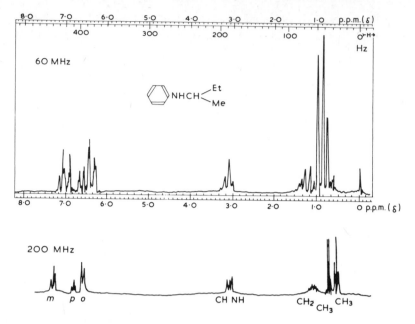

Fig. 210. NMR spectra of *N-sec*-butylaniline obtained at 60 and 200 MHz show that greater resolution is obtained at the higher frequency (lower spectrum). For example, the spectrum taken at 200 MHz shows well separated resonance peaks for the ortho, meta, and para aromatic protons

volume of 0.35–0.5 ml is needed, 35–50 mg of material must be available. Solutions of 20–25 % afford excellent spectra with a high signal to noise ratio. Paramagnetic impurities in either sample or solvent cause resonance line broadening and efforts must be made to preclude all extraneous particles. Small amounts of oxygen dissolved in a liquid sample or solution may cause considerable loss of spectrum resolution (line broadening). Oxygen may be conveniently removed by vigorously shaking the sample with about 0.3 ml of 10 % sodium dithionite solution; a marked improvement in spectrum resolution results (Brophy, Laing and Sternhill, 1968). Other methods of deoxygenation include vacuum distillation, repeated freezing, evacuation, or bubbling argon or nitrogen through the sample immediately before recording the spectrum.

The properties required of solvents for PMR spectroscopy are:

(1) chemical inertness
(2) magnetic isotropy
(3) volatility, to facilitate sample recovery
(4) absence of hydrogen atoms (this is not always possible)

Solvents commonly employed are CCl_4, $CDCl_3$, D_2O, CD_3OD, $(CD_3)_2SO$, CD_3COOD and CF_3COOH. (See solvent effects p. 379.)

An excellent chart showing characteristic resonance bands of solvents used in NMR spectroscopy has been published. (Henty and Vary, 1967).

Special Techniques for Small Samples

Often, sufficient sample for normal NMR investigation is lacking. Special techniques must then be used to obtain satisfactory spectra.

Microcells. Several types of microcells are available requiring as little as 1 mg of sample dissolved in 0.025 ml of solvent. The cell confines a small amount of sample to the locality of the probe receiver coil. Microcells may be either spherical or cylindrical in design.

Time-averaging computer (Computer of Average Transients or CAT). Signal to noise ratio enhancement can be effected by a computer, which stores the resonance signals from many passes of the spectrum and averages out the random noise. Sensitivity is increased by the square root of the number of scans.

An acceptable spectrum may be obtained from a fraction of a milligram sample. Either a whole spectrum or any desired portion of it may be studied in this way (Fig. 211).

Fourier Transform (FT) NMR. A disadvantage of the CAT technique is the length of time needed for the summation of numerous standard spectrum scans. In recent years this has been overcome by employing a strong r/f pulse of 50 μsec duration over a broad bandwidth and recording the total spectral response in the memory of a computer. Each nucleus in the sample absorbs its characteristic frequency component and the resultant pattern recorded is called the *free induction decay.* From the total series of overlapping decay patterns (one from each absorbing nucleus) the spectrum is obtained *via* a computer by a mathematical operation known as *Fourier Transformation.* Thus by the accumulation of several hundred or even several thousand pulse responses spaced 0.1 sec to several seconds apart, the recording time for a spectrum may be reduced at least one hundredfold. A small storage computer is employed to add the spectra from a series of pulse irradiations. Enhancement of signals by a factor of about 10 has been achieved practically.

Fourier Transform NMR is of particular value in ^{13}C studies (p. 412).

Where low solubility or a limited quantity of sample are restricting factors the following should be considered:

Insoluble sample	Small quantity of sample
(i) Use a more sensitive instrument, e.g. 100 MHz	(i) Use a more sensitive instrument
(ii) Use a CAT or FT	(ii) Use a CAT or FT
(iii) Increase temperature	(iii) Use microcell
(iv) Change solvent	

Temperature variation is often necessary in NMR spectroscopy and many instruments have a variable temperature probe attachment able to operate between -100 and $+200°$. Devices for wider temperature ranges have been described by Pope, Schneider and Bernstein (1959). Kinetic studies, conformational equilibria and hydrogen bonding investigations are facilitated by temperature variation.

Resolution

The resolution of a spectrum is its clarity of division into distinct bands corresponding to atomic or electronic transitions. Resonance line width is governed,

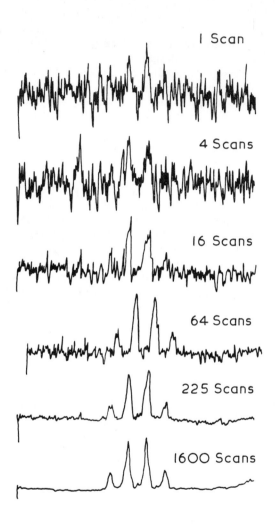

Fig. 211. Weak NMR signals may be greatly enhanced by retracing the spectra many times and summing the traces with a multi-channel analyser, essentially a small computer which can 'remember' several hundred traces. These are spectra of the methylene quartet of ethylbenzene in very dilute solution (0.1% in carbon tetrachloride) shown after 1, 4, 16, 64, 225, and 1600 scans have been summed by a 1024-channel computer of average transients (CAT) (Varian Associates Inc., Palo Alto, California)

in part, by magnetic field homogeneity, and therefore in assessing any spectrum an index of resolution is desirable. The resolution of a resonance band is conveniently expressed by its width, in hertz, at half height. The acetaldehyde, CHO quartet (Fig. 212), in the absence of atmospheric oxygen which causes paramagnetic line broadening, is often used as a standard for resolution in PMR studies.

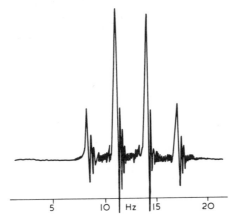

Fig. 212. Spectrum of the CHO group of acetaldehyde

CHEMICAL SHIFT

Dissimilar isotopes will require different amounts of energy to undergo transitions to a higher energy level and therefore they will absorb r/f energy at different frequencies. Nuclei of the same isotope, depending upon their molecular environments, also absorb energy at different frequencies. The foregoing discussion has been restricted to nuclei stripped of their electrons (unshielded nuclei). They must now be considered with regard to their environment. The electrons around each nucleus, when influenced by a magnetic field (H_0), circulate in such a way as to create a small local magnetic field of their own which opposes H_0. Thus, the resultant field experienced at the nucleus is H_0 less a factor for the electron cloud *diamagnetic shielding*. The nature of the electron cloud about the nucleus will, therefore, govern the region of the radiofrequency portion of the electromagnetic spectrum from which the nucleus absorbs energy. Protons in different parts of an organic molecule will usually give rise to separate resonance bands. The difference between the absorption position of a proton and the absorption position of the protons of a reference compound is known as the *chemical shift* of that particular proton. The most commonly employed reference compound is tetramethylsilane (p. 382).

The field (H) experienced at a particular atomic nucleus in a molecule, where σ is the *shielding* or *screening* constant of the atom, subjected to an applied field H_0 is given by:

$$H = H_0(1-\sigma)$$

and the resonance frequency for that particular atom, from eq. (3) becomes:

$$v = \frac{\gamma H_0(1-\sigma)}{2\pi} \tag{5}$$

Thus the amount of shielding determines how much the applied field has to be increased in order to attain that field (H) which induces a transition in the nucleus. For protons, screening constants are smaller than for most other nuclei, since the electron densities about protons are usually smaller than about other nuclei. The relationship between the field independent scale for chemical shift (δ) and the screening constant is given by the expression:

$$\delta = \sigma - \sigma_{ref} = \frac{H_0 - H_{ref}}{H_{ref}} \qquad \text{See p. 382.}$$

Chemical shift values are proportional to the applied magnetic field (H_0). They cannot be calculated but must be determined empirically. Examples of chemical shift values for protons in different environments are found in Figs. 213, 214, 215, and 216 (pp. 373–376).

Factors Influencing Chemical Shift

Often it is incorrectly assumed that only the electron density about a proton is responsible for the shielding effect, although it is true that a proton on a carbon H—C—X resonates to lower field as X becomes more electronegative. In the series where X is Si, C, N and O respectively, then that is the order of decreasing magnetic field required to induce a nuclear transition. The proton becomes progressively more deshielded or has a diminished electron cloud density. Similarly, as the nature of carbon moves from sp^3 to sp^2, attached protons resonate at lower field. This tendency is not adhered to, however, since protons on sp hybridised carbon, acetylenic protons, resonate upfield of olefinic protons; and cyclopropane protons which possess some sp^2 character are amongst the most shielded of protons.

In gases, liquids and solutions, a nucleus is only affected by localised diamagnetic fields from its own molecule; it is not usually influenced by those fields from neighbouring molecules, the effects of which are averaged by random motion. However, where association is possible, particularly in solutions where solute and solvent may form loose complexes, such effects may influence chemical shifts considerably (p. 379).

Localised diamagnetic fields are induced about an individual nucleus by the *intra-atomic* circulation of its electron cloud. In a molecule having groups of atoms with multiple bond character (π-electrons), *inter-atomic* electron circulation is possible. Here the circulation of electrons is only in certain preferred directions about the bond and, thus, such bonds exhibit *anisotropy* of *diamagnetic susceptibility* (diamagnetic anisotropy). The effect, either shielding or deshielding, on a nearby proton depends (a) upon its distance from the multiple bond, and (b) upon its orientation with respect to the bond. Ethylene, acetylene and benzene show this phenomenon.

text continued on p. 377

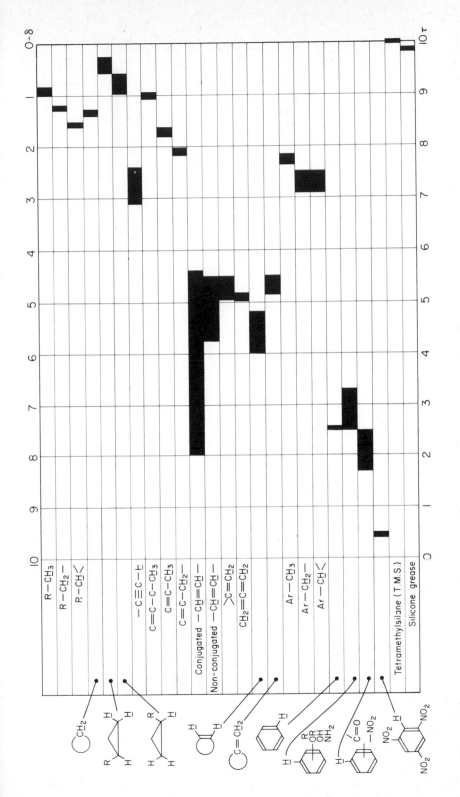

Fig. 213. Characteristic proton shift positions

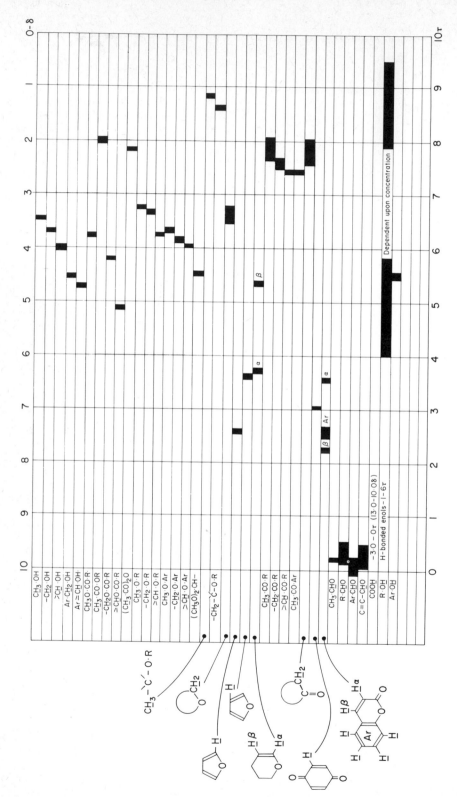

Fig. 214. Characteristic proton shift positions

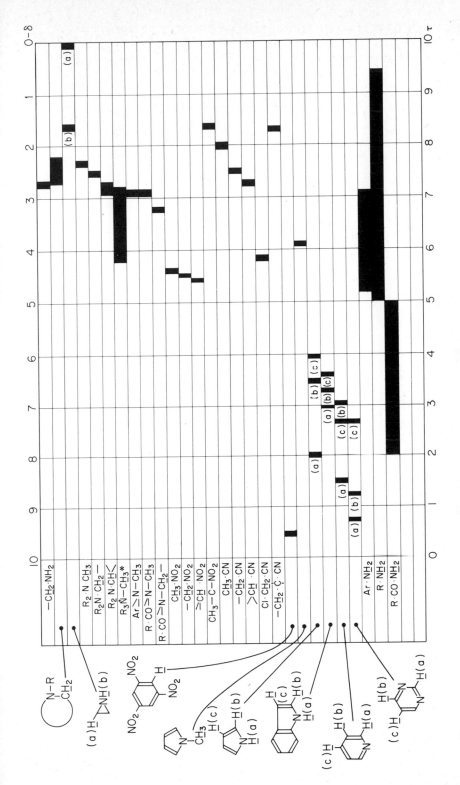

Fig. 215. Characteristic proton shift positions

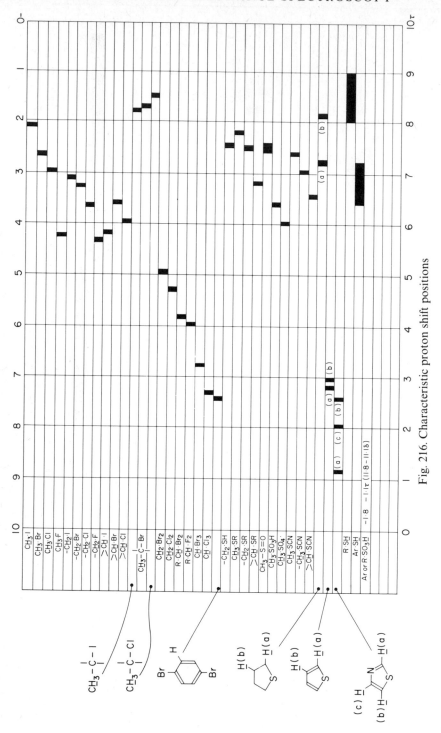

Fig. 216. Characteristic proton shift positions

Ethylene, when its double bond is orientated at right-angles to the applied field, has a π-electron circulation about the bond (Fig. 217a). The induced magnetic field reinforces the applied field at the protons, which are consequently deshielded. Olefinic protons occur in the region of 7.6–4.5δ compared with 0.9δ for CH_3—C$\leqslant$. Figure 217b illustrates the shielding cones (+) and regions

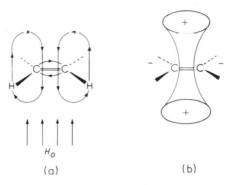

(a) (b)

Fig. 217. Ethylene

of deshielding (−); any proton held in these regions will be shielded or de-shielded accordingly.

Carbonyl groups ($>$C$=$O) behave similarly.

Acetylene is a linear molecule with an axis of symmetry passing through the triple bond. Orientation of the acetylene molecule with its longitudinal axis parallel to the magnetic field causes a diamagnetic circulation of π-electrons in such a way that its protons experience an induced magnetic field opposing the applied field (Fig. 218). The protons are therefore shielded by the anisotropy of the triple bond and resonate at 2.6δ.

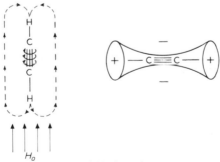

Fig. 218. Acetylene

Benzene and aromatic protons. At the aromatic protons the induced magnetic field reinforces the applied field and they therefore resonate at low field, about 7δ. Any proton lying above the plane of an aromatic ring will be shielded and similarly a proton in the plane of the ring will be deshielded (Bovey and Johnson, 1958) (Fig. 219).

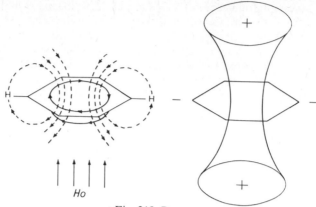

Fig. 219. Benzene

Ring current effects are well illustrated by the 60 MHz spectrum of C-18-annulene (I). The outer twelve protons experience a ring-current-induced reinforcing field and absorb at 530 Hz downfield from tetramethylsilane; whereas the inner six protons are subject to an H_0-opposing field and resonate at 115 Hz upfield from tetramethylsilane.

The deduction of the absolute configuration about C-9 of the analgesic 2'-hydroxy-5,9-dimethyl-6,7-benzomorphans (II) (Fullerton, May and Becker,

(I)

(II)

1962) was aided by a consideration of ring current effects. In the α-isomer where the 5- and 9-methyl groups are *cis*, the 9-methyl group overhangs the aromatic ring, and the ring diamagnetic anisotropy produces a shift of 25 Hz (60 MHz) upfield from the 9-β-methyl signal. Resonance signals from the 5-methyl groups are little affected. These observations enabled absolute structures to be assigned to the α- and β-5,9-dialkyl-6,7-benzomorphans.

An aromatic compound has been defined as a compound that can support a ring current. This may not be so and the anisotropic nature of many cyclic systems could be due to other factors not yet understood (Jones, 1968).

It should be noted that significant diamagnetic anisotropy is exhibited by C—C single bonds.

Hydrogen bonds. Dilute solutions of compounds bearing protons capable of intermolecular hydrogen bonding have resonance lines for such protons which on increasing solute concentration, shift downfield. That is, increased hydrogen bonding leads to greater proton deshielding. Similarly, the weakening of H-bonds by raising the sample temperature causes an upfield shift of the proton resonance line. The chemical shifts of intra-molecular hydrogen bonded protons are independent of temperature and solute concentration. Nuclear magnetic resonance spectroscopy is an excellent means of studying hydrogen bonding (p. 401).

Solvent Effects. (Bhacca and Williams, 1964; Ronayne and Williams, 1966). Solvent variation often results in dramatic chemical shift changes of certain groups of protons, a phenomenon which may be exploited in structural investigations. Magnetic non-equivalence may be induced where previously spectrum interpretation was severely impaired by resonance band overlap. In solutions of polar solutes, solvent molecules often adopt a specific orientation with respect to those of the solute.

Non-aromatic solvents, such as CS_2, CCl_4 and CH_3CN, in close proximity to a molecule of polar solute are polarised; a local 'reaction field' is induced and the chemical shift values of solute protons are influenced to varying degrees. Although these effects are small they may be used to determine the proximity of a proton to a polar centre, such as a carbonyl or ether oxygen.

Aromatic solvents often induce large 'solvent shift' effects, particularly in proton groups adjacent to a carbonyl or other polar function. The anisotropy of magnetic susceptibility of an aromatic solvent such as benzene will affect the proton groups of a solute differently, depending upon the mode of solute-solvent alignment. *N*-Methylformamide, represented by the resonance forms (III) $\rightleftharpoons$ (IV), shows restricted rotation about the C—N bond and gives both *cis* (IIIa and IVa 92%) and *trans* (IIIb and IVb 8%) N—CH_3 PMR signals in a spectrum of the neat liquid. In benzene solution the *trans* N-CH_3 signal is

(IIIa)   (IIIb)

(IVa) *cis*   *trans* (IVb)

shifted upfield far more than that of the *cis* form, because of the alignment with the benzene solvent illustrated in (IVb). Here the partially positive nitrogen atom is close to the high π-electron density region of the benzene ring, which in turn is as far away as possible from the carbonyl oxygen bearing a fractional negative charge. In other words, orientation IVb is that preferred in the *trans* isomer and orientation IVa in the *cis* isomer. In the former orientation, the *N*-methyl group falls above the plane of the aromatic ring and is therefore shielded. In the latter no such shielding results. Polar compounds often exhibit good solvent shift effects with benzene; with pyridine somewhat larger spectral changes result. Solvent shift effects are best studied in weak (5%) solutions. Acidic solvents, such as CF_3COOH, CH_3COOH and CD_3COOD, also show useful solvent shift effects, particularly with amines.

Lanthanide Shift Reagents (Sanders and Williams, 1972; Mayo, 1973). Chemical shift changes induced by solvents are relatively small. A very much more effective method of selectively altering the magnetic environments of nuclei and thus changing their chemical shifts is by the addition of paramagnetic metal complexes to the sample.

The presence of paramagnetic ions usually causes rapid spin-lattice relaxation in nearby nuclei and this results in excessive line broadening of the NMR signal. However the europium ion (Eu^{3+}) has anomalously inefficient nuclear spin-lattice relaxation properties and in the form of suitable complexes provides a valuable range of *shift reagents* affording well resolved spectra.

The most useful complexes are β-diketone derivatives in which the lanthanide ion may increase its co-ordination number by interaction with lone pairs of electrons. An example of a commonly used shift reagent is tris(dipivalo methanato)europium, Eu $(DPM)_3$.

$$\left[Eu \underset{O=}{\overset{O-}{\Big\langle}} \underset{n\text{-}Bu}{\overset{n\text{-}Bu}{\Big\rangle}} \right]_3$$

As a result of the formation of a new complex between the shift reagent and a sample molecule possessing suitable donor atoms, the chemical shift of protons of the sample is altered. The extent of the change is related to the proximity of the protons to the donor functional group (Fig. 220).

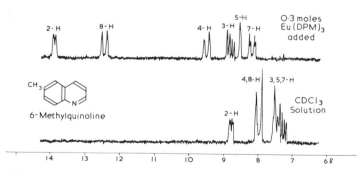

Fig. 220. 60 MHz spectra showing the effect of a lanthanide reagent on the chemical shifts of the aromatic protons of 6-methylquinoline (Courtesy Perkin-Elmer Ltd)

The paramagnetic metal species induces the shifts described by two mechanisms:

(1) *Contact Shifts* arising from delocalisation of unpaired electron spin, *via* covalent bonds, to the affected nuclei.
(2) *Pseudocontact Shifts* from secondary magnetic effects generated by the magnetic moment of the paramagnetic ion and transmitted through space.

Shift reagents of other lanthanides are also valuable in structural studies, for example the dipivalomethanato complex of praseodymium, Pr $(DPM)_3$. In contrast to shifts to low field induced by Eu $(DPM)_3$, the praseodymium reagent causes somewhat larger shifts generally to high field. Ytterbium complexes, e.g. Yb $(DPM)_3$ are often useful reagents for N-heterocycles but they give rise to moderate line broadening.

Fluorinated β-diketone complexes of lanthanides often exhibit greater shifting power than non-fluorinated reagents. Examples of such complexes are:

Eu$(FOD)_3$ Tris (1,1,1,2,2,3,3-heptafluoro-7,7-dimethyloctan-4,6-dionato)-europium.

Eu$(PFD)_3$ Tris (1,1,1,2,2-pentafluoro-6,6-dimethylheptane-3,5-dionato) europium.

Eu$(FHD)_3$ Tris (1,1,1-trifluoro-5,5-dimethylhexane-2,4-dionato)europium.

In practice known amounts of the lanthanide shift reagents are added successively to the sample dissolved, preferably, in a non-polar solvent. The chemical shift of each proton or proton group of the sample molecule changes with each addition of reagent and the extent of the induced shift is measured. A plot of the induced shift against the ratio of shift reagent to substrate will give a straight line at low values of the ratio. From plots for each proton group in a molecule valuable structural information may be deduced.

Clearly the application of lanthanide shift reagents extends the utility of NMR spectroscopy. They may be used simply to resolve overlapping signals from different proton groups in a molecule or by more quantitative studies to provide information concerning molecular configuration. Chiral lanthanide shift reagents are of value in the quantitative determination of mixtures of enantiomers.

Shoolery's rules (Dailey and Shoolery, 1955). The approximate chemical shifts of protons in aliphatic methylene groups may be predicted by the application of Shoolery's additive constants. Additive constants for many other groups are also available.

Protons are described as *magnetically equivalent* when they have identical screening constants (σ). Such protons are almost always also chemically equivalent, as for example in a —CH_3 group. In methylacetylene magnetically equivalent nuclei are not chemically equivalent; its spectrum consists only of a four proton singlet at 1.8δ.

Much information concerning the chemical shifts of protons in organic molecules has been collected and compiled into correlation charts, examples are found in Figs. 213, 214, 215 and 216 (pp. 373–376). These charts should be used empirically like the corresponding compilations in infra-red spectroscopy. Collections of spectra (Bhacca, Johnson and Shoolery, 1962; Bhacca, Hollis, Johnson and Pier, 1963) are also of value in structural studies.

Scales of Measurement

Having established that protons in different environments in an organic compound resonate at different frequencies, a method of expressing the positions of resonance lines is required. Most proton absorption arises within a 600 Hz range on a frequency scale of 60×10^6 Hz (i.e. 60 MHz) and measurements are often required to 0.2 Hz. To measure frequency or magnetic field strength to this degree of accuracy on an absolute scale is not possible. A comparative scale is therefore employed. Resonance lines are measured with respect to the absorption from a reference compound, usually tetramethylsilane, $Si(CH_3)_4$, (TMS) (p. 383). Chemical shifts are expressed in units downfield from TMS, usually at 60 MHz. In the PMR spectra of organic compounds, shift values upfield of TMS are rarely encountered.

There are three scales for expressing chemical shift values:

(1) *ν-scale* (Hertz, Hz, cycles per second, c/sec, cps). The Hertz is the internationally accepted radiofrequency scale notation, where 1 Hz = 1 c/sec. A chemical shift value of 120 Hz at 60 MHz is equivalent to 120 c/sec at 60 Mc/sec. Chemical shifts quoted in Hz from a reference signal must be accompanied by a statement of the operating frequency of the instrument. The chemical shift is proportional to both the applied field (H_0) and the applied frequency ($ν_0$).

Consider a chemical shift of 80 Hz downfield from TMS measured at 40 MHz; at 60 MHz this becomes $\frac{60}{40} \times 80 = 120$ Hz, and at 100 MHz it becomes $\frac{100}{40} \times 80$ or $\frac{100}{60} \times 120 = 200$ Hz. Thus, at 100 MHz, line separation and therefore resolving power is greater than at 60 MHz.

(2) *δ-Values*. It became necessary to provide a field-independent scale, thus δ-values are used where:

$$\delta = \frac{H_0 - H_{ref}}{H_{ref}} \times 10^6 \text{ ppm}$$

H_0 = field strength for resonance of protons of the sample, H_{ref} = field strength for resonance of protons of the reference compound. This provides a 0 to -10 scale with TMS as the zero marker (Fig. 221). δ is a dimensionless expression

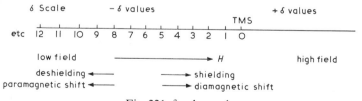

Fig. 221. δ-value scale

which is negative for most protons and is often referred to as parts per million (ppm).

The International Union of Pure and Applied Chemistry define the δ-scale thus:

'Whenever possible the dimensionless scale should be tied to an internal reference which should normally be tetramethylsilane. The proton resonance of tetramethylsilane should be taken as zero; if some other internal reference is used that reference, and the conversion shift used to convert the measured shifts to tetramethylsilane reference scale, should be explicitly stated. The dimensionless scale should be defined as positive in the high frequency (low field) direction. The scale in parts per million (ppm) based on zero for tetramethylsilane should be termed the δ-scale.'

The relationship between the δ-scale and the v-scale is expressed by:

$$\delta = \frac{\text{Hz}}{\text{instrument frequency } v_0 \text{ (MHz)}}$$

(3) τ-*Values.*

$$\tau = 10 - \delta$$

By definition the TMS signal occurs at 10. Use of the τ-scale is now discouraged.

Spectrum calibration is accomplished by aligning the single sharp resonance line of tetramethylsilane, in which all twelve protons are chemically and magnetically equivalent, with the 0δ mark on a precalibrated chart. TMS is the most commonly used internal standard, being dissolved to the extent of 0.5% in a solution of the sample. External references are necessary under certain circumstances, and then precision co-axial sample tubes are employed.

An internal reference should possess the following properties:

  (i) chemical inertness,
  (ii) magnetic isotropy,
  (iii) it should give a single, sharp, easily recognised absorption signal,
  (iv) miscibility with a wide range of solvents and organic liquids,
  (v) it should be volatile, facilitating its removal from valuable samples.

TMS has all of these properties, and is an excellent reference for organic solvents and liquids. For solutions in water or deuterium oxide (D_2O), sodium 2,2-dimethyl-2-silapentane-5-sulphonate or DSS (V) is a reasonable internal reference, although no standard for aqueous solutions is really satisfactory.

$$CH_3-\underset{\underset{\displaystyle CH_3}{|}}{\overset{\overset{\displaystyle CH_3}{|}}{Si}}-CH_2.CH_2.CH_2.SO_3^-Na^+$$

(V)

SPIN-SPIN COUPLING

A high resolution spectrum is distinguished from one at low resolution by the presence, in the former, of fine structure. Many PMR spectra have resonance bands split into doublets, triplets, quartets, etc. This phenomenon is known as spin-spin coupling or spin-spin splitting.

Consider two protons on adjacent carbons:

$$\begin{array}{cc} \text{(A)} & \text{(X)} \\ \text{H} & \text{H} \\ | & | \\ \text{R}'''\!-\!\text{C}\!-\!\text{C}\!-\!\text{R} \\ | & | \\ \text{R}'' & \text{R}' \end{array}$$

The R groups are not protons and do not bear protons on nuclei adjacent to those under consideration. Then, provided that the R groups are not all the same, the signal for each proton will appear as a doublet (Fig. 222).

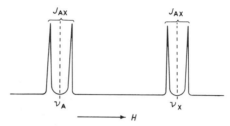

Fig. 222. Coupling between protons with widely different chemical shifts

The separation between the lines of each doublet, J_{AX}, is identical for each signal and is referred to as the coupling constant (J). The J value, which is usually measured in Hz, is independent of field strength. An increase in field strength causes v_A and v_X to shift further apart, but J_{AX} remains the same for any particular system.

By a consideration of the spin state of each proton with respect to that of its neighbour in a population of identical molecules, an explanation of spin-spin coupling is forthcoming. Proton **A** can have either a spin parallel or antiparallel with respect to that of **X**, and likewise proton **X** is aware of two spin forms of **A**. Each proton being aware of two spin states of its neighbour experiences two effective fields from it, and generates two corresponding resonance signals. Since the populations in each spin state are the same, resonance lines of equal intensity result. The 'awareness' of the spin state of one nucleus by an adjacent nucleus may be considered as being relayed by interceding bonding electrons. For further study, an excellent account of the theory of spin-spin interactions has been published by Roberts (1961).

Run under conditions of low resolution, the PMR spectrum of ethanol appears as three broad bands integrating for one, two and three protons respectively (Fig. 223a). In a high resolution spectrum, fine structure is apparent (Fig. 223b). Here the $-CH_2-$ group is a quartet with an area-ratio of lines $1:3:3:1$, and the $-CH_3$ signal a triplet of area-ratio $1:2:1$. The spacing between the lines of the $-CH_2-$ and CH_3- patterns is identical and is the *coupling constant J*. Figure 224 shows the spin arrangements possible for the three methyl protons and the two methylene protons. Four spin combinations of the CH_3 protons are possible with their statistical probabilities of occurrence being $1:3:3:1$, and similarly three combinations, of probability $1:2:1$, are

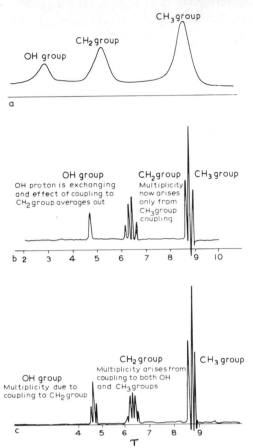

Fig. 223. (a) Low resolution spectrum of ethanol; (b) High resolution spectrum of ethanol (trace of OH⁻ or H⁺); (c) High resolution spectrum of highly purified ethanol

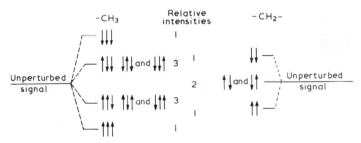

Fig. 224. Possible spin arrangements for three methyl protons and two methylene protons

possible for the —CH_2— protons. Each proton group splits the adjacent group signal to its own tally of possible spin states.

The constitution of multiplets arising from spin-spin coupling is governed by the following rules:

(a) Magnetically equivalent nuclei do not interact, e.g. methane and ethane will each give only a single resonance line.

(b) Band multiplicity is determined by the neighbouring groups of equivalent nuclei, and is given by the expression:

$$\text{Number of lines} = (2nI + 1)$$

where

$$I = \text{the spin quantum number}$$

and

$$n = \text{number of adjacent equivalent nuclei}$$

The integration of the whole multiplet is equal to the number of protons responsible for the signal, e.g., the $1:3:3:1$ quartet in the ethanol spectrum integrates for two protons and the $1:2:1$ triplet for three protons.

(c) If there are more than two interacting groups, the multiplicity of **A** as split by **B** and **C** is expressed by $(2n_B I_B + 1)(2n_C I_C + 1)$ etc. With protons of spin $1/2$ this simplifies to $(n_B + 1)(n_C + 1)$ etc. For example, under first order coupling rules (p. 387) consider

$$R-CH_2-CH_2-CH_2-R'$$

$$\textbf{B} \qquad \textbf{A} \qquad \textbf{C}$$

the expression becomes $(2+1)(2+1) = 9$. Proton **A** resonance will therefore be seen as a pattern with a maximum of nine lines (three triplets), but fewer than nine lines may ensue from band overlap. If, however, R and R′ are identical, then the **B** and **C** protons are chemically and magnetically equivalent and give rise to a five line pattern for the **A** protons, i.e. $n+1 = 4+1 = 5$. This is not due to band overlap.

Note that the coupling constants J_{AB} and J_{AC} will most probably be different. Also, with nuclei other than protons the value of I may vary and, therefore, so will the band multiplicity.

(d) The intensities of lines in a first order multiplet are theoretically symmetrical about the mid-point of the absorption band, and follow the coefficients of a binomial expansion $(1+x)^n$. The intensity ratios are best remembered by use of the Pascal triangle:

												n
					1						--------	0
				1		1					--------	1
			1		2		1				--------	2
		1		3		3		1			--------	3
	1		4		6		4		1		--------	4
1		5		10		10		5		1	--------	5
1	6		15		20		15		6	1	--------	6

where n is the number of interacting nuclei

Coupling is described as **first order** when:

(1) the chemical shift separation of the nuclei, $(\Delta v = v_A - v_X)$, is at least six times the coupling constant, J_{AX},

(2) each proton of a group is coupled equally to every proton of the other group.

Under these circumstances the simple multiplicity rules outlined above apply and the pattern is described as first order. The splitting between the CH_3- and $-CH_2-$ groups of ethanol is an example of first order coupling.

In the two proton picture referred to earlier (p. 384), under ideal conditions of first order coupling all four line intensities will be the same. As the chemical shift difference between **A** and **X** is reduced, then deviation from the strict binomial intensities results, and the quartet collapses, ultimately, to a singlet. (Fig. 225). This occurs when the chemical shifts of **A** and **X** are the same; in other words the protons are magnetically equivalent and resonate at the same frequency.

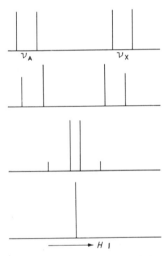

Fig. 225. Collapse of quartet to singlet with reduction of chemical shift difference between **A** and **X**

Exchange Effects

The spectrum of ethanol run under normal circumstances, exhibits a four-line pattern for the $-CH_2-$ protons (Fig. 223b). This behaviour appears anomalous since further splitting by the hydroxyl proton would be expected. Such coupling becomes apparent in the spectrum of very pure ethanol where the $-CH_2-$ signal is a double split quartet (8 lines) and the hydroxyl proton is the expected triplet (Fig. 223c). The addition of a trace of either acid or base to pure ethanol restores the simplified spectrum (Fig. 223b). Laboratory ethyl alcohol normally contains acid or base impurities which catalyse the exchange of hydroxyl protons between ethanol molecules. Thus in a given time, any individual hydroxyl proton experiences a wide variety of spin arrangements from different $-CH_2-$ group protons. The spin arrangements of the $-CH_2-$ group protons as seen by the —OH protons, and *vice versa*, are effectively averaged, and no spin coupling is observed. The catalytic effect of H^+ and OH^- is apparent in the spectrum of a mixture of ethanol and water which is acid and base free. Here separate resonance signals are seen for C_2H_5—O$\underline{H}$ and $\underline{H}_2O$, suggesting very slow exchange. The addition of acid causes coalescence of the two resonance lines to a sharp singlet, the signals being averaged by rapid exchange.

Chemical exchange may be demonstrated further by the addition of deuterium oxide (D_2O) to ethanol. Resonance from the ethanol hydroxyl proton disappears since the —OH proton is replaced by D; D$\underline{H}$O resonance occurs at about 4.88δ

$$C_2H_5OH + D_2O \rightleftharpoons C_2H_5OD + DHO$$

Protons bonded to O, N and S are exchangeable by deuterium, thus providing a good means of characterisation.

However, the rate of exchange of NH and SH protons in non-aqueous media is much slower than that of —OH protons.

Spin-Spin Coupling Notation

An alphabetical notation for referring to types of nuclei engaged in spin-spin coupling has evolved. The following is a guide to this notation.

(1) Equivalent nuclei are referred to by the same letter of the alphabet, with sub-numbers indicating the number of equivalent nuclei.

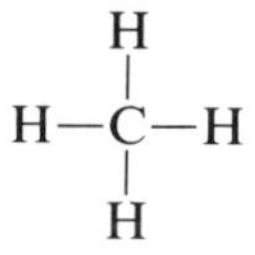

$$\begin{array}{c} H \\ | \\ H-C-H \\ | \\ H \end{array} \quad \text{is an } A_4 \text{ system}$$

and

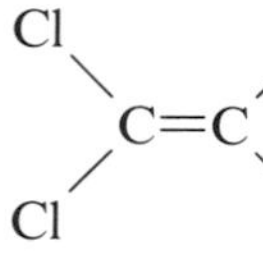

$$\begin{array}{c} Cl \\ \diagdown \\ \quad\quad C=C \\ \diagup \\ Cl \end{array}\begin{array}{c} H \\ \diagup \\ \\ \diagdown \\ H \end{array} \quad \text{an } A_2 \text{ system}$$

(2) Nuclei with resonance signals separated by only a small chemical shift difference are designated by letters from the same region of the alphabet.

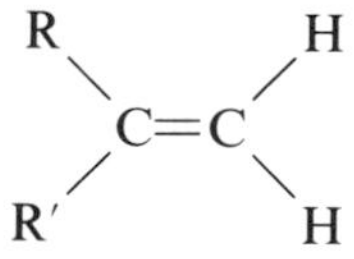

$$\begin{array}{c} R \\ \diagdown \\ \quad\quad C=C \\ \diagup \\ R' \end{array}\begin{array}{c} H \\ \diagup \\ \\ \diagdown \\ H \end{array} \quad \text{is AB}$$

and

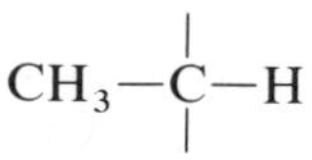

$$CH_3-\overset{|}{\underset{|}{C}}-H \quad \text{is } A_3B$$

(3) Where chemical shifts are widely separated, letters from quite different parts of the alphabet are employed.

$$H-\underset{\underset{F}{|}}{\overset{\overset{F}{|}}{C}}-H \quad \text{is } A_2X_2$$

(A)

$$\underset{\underset{(X)}{H}}{\overset{Ph}{\diagdown}}C=C\underset{(B)}{\overset{H}{\diagup}}\overset{H}{\diagdown}$$

is ABX

$$\underset{CH_3-\underset{(A_3)}{C}=C}{\overset{(M)\quad(N)}{\overset{H\quad H}{\diagdown}}}$$

$$\underset{F(X)}{}\quad \text{is } A_3MNX$$

(4) Where more than one coupling constant between equivalent groups is possible, primed letters are resorted to:

$$\underset{H}{\overset{H}{\diagdown}}C=C\underset{F}{\overset{F}{\diagup}}$$

is an AA′XX′ system, not A_2X_2

Each proton does not couple equally to each fluorine, and *vice-versa*, thus the J_{cis} and J_{trans} values will be different. Similarly, in *para*-substituted aromatic compounds, e.g.

is AA′BB′   and   is AA′XX′

In both cases, *meta* and *para* couplings occur, as well as *ortho* coupling.
    Coupling to the C$\underline{H}_3$ protons has been ignored.

Higher Order Spin-Spin Coupling

Values for chemical shifts and coupling constants are readily arrived at for spectra or portions of spectra arising from first order coupling, however when first order coupling rules are not obeyed, these values are more difficult to deduce. Higher order spin systems, for example ABX, A_2B_2, etc, may be solved by recognising certain band spacings and substituting these values in established equations. In more complex cases the analysis of a spectrum may get very involved. For further reading on the analysis of non-1st order spin systems reference should be made to specialist publications (Pople, Schneider and Bernstein, 1959; Roberts, 1961; Mathieson, D. W., 1967; Bible, 1965; Becker, 1965). Computer programmes are available for the calculation of certain complex spectra and Wiberg and Nist (1962) have published a compilation of calculated spectra.

Factors influencing the Value of Coupling Constants

It is not fully understood why coupling constants vary considerably with proximate structure, and at present the prediction of approximate J-values can only be made on an empirical basis. The following are some of the factors affecting the size of coupling constants.

(1) When only single bonds intercede between nuclei, coupling usually occurs only through three bonds.

In compounds such as

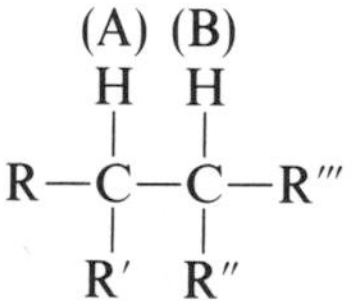

$$
\begin{array}{cc}
\text{(A)} & \text{(B)} \\
\text{H} & \text{H} \\
| & | \\
\text{R}-\text{C}-\text{C}-\text{R}''' \\
| & | \\
\text{R}' & \text{R}''
\end{array}
$$

J_{AB} is in the order of 5–8 Hz.

Occasionally, when molecular geometry is favourable, long range low order coupling can occur through four or even five bonds.

(2) Couplings through multiple bonds can occur over greater distances, presumably because of the mobility of π-electrons. Coupling values of olefinic protons decrease in the order:

trans > cis > geminal

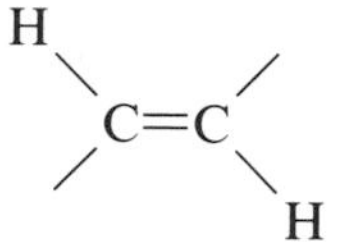

$J_{AB\,trans}$ 11–18 Hz

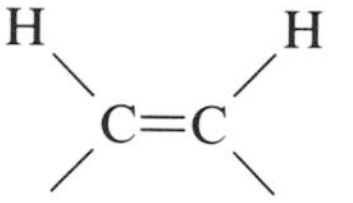

$J_{AB\,cis}$ 6–14 Hz

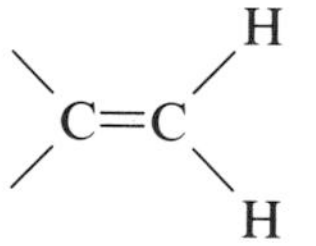

$J_{AB\,gem}$ 0–3.5 Hz

In unsaturated compounds coupling of the order of 1 Hz through four bonds is often observed.

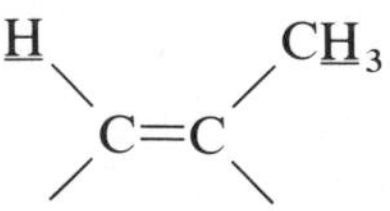

$J_{A_3X} \simeq 1$ Hz

(3) In aromatic systems *ortho* coupling ($J \simeq 8$ Hz) > *meta* ($J \simeq 2$–3 Hz) > *para* ($J \simeq 0$–1 Hz).

(4) Coupling between geminal protons on saturated carbon is dependent, in part, upon the bond angle θ. J_{AB} decreases from about 20 Hz when θ is 105° to zero when θ is 125°.

(A)

R   H

C θ

R′   H

(B)

It must be emphasised that this is an approximate relationship since other factors influence J_{AB}. Cookson, Crabb, Frankel and Hudec (1966) have published a comprehensive account of geminal coupling constants and factors influencing them.

(5) The size of the vicinal proton coupling constant is dependent upon the dihedral angle θ. A plot of θ against J (Fig. 226) shows that J_{AB} is maximum ($\simeq 10$ Hz) when the dihedral angle is 180°, and a minimum when it is 90° and about 8 Hz when θ is 0°. In cyclohexane axial-axial coupling has $J_{ax\,ax} \simeq 8$–12 Hz, whereas $J_{eq\,ax} \simeq 3$–4 Hz. Karplus (1959), who has studied vicinal coupling extensively, warns against attempting to determine dihedral angles with any degree of accuracy from vicinal coupling constants.

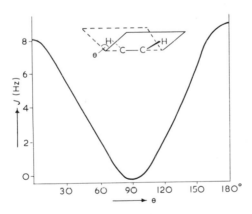

Fig. 226. Karplus relationship between the dihedral angle and coupling constant for vicinal protons

(6) The electronegativity of substituent X in a system:

H H

—C—C—X

has little effect on the vicinal coupling constant (< 2 Hz). In the limited number of cases investigated increasing substituent electronegativity gives a decrease in the J-value.

(7) Coupling values for splitting between protons and other nuclei, for example ^{15}N, ^{31}P, ^{13}C, etc., are often very large; an exception is proton-deuterium coupling, where the J-values are about one-seventh of those between protons. Advantage is taken of this in *spin-decoupling* by deuterium replacement of hydrogen.

Measurement of Coupling Constants

A major problem encountered in comparing NMR data from different published sources is the variability of conditions under which spectral parameters are measured. To minimise errors in the measurement of coupling constants the use of scale expanded spectra is recommended.

In the spectrum of some molecules where certain H—H coupling constants are inaccessible from direct observation, ^{13}C satellite signals may be exploited. The 1.1% natural abundance of ^{13}C and the large $^{13}C—^{1}H$ coupling constant gives rise to weak '^{13}C satellites' almost symmetrically positioned about the proton signal being investigated. The signals are due to protons bonded to and therefore coupled to ^{13}C $(I = \frac{1}{2})$ rather than ^{12}C. Although such satellite signals are weak, the sensitivity of modern NMR spectrometers renders them easily observable and H—H couplings of symmetrical molecules may often be extracted (see Becker (1969) and Batterham (1973)). For tables of coupling constants for protons see Mathieson (1967), Becker (1969), Batterham (1973), and Silverstein and Bassler (1968).

Spin-Spin Decoupling

A spectrum is often rendered very complex and difficult to interpret by spin-spin coupling. Simplification may be achieved in two ways.

(1) By using an instrument with a more powerful homogeneous magnetic field, for example, a 100 MHz instrument in preference to one operating at 60 MHz. A greater line separation (higher resolution) results, but there is a practical limit to the magnet strength available (p. 367).

(2) By a spin-spin decoupling technique.

Such techniques not only reduce the amount of fine structure and clarify the spectrum, but enable interacting protons to be identified.

Isotope exchange. The hydroxyl protons of ethanol, as we have seen, are able to exchange with other protons or deuterium (^{2}H) causing decoupling and simplification of the —CH_2— proton resonance. Dimethylsulphoxide, as a solvent, facilitates the study of coupling through exchangeable protons of this sort by hydrogen bonding to them and slowing the exchange rate.

Isotopic exchange may be used to advantage in the deuteration of compounds with active (acidic) hydrogens bonded to carbon, as for example, in protons *alpha* to a ketone or nitrile.

$$R—CH_2—\overset{\overset{O}{\|}}{C}—R \xrightarrow[OH^-]{D_2O} R—\underset{\underset{D}{|}}{\overset{\overset{O}{\|}}{C}H}—\overset{\overset{O}{\|}}{C}—R \xrightarrow[OH^-]{D_2O} RCD_2—\overset{\overset{O}{\|}}{C}—R$$

and

$$NC-CH_2-CH_2-OH \xrightarrow[OH^-]{D_2O} NC-\underset{\underset{D}{|}}{CH}-CH_2-OD \xrightarrow[OH^-]{D_2O}$$

$$NC-CD_2-CH_2-OD$$

(VI)   (VII)   (VIII)

The scheme (VI–VIII) illustrates isotopic exchange in 2-hydroxypropionitrile (Lapidot, Reuben and Samuel, 1964). In D_2O its spectrum appears as an A_2X_2 pair of triplets, the C-1 and C-2 proton signals being at 2.68δ and 3.80δ respectively (Fig. 227). The addition of hydroxide ions to this solution causes the

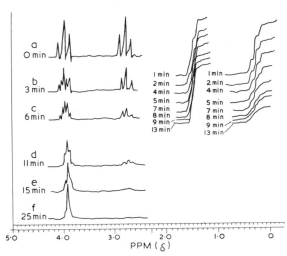

Fig. 227. Deuterium exchange of 2-hydroxypropionitrile with D_2O in the presence of sodium hydroxide. Left: Successive spectra. Centre: Integrated intensities of β-proton triplets. Right: Integrated intensities of α-proton triplets (Courtesy *J. Chem. Ed.*)

activated C-1 protons to exchange successively (VII and VIII) with deuterium and show the corresponding changes in the PMR spectrum. The rate of exchange is dependent upon the concentration of OH^-. Figure 227 illustrates the change in the spectrum of a solution of 2-hydroxypropionitrile (VI) in *1.44N* NaOH in D_2O with time:

(*a*) At zero time only the A_2X_2 pattern of (VI) is present.

(*b*) and (*c*) A doublet due to the C-2 protons split by the single C-1 proton (VII) is superimposed upon the original lowfield triplet giving a five line pattern.

(*d*) Here there are contributions from structures (VI), (VII) and (VIII), a triplet, doublet and singlet respectively appear superimposed for the C-2 protons.

(*e*) Contributions from only (VII) and (VIII) are apparent in the resonance of the C-2 protons.

(*f*) Finally, after 25 min almost complete exchange has occurred, the C-2 protons are almost entirely decoupled to yield a singlet and C-1 proton resonance is lost. Integration curves for C-2 and C-1 protons respectively are shown to the right of the spectra.

The synthesis of deuterated compounds from appropriately deuterated starting materials also aids the elucidation of molecular fine structure. An example is the investigation of the conformational preference of the non-bonding electron pair in piperidines by the synthesis, and PMR examination of the α-protons, of 3,3,5,5-tetradeuteropiperidines (IX) (Lambert, and Keske, 1966).

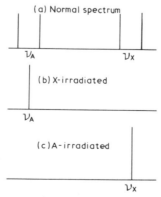

(IX)

Nuclear Magnetic Double Resonance (NMDR) (von Philipsborn, 1971)

During a normal NMR experiment the sample is subjected to a single radio-frequency field. If a second (or sometimes more) radiofrequency field is applied in a specific manner, then a series of techniques collectively known as nuclear magnetic double resonance results. From the careful application of these techniques, particularly in complex spectra with overlapping resonance bands, much valuable information is forthcoming.

Spin-Spin decoupling by NMDR. If during an NMR experiment on a simple AX system (Fig. 228a), the X nucleus is irradiated with a second radiofrequency signal of frequency (v_x) equal to the precessional frequency of X, then the X-nuclei will undergo rapid changes of spin state. Nucleus A is aware of only one equivalent (averaged) spin state of X and therefore its resonance signal appears, not as a doublet, but as a sharp singlet. Irradiation of X effectively decouples the signal from A (Fig. 228b). Similarly if A is irradiated with a second r/f signal equal to its own resonance frequency (v_A), then the signal from X will collapse to a singlet (Fig. 228c).

(a) Normal spectrum

v_A v_x

(b) X-irradiated

v_A

(c) A-irradiated

v_x

Fig. 228. Double irradiation

Double resonance experiments may be performed either by: (1) keeping both the applied frequency (v_1) and the irradiation frequency (v_2) constant, and varying the magnetic field (*field sweep*), or (2) by having a constant magnetic field (H_0) and irradiation frequency (v_2), and sweeping with the applied frequency (v_1) (*frequency sweep*). The irradiating frequency (v_2) may be changed after each scan, permitting an extensive investigation of spin-spin coupling. Nuclear triple resonance, where two irradiation frequencies are applied simultaneously and the signal due to a third nucleus examined, is valuable in the investigation of multiple and long-range coupling.

The irradiation frequency (v_2), in order to decouple the nuclei completely, is a powerful r/f signal; v_1 is weak in comparison.

Heteronuclear decoupling is the decoupling of different isotopes where the difference in resonance frequencies is large, e.g., ^1H and ^{19}F. The experiment is expressed as

$$\overset{\text{Nucleus being studied}}{\overbrace{}}\ ^1H[^{19}F]\ \overset{\text{Nucleus irradiated}}{\overbrace{}}$$

Homonuclear decoupling involves nuclei of the same isotope, where the difference in resonance frequencies is small. It is expressed by $^1H\,[^1H]$. Similarly, the irradiation of the X nucleus of an ABX system is expressed as AB [X].

For effective double irradiation a good chemical shift difference between the interacting nuclei is required.

Spin decoupling techniques, as exemplified by ethyl alcohol spectra, are summarised in Fig. 229.

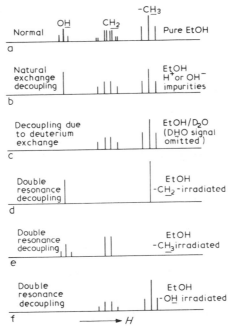

Fig. 229. Decoupling—illustrated by ethanol spectra

Decoupling experiments by NMDR are, in general, easy to conduct, however complete decoupling only occurs under rather restricted experimental conditions. True decoupling is only seen when the chemical shift separation is much greater than the coupling constant.

Spin Tickling

When the second irradiating frequency (v_2) is relatively weak and spectrum collapse is incomplete extra resonance lines or 'residual splittings' are observed. These lines may be related to energy transitions of the spin system observed, and may aid the interpretation of complex spectra.

Nuclear Overhauser Effect (NOE) (Backers and Shaefer, 1971)

Unlike spin decoupling and spin tickling which are based upon changes in energy levels, the nuclear Overhauser effect results from population changes in the energy levels which cause corresponding intensity changes in the spectra. The NOE experiment consists of saturating one signal and observing the intensities of other signals. The resonance bands due to protons spatially close to the group being saturated will show an increase in intensity. Such intensity changes are best measured by integration rather than peak height. The theory of the NOE is dependent upon relaxation mechanisms, and the relaxation of any nucleus is affected by all surrounding magnetic nuclei. Since the contribution of any nucleus to the relaxation of a second nucleus is dependent upon the mean square value of the magnetic field produced by the former, then the magnitude of the nuclear Overhauser effect will relate to the internuclear distance.

In the spectrum of dimethyl formamide (X) (Fig. 231) it is not possible from shift or coupling parameters to assign the two methyl signals. Saturation of the low field methyl signal causes a 17% increase in the intensity of the CHO signal, whereas saturation of the high field methyl signal gives no change in the formyl proton band. Clearly the low field line is *cis* to the formyl hydrogen.

Interactions giving rise to nuclear Overhauser effects may be intermolecular or intramolecular. In dilute solutions in aprotic solvents only the latter will be observed. It is important from a experimental standpoint in all NOE experiments to de-gas samples. Oxygen, being paramagnetic, will affect and often dominate the relaxation process and its presence will invalidate results.

INDOR Spectroscopy

Probably the most powerful double resonance technique for the observation of the multiplicity of signals hidden by band overlap is INternuclear DOuble Resonance (INDOR). Here signals are only observed from coupled protons.

In the INDOR experiment the magnetic field (H_0) and the observation frequency (v_1) are held fixed on one of the A signal lines in a spectrum where, for example, the X signal is obscured. The second weak perturbing frequency v_2 is now swept through the spectrum. As v_2 passes in turn through the hidden X lines a change in intensity of the A line may result, leading to a vertical movement of the recorder pen. Thus a negative or positive signal will result at positions corresponding to the X resonance frequencies. The signals overlapping X will not register unless they too are coupled to proton A (Fig. 230).

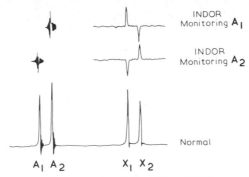

Fig. 230. 90 MHz spectra of $C_6H_5.CHBr.CHBr.NO_2$, 300 Hz scan, demonstrating the INDOR effect (Courtesy Perkin-Elmer Ltd)

APPLICATIONS

Structure Elucidation

The versatility of nuclear magnetic resonance spectroscopy as an analytical technique is illustrated here by a selection of its many applications. PMR is particularly useful in view of the abundance of ^1H in organic compounds, its strong resonance signal and convenient spin value of $\frac{1}{2}$. It has been applied extensively in structural elucidation.

The unequivocal proof of an organic structure by PMR spectroscopy is not often possible, but in conjunction with ultraviolet, infrared and mass spectrometry, it is an extremely valuable tool. Theoretically, all the resonance lines of a spectrum can be assigned to some molecular function, but in practice, for complex spectra, this proves difficult and often unnecessary.

An examination of a PMR spectrum must entail careful consideration of the following features.

(1) Resonance line positions, or *chemical shift* values, indicate the electronic environment of groups of equivalent protons. Correlation charts should be consulted to establish the probable nature of each group. Resonance line overlap occurs frequently, particularly at high field, and spectrum simplification by spin-spin decoupling may be necessary. Examination of a spectrum in the presence of D_2O or CD_3OD will establish the identity of bands due to $-OH$ or NH, bases exhibit characteristic shifts in CF_3COOD solution.

(2) *Line intensity*, as determined by the electronic integrator, is proportional to the number of protons responsible for the signal.

(3) *Spin-spin coupling* patterns give the number of protons in interacting equivalent groups. Coupling patterns may be clarified by employing a lanthanide shift reagent or by INDOR spectroscopy.

A detailed account of structure elucidation by NMR is beyond the scope of this chapter and reference should be made to the excellent coverage of the subject by Bible (1965), Silverstein and Bassler (1968), Chapman and Magnus (1966), Casy (1971), Bhacca and Williams (1964) and Mathieson (1965).

INVESTIGATION OF DYNAMIC PROPERTIES
OF MOLECULES

Conformational Isomerism

At room temperature the cyclohexane molecule undergoes a rapid interconversion of chair forms, where the hydrogen atoms alternate between axial and equatorial conformations.

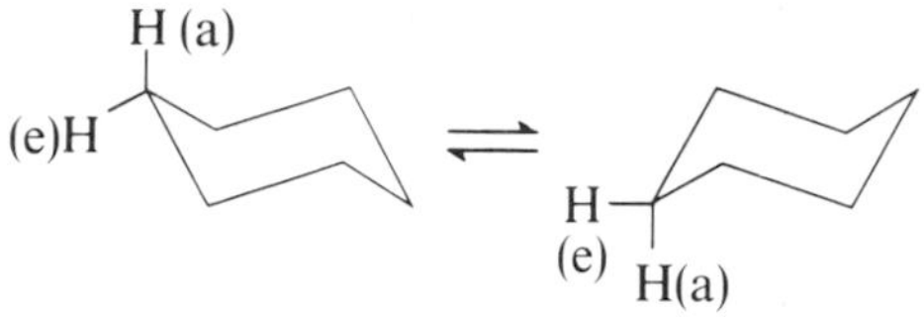

Above $-50°C$ the PMR spectrum of cyclohexane in carbon disulphide exhibits a single resonance peak, a consequence of the average environment experienced by each proton. Below $-50°$ the signal broadens, as the conformational interconversion slows down, to a very broad band at $-65°$. At $-70°$ two ill-defined peaks are apparent and these develop until at $-100°$ they are seen as a pair of doublets ($J = 5–7$ Hz) with a chemical shift difference of 27 Hz. As the temperature is lowered the protons spend increasingly longer times in axial and equatorial positions until ultimately the NMR spectrometer records signals due to each conformation. The low field doublet is assigned to the equatorial protons and that at high field to the more shielded axial protons. The inversion rate and the energy of inversion of such systems may be calculated from NMR data.

Further examples of the application of NMR in the investigation of conformational isomerism are given in Anderson (1965), Roberts (1966) and Jensen, Noyce, Sederholm and Berlin (1960).

Restricted Rotation

Free rotation about the $C—N$ bond of amides is impaired by the partial double bond character bestowed by resonance forms; dimethylformamide, for example, may be formulated thus:

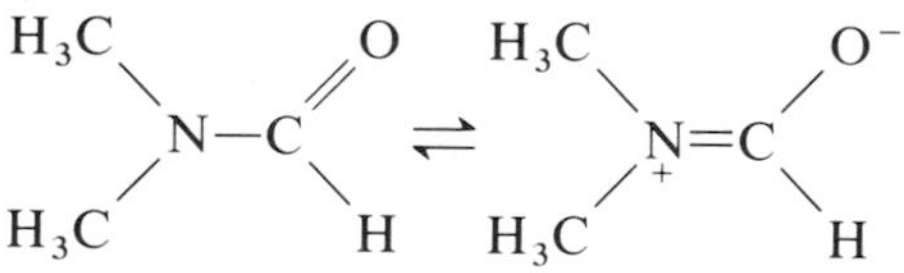

In the absence of a contribution from structure (XI) free rotation about the $C—N$ bond would render the $N—CH_3$ groups equivalent and their hydrogen resonances coincident. However below $64°$ the N-groups are magnetically non-equivalent and two sharp resonance lines result (Fig. 231). Temperature increase enhances the rotation rate about the $C—N$ bond until the methyl groups achieve magnetic equivalence. Both the rotation rate and the energy

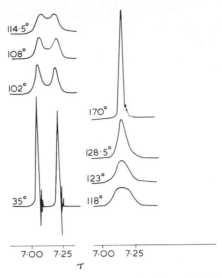

114·5°

108°

102°

170°

128·5°

123°

118°

35°

7·00   7·25   7·00   7·25

τ

Fig. 231. The spectrum of *N,N*-dimethylformamide (neat liquid) at room temperature shows a sharp peak for each methyl group. The peak at high field is trans to the formyl proton and is slightly broadened by coupling to it. When the temperature is raised, the peaks broaden and coalesce because the methyl groups exchange their environments by rotation about the C—N bond. This bond has partial double-bond character, as indicated by the 12-kcal activation energy required for rotation

barrier to rotation may be calculated from PMR data. (Pople, Schneider and Bernstein, 1959). A more detailed discussion of the detection of hindered rotation and inversion by NMR spectroscopy may be found in Kessler (1970).

Molecular Asymmetry

Under conditions of unrestricted rapid rotation about a C—C single bond in an open chain system, the NMR spectrum reflects an average of the environments experienced by the nuclei concerned. In $CH_3.CH_2.X$, the CH_3 protons form one equivalent group and the $-CH_2-$ another giving an A_3X_2 resonance pattern. A fixed configuration (XII) or slow rotation, on the other hand, would be observed in the PMR spectrum by the non-equivalence within the proton groups and thus a more complex splitting pattern would result. 1,2-Difluorotetrachloroethane, at room temperature, yields a single line ^{19}F spectrum: rotational isomers (XIII), (XIV) and (XV) appear as averaged.

(XII)

(XIII)   (XIV)   (XV)

Rotation about the C—C bond decreases as the temperature is lowered until at $-120°$ the fluorine nuclei exhibit two distinct signals separated by 51 Hz at 56.4 MHz.

Consider now, the rotational isomers (XVI), (XVII) and (XVIII) of the compound

$$
\begin{array}{c}
\ \ X\ \ H \\
\ \ |\ \ \ | \\
Y-C-C-R \\
\ \ |\ \ \ | \\
\ \ Z\ \ H
\end{array}
$$

where the methylene group is attached to an asymmetrically substituted carbon. The protons H′ and H″ in the three rotational isomers do not experience equivalent averaged environments even when rapid rotation about the carbon–carbon bond occurs and they give rise to a geminal AB quartet. A similar splitting pattern would also result where the configuration is frozen by steric hindrance, even in the absence of an adjacent asymmetric carbon.

(XVI)   (XVII)   (XVIII)

Differentiation between *cis* and *trans* 2,6-dimethyl-1-benzylpiperidines (XIX and XX) exploits the principle of the non-equivalence of methylene protons through molecular asymmetry. The *trans* isomer, lacking a plane of symmetry, has magnetically non-equivalent benzylic methylene protons which appear as an AB quartet ($J = 13.2$ Hz) centred at 3.41δ (Fig. 232a). On the other hand, the benzylic methylene protons of the symmetrical *cis* isomer appear (Fig. 232b) as a sharp singlet at 3.41δ, Hill and Chan (1965).

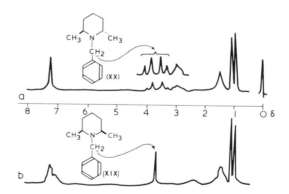

Fig. 232. NMR spectra of *cis* and *trans*-2,6-dimethyl-1-benzylpiperidines

The term *intrinsic diastereotopism* (Mislow and Rabin, 1967) is used to describe this phenomenon which is seen not only in the methylene hydrogens of a benzylic group but also in sterically constrained methylene groups (Morris, 1973).

A knowledge of the preferred conformation of a drug in aqueous solution may provide a useful insight into structure-biological activity relationships. In the case of acetylcholine (Culvenor and Ham, 1966) a consideration of the size of vicinal coupling constants from the A_2B_2 pattern of the $-CH_2.CH_2-$ group, favours a *gauche* conformation (XXI*a*), best represented by (XXII), rather than a *trans* conformation (XXI*b*).

(XXI*a*)   *gauche*   (XXI*b*) *trans*   (XXII)

Hydrogen Bonding

Protons able to participate in hydrogen bonds can be identified and studied by means of PMR spectroscopy (Emsley, Feeney and Sutcliffe, 1965). The position of the hydroxyl proton resonance signal of ethanol is temperature dependent, temperature elevation causing a diamagnetic shift, whereas the chemical shifts of the $-CH_2-$ and $-CH_3$ signals are unaffected. A consideration of the nature and origin of the hydroxyl proton signal provides an explanation of this phenomenon. Owing to rapid chemical exchange the O—H proton experiences many magnetic environments from both dissociated and hydrogen-bonded molecules, and its resonance signal reflects an average of these. As the temperature is raised dissociation increases and the —OH proton's 'average' environment is changed. If H-bonding is thought of as depleting the electron density around the bonded proton, then shielding, and a consequent upfield shift, results from dissociation. Similarly a shift of the ethanol OH proton signal towards TMS (0.5δ–5δ) is observed when a sample is gradually diluted with a non-polar solvent, such as carbon tetrachloride. Dilution, like temperature elevation, favours the dissociated species and intermolecular H-bonded protons may be recognised by the dependence of their chemical shifts on these variables. Assignments of hydroxyl protons however, should be verified by deuterium exchange. Protons engaged in intramolecular H-bonds exhibit only small shifts on dilution or temperature elevation. Quantitative relationships between H-bonding and the magnitudes of chemical shift changes have not been established.

A solution of chloroform in benzene provides an interesting example of the H-bonding to the benzene π-cloud. The chloroform proton resonance signal occurs *upfield* of the normal position (7.3δ), because of shielding by the aromatic diamagnetic anisotropy.

High resolution PMR admirably complements infra-red spectroscopy in the investigation of H-bonding. However unlike the latter, because of chemical exchange, it cannot detect monomers, dimers, polymers, etc., individually.

Keto-Enol Tautomerism

Nuclear magnetic resonance spectroscopy is probably the most powerful physical analytical method for qualitative and quantitative investigations of keto-enol equilibria (Smith, 1964). During such equilibria a specific proton experiences two distinct magnetic environments, which, if enough of each tautomer is present, will show up as separate resonance signals.

Acetylacetone (XXIII) at room temperature is a mixture of four parts enol and one part keto. Strong intramolecular hydrogen bonding stabilises the

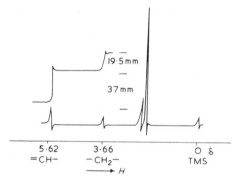

(XXIII)

enolic form, and because chemical exchange is slow, both forms are seen in the PMR spectrum (Fig. 233). In the presence of triethylamine only the enol tautomer occurs; it is stabilised by strong hydrogen bonding between the

Fig. 233. NMR spectrum of liquid acetylacetone at 43° taken at 60 MHz

amine nitrogen and the hydroxyl proton. The concurrent upfield shift of the hydroxyl resonance reflects the strength of the intermolecular attraction. Tautometer ratios in keto-enol mixtures may be calculated from the integration curve. In the spectrum of acetylacetone taken at 43° (Fig. 233), for example the band at 5.62δ is assigned to the enol olefinic hydrogen and that at 3.66δ to the keto $-CH_2-$ group which correspond to integration values of 37 mm and 19.5 mm respectively. Thus

$$\text{enol} \equiv =CH- \equiv 1H \equiv 37\,\text{mm}$$

$$\text{keto} \equiv -CH_2- \equiv 2H \equiv 19.5\,\text{mm}$$

$$\text{or} \quad 1H \equiv 19.5/2\,\text{mm}$$

Therefore

$$\% \text{ enol} = \frac{37}{37 + 19.5/2} \times 100$$

$$= 79.1\%$$

The enthalpy of keto-enol conversion may be established from variable temperature studies (Reeves, 1957).

Determination of Optical Purity

In pharmaceutical chemistry and particularly in drug-structure-activity studies, it is often desirable to determine the ratio of enantiomorphs in a mixture or to assess the optical purity of a compound without resorting to the physical separation of diastereoisomers. Since it is known that diastereoisomers differ in their NMR characteristics (Mateos and Cram, 1959), the technique has been investigated for practical applications. (Raban and Mislow, 1965, 1966). If the mixture of enantiomorphs is quantitatively converted to the corresponding mixture of diastereoisomers by reaction with an optically pure reagent, examination of the NMR spectrum of the mixture often permits assignment of bands to each isomer. Enantiomorph ratios can be calculated from integral values in the manner described for keto-enol tautomers. An instrument in good operating condition is able to detect less than 1% of optical impurity.

Examination under normal solvent conditions fails to reveal differences in the spectra of enantiomers. However the fact that strong solvent-solute interactions such as hydrogen bonding and dipolar forces often cause a specific preferred alignment of solute with solvent (p. 379) may be exploited. In dilute solution in an optically pure chiral solvent such as $(+)$ or $(-)$ α-phenylethylamine, each of a pair of enantiomers may align itself differently with the solvent molecules, and chemical shift differences between corresponding groups in each isomer are then seen (Pirkle, 1966). Effectively the solute molecules combine with solvent molecules to form 'loose' diastereomeric complexes. The use of a racemic solvent does not differentiate between optical isomers. Similarly the addition of an optically pure chiral solute to a solution of enantiomers may result in a formation of diastereomeric complexes with observable shift differences (Anet, 1968). The enzyme α-chymotrypsin, for example, interacts in solution with racemic N-trifluorophenylalanine giving rise to two ^{19}F signals (Zeffren and Reavill, 1968). Chiral lanthanide shift reagents (p. 381) often accentuate chemical shift differences in diastereomeric complexes (Whiteside and Lewis, 1970).

An interesting method for determining optical purity has been described for the schistosomicidal tetrahydroisoquinoline (XXIV) where the $(+)$ isomer is the more biologically active. A salt is made of the base isomeric mixture with an optically pure acid, $(+)$ α-methoxy-α-trifluoromethylphenylacetic acid (XXV). In the salts of the enantiomers there is a 0.05 ppm shift difference seen for the 8-position aromatic protons. The anisotropic effect of the acid aromatic ring in the complex (XXVI) appears to be responsible for this difference (Baxter and Richards, 1972).

Molecular Interactions

Chemical shift differences arising from the interaction of solute and solvent, and solute and solute have been described, illustrating that NMR may provide an insight into the nature of intermolecular complexes. Thus, it should also be possible to explore drug–drug and drug–macromolecule complexation in a similar manner. Such phenomena need to be studied in dilute solution and,

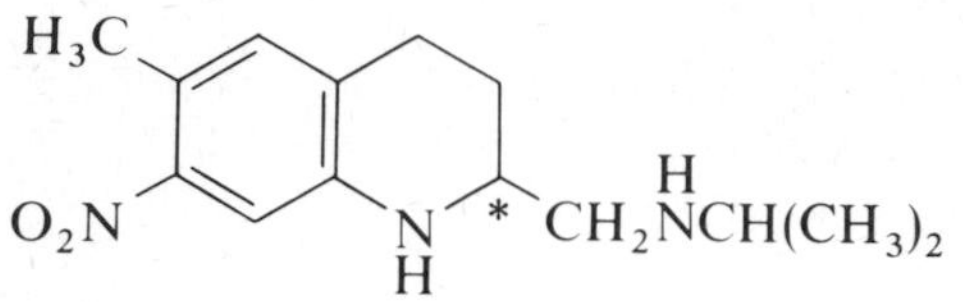

(XXIV)

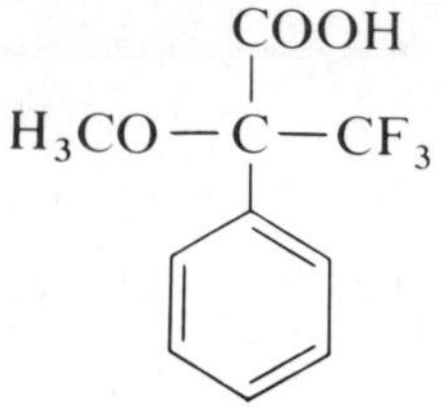

(XXV)

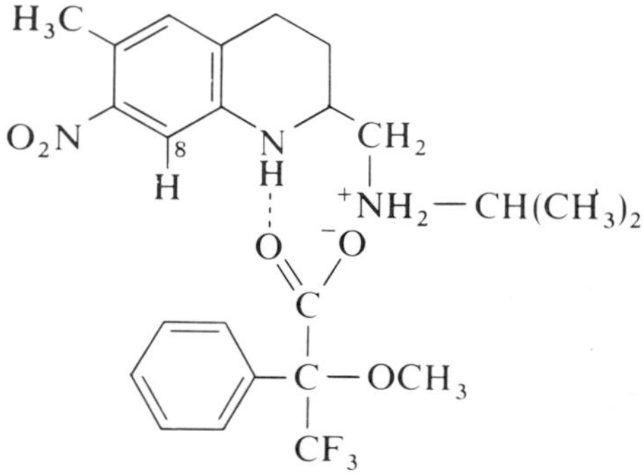

(XXVI)

therefore, to overcome the limited sensitivity of NMR, a spectrum accumulation technique is often employed.

Micelle Formation. Micelle formation by drugs in solution is a widespread phenomenon. With ionic compounds in aqueous solution, micellisation results in the removal of hydrophobic portions of the monomer from contact with water and concentration of the hydrophilic head groups at the surface of the micelle. The changes in the environment of these groups leads to changes in their PMR absorption characteristics. A series of phenothiazine drugs, which micellise at relatively high concentrations, for example $4 \times 10^{-2} M$ (1.33%) for promethazine hydrochloride (XXVII) have been studied at concentrations above and below the critical micelle concentration (CMC) (Florence and

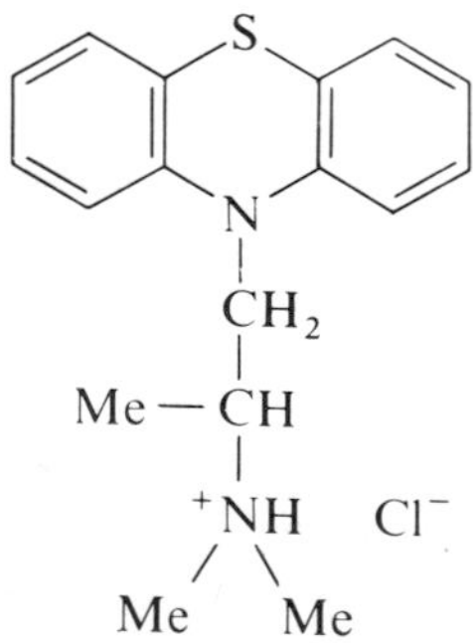

(XXVII)

Parfitt, 1970, 1971). Chemical shifts were measured relative to appropriate signals in an external standard of a 20% promethazine solution (micellar). The shifts of the resonance line of the hydrophilic $\overset{+}{N}H(CH_3)_2$ group to higher fields on increasing concentration were explained by the increased dissociation of the group at the micelle surface. Diamagnetic shifts of the aromatic protons suggested a parallel stacking of the phenothiazine rings in the interior of the micelle (Figs. 234 and 235).

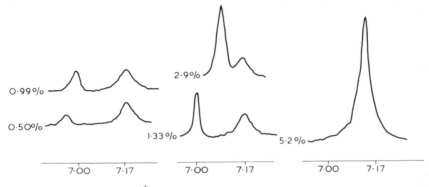

Fig. 234. Pmr signal of the $\overset{+}{N}(CH_3)_2$ protons of promethazine hydrochloride in D_2O. The high-field line of constant intensity is due to the external reference signal of 20% promethazine HCl in D_2O. The concentrations of the solution are marked (Courtesy of *J. Pharm. Pharmacol.*)

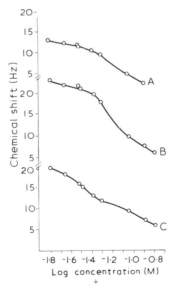

Fig. 235. Chemical shifts (Hz) of signal of A: $\overset{+}{N}(C\underline{H}_3)_2$, B; $CH—C\underline{H}_3$ and C: aromatic ring protons, as a function of the concentration of solution, relative to the position of 20% promethazine HCl signals from the respective groups (Courtesy of *J. Pharm. Pharmacol.*)

Drug-macromolecule Interactions (Jardetsky, 1964; Burgen and Metcalfe, 1970). When considering interactions between molecules of low molecular weight and macromolecules, two NMR parameters, chemical shift and line width are usually observed. If a small drug (or substrate) molecule interacts and binds with some portion of a macromolecular surface then it effectively becomes a part of the macromolecule. Dependent upon the strength of the binding, the drug will only reside for a portion of any given time in the complex and thus during an NMR experiment an average picture is seen. Because the small molecule has become, for a time, part of a more rigid lattice, the spin-spin relaxation time (T_2) of proton groups will change giving rise to changes in line width.

Drug or Substrate	Enzyme or Macromolecule	Complex
Large T_2	Small T_2	Bound substrate has smaller T_2 than free substrate
Narrow line	Broad line	Broader line than substrate alone

The magnetic environment of the bound substrate will also be different from that of the free molecule, thus chemical shift changes will also occur. A consideration of the changes in chemical shift and relaxation times of specific proton groups in the drug when it moves from the free to the bound state will provide valuable information concerning the mode of binding.

A simple example is the broadening of the benzyl alcohol aromatic signal on moving from an aqueous solution to the same strength solution in a 1% suspension of erythrocyte membranes. The broadening is due to the partitioning of benzyl alcohol into the lipid membrane.

The storage of catecholamines, for example adrenaline in the adrenal medulla, is believed to be in the form of weak complexes with nucleotides. Relaxation time differences between solutions of adrenaline and of 1:1 adrenaline-adenosinetriphosphate complex in D_2O indicated that the complex is formed by attachment of the catecholamine side-chain to, probably, the phosphate of the nucleotide. Similar studies suggest that protein binding of benzylpencillin occurs through the pencillin benzyl group interacting with some aromatic groups of the protein.

QUANTITATIVE ANALYSIS

Attention has been focused largely on nuclear magnetic resonance spectroscopy as an aid to structure elucidation and to the study of molecular dynamic processes. Far less heed has been paid to its applications to quantitative analysis, though this is now beginning to receive attention (Kasler, 1973; Perkin-Elmer NMR Quarterly, 1972). Automatic integration of resonance bands affords an easy and rapid quantitative means of determining the ratio of compounds in a mixture, provided that at least one resonance band from each constituent is free from extensive overlap by other absorption. The estimation of the keto-enol ratio in acetylacetone (p. 402) is an example of the quantitative analysis of a mixture. An example (Smith, 1964) where a limited amount of band overlap

Broad 3-proton
singlet at 1.92δ

$$H_3C \diagdown C=O$$
$$H_3C \diagup$$

6-proton
singlet at
2.10δ

(XXVIII)

$$H_3C \diagdown CH.OH$$
$$H_3C \diagup$$

6-proton
doublet centred
at 1.12δ

(XXIX)

$$H_3C \diagdown C-O.C-CH_3$$
$$H_2C \diagup \quad \overset{O}{\underset{\|}{}}$$

3-proton singlet
at 2.10δ

(XXX)

has been surmounted, is the analysis of a mixture of acetone (XXVIII), isopropanol (XXIX) and isopropenyl acetate (XXX), the high field spectrum of
which is illustrated in Fig. 236.

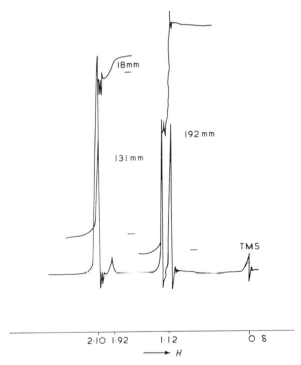

Fig. 236. NMR spectrum of a mixture of isopropenyl acetate, isopropanol and acetone

The six-proton singlet of acetone at 2.10δ, coincides with the acetate methyl
three proton singlet of (XXX). Fortunately the broad apparent singlet at 1.92δ
is assignable to the methyl group of isopropenyl acetate attached to the double
bond (low order coupling to the olefinic protons produces line broadening).
The band at 1.92δ integrates for 18 mm, therefore the acetone signal corresponds

to an integral of (131–18) mm. Thus for the mixture:

$$\text{Acetone (XXVIII)} \equiv 6H\ (2.10\delta) \equiv (131-18)\ \text{mm} \equiv 113\ \text{mm}$$

$$\text{Isopropanol (XXIX)} \equiv 6H\ (1.12\delta) \equiv 192\ \text{mm}$$

$$\text{Isopropenyl acetate (XXX)} \equiv 3H\ (1.92\delta) \equiv 18\ \text{mm}$$

therefore

$$6H \equiv 36\ \text{mm}$$

$$\% \text{ Acetone} = \frac{113}{113+36+192} \times 100 = \frac{113}{341} \times 100 = 33\%$$

$$\% \text{ Isopropanol} = \frac{192}{341} \times 100 \qquad\qquad = 56\%$$

$$\% \text{ Isopropenyl acetate} = \frac{36}{341} \qquad\qquad = 11\%$$

Similarly, mixtures of aspirin, phenacetin and caffeine and A.P.C. tablets have been assayed using the aspirin methyl ester protons (2.3δ), the protons of the phenacetin acetyl group (2.1δ) and for caffeine the 1- and 3-N—CH_3 protons (3.4δ and 3.6δ). The procedure takes 15–20 minutes and is reasonably accurate (Hollis, 1963).

The absolute concentration of a constituent in a pharmaceutical raw material or formulation may be obtained by adding a reference compound to the solution for NMR examination. Weighed amounts of both sample and reference are dissolved in an appropriate solvent and the spectrum and integrals are obtained.

Corticosteroids both in bulk and in formulations have been assayed in this manner (Avdovich 1970). Prednisone (XXXI), Prednisolone and Triamcinolone are 1,4-dien-3-ones and exhibit a characteristic olefinic-region spectrum (Fig. 237). The C-1 proton is seen as part of an AB quartet at 7.9δ, coupled to the C-2 proton at 6.4δ, which in turn is partly obscured by the C-4 (1H) resonance at 6.3δ. To the 1,4-dien-3-one sample in dimethylsulphoxide solution is added a known weight of a pure stable reference standard, fumaric acid, the olefinic resonance line of which is seen at 7.1δ. At 5.9δ a peak due to the 4-position proton of a 4-ene-3-one steroid is observed, but does not interfere with the assay.

Several integrals are run for each determination and the integrals from the 7.9δ sample signal and the reference integral are measured. The amount of corticosteroid present is calculated from the simple expression:

Amount of Steroid (mg)

$$= \frac{\text{EW (Steroid)}}{\text{EW (Standard)}} \times \frac{\text{Integral (mm) Steroid}}{\text{Integral (mm) Standard}} \times \text{Weight of Standard}$$

where

$$\text{EW} = \frac{\text{Molecular Weight}}{\text{Number of hydrogens in signal chosen}}$$

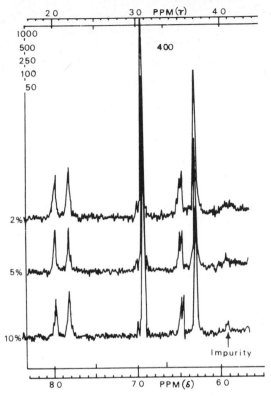

Fig. 237. Partial NMR spectrum of steroidal 1,4-dien-3-one in dimethyl sulphoxide containing fumaric acid and 2, 5, or 10% of added steroidal 4-en-3-one (Courtesy of *J. Pharm. Sciences*)

Advantages of the NMR method over the UV and colorimetric methods in this determination are speed and specificity. Although the 4-en-3-one impurities may be detected, they are not measured.

(XXXI)

During quantitative studies the conditions under which spectra are run must be considered carefully. Saturation and relaxation effects are dependent upon the operating H_1 level and the speed of scan. The fastest practical speed of scan is limited not simply by the resolution required, but also by the necessity to allow sufficient time between scans to permit complete relaxation. Thus instrumental conditions must be carefully optimised for each type of assay.

Other spectrum factors which must be considered in quantitative studies are spinning side-bands and ^{13}C satellites. These are part of the main proton signal and must be treated consistently in each assay.

Surfactant Chain-Length Determination

To establish the ethylene ether chain length of a non-ionic surface-active agent by chemical methods is a lengthy procedure. Nuclear magnetic resonance provides a rapid, accurate and non-destructive method of analysis.

A spectrum and at least five repeat integrals are obtained for a 10% solution of the surfactant in carbon tetrachloride. The ratio of the ethylene ether proton integral to the cetyl terminal chain of the surfactant is determined. From the knowledge that the cetyl end group integral is equivalent to 28H, the value of n for the $-O-(CH_2-CH_2-O-)_n$ chain is obtained from a simple proportion calculation. Table 50 illustrates how the NMR results compare with the manufacturers specifications.

Table 50

Detergent	Chain Length (n)	
	Specification	NMR Value
Texofor A10	10	10
Cetomacrogol 100	22–25	23
Texofor A45	~45	47
Texofor A60	~60	62

Hydrogen Analysis

The percentage of hydrogen in an unknown sample may be determined easily and rapidly from the total integral of its PMR spectrum. Results as accurate as those obtained by conventional microanalytical means are claimed. The hydrogen percentage (H%) is given by the expression:

$$H\% = \frac{\text{total integral area}}{kCV} \tag{6}$$

where

k = proportionality constant for the sample tube
C = weight of sample per unit volume
V = effective volume of solution

Two sample tubes of identical diameter are required and both are charged with a known volume (0.532 ml) of a solution of n-octane in carbon tetrachloride (0.2040 g/ml). A volume calibration mark is made on each tube and one tube is sealed as a permanent reference standard. Total integral values for each sample should be the same if the tube diameters are identical. The unknown compound (10–150 mg) is weighed directly into the matched sample tube and made up to the calibrated volume with carbon tetrachloride. The percentage hydrogen is calculated from the total integral area, using eq. (6). Samples, which may be recovered, should be as large as possible in order to minimise errors introduced by the integration of background noise.

Iodine Value

Triglycerides have in their PMR spectra four characteristic sets of signals from the resonance of olefinic protons, the four C-1 and one C-3 glyceride methylene protons, methylene protons directly linked to a double bond, and the remaining protons on saturated carbons. The integration curve of the combined C-1 and C-3 glyceride methylene protons, occurring in isolation around 4δ, can be measured accurately, and with these as an internal calibration, the olefinic protons (the degree of unsaturation) and the total number of protons (a measure of the average molecular weight) may be determined. Iodine values, calculated from the olefinic proton integration (Johnson and Shoolery, 1962) and the molecular weight, agree well, in general, with those estimated by Wijs method (Table 51). An exception is Tung oil which, because it possesses conjugated unsaturation, a function of a high oleostearic ester content, gives a low iodine value by Wijs method. The value calculated from PMR data corresponds well with that determined by vapour phase chromatography.

Table 51. Iodine Values for a Selection of Natural Oils

Oil	NMR method	Wijs method
Coconut	10.5 ± 1.3	8.0–8.7
Olive	80.8 ± 0.9	83.0–85.3
Peanut	94.5 ± 0.6	95.0–97.2
Soybean	127.1 ± 1.6	125.0–126.1
Sunflower seed	135.0 ± 0.9	136.0–137.7
Safflower seed	142.2 ± 1.0	140.0–143.5
Whale	150.2 ± 1.0	149.0–151.6
Linseed	176.2 ± 1.2	179.0–181.0
Tung	225.2 ± 1.2	146.0–163.5

Moisture Analysis

Water adsorbed on biological materials, such as food products, appears in an NMR spectrum as a relatively sharp band superimposed on the very broad proton absorption of the solid. The moisture content of a powdered solid may be determined from the adsorption curve (Pople, Schneider and Bernstein, 1959), which is rather broader than that of pure liquid water, or more usually from the derivative curve. Comparison is made with a calibration plot of signal amplitude against percentage moisture for samples of known moisture content. The technique is non-destructive and is independent of the particle size of the powder.

Carbon-13 NMR (CMR) Spectroscopy

During the last ten years natural abundance ^{13}C-NMR spectroscopy has progressed from infancy to maturity (Levy and Nelson, 1972; Grutzner, 1972). Instrumentation has developed to such an extent that the technique is rivalling ^{1}H-NMR in versatility.

Almost all the foregoing discussion for ^{1}H-NMR may be applied to CMR, but there are important differences, Table 52.

Table 52. Nuclear Properties of ^{1}H and ^{13}C

	^{1}H	^{13}C
Nuclear Spin (I)	$\frac{1}{2}$	$\frac{1}{2}$
Resonance Frequency at 23.5 Kgauss	100 MH$_2$	25.2 MH$_2$
Natural abundance	99.9%	1.1%
Sensitivity for an equal number of nuclei	1.00%	0.016
Shift Range	20 ppm	600 ppm

Sensitivity

A natural abundance of 1.1% for ^{13}C is clearly a factor that renders CMR less sensitive than PMR. However, because of its low abundance, the probability of a ^{13}C atom residing adjacent to another ^{13}C atom is low. Complications arising from ^{13}C–^{13}C coupling are, therefore, negligible.

A second factor affecting sensitivity is the low magnetogyric ratio (γ) of ^{13}C, about $\frac{1}{4}$ that of ^{1}H. Since sensitivity is proportional to γ^3, then ^{13}C affords about $\frac{1}{60}$ the sensitivity of ^{1}H in an NMR experiment. Effectively, ^{1}H is about 6000 times more sensitive than ^{13}C in NMR terms, i.e. $\simeq 60$ (γ factor) $\times 100$ (abundance factor).

The low sensitivity of CMR is overcome by the use of large samples, up to 2 ml in 15 mm tubes, and by enhancement and decoupling techniques in conjunction with highly stable spectrometers operating at high fields.

Both CAT and *Fourier Transform* (p. 369) have been described and are used to obtain ^{13}C spectra. The latter is particularly useful because of the speed of spectrum accumulation.

Coupling of ^{13}C to ^{1}H not only complicates spectra from the standpoint of interpretation but it gives, instead of a single sharp resonance signal, a multiplicity of bands often of low intensity. Heteronuclear decoupling of each ^{1}H individually from each ^{13}C signal would be a lengthy and tedious process. The problem is overcome by *proton noise decoupling*. During a noise decoupling experiment, all the protons in the sample are decoupled simultaneously and each ^{13}C resonance band is seen as a single line. Proton noise decoupling also disturbs the ^{13}C energy level populations and results in a nuclear Overhauser effect affording an enhancement of each ^{13}C signal intensity by up to a factor of three.

By employing a combination of the above techniques, a ^{13}C natural abundance spectrum may be obtained from as little as 10 mg of sample. Much smaller amounts of compound are needed if the sample is ^{13}C-enriched.

Chemical Shift

Chemical shift is usually the most important spectral parameter in CMR spectroscopy. Whereas ^{1}H resonances occur over a relatively narrow range (10–20 ppm), ^{13}C shifts cover a range of about 600 ppm, with most resonances falling within a 200 ppm range.

The signal spread and the application of proton noise decoupling results in spectra with each carbon atom in a molecule being displayed as a single discrete sharp band. As in PMR, ^{13}C shifts are measured relative to a TMS internal standard. In general, ^{13}C shifts fall into well defined ranges according to the electronic and magnetic environment of the carbon (Fig. 238).

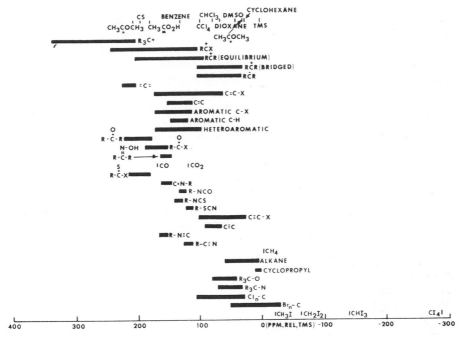

Fig. 238. General ^{13}C chemical shift chart (Courtesy of Wiley Interscience)

Lanthanide shift reagents are occasionally employed in CMR spectroscopy, but because of the low incidence of resonance band overlap, they are much less useful than in ^{1}H investigations.

Spin-Spin Coupling

For the reasons outlined, about ^{13}C–^{13}C and ^{13}C–^{1}H couplings are not usually seen in CMR spectra. The former may be examined in ^{13}C-enriched samples and the latter can sometimes be extracted from ^{13}C satellites in ^{1}H spectra ($J_{^{13}C-^1H} \simeq 150$ Hz). In the absence of proton noise decoupling, ^{13}C spectra are very complex. However, selective decoupling techniques may help to determine specific ^{13}C–^{1}H coupling constants.

Integration

At present in normal CMR spectra, there is no correlation between integrated peak areas and the number of ^{13}C nuclei in each signal. Integrals, are therefore, not generally measured. A Fourier Transform derived spectrum is based upon the spin-lattice relaxation of each ^{13}C atom and the intensity of each band is not dependent solely upon the number of ^{13}C atoms responsible for the signal. Similarly, because of the variation in relaxation times, proton noise decoupling results in variable nuclear Overhauser enhancement.

Modification of the Fourier Transform technique is being explored as a means of quantitating ^{13}C signals.

Applications

Structure elucidation is the most common application of ^{13}C spectroscopy (Gray, 1973). PMR and CMR are often employed as complementary techniques. CMR is particularly useful in the investigation of the structures of complex natural products and biopolymers. Other applications in common with PMR include the study of the dynamic properties of molecules and molecular geometry.

The investigation of biosynthetic pathways, particularly in micro-organisms, is being assisted by the increasing availability of ^{13}C-enriched substrates. Detection and identification of labelled sites may be accomplished by CMR directly or by examination of $^{13}C-^1H$ satellite signals, e.g. Cephalosporin C (Neuss 1971) and the rifamycins (Martinelli, 1973).

References

Anderson, J. E., *Q. Rev.* (1965) **19**, 426.

Anet, F. A. L., Sweeting, L. M., Whitney, T. A. and Cram, D. J. *Tetrahedron Letters* (1968) 2617.

Avdovich, H. W., Hanbury, P. and Lodge, B. A., *J. Pharm. Sci.* (1970) **59**, 1164.

Backers, G. E. and Shaefer, T., *Chem. Rev.* (1971) **71**, 617.

Batterham, T. J., *NMR Spectra of Simple Heterocycles*, Wiley, New York, 1973.

Baxter, C. A. R. and Richards, H. C., *Tetrahedron Letters* (1972) 3357.

Becker, E. D., *J. chem. Educ.* (1965) **42**, 591–96.

Becker, E. D., *High Resolution NMR*, Academic Press, New York, 1969.

Bhacca, N. S., Johnson, L. F. and Shoolery, J. N., *NMR Spectra Catalog.*, Vol. 1, Varian Associates, Palo Alto, California, 1962.

Bhacca, N. S., Hollis, D. P., Johnson, L. F. and Pier, E. A., *NMR Spectra Catalog.*, Vol. 2, Varian Associates, Palo Alto, California, 1963.

Bhacca, N. S. and Williams, D. H., *Applications of NMR Spectroscopy in Organic Chemistry*, Holden-Day, San Francisco, 1964.

Bible, R. H., *Interpretation of NMR Spectra*, Plenum Press, New York, 1965.

Bovey, F. A. and Johnson, C. E., *J. chem. Phys.* (1958) **29**, 1012.

Brophy, G. C., Laing, O. N. and Sternhill, S., *Chem. and Ind.* (1968) 22.

Burgen, A. S. V. and Metcalfe, J. C., *J. Pharm. Pharmac.* (1970) **22**, 153.

Casy, A. F., *PMR Spectroscopy in Medicinal and Biological Chemistry*, Academic Press, London, 1971.

Chapman, D. and Magnus, P. D., *Introduction to Practical High Resolution Nuclear Magnetic Resonance Spectroscopy*, Academic Press, London, 1966.

Cookson, R. C., Crabb, T. A., Frankel, J. J. and Hudec, J., *Tetrahedron* (1966) Supplement No. 7, 355–90.

Culvenor, C. C. J. and Ham, N. S., *Chem. Comm.* (1966) 537.

Dailey, B. P. and Shoolery, J. N., *J. Am. chem. Soc.* (1955) **77**, 3877.

Emsley, J. W., Feeney, J. and Sutcliffe, L. H., *High Resolution Nuclear Magnetic Resonance Spectroscopy*, Vol. 1, Pergamon Press, Oxford, 1965.

Farrar, T. C. and Becker, E. D., *Pulse and Fourier Transform NMR*, Academic Press, New York, 1971.

Florence, A. T. and Parfitt, R. T., *J. Pharm. Pharmac.* (1970) **20**(*suppl.*), 121.

Florence, A. T. and Parfitt, R. T., *J. phys. Chem.* (1971) **75**, 3554.

Fullerton, S. E., May, E. L. and Becker, E. D., *J. org. Chem.* (1962) **27**, 2144.

Gray, G. A., *Applications of ^{13}C Nuclear Magnetic Resonance in Biochemistry—Critical Reviews in Biochemistry*, Chemical Rubber Company, Spring, 1973.

Grutzner, J. B., *Lloydia* (1972) **35**, 375.

Henty, D. N. and Vary, S., *Chem. and Ind.* (1967) 1782.

Hill, R. K. and Chan, T. H., *Tetrahedron* (1965) **21**, 2015.

Hollis, D. P., *Analyt. Chem.* (1963) **25**, 1682.

Jardetzky, O., *Adv. chem. Phys.* (1964) **7**, 499.

Jensen, F. R., Noyce, D. S., Sederholm, C. H. and Berlin, A. J., *J. Am. chem. Soc.* (1960), **82**, 1256.

Johnson, L. F. and Shoolery, J. N., *Analyt. Chem.* (1962) **34**, 1136.

Jones, A. J., *Rev. Pure Appl. Chem.* (1968) **18**, 253.

Karplus, M., *J. chem. Phys.* (1959) **30**, 11.

Karplus, M. and Anderson, D. H., ibid. (1959) **30**, 6.

Kasler, F., *Quantitative Analysis by NMR Spectroscopy*, Academic Press, New York, 1973.

Kessler, H., *Angew. Chem.* (*Int. Ed.*) (1970) **9**, 219.

Lambert, J. B. and Keske, R. G., *J. Am. chem. Soc.* (1966) **88**, 620.

Lapidot, A., Reuben, J. and Samuel, D., *J. chem. Educ.* (1964) **41**, 570.

Levy, G. C. and Nelson, G. L., *Carbon-13 Nuclear Magnetic Resonance for Organic Chemists*, Wiley-Interscience, 1972.

Martinelli, E., White, R. J., Gallo, G. G. and Beynon, P. J., *Tetrahedron* (1973) **29**, 3441.

Mateos, J. L. and Cram, D. J., *J. Am. chem. Soc.* (1959) **81**, 2756.

Mathieson, D. W. (ed.), *Interpretation of Organic Spectra*, Academic Press, London, 1965.

Mathieson, D. W. (ed.), *Nuclear Magnetic Resonance for Organic Chemists*, Academic Press, London, 1967.

Mayo, B. C., *Chem. Soc. Revs.* (1973) **1**, 49.

Mislow, K. and Raban, M., *Topics in Stereochemistry* (1967) **1**, 1.

Morris, D. G., Murray, A. M., Mullock, E. B., Plews, R. M. and Thorpe, J. E., *Tetrahedron Letters* (1973) 3179.

Neuss, N., *et al.*, *J. Am. chem. Soc.* (1971) **93**, 2337.

Perkin-Elmer, *NMR Quarterly*, 1972, No. 3.

Pirkle, W. H., *J. Am. chem. Soc.* (1966) **88**, 1837.

Pope, J. A., Schneider, W. G. and Bernstein, H. J., *High-Resolution Nuclear Magnetic Resonance*, McGraw-Hill, London, 1959.

Raban, M. and Mislow, K., *Tetrahedron Letters* (1965) 4249.
idem, ibid. (1966) 3961.

Reeves, L. W., *Canad. J. Chem.* (1957) **35**, 1351.

Roberts, J. D., *An Introduction to Spin-Spin Splitting in High Resolution NMR Spectra*, Benjamin, New York, 1961.

Roberts, J. D., *Chemistry in Britain* (1966) **2**, 529.

Ronayne, J. and Williams, D. H., *Chem. Comm.* (1966) 712.

Sanders, J. K. M. and Williams, D. H., *Nature, Lond.* (1972) **240**, 385.

Silverstein, R. M. and Bassler, G. C., *Spectrometric Identification of Organic Compounds*, Wiley, New York, 1968.

Smith, W. B., *J. chem. Educ.* (1964) **41**, 97.

von Philipsborn, W., *Angew. Chem.* (*Int. Ed.*) (1971) **10**, 472.

Whiteside, G. M. and Lewis, D. W., *J. Amer. chem. Soc.* (1970) **92**, 698.

Wiberg, K. B. and Nist, B. J., *The Interpretation of NMR Spectra*, Benjamin, New York, 1962.

Zeffren, E. and Reavill, R. A., *Biochem. Biophys. Res. Commun.* (1968) **32**, 73.

13 Mass Spectrometry

R. T. PARFITT

Many excellent reviews* have been written on the theory of mass spectrometry and its applications in organic chemistry and reference should be made to these for a detailed treatment of the subject. The objective of this chapter is to make the student aware of its potential in qualitative and quantitative pharmaceutical analysis.

The mass spectrometer produces positive ion spectra which, unlike the overlapping band spectra from most other spectrometric methods, are line spectra. When an organic molecule is bombarded with electrons of sufficient energy ($> 10\,eV$), it may lose an electron and so yield a positive ion:

$$M + e \rightarrow M^+ + 2e$$

M^+ is often unstable, having an energy excess, and in the mass spectrometer it fragments in a specific manner. During electron bombardment one of the molecules' valence electrons can be removed and the imparted energy excess surges through the molecule. Whenever sufficient energy accumulates in a particular bond, then that bond will cleave. Since different bonds require different amounts of energy to break them, each molecule will give rise to a unique fragmentation or 'cracking' pattern. Among the fragments produced are further positive ions which are separated and recorded by the mass spectrometer according to their mass-to-charge ratios (m/e). Since the charge is usually unity, $m/e = m$, the mass of the fragment; occasionally, however, fragments of higher charge are encountered.

Wien, in 1898, produced the first crude mass spectra when he demonstrated that positive ions could be deflected according to their masses in electric or magnetic fields. This observation was developed by Thomson (1910) who used a combined electrostatic and magnetic field to observe the mass spectrum of a mixture of rare gases and thereby identified two neon isotopes. Instrumentation was developed further by Dempster (1918) and Aston (1919), the latter confirming the nature of Thomson's neon isotopes. By 1924, Aston had determined the isotopic constitution of about fifty elements.

The potential of mass spectrometry in the study of organic compounds was quickly realised, but it was not until 1940 that a commercial modification of the Dempster instrument was employed in the quantitative estimation of a complex mixture of hydrocarbons. Computer-assisted quantitative analysis of hydrocarbon fractions is now of considerable importance in the petroleum

---

* Beynon (1960); Biemann (1962); Budzikiewicz, Djerassi and Williams (1964); Hill (1966); McLafferty (1963); Milne (1971); Spiteller and Spiteller-Friedmann (1965); Waller (1972).

industry. During the past twenty years, with the advent of high resolution mass spectrometry, the technique has been employed extensively for the elucidation of complex organic structures, by the rationalisation of their fragmentation patterns.

Theory

Figure 239 is a diagram of a single focusing mass spectrometer. The sample is introduced into the instrument in such a way that its vapour is bombarded by

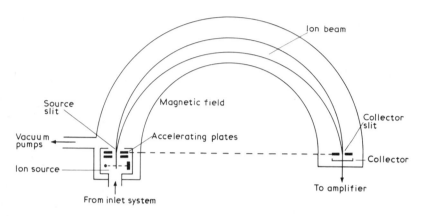

Fig. 239. Single focusing mass spectrometer

electrons having an energy of about 70 eV. Positive ions formed in the ion-source are accelerated between two plates by a potential difference of a few thousand volts (V). The ions pass through the source slit and are deflected by a magnetic field (H) according to their mass/charge ratios. They then pass through the exit or collector slit, impinge upon the collector, and the signal received is amplified and recorded (usually on photographic paper). The height or intensity of the resulting peak is proportional to the ion abundance, that is, the number of ions of identical mass received by the collector.

For an ion of unit charge e, mass m and velocity after acceleration v, the potential energy of the particle eV, is equal to its kinetic energy: i.e.

$$eV = \tfrac{1}{2}mv^2 \tag{1}$$

where V = acceleration voltage.

In a magnetic field the ion experiences a force Hev at right angles both to the direction of the field and its direction of motion. It therefore moves in the arc of a circle, where the radius is given by the expression:

$$Hev = \frac{mv^2}{r}$$

where

$$r = \text{radius of the ion path}$$

and

$$H = \text{strength of the magnetic field}$$

Therefore

$$r = \frac{mv}{eH} \tag{2}$$

Eliminating v between eqs. (1) and (2) gives

$$m/e = \frac{H^2 r^2}{2V} \tag{3}$$

Thus the radius of the ion path may be changed by varying either the magnetic field (H) or the accelerating voltage (V). By either method, ions of different mass-to-charge ratios (m/e) can be made to impinge upon the collector in turn thus giving rise to a spectrum (Fig. 240). Magnetic field variation enables a wide range to be covered in a single sweep, but, for very rapid scanning a voltage sweep must be employed.

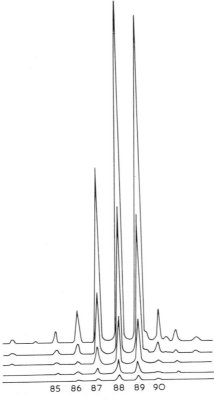

85 86 87 88 89 90

Fig. 240. A portion of a mass spectrum traced by a five-element galvanometer

The ion-source, ion path and collector of the mass spectrometer must be under high vacuum (10^{-7} mm Hg) for optimum operation. Any extraneous materials in the source, including atmospheric gases, will themselves ionise in the electron beam and their spectra will be recorded.

Instrumentation

The above description is that of a typical single focusing mass spectrometer, where a maximum resolution of 3000 is attainable (p. 423). The principle of magnetic focusing falsely assumes that all ions are formed with zero kinetic energy, and in the magnetic field are focused only according to their mass. Velocity or kinetic energy focusing is achieved by the introduction of an electrostatic analyser between the ion accelerating chamber and the magnetic analyser. The electrostatic analyser selects a beam of ions of very narrow energy range for final deflection in the magnetic analyser. The resolving power of this *double-focusing mass spectrometer* (Fig. 241) is of the order of 30 000. Such a resolving capability enables high molecular weight fragments, which differ by only one mass unit, to be distinguished.

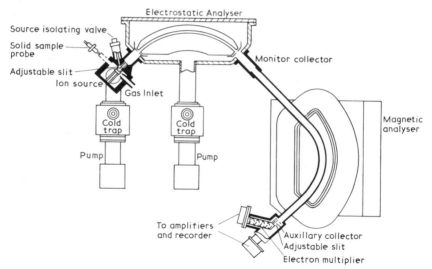

Fig. 241. Double focusing mass spectrometer

Double beam instruments where two ion beams from independent sources pass side by side through a common mass analyser and are detected by separate collectors are available. Such instruments may be used to compare samples directly, to investigate a single sample under different ionising conditions or to compare a sample with a standard (e.g. perfluorokerosene) as a mass marker.

The *Quadrupole Mass Spectrometer*, initially devised to separate uranium isotopes, has recently been adopted in organic mass spectrometry, particularly when in combination with a gas chromatograph. Focusing of ions, after acceleration from the ion source, is effected by a quadrupole mass filter where they are separated according to mass (Fig. 242), and detected by an electron

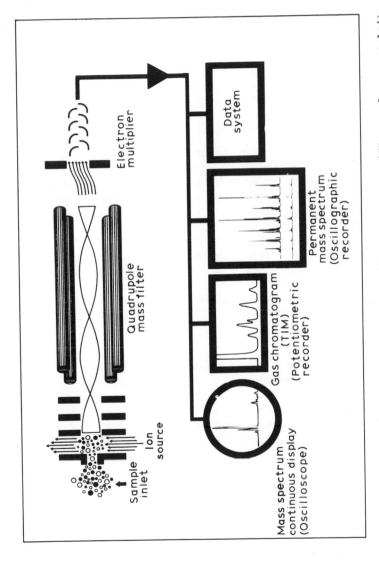

Fig. 242. Functional diagram of quadrupole mass spectrometer (Courtesy of Finnegan Instruments Ltd.)

multiplier. The mass filter consists of a quadrant of four parallel circular (or ideally hyperbolic) tungsten rods which focus ions by means of an oscillating and variable radiofrequency field.

Advantages of the quadrupole mass spectrometer are:

(1) The mass of ions focused is directly proportional to the voltage applied to the quadrupole filter. Since the voltage is a linear function, the mass spectrum display is also linear. Conventional sector instruments usually yield logarithmic spectral displays.

(2) Accelerating potentials in quadrupole instruments are in the order of 5–30 V, whereas a sector mass spectrometer generally requires 3000–6000 V. A high accelerating voltage requires a very low ion source pressure ($\simeq 10^{-7}$ mm Hg) to prevent arcing, whereas a low accelerating potential permits the use of a relatively high source pressure (10^{-4}–10^{-5} mm Hg). The latter is of particular advantage in gas chromatograph–mass spectrometer (GC/MS) combinations.

Quadrupole instruments are able to scan up to about 1000 atomic mass units (amu), but they do not give rise to metastable ions (p. 431).

The *linear time-of-flight mass spectrometer* does not require a magnet. Positive ions are produced in the conventional manner but instead of constant electron bombardment of the sample, there is pulse bombardment in short bursts lasting about 0.25 μsec. The ions are accelerated into an evacuated drift tube about 1 m long, in the absence of a magnetic field, and they separate or focus according to the momentum received during acceleration. The velocity achieved by each ion in the drift tube is dependent upon the mass-to-charge ratio. Maximum resolving power for a 'time-of-flight' instrument is low, about 1500, but scanning speeds are very fast rendering the instrument useful for combination with a gas-liquid chromatograph.

PRACTICAL CONSIDERATIONS

Sample Introduction

Physical and chemical properties govern the way in which a sample is introduced into the mass spectrometer. Comprehensive accounts of sample handling techniques are given by Beynon (1960), Biemann (1962) and Hill (1966). The sample must be in the vapour state when bombarded by electrons. This requires it to have a minimum vapour pressure of about 10^{-6} mm Hg since the ion-source is kept at an operating pressure of 10^{-7} mm Hg. Even compounds of low volatility, such as amino acids and peptides, may be investigated with the aid of modern inlet systems. Gases and volatile liquids are admitted to the source through a small leak from a *gas reservoir*. Solid samples and liquids of low volatility are introduced to within a short distance of the electron beam on the bakelite tip of a *direct-insert-probe*, which is inserted into the ion-source through a system of vacuum locks. The direct-insert-probe may be heated by platinum wire heating elements to aid sample vaporisation.

The *sample size* required to record a mass spectrum is dependent upon the method of sample introduction. As little as 0.1 μg is required for the direct-insert-probe but as much as 1 mg may be necessary for introduction *via* the gas reservoir. The size of samples from vapour phase chromatographs is of the order of 0.01–1 μg.

Sample Purity

Small amounts of impurities in the sample need not seriously affect the interpretation of a mass spectrum, although this depends largely upon the nature of the problem. The appearance of additional or more intense peaks facilitates the recognition of impurities. If the spectrum of a pure compound is available, then inspection of the mass spectra of other samples of that compound will show the presence of impurities by the additional peaks. Subtraction of the known spectrum from the spectrum of the impure sample furnishes a mass spectrum of the impurity (or impurities), and may lead to its identification. High molecular weight contaminants are readily detectable (provided they are sufficiently volatile) by peaks at higher mass than the known M^+.

The mass spectrometer does not distinguish readily an impurity which is stereoisomeric with the main component, since stereoisomers have very similar fragmentation patterns.

Temperature effects. Chemical changes induced in the sample by the elevated temperature of the inlet system may lead to variations in mass spectra. Alcohols, for example, may be dehydrated to olefins, resulting in the absence of a mass peak for the alcohol, and a spectrum characteristic of the corresponding olefin. To overcome this either the spectrum is run at a lower temperature or the sample is converted to a less sensitive derivative. Alcohols are often converted to their methyl ethers.

Spectra for comparison should be taken at, or as near as possible to the same operating temperature and voltage.

Spectrum Resolution

The resolving power of a mass spectrometer is its ability to separate ions of different mass-to-charge ratio (m/e). For two peaks corresponding to ions of mass m_1 and m_2 respectively, separated by Δm then:

$$\text{resolving power} = m_1/\Delta m$$

An illustration of definitions of resolution is given in Fig. 243.

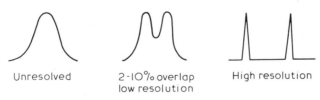

| Unresolved | 2–10% overlap low resolution | High resolution |

Fig. 243. Spectrum resolution

For a single focusing instrument of resolving power 1000, ions of mass 999 and 1000 are just discernable as separate peaks. A high resolution instrument has a resolving power of over 20 000, but this does not imply that molecular weights of the order of 20 000 can be obtained since such compounds are generally non-volatile. A high resolution instrument does, however, give a wider spread of peaks from ions of lower mass, and the fragment ions $C_2H_6^+$ (30.0469), CH_2O^+ (30.0105) and CH_4N^+ (30.0344), for example, give rise to three distinct lines.

COMBINED GAS CHROMATOGRAPHY—MASS SPECTROMETRY (GC/MS)

Gas-liquid chromatography (GC) is a very efficient method for separating a complex mixture into its components. The high sensitivity of mass spectrometry provides the necessary information for either identification of compounds by comparison with available spectra, or structure elucidation of a small quantity of compound. A combination of these techniques, with the introduction of GC effluents, after removal of most of the carrier gas, into a mass spectrometer is finding increasing use in analytical and structural organic chemistry and bio-chemistry (Dijkstra, 1965; McFadden, 1973). Efforts to combine the two techniques have been in progress for many years. The most obvious method of combination is to condense the fraction emerging from the GC column into a capillary or onto a small metal surface; the fractions collected in this way are introduced, in the normal manner, into the mass spectrometer source. Many operations are required for a multicomponent mixture and losses are likely to occur during collection of the fraction, however, the mass spectrometer may be operated at high resolution and no GC carrier gas is admitted to the instrument.

A second method is to feed the column effluent directly into the ion source after passage through an *interface* or *separator* which enhances the concentration of the sample by removing some of the carrier gas (Simpson, 1972).

The interface between the GC and MS has an important role to play in the overall efficiency of the instrument. It must be capable of providing an inert pathway from the column to the ion source without loss of chromatographic resolution, whilst at the same time removing the carrier gas and reducing the pressure from about one atmosphere at the column outlet to $10^{-5}-10^{-6}$ mm Hg in the ion source. It is hardly surprising that the separator has been a major cause of technical problems encountered in GC/MS.

The Watson–Biemann effusion separator consists of a sintered glass tube, the surrounds of which are evacuated. The carrier gas, usually helium, passes preferentially through the sintered glass and the effluent is concentrated by a factor of up to 100. Two stage separators may enrich the effluent by a factor of 400 and be capable of dealing with gas flow rates in the order of 20–60 ml per min.

The Ryhage jet separator is based upon the differing rates of diffusion of different gases in an expanding supersonic jet stream. In the centre of the gas jet the heavier (sample) component concentrates. The gases pass at high speed through an orifice aligned with a second orifice a short distance away. The concentrate passes through the second orifice, and then on to the ion source whilst the carrier gas is pumped away. Usually a Ryhage separator is two stage, although all glass single stage units (Fig. 244) are used, particularly in combination with quadrupole mass spectrometers.

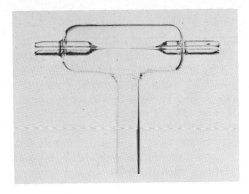

Fig. 244. The Ryhage jet separator

In the *Llewellyn–Littlejohn* separator (Fig. 245), separation of organic molecules from carrier gas molecules is achieved by means of the selective permeability of an elastomer membrane. Permeability is a function of both the solubility of the gas molecules in the membrane and their ability to diffuse through it. Gases such as hydrogen, helium, argon and nitrogen having a very low solubility and high diffusion rate, pass through the membrane much more slowly than organic vapours where the reverse is true. The vapour, therefore, is concentrated and the remaining carrier gas expelled into the atmosphere. The concentrate is further enriched by passage through a second semipermeable membrane. Enrichment of the organic phase by a factor of greater than 10^5 has been achieved.

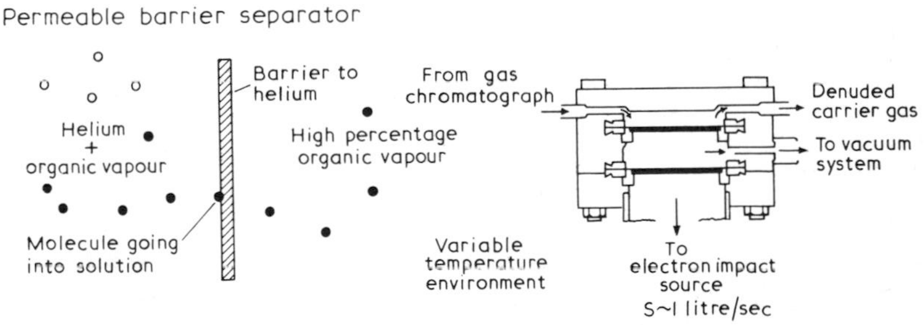

Permeable barrier separator

Fig. 245. The Llewellyn separator

No single separator is likely to give satisfactory results over a wide molecular weight spread.

After passage through an interface each component of the mixture is fed directly into the mass spectrometer in turn, and its mass spectrum is recorded. Overlapping spectra do occur, but by a process of spectrum subtraction the individual spectra of a partially resolved GC peak may often be deduced. In GC/MS analysis, metal chromatography columns and metal parts and connections between the chromatograph and the ion source are to be avoided, since they may catalyse thermal breakdown of certain compounds.

Selected Ion Monitoring—Mass Fragmentography

In a GC/MS experiment, the gas chromatogram may be obtained by simply splitting the GC effluent, part of which goes to a GC detector and the remainder passes into the MS. It is, however, more convenient to use the mass spectrometer as a GC detector, the total ion current from the MS collector being plotted against time to yield a chromatogram. Even more valuable is the use of *multiple ion detection* or specific ion detection. Here from one to eight ions from the MS are monitored simultaneously and the resultant chromatograms displayed by a multipen recorder. This technique is known as *selected ion monitoring* or *mass fragmentography* (Waller, 1972; McFadden, 1973).

Multiple ion detection is facilitated by an accelerating voltage alternater which permits a 'jump' scan where only ions of specific (selected) *m/e* are recorded. Several ions from a single compound may be monitored, affording a high confidence in assigning identity or structure, or, alternatively, ions from several components of a mixture may be monitored simultaneously.

Quadrupole or time-of-flight mass spectrometers are particularly suited to multiple ion detection, since up to eight selected ions may be monitored over any mass range. Most magnetic sector instruments, however, monitor fewer ions over a restricted (20% of total) mass range.

Mass fragmentography is a technique endowed with a high level of specificity. It is most unlikely that two compounds would have not only the same GC retention time but also the same MS fragmentation with identical peak ratios. Perhaps the greatest advantage, however, is the sensitivity of the technique, particularly in the detection of trace compounds in biological fluids. By employing an internal standard, either a known amount of isotopically labelled isomer or even an unrelated compound added to the sample, quantitation down to nanomole or picomole levels may be achieved.

Mass fragmentography is of particular value in drug metabolism investigations (Maume 1973; Ebbighausen, 1973).

Data Acquisition and Processing

Mass spectrometers and GC/MS combinations produce a wealth of data rapidly. To process and interpret all of this data manually would be excessively time consuming. Clearly, in order to realise the maximum potential of mass spectrometry extensive use must be made of computers. An account of mass spectrometer data acquisition and processing systems is beyond the scope of this chapter but receives comprehensive treatment elsewhere (Waller, 1972; Henneberg, 1972).

Chemical Ionisation Mass Spectrometry (CIMS)

The discussion so far has concerned spectra generated by electron impact (EI). Such spectra may suffer from the disadvantages of excessive fragmentation and the lack of a molecular ion; problems which can sometimes be surmounted by lowering the electron impact energy.

Chemical ionisation (Waller, 1972; McFadden, 1973; Beggs and Yergey, 1973) has recently emerged as a valuable technique that generates simple positive ion spectra from a sample by low energy ion–molecule reactions. Chemical ionisation mass spectrometry evolved from fundamental physico-chemical studies of the very rapid reactions that occur when ions collide with neutral molecules.

In a mass spectrometer chemical ionisation source, a reactant (or ionising) gas is admitted in several thousandfold excess over the gaseous sample. The pressure within the ion chamber is about 1 mm Hg (1 Torr). The mixture is subjected to electron bombardment (100 eV) in the usual manner, whereon, because of the relatively low sample concentration, almost all primary ionisation occurs to the reactant gas. Reagent ions so formed undergo rapid reactions with their own neutral species to form a steady-state ion plasma, which in turn interacts in a specific manner with molecules of the sample.

If we consider methane as a typical reactant gas, electron impact first removes an electron from the molecule to give $CH_4^{+\cdot}$ which is then involved in ion-molecule reactions to yield the reagent plasma:

$$CH_4 + e \rightarrow CH_4^{+\cdot} + 2e$$

$$CH_4^{+\cdot} \rightarrow CH_3^+ + H^{\cdot}$$

$$CH_4^{+\cdot} + CH_4 \rightarrow CH_5^+ + CH_3^{\cdot}$$

$$CH_3^+ + CH_4 \rightarrow C_2H_5^+ + H_2$$

$$CH_3^+ + 2CH_4 \rightarrow C_3H_7^+ + 2H_2$$

$$CH_2^{+\cdot} + 2CH_4 \rightarrow C_3H_5^+ + 2H_2 + H^{\cdot}$$

and so on.

Most sample molecules, particularly in biological studies, possess oxygen or nitrogen in the form of nucleophilic sites ideal for the acceptance of a proton. Reaction between reagent ions and sample may be summarised thus:

$$CH_5^+ + BH \rightarrow BH_2^+ + CH_4$$

$$C_2H_5^+ + BH \rightarrow BH_2^+ + C_2H_4$$

$$C_2H_5^+ + BH \rightarrow B^+ + C_2H_6$$

In CI spectra the molecular ion (M^+) is often weak. However, the spectrum base peak is usually $(M+1)^+$, and is known as the *quasi-molecular ion*. A comparison of the EI and CI spectra of ephedrine is seen in Fig. 246 (Fales, Lloyd and Milne, 1970). From an interpretative standpoint ions occurring in methane CI spectra at $(M+29)^+$ $(+C_2H_5)$ and $(M+41)^+$ $(+C_3H_5)$ are often useful.

When the sample molecule is not a good proton acceptor, hydride ion abstraction occurs affording a $(M-1)^+$ quasi-molecular ion.

The most commonly used reagent gases in CIMS are methane, isobutane and ammonia, the ions from each of which have a specific proton affinity. Ion plasma with a high proton affinity, e.g. that from isobutane, causes little or no fragmentation and gives an uncomplicated spectrum of the sample with an intense quasi-molecular ion. Consequently CI spectra of mixtures are relatively easy to interpret (Fig. 247).

Combined gas chromatography and mass spectrometry may be adapted to CI studies (GC/CIMS) thus providing a valuable tool for the investigation of

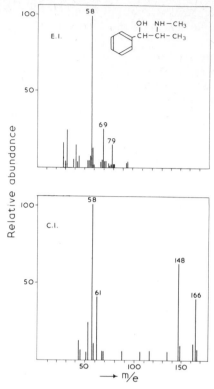

Fig. 246. A comparison of the EI and CI spectra of ephedrine. (Courtesy of *J. Am. Chem. Soc.*)

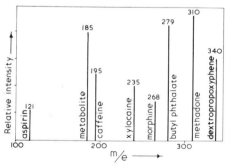

Fig. 247. Chemical ionisation spectrum of compounds detected in a chloroform extract of a urine sample from drug overdose patient. Reagent gas: isobutane. (Adapted from *Industrial Research*)

complex mixtures of compounds particularly from biological fluids. Quadrupole mass spectrometers are well-suited to this combination. The GC carrier gas and the CI reactant gas may be the same compound, e.g. methane. Once the GC effluent has left the column, it passes directly into the CI source without the intervention of a separator. The CI source is able to handle a high gas input and the pressure gradient between the column outlet and the source is relatively shallow. If methane has an adverse effect on gas chromatographic resolution, when compared with, say, helium, the latter may be employed as carrier with the reagent gas being admixed with the column effluent before passage into the CI source.

Mass fragmentography is facilitated by the simplicity of spectra produced by GC/CIMS.

Field Ionisation (FI) Mass Spectrometry

Field ionisation (Waller, 1972; McFadden, 1973; Beckey, 1969) is another low energy method of generating positive ions for mass spectrometric analysis, providing information complementary to EI ionisation. 'Soft' ionisation of molecules is induced by a very high positive electric field ($10^7–10^8$ V cm^{-1}) produced at a fine metal point, sharp metal edge or thin wire. The high electric field gradient between a sample molecule and the metal, usually platinum or tungsten, results in the loss of an electron by the molecule to the anode. The ions formed are repelled by the anode and accelerated into the MS analyser.

In field ionisation only sufficient energy (12–13 eV) is available to just ionise the molecule. Thus an intense parent ion with little or no fragmentation usually results. Spectrum simplicity may be of value when operating in a GC/MS mode or in mass fragmentography. Disadvantages of FI are the relatively low (about $\frac{1}{10}–\frac{1}{100}$th EI) sensitivity of the source, its fragility and its susceptibility to being affected by previous samples.

Field ionisation is particularly useful in the determination of the structures of large molecules or molecules lacking a parent ion by EI, e.g. carbohydrates. Often a quasi-molecular ion $(M+1)^+$ is observed due to interaction on the metal source surface between the sample and adsorbed water.

A modification of FI is *field desorption* (FD), a technique showing great promise for the MS analysis of non-volatile or thermolabile compounds. The sample is applied in solution to the field ion emitter and the solvent allowed to evaporate. Ionisation ensues by application of the high electric field in the manner described above. Nucleotides and quaternary ammonium salts are among the compounds reported to exhibit a molecular ion by this technique (Brent, 1973).

Figure 248 offers a comparison of the EI, FI and FD mass spectra of glutamic acid (Beckey, Heindricks and Winkler, 1970).

APPLICATIONS

Structure Elucidation*

The mass spectrum is a line spectrum, each line corresponding to a positive ion of specific mass. In the case of a compound which has simply lost one electron, the parent (P^+) or mass (M^+) peak corresponds to the exact molecular

*Biemann (1962); Hill (1966); Budzikiewicz, Djerassi and Williams (1964); Waller (1972); Spiteller and Spiteller–Friedmann (1965).

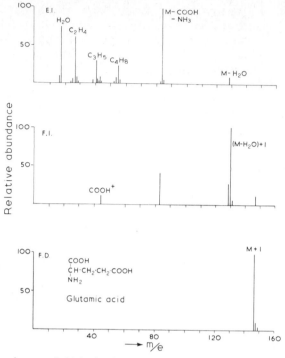

Fig. 248. Electron impact, field ionisation and field desorption mass spectra of glutamic acid. (Courtesy of Wiley-Interscience)

weight of the compound. The 'nearest-whole-number' value for a molecular weight is arrived at rapidly by counting the number of lines (Fig. 240) from a reference line. Traces of air introduced into the instrument give rise to peaks corresponding to H_2O, m/e 18; N_2, m/e 28; O_2, m/e 32; Ar, m/e 40; and CO_2, m/e 44, which may act as references. The final ion recorded on the spectrum, C^+, m/e 12, may also be used.

By a process known as *peak matching*, the molecular weight of an ion may be determined to six decimal places on a double-focusing mass spectrometer. The peak, corresponding to an unknown mass, is matched with a known fluoro-carbon peak in that mass region, on an oscilloscope. Accuracy of this order permits the absolute identification of ions of identical 'rough' mass. For example, the ions $[CO]^+$, $[N_2]^+$, $[CH_2N]^+$ and $[C_2H_4]^+$ each correspond to a mass of 28; mass measurement to six decimal places establishes the identity and ele-mental composition of each ion:

$$
\begin{array}{ll}
CO & 27.994\,914 \\
N_2 & 28.006\,154 \\
CH_2N & 28.018\,723 \\
C_2H_4 & 28.031\,299 \\
\end{array}
$$

Another method of determining the element make-up of an ion is by a consideration of isotope peaks. Isotopes of elements, which differ only by the number of neutrons in their nuclei, give rise to additional peaks in the mass spectrum of a compound. In a compound possessing only one carbon atom both ^{12}C and ^{13}C occur in the ratio 98.982 to 1.108. Compounds with more than one carbon atom have a correspondingly greater chance of possessing a ^{13}C nucleus. Thus a molecular ion having a single carbon atom affords a mass peak (M) and mass plus 1 peak ($M+1$) in the ratio of approximately 99:1, that is, corresponding to the natural isotopic abundance ratio of carbon. Important contributions to an $M+1$ (or fragment ion $+1$) peak are made by ^{13}C, ^{2}H, ^{15}N and ^{33}S, and to an $M+2$ (or fragment ion $+2$) peak by ^{18}O, ^{34}S, ^{37}Cl and ^{81}Br (Fig. 249). Because the number of each type of nucleus present

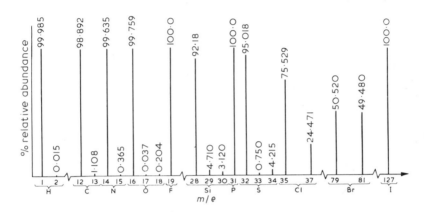

Fig. 249. Relative natural abundances of isotopes

in a molecule governs its contribution to the $M+1$ and $M+2$ peaks, the ratios of the intensities of $M+1$ and $M+2$ to the parent peak correspond to a specific elemental composition of the molecular ion. The intensities of the $M+1$ and $M+2$ (or fragment ion $+1$ and fragment ion $+2$) peaks as a ratio of the parent or fragment ion should be established and the corresponding elemental composition of the molecule or fragment ascertained by reference to a table of *isotope abundance ratios* (Beynon, 1960; Beynon and Williams, 1963; Silverstein and Bassler, 1968). It should be remembered that fluorine and iodine consist of single atomic species of masses 19 and 127 respectively.

Nitrogen rule. If a molecular ion has an even molecular weight it must possess either no nitrogen or an even number of nitrogen atoms. An odd molecular weight compound requires an odd number of nitrogen atoms.

The *base peak* is the most intense line of the mass spectrum.

Metastable Ions. Fragmentation ($m_1^+ \rightarrow m_2^+ +$ a neutral fragment), in general, occurs in the ion source of the mass spectrometer before the positive ions are accelerated and therefore a distinct peak results for each fragment ion. If an ion breaks up slowly during acceleration and flight then m_2^+, which has been generated in the ion source, will appear as a normal peak, whereas the same ion formed during flight occurs at a lower mass than m_2^+ and is designated m^*, a metastable ion. The fragment ion m_2^+ formed after acceleration has less kinetic energy than one formed in the source because some of the kinetic energy received by m_1^+ during acceleration is carried off by the neutral fragment. The peak (m^*) due to such fragmentation, therefore, occurs at lower mass than m_2^+ and is generally broad and of low intensity (Fig. 250). The relationship of the position in the spectrum of m^*, with respect to the ions m_1^+ and m_2^+ is given by the expression:

$$m^* = (m_2)^2/m_1 \qquad (4)$$

Thus if a fragment m_2^+ is suspected as having arisen from an ion m_1^+ the observation of a metastable ion (m^*) at the mass value calculated from eq. (4) lends support to the fragmentation pathway. The absence of a metastable ion does not preclude the pathway.

The utility of mass spectrometry is not restricted to molecular and fragment weight determinations. The manner in which molecular ions fragment in the mass spectrometer is of the utmost importance and yields much information concerning the structure of the parent compound.

Some of the most important aspects of molecular fragmentation on electron impact are summarised below. A one electron shift is denoted by a 'fish-hook' arrow⌒ and a two electron shift in the usual manner by⌒following the convention proposed by Budzikiewicz, Djerassi and Williams (1964).

(1) The parent peak (M^+) is the most intense in straight chain compounds, the intensity diminishes with increased chain branching.

(2) In a homologous series the intensity of M^+ decreases with increase in molecular weight.

(3) In branch-chain hydrocarbons cleavage is favoured at the bond adjacent to the branch (Fig. 251), thus giving rise to tertiary ions rather than secondary and secondary ions in preference to primary. This follows the order of stability of carbonium ions, tertiary > secondary > primary. For example, in the following methyl pentanes:

$$\left[\begin{array}{c} \overset{\displaystyle CH_3}{\underset{\displaystyle CH_3}{\overset{|}{\underset{|}{CH_3-C}}}}\,\overset{\displaystyle CH_3}{\overset{|}{CH}}-CH_2-CH_3 \end{array} \right]^{+\cdot} \rightarrow \underset{\displaystyle CH_3}{\overset{\displaystyle CH_3}{\overset{|}{\underset{|}{CH_3-C^+}}}} + \overset{\displaystyle CH_3}{\overset{|}{\cdot CH}}-CH_2-CH_3$$

m/e 57

and

$$\left[\overset{\displaystyle CH_3}{\overset{|}{CH_3-CH}}\,CH_2-CH_2-CH_3 \right]^{+\cdot} \rightarrow \overset{\displaystyle CH_3}{\underset{\displaystyle H}{\overset{|}{\underset{|}{CH_3-C^+}}}} + \cdot CH_2-CH_2-CH_3$$

m/e 43

(4) The probability of the existence of a strong M^+ peak is high when unsaturated or cyclic systems are present in the molecule. As in carbonium ion chemistry, the ions formed are stabilised.

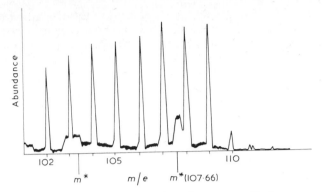

Fig. 250. Metastable peaks. m^* = metastable ion

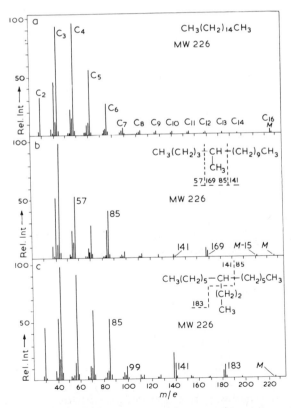

Fig. 251. Mass spectra of isomeric C_{16} hydrocarbons

(5) Alkylbenzenes cleave at the C—C bond β to the aromatic ring resulting in a highly stabilised carbonium ion. Most alkylbenzenes have been shown to give the more stable tropylium ion (I) rather than the benzyl cation (II).

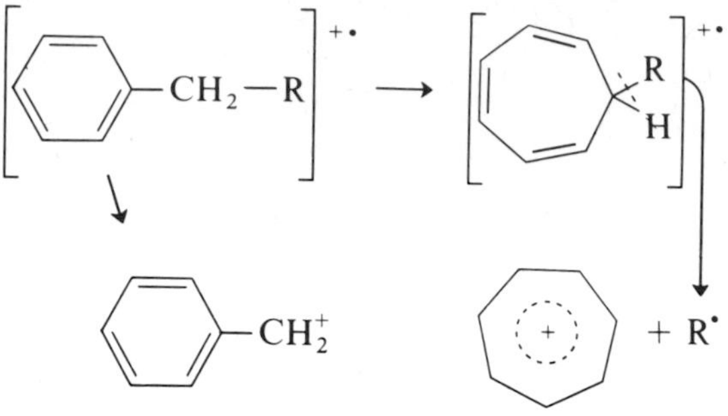

(II)   (I)

Similarly, 3-alkylpyridines give an azatropylium ion (III):

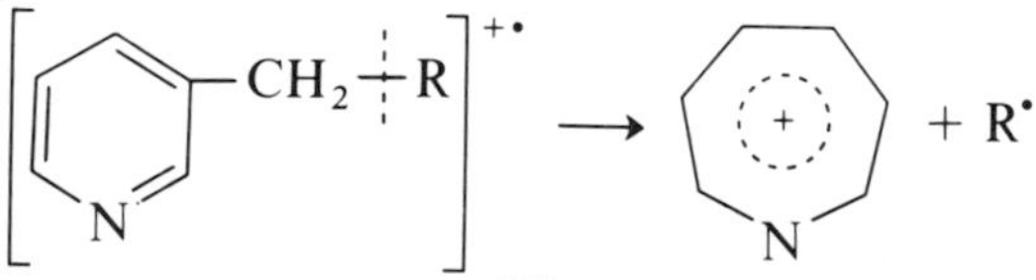

(III)

(6) The preferred cleavage in compounds possessing a carbon–carbon double bond is β to that bond, to yield the resonance stabilised allyl cation (IV).

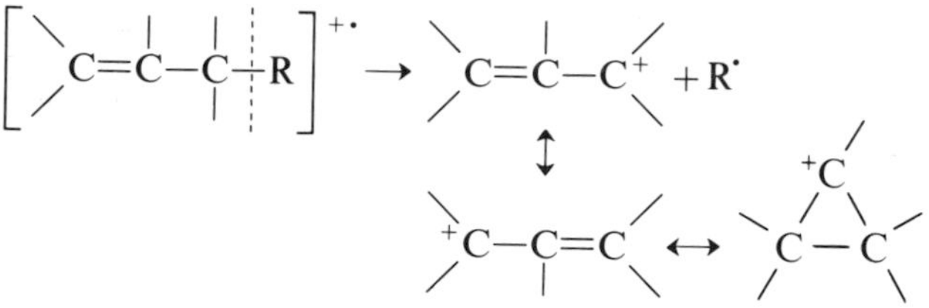

(IV)

Fragmentation of this sort should lead to considerable differences in the mass spectra of isomeric olefins; however, in the molecular ion the double bond appears to migrate readily. Isomeric olefins therefore give very similar mass spectra.

Cyclic olefins, in which the double bond is fixed, often undergo a characteristic retro-Diels–Alder fission:

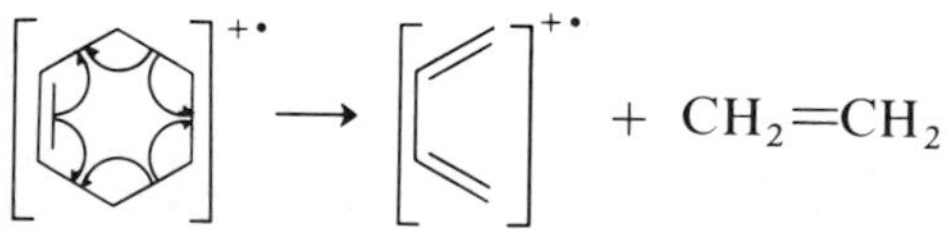

This fragmentation pathway is of importance in the rationalisation of the spectra of terpenes and unsaturated steroid molecules.

(7) Compounds such as alcohols, mercaptans, amines and esters cleave at the carbon-carbon bond β to the hetero-atom. Again, resonance stabilisation of the positive charge is possible by virtue of the hetero-atom lone pair.

$$\left[R\!\!\!-\!\!\!\overset{|}{\underset{|}{C}}\!\!-\!\!\ddot{X}H\right]^{+\cdot} \longrightarrow R^{\cdot} + \overset{+}{\underset{/}{\diagdown}}C\!\!-\!\!\overset{\frown}{\ddot{X}}H$$

$$\updownarrow$$

$$\underset{/}{\diagdown}C\!\!\overset{\equiv}{=}\!\!\underset{+}{X}\!\!-\!\!H$$

Where two hetero-atoms are present in the same molecule the charge is retained predominantly by the fragment bearing the more electronegative hetero-atom (Fig. 252). Although cleavage is very much favoured in one direction, the reverse electron flow does give rise to some of the less stable ion.

$$\left[\begin{array}{c} CH_2\!-\!CH_2 \\ | \qquad | \\ HO \qquad NH_2 \end{array}\right]^{+\cdot}$$

$$CH_2\!\!=\!\!\overset{+}{O}H + \underset{\underset{NH_2}{|}}{\overset{\cdot}{C}H_2} \qquad\qquad \overset{\cdot}{C}H_2 + H_2\overset{+}{N}\!\!=\!\!CH_2$$
$$\underset{OH^+}{|}$$

$$m/e \; 31 \qquad\qquad\qquad m/e \; 30$$
$$30\% \text{ abundance} \qquad\qquad 57\% \text{ abundance}$$

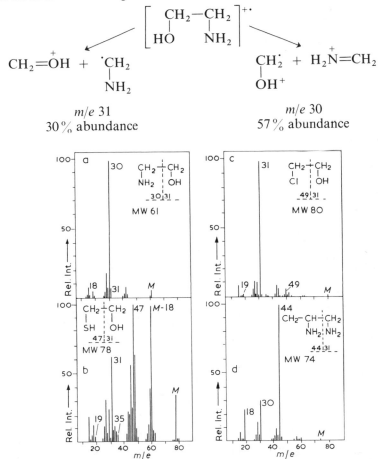

Fig. 252. Mass spectra of various simple, bifunctional molecules. (a) 2-aminoethanol; (b) 2-thioethanol; (c) 2-chloroethanol; (d) 1,2-diaminopropane

In ethers, alkyl halides and related compounds cleavage of the carbon-hetero-atom bond may occur leading to the more stable alkyl carbonium ion rather than the X^+ ion:

$$\left[-\overset{|}{\underset{|}{C}} \!\vdots\! X \right]^{+\cdot} \rightarrow -\overset{|}{\underset{|}{C}}{}^+ + X^\cdot \qquad \text{(V)}$$

where X = halogen, O—R, S—R, N—R_2. R can only be hydrogen when a tertiary carbonium ion is possible as a stable positive ion (V).

Carbonyl compounds cleave at the carbon–carbon bond α to the carbonyl group. In an asymmetrical ketone two such modes of fission are possible, the predominant positive ion produced being that which favours the elimination of the largest alkyl chain as the neutral fragment:

$$\left[\begin{matrix} \text{R} & & \text{R}' \\ & \overset{\times}{\underset{\underset{\text{O}}{\|}}{\text{C}}} & \\ \text{(b)} & & \text{(a)} \end{matrix} \right]^{+\cdot}$$

$$\text{R}^\cdot + \text{R}'\!-\!\overset{+}{\text{C}}\!\!\overset{\frown}{=}\!\!\overset{\cdot\cdot}{\text{O}} \qquad\qquad \overset{\text{R}''}{+\text{R}\!-\!\overset{+}{\text{C}}\!\!\overset{\frown}{=}\!\!\overset{\cdot\cdot}{\text{O}}}$$

$$\updownarrow \qquad\qquad\qquad\qquad \updownarrow$$

$$\text{R}'\!-\!\text{C}\!\!\equiv\!\!\overset{+}{\text{O}} \qquad\qquad \text{R}\!-\!\text{C}\!\!\equiv\!\!\overset{+}{\text{O}}$$

If R = CH_3 and R' = C_3H_7 then CH_3—C≡O^+ (m/e 43) will occur in greater abundance than C_3H_7—C≡O^+ (m/e 71). In the case of acetaldehyde, H—C≡O^+ (m/e 29) is formed in preference to CH_3—C≡O^+ (m/e 43), and the former appears as the base peak of the spectrum.

(8) Molecular or fragment ions may also rearrange, often with the elimination of a neutral fragment. Such rearrangements involve the transfer of hydrogen from one part of the molecular ion to another via, preferably, a six-membered cyclic transition state. The process is favoured energetically because as many bonds are formed as are broken.

Molecular ions possessing electronegative hetero-atoms often eliminate stable neutral molecules such as H_2O, CO, NO, HCl, H_2S, NH_3 or olefins. The neutral fragments are not detected by the mass spectrometer but their loss is indicated by the nature of the positive ions formed:

$$\left[\begin{matrix} \diagdown\diagup \\ \text{C}\!\!\diagdown\!\!\text{X} \\ \text{(C)}_n \diagup \quad \diagdown \\ \diagdown\diagup \\ \text{C}\!\!\diagup\!\!\text{H} \\ \diagup\diagdown \end{matrix} \right]^{+\cdot} \longrightarrow \left[\begin{matrix} \diagdown\diagup \\ \text{C} \\ \text{(C)}_n \diagup \\ \diagdown \\ \text{C} \\ \diagup\diagdown \end{matrix} \right]^{+\cdot} + \text{HX}$$

where X = —OH, —SH, ester, halogen and sometimes —C≡N.

A rearrangement involving the transfer of a hydrogen atom from one part of an ion to another *via* a six-membered ring transition state is the McLafferty rearrangement. An example of this process is the elimination of ethylene from *n*-propylmethyl ketone:

$$\left[\begin{array}{c} H_2C \\ H_2C \quad \underset{CH_2}{\overset{H}{\diagdown}} \quad \overset{O}{\underset{\parallel}{C}}-CH_2 \end{array} \right]^{+\cdot} \longrightarrow \quad \underset{CH_2}{\overset{H\diagdown O^{\cdot+}}{\underset{\diagdown}{C}}} \underset{CH_3}{\diagup} \quad + CH_2{=}CH_2$$

Many aldehydes, ketones, unsaturated hydrocarbons, amides, nitriles, and esters, will exhibit rearrangement peaks in their mass spectra when the stereochemistry of the molecule favours hydrogen transfer.

For a detailed treatment of fragmentation pathways the reader is referred to the works of Biemann (1962), Hill (1966), and McLafferty (1966).

Detection of Impurities

The mass spectrometer is able to detect as little as a few parts per million of an impurity in a compound, particularly if the structure of the impurity is quite different from that of the main component. A good example is the detection of trace amounts of xylene in acetylacetone (Beynon, 1960; Nicholson, 1957) which had been purified for heat of combustion investigations. In the mass spectrum of the acetylacetone sample ($M^+ = 100$) extra peaks were observed at m/e 106, 105 and 91 suggesting the presence of a higher molecular weight impurity ($M^+ = 106$). Fragmentation of the acetylacetone molecular ion cannot give rise to the 91 peak since this would involve the loss of nine hydrogen atoms. The anomalous peaks may be explained by the presence of xylene as a contaminant, the parent ion (VI) of which through loss of a proton may rearrange to a tropylium ion (VII), or by loss of a methyl fragment yields the 91 peak (VIII) ($M^+ - 15$). The presence of xylene (the isomers could not be

$$\left[\begin{array}{c} CH_3 \\ \\ \\ CH_3 \end{array} \right]^{+\cdot} \quad \overset{-H^+}{\longrightarrow} \quad \overset{CH_3}{\bigcirc{+}}$$

$$\begin{array}{cc} m/e\ 106 & m/e\ 105 \\ (VI) & (VII) \end{array}$$

$$\overset{-CH_3^{\cdot}}{\searrow} \quad \left[\begin{array}{c} CH_3 \\ \\ \\ + \end{array} \right]$$

$$\begin{array}{c} m/e\ 91 \\ (VIII) \end{array}$$

distinguished) was traced to the use of sodium, which had been stored over xylene, during the preparation of the sample. Since acetylacetone and xylene have approximately the same boiling point, distillation did not effect separation. The amount of xylene present in the sample was measured quantitatively, enabling the true heat of combustion to be calculated by the application of a correction factor. Contaminants of molecular weight higher than the major component are those most readily detected. The presence of impurities in a sample may also be established by lowering the intensity of the ionising electron beam. If all peak heights do not decrease proportionally then a contaminant (or contaminants) is present.

The advantage of such a sensitive means of impurity detection has obvious implications in quality control and in forensic science. From a consideration of the nature of the impurities present in a compound, it may be possible to rationalise its mode of manufacture.

Quantitative Analysis

The main stimulus for the development of early mass spectrometers came from the value of the technique in the quantitative analysis of multicomponent mixtures. The petroleum industry, in particular, used mass spectrometry for the analysis of complex hydrocarbon mixtures. A prerequisite of the technique is that spectra of pure samples of each component of the mixture must be available. The intensities of major peaks from each spectrum are calculated as a percentage of its base peak, and an intense peak from each component is chosen for the analysis. The partial pressure of each component in the gas reservoir is proportional to its amount in the mixture (Dalton's law of partial pressures) and therefore peak height is a quantitative measure of the constituents of the mixture. In a mass spectrometric assay of a mixture account is taken of contributions of fragments from other components to the chosen peak. Fortunately, peaks are quantitatively additive and corrections are therefore easily applied. To-day, spectra from multicomponent mixtures are often programmed for feeding into a computer, permitting rapid and accurate analysis.

In the spectrum of a simple mixture of n-butane, iso-butane, propane, ethane and methane, the butanes possess peaks at m/e 58 and 57 (due to $C_4H_{10}^+$ and $C_4H_9^+$). n-Propane has a mass peak at m/e 44 which has to be corrected for contributions from the butanes. At m/e 30 and 16 are the mass peaks of ethane and methane respectively and these, in turn, must be corrected for contributions from the higher homologues. Problems such as this may be solved by the application of linear simultaneous equations (Kiser, 1965).

A similar method may be employed for compounds other than hydrocarbons. Amino acids (Biemann, 1960), for example, may be analysed successfully by their prior conversion to ethyl esters by treatment with boiling ethanol/hydrochloric acid. Less than 3 mg of the amino acid mixture is required but care must be taken to avoid the formation of diketopiperazines. Amino acid ethyl esters exhibit only a few intense peaks in their mass spectra, (Fig. 253) and for different amino acids these fall at quite different mass values with a minimum of interference from contributions of other amino acid fragments. The base peak, and often the most valuable peak, of an amino acid ethyl ester spectrum

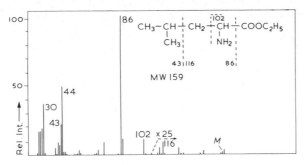

Fig. 253. Mass spectrum of leucine ethyl ester

is usually due to the fragment $R-CH=\overset{+}{N}H_2$, which corresponds in glycine to m/e 30, alanine, m/e 44 and valine, m/e 72 etc. Thus, provided that the mass spectra of pure samples of all the amino acid ethyl esters which constitute the mixture are available, the quantitative composition of the mixture can be calculated. The instrument time involved is about 30–40 min which permits rapid and accurate analysis of many mixtures during metabolism studies and protein structure investigation.

A more elegant method of determining quantitatively the components in a mixture of similar compounds is known as the *isotope dilution technique*. For this the compound or compounds to be assayed must be available in an iso-topically labelled form, e.g. with ^{13}C, ^{15}N or ^{131}I. A known amount of labelled compound of known purity is added to the mixture to be estimated and a small quantity of compound(s) isolated, the isotope ratio of compound isolated is determined by MS. Isolation may be effected by standard chemical means or far better the complete exercise is performed by GC/MS. From the ratio of labelled compound to non-labelled, taking into account natural abundance and the purity of the labelled standard, the amount of compound(s) in the unknown mixture may be calculated. The technique has been applied widely in amino acid studies.

An interesting study has been reported by Horning (1973) where diphenyl-hydantoin, caffeine, pethidine and several barbiturates and barbiturate meta-bolites have been measured in the picogram and nanogram range in human body fluids, with the aid of stable isotope labelled drugs. Her objective was to establish analytical procedures that afforded rapid drug-body fluid profiles of patients under multiple drug therapy. Gas chromatography, GC/MS with CI and EI, selective or multiple ion detection and isotope dilution were used separately or in combination to examine samples of urine, plasma, breast milk and amniotic fluid in mother-infant pairs shortly after birth. Diphenylhydantoin-2,4,5-^{13}C and pentobarbitone-2,4(6),5-^{13}C were the internal standards. Body fluid samples as low as 50–200 μl were employed successfully to monitor profiles. Additionally three metabolites of *phenobarbitone*, 5-ethyl-5-(3,4-dihydroxy-1,5-cyclohexadien-1-yl)barbituric acid, 5-ethyl-5-(4-hydroxyphenyl)barbituric acid and 5-ethyl-5-(3,4-dihydroxyphenyl)-barbituric acid, and two metabolites of *quinalbarbitone*, 5-allyl-5-(3-hydroxy-1-methylbutyl) barbituric acid and 5-(2,3-dihydroxypropyl)-5-(1-methylbutyl) barbituric acid were identified.

Amino Acid Sequence Analysis in Peptides*

The determination of the structures of peptides and proteins is of major importance, and is often performed by a process of stepwise hydrolysis and much column chromatography. A method for overcoming the lack of appreciable volatility of peptides, a disadvantage in mass spectrometry, is to reduce them with lithium aluminium hydride to the corresponding polyamino alcohols.

$$H_2N-CH-C-NH-CH-C-OH \xrightarrow{\text{LiAlH}_4}$$

$$H_2N-CH-CH_2-NH-CH-CH_2OH$$

Polyamino alcohols are volatile and upon electron impact fragment in a characteristic manner. The problem of determining an amino acid sequence may be simplified further by the use of gas chromatography. If a polyamino alcohol mixture derived from a protein-hydrolysate peptide mixture is subjected to gas chromatographic separation, then the effluent fractions may be introduced into the mass spectrometer individually. The development of GC/MS has obviated the effluent isolation step and speeded up the process considerably. Derivatisation of amino alcohols also aids sequencing. Hydroxyl groups may be converted to trimethylsilyl ethers by treatment with N-(trimethylsilyl)diethyl-amine, a reagent that does not react with amino groups. Specific O-silyl deriva-tives exhibit a considerable enhancement of fragment ions from the C-terminal amino acid.

Mass spectrometric examination of peptides may be performed without prior reduction of the amide linkages. Compounds of sufficient volatility and stability can be prepared by esterification of the free carboxylic acid end-group and the conversion of the amine end-group to a characteristic derivative, for example, acetyl, trideuteroacetyl, trifluoroacetyl and carbobenzoxy.

Amino acid sequencing by MS is being employed increasingly. The inter-pretation of spectra, however, although based upon simple arithmetic summa-tions, is tedious. Fortunately MS spectral data are amenable to computer analysis, thus speeding the sequencing process considerably (McFadden, 1973).

Drug Metabolism†

Drug metabolites usually arise from relatively minor structural modifications of the parent molecule. Oxidation to hydroxyl derivatives or O-demethylation, for example, often renders a foreign molecule more readily eliminated from the body. Mass spectrometry is a technique affording considerable structural information from a small sample and is consequently ideal for metabolism studies. The scope of the technique is extended further by its ability to detect drugs and metabolites that have been cold labelled, for example with ^2H, ^{13}C or ^{18}O.

---

* Waller (1972); Biemann (1960); Senn, Venkataraghavan and McLafferty, 1966; Biemann, Cone, Webster and Arsenault (1966); Jones (1968).

† Waller (1972); Fenselau (1972).

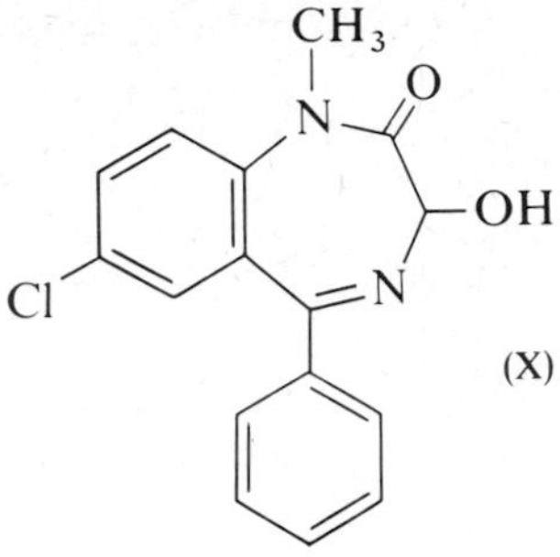

(X)

$C_{16}H_{13}N_2O_2Cl$
$+O$

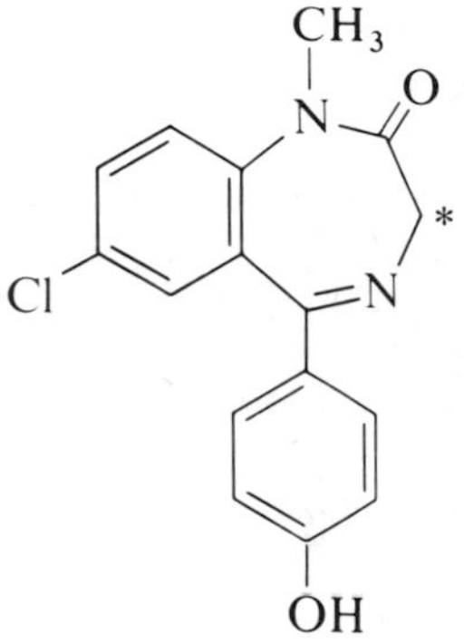

(XI)

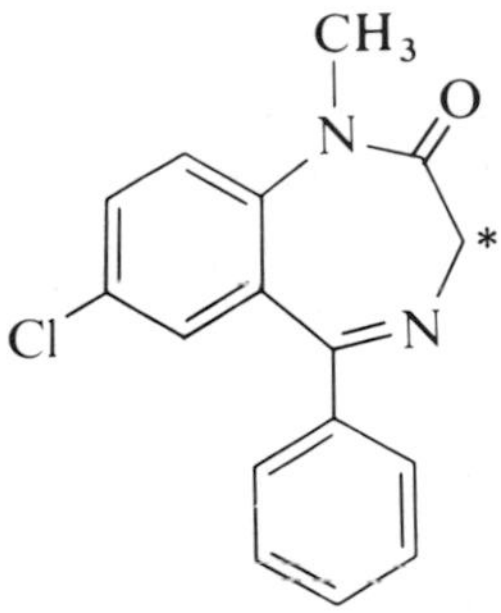

Diazepam (IX)

$C_{16}H_{13}N_2OCl$

$\longrightarrow \left\{ \begin{array}{l} C_{16}H_{13}N_2O_2Cl \\ +O \end{array} \right.$

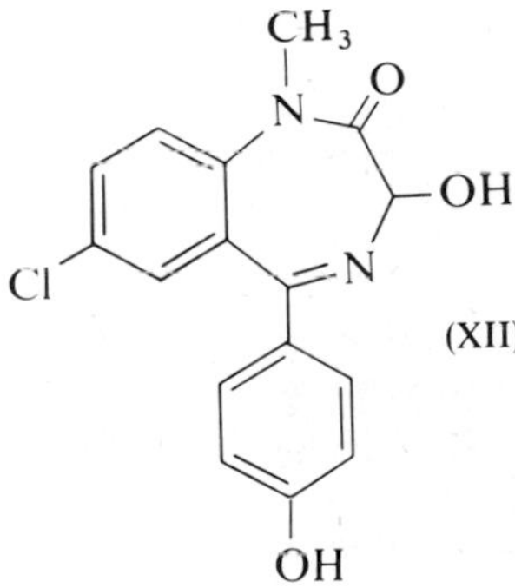

(XII)

$C_{16}H_{13}N_2O_3Cl$
$+2O$

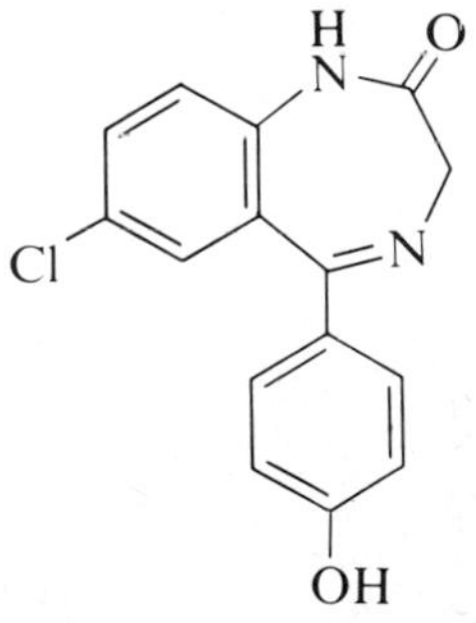

(XIII)

$C_{16}H_{11}N_2O_2Cl$
$+O; -CH_2$

The most obvious approach to identifying metabolites is to compare directly spectra of the pure drug with those of its biotransformation products. Direct insertion of metabolites, after isolation by column chromatography, preparative TLC or preparative GC, into the ion source of a high resolution instrument, followed by accurate mass measurement of each molecular ion, affords their molecular formulae. Differences in molecular formula indicate the transformation involved and the fragmentation pattern of a metabolite will often yield information concerning the position of gain or loss of a molecular unit. Identification of the major metabolites of diazepam (IX) was established in this manner (Schwarts, Bommer and Vane, 1967). Initial oxidation occurs at the positions indicated(*) in (X) and (XI). Nuclear magnetic resonance spectroscopy was used to assign unambiguously the positions of substitution. Secondary metabolism occurs by N-demethylation of (XI) to give (XIII) and by inserting a second oxygen (XII).

Combined GC/MS is becoming an increasingly common technique for the identification of organic compounds in admixture and in high dilution. It is now the method of choice in most metabolism studies.

For many years, 1-(2-acetylhydrazino)phthalazine (XV) was presumed to be a major metabolite of hydralazine (XIV). However, attempts to synthesise it resulted in the isolation of the triazolophthalazine (XVI) only. Paper chromatography of a metabolite, previously identified as the acetyl derivative (XV), was confirmed by a GC/MS study of the urine from three patients to be (XVI) (Zimmer, Kokosa and Garteiz, 1973).

NH.NH$_2$	NH.NH.C.CH$_3$	
	$\parallel$
	O
Hydralazine (XIV)	(XV)	(XVI)

A total ion current gas chromatogram is illustrated in Fig. 254, and Fig. 255 demonstrates the mass spectral characterisation of compound (XVI).

A fully computerised method for the quantitative determination of picogram quantities of drugs by GC/MS has been described by Baczynskyj (1973).

The first application of mass fragmentography to the determination of drug metabolites was described for chlorpromazine (XVII) (Hammer, Holmstedt and Ryhage, 1968).

A mass fragmentogram where the intensity of ions m/e 232 (A), m/e 234 (A^1) and m/e 246 (B) have been monitored with time is shown in Fig. 256. Fragmentation at bond X gives rise to ions A, A^1 and fragmentation at Y ion B. The intensity of these ions in the mass spectrum of a compound related to chlorpromazine will vary according to the nature of the basic side-chain. Preparation of trifluoroacetate (TFA) derivatives of the mixture of metabolites increases the specificity of the method. By a consideration of their GC retention times and the relative intensities of the three ions monitored as seen in the mass fragmentogram, desmethylchlorpromazine (XVIII) and didesmethylchlorpromazine (XIX) were identified as chlorpromazine metabolites.

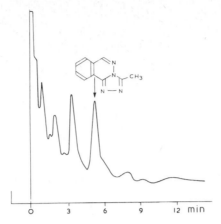

Fig. 254. Total ion current recording after injection of a 3 μl aliquot of a urine extract from a patient treated with hydralazine (Courtesy of *Arzneim. Forsch.*)

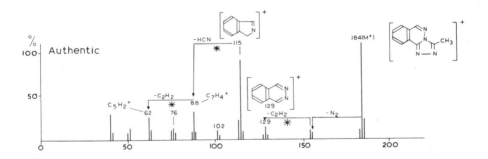

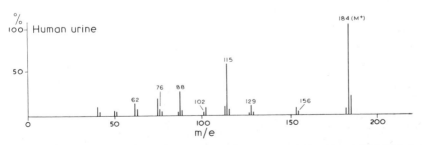

Fig. 255. Upper panel: Mass spectrum of authentic 3-methyl-s-triazolo[3,4-a]phthalazine. Lower panel: Mass spectrum of a compound with retention time of 5.5 min in a urine extract from a patient treated with hydralazine (Courtesy of *Arzneim. Forsch.*)

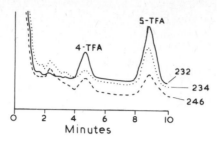

Fig. 256. Mass fragmentogram, chlorpromazine metabolism study. Fragments monitored *m/e* 232, 234 and 246. TFA = trifluoroacetate derivative. (Adapted from *Biochemical Applications of Mass Spectrometry*, G. R. Waller, courtesy of Wiley-Interscience)

$R = R' = CH_3$ = Chlorpromazine (XVII)

$R = H; R' = CH_3$ (XVIII)

$R = R' = H$ (XIX)

$A = {}^{35}Cl$

$A' = {}^{37}Cl$

$B = {}^{35}Cl$

Clinical and Forensic Applications*

A delicate balance of biochemical reactions is necessary to ensure the health of the human organism. Specific disease states cause changes in body chemistry and excretion products and the latter may be detected in body fluids, expired air or sweat. Mass spectrometry and the GC/MS combination may be employed to monitor these changes and aid diagnosis.

The health of an individual may also be impaired by accidental or deliberate abuse or overdosage of drugs or toxic chemicals. Forensic science and toxicology, therefore, are areas where MS and GC/MS are finding increasing application.

* Waller (1972); McFadden (1973); Fenselau (1972).

Finkle (1973) for example, has reported the case of a male with a history of drug abuse found dead with narcotic overdosage being suspected as the cause.

Analysis by GC/MS showed that his blood and urine contained codeine, and that bile, and veins at the injection site, contained morphine, supporting the probable cause of death.

Chemical ionisation (p. 246) often gives peaks of much higher intensity in the molecular ion region than do corresponding electron impact spectra. In the identification of mixtures of drugs in body fluids this enhanced sensitivity coupled with spectrum simplicity is of special value.

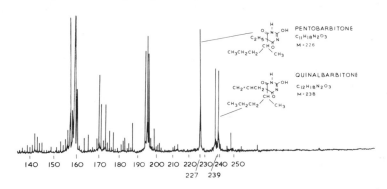

Fig. 257. Chemical ionisation mass spectrum of extract of gastric contents. (Courtesy of *Analytical Chemistry*)

Barbiturates have been prescribed so widely as sedatives that accidental or deliberate overdosage is relatively common. Because of the multiplicity of barbiturate analogues, identification by standard procedures is difficult, if not impossible. Fales, Milne and Axenrod (1970) have reported a method of barbiturate identification by GC/MS. The detection of quinalbarbitone and pentobarbitone in an extract of gastric contents is illustrated in Fig. 257.

Mass spectrometric techniques are advancing rapidly and are exploited in an increasing number of medical and related disciplines. Identification and quantitation of drugs, metabolites and toxins will become more sophisticated, rapid medical diagnosis by MS will spread, and the monitoring of the environment will develop.

References

Baczynskyj, L., Duchamp, D. J., Zieserl, J. F. and Axen, U., *Anal. Chem.* (1973) **45**, 479.

Beckey, H. D., *Angew. Chem.* (Int. Ed.) (1969) **8**, 623.

Beckey, H. D., Heindricks, A. and Winkler, H. W., *Int. J. Mass Spectrom. Ion. Phys.* (1970) **3**, 9.

Begg, D. and Yergey, A., *Industrial Research* (1973) 46.

Beynon, J. H., *Mass Spectrometry and its Applications to Organic Chemistry*, Elsevier, Amsterdam, 1960.

Beynon, J. H. and Williams, A. E., *Mass Abundance Tables for use in Mass Spectroscopy*, Elsevier, Amsterdam, 1963.

Biemann, K., *Chimia (Switz.)* (1960) **14**, 393.

Biemann, K., *Mass Spectrometry*, McGraw-Hill, New York, 1962.

Biemann, K., Cone, C., Webster, B. R. and Arsenault, G. P., *J. Am. chem. Soc.* (1966) **88**, 5598.

Brent, D. A., Rouse, D. J., Sammons, M. C. and Bursey, M. M., *Tetrahedron Letters* (1973) 4127.

Budzikiewicz, H., Djerassi, C. and Williams, D. H., *Interpretation of Mass Spectra of Organic Compounds*, Holden-Day, San Francisco, 1964.

Budzikiewicz, H., Djerassi, C. and Williams, D. H., *Structure Elucidation of Natural Products by Mass Spectrometry*, 2 vols., Holden-Day, San Francisco, 1964.

Dijkstra, G., in *Advances in Mass Spectrometry* (1965) **3**, 441, Pergamon Press, Oxford.

Ebbighausen, W. O. R., Mowat, J. and Vestergaard, Per, *J. Pharm. Sci.* (1973) **62**, 146.

Fales, H. M., Milne, G. W. A. and Axenrod, R., *Anal. Chem.* (1970) **42**, 1432.

Fales, H. M., Lloyd, H. A. and Milne, G. W. A., *J. Am. chem. Soc.* (1970) **92**, 1590.

Fenselau, C., in *Methods in Pharmacology*, Vol. 2, *Physical Methods*, ed. Chignell, C. F., Appleton-Century Crofts, New York, 1972.

Finkle, B. S., in *Techniques of Combined Gas Chromatography/Mass Spectrometry*, Wiley, New York, 1973, p. 385.

Hammer, C. G., Holmstedt, B. and Ryhage, R., *Anal. Biochem.* (1968) **25**, 532.

Henneberg, D., Casper, K., Ziegler, E. and Wiemann, B., *Angew. Chem.* (*Int. Ed.*) (1972) **11**, 357.

Hill, H. C., *Introduction to Mass Spectrometry*, Heyden, London, 1966.

Horning, M. G., Nowlin, J., Lertratanangkoon, K., Stillwell, W. G., and Hill, R. M., *Clini. Chem.* (1973) **19**, 845.

Jones, J. H., *Q. Revs.* (1968) **22**, 302.

Kiser, R. W., *Introduction to Mass Spectrometry and its Applications*, Prentice-Hall, Englewood Cliffs, N.J., 1965.

McFadden, W., *Techniques of Combined Gas Chromatography/Mass Spectrometry*, Wiley, New York, 1973.

McLafferty, F. W., Mass *Spectrometry of Organic Ions*, Academic Press, New York, 1963.

McLafferty, F. W., *Chem. Comm.* (1966) 78.

Maume, B. F., Bournot, P., Lhuguenot, J. C., Baron, C., Barber, F., Maume, G., Prost, H. and Padieu, P., *Anal. Chem.* (1973) **45**, 1073.

Milne, G. W. A., *Mass Spectrometry*, Wiley, New York, 1971.

Nicholson, G. R., *J. Chem. Soc.* (1957) 2431.

Schwarts, M. A., Bommer, P. and Vane, F. M., *Arch. Biochem. Biophys.* (1967) **121**, 508.

Senn, M., Venkataraghavan, R. and McLafferty, F. W., *J. Am. chem. Soc.* (1966) **88**, 5593.

Silverstein, R. M. and Bassler, G. C., *Spectrometric Identification of Organic Compounds*, Wiley, New York, 1968.

Simpson, C. F., *Critical Reviews in Analytical Chem.* (1972, Sept.), 1.

Spiteller, G. and Spiteller-Friedmann, N., *Angew. Chem.* (*Int. Ed.*) (1965) **4**, 383.

Waller, G. R., *Biochemical Applications of Mass Spectrometry*, Wiley, New York, 1972.

Zimmer, H., Kokosa, J. and Gorteiz, D. A., *Arzneim. Forsch.* (1973), **23**, 1028.

14 Radiochemical Techniques*

N. D. HARRIS

* No attempt has been made to deal with theoretical material except in so far as it relates to the experiments under consideration, and it is assumed that the reader has an adequate theoretical background.

COUNTING EQUIPMENT

The usual basic components of a set of counting equipment are shown in Fig. 258.

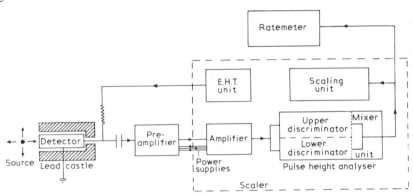

Fig. 258. Block diagram of the basic components of a set of counting equipment

Castles

The castle serves three functions; it supports the detector, shields the detector from background radiation and from extraneous radioactive sources (usually with a minimum of 2.5 cm of lead), and supports the source in a suitable position relative to the detector. Also, it may protect the operator from the radiation emitted by the source. Castles convenient for student use with Geiger-Müller counters are illustrated in Fig. 259.

Detectors

Detectors are commonly misnamed 'counters'. It should be noted that 'counters' do not themselves count, but merely respond to the passage of radiation. The resultant pulses are enumerated by electronic equipment (scalers, ratemeters or electrometers).

Counters Dependent on Ionisation

When radiation passes between the electrodes of a detector as shown in Fig. 260, a number of ion pairs (electrons plus cations) is produced. If there is no

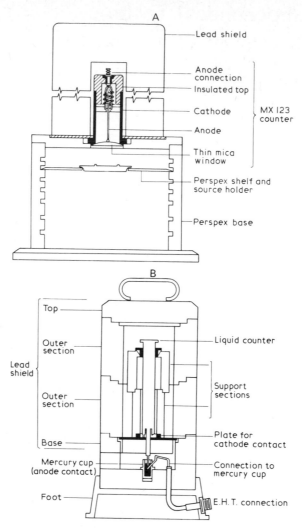

Fig. 259. Types of castle (A) Labgear*; (B) Panax Type LC3, set up for a liquid Geiger-Müller counting

potential difference between the electrodes the ions recombine, but if a potential is applied the ions separate and move towards the electrodes. The relationship between the applied potential and the charge collected is shown in Fig. 261.

As the applied potential is increased, a point is reached at which all the primary ions are collected (*saturation collection*) and change of potential has little effect, i.e. there is a plateau (Region II). If the potential is further increased, the primary ions give secondary ions by collision with molecules of the filling gas, and *internal gas multiplication* occurs (Region III). This is the *proportional*

* See section on 'source mounts' on pages 467 and 470.

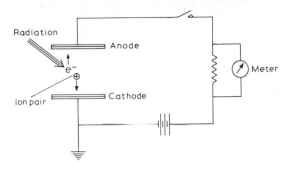

Fig. 260. Schematic radiation detector dependent on ionisation

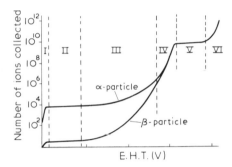

Fig. 261. The relationship between applied potential (EHT) and the charge collected in a radiation detector dependent on ionisation

region in which the *relative* increase in charge depends on the applied potential. The charge collected may be increased up to 10^6 times. With further potential increase another plateau is reached (Region V). This is the *Geiger region* in which the number of secondary ions produced saturates the gas space and the same charge is collected irrespective of the number of primary ions formed initially. In Region VI the gas ionises due to the applied potential and a condition of *continuous discharge* occurs. The charge collected gives a pulse at the electrode, which is amplified and passed to the recording equipment.

Ionisation Chambers

These work in Region II and, being very stable, are often used for accurate work. However, they require highly sensitive and stable associated equipment. A form of ionisation chamber used as a sub-standard instrument and suitable for the assay of the pharmacopoeial radionuclides is shown in Fig. 262.

Proportional Counters

These counters operate in Region III. A common form is the *Gas Flow Counter* (Fig. 263) in which argon or methane is passed through the counter at just above atmospheric pressure to keep out oxygen. Although internal gas multiplication occurs the pulses are small and considerable amplification is needed

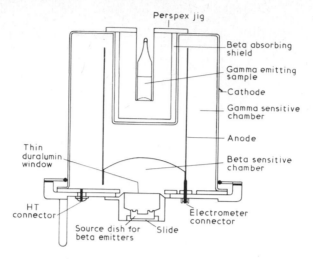

Fig. 262. Sub-standard ionisation chamber Type 1383A

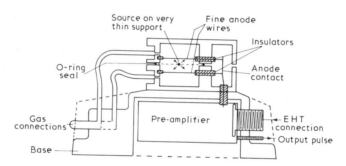

Fig. 263. A gas flow proportional counter

before recording. The applied potential must be very stable since the charge collected varies considerably with applied potential (see Fig. 261). Proportional counters have a number of advantages. They can count at a high rate, and since the pulse size depends on the energy of the radiation, they can be used to measure particle energy. Gas flow counters have a very high efficiency for low energy β-particles since the sample is inside the counter.

Geiger-Müller Counters

Mechanism. When a particle enters the counter it produces ion pairs. The electrons are accelerated to the anode and produce an avalanche of electrons, which are collected there in about 0.05 to 0.3 μsec giving a negative pulse which is recorded as one count in the scaler. This electron collection reduces the field strength and the counter is unable to function. The large positive ions move slowly to the cathode where they are discharged during a period of 100 to 500 μsec. Excitation of the cathode results and this would give spurious discharges and multiple pulses in the counter if not controlled. Control is achieved by including in the filling gas a small proportion of an organic vapour (ethanol or ethyl formate) or a halogen (chlorine or bromine). Such counters are described as being *internally quenched*, dissociation of the quenching agent dissipating the excitation energy of the cathode. Organic quenching agents become used and such counters have a finite life; halogen molecules on the other hand merely dissociate and recombine. Typical fillings for organic and halogen quenched counters are given in Table 53, which also lists some of their characteristics.

Table 53. Some Properties of Geiger-Müller Counters

	Typical Fillings and Characteristics			
	Organic Quenched		Halogen Quenched	
End-window counters	Helium	60 cm Hg	Neon	19.6 cm Hg
	Argon	4 cm Hg	Argon	0.4 cm Hg
	Ethyl formate	1 cm Hg	Bromine	0.4 mm Hg
Liquid counters	Argon	9 cm Hg	Neon	19.5 cm Hg
	Ethyl alcohol	1 cm Hg	Argon	0.4 cm Hg
			Chlorine	0.1 cm Hg
Working potential (volts)	1200–1600		350–700	
Plateau length (volts)	250–300		100–200	
Plateau slope (% per volt)	0.05		0.07	
Life (counts)	5×10^8		$> 10^{10}$	
Temperature range (°C)	-20 to $+50$		-55 to $+75$	

Characteristics of Geiger-Müller (G-M) counters. Following irradiation and the emission of a pulse there is a period of about 100 μsec during which the positive ions begin to be collected, and the field strength is so reduced that the counter cannot operate. This is the *true dead time*. The field strength is gradually restored and after about a further 100 μsec *recovery time* the counter will again yield full-size pulses (Fig. 264). The total period of about 200 μsec

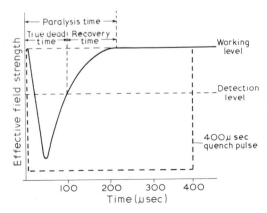

Fig. 264. Paralysis and electronic quenching in a Geiger-Müller counter

is the *paralysis time*, i.e. the time interval required to count separate pulses, and if a particle enters the counter during this interval it is not counted.

Losses due to paralysis can be corrected for since

$$N = N_0/(1 - N_0 T)$$

where N is the true counting rate, N_0 is the observed counting rate and T is the paralysis time. However, paralysis times vary for different G-M tubes and with different circuits, so that losses cannot be corrected unless the paralysis time of a particular counting set-up is known. This situation is dealt with by *electronic quenching* using a *probe unit* which reduces the anode potential for an exact predetermined time from the start of the pulse, and so paralyses the counter. The duration of the quench pulse may be set at will to 200, 300, 400, 500, or 800 μsec. The situation with a 400 μsec quench pulse is shown in Fig. 264.

Using a probe unit, corrections for paralysis, sometimes referred to as 'coincidence corrections', may be made by reference to prepared tables or graphs (see Tables 54 and 55).

End window counters (Fig. 265A) are used for counting solid sources. They consist usually of a cylindrical metal cathode about 2 cm in diameter closed at one end, along the centre of which is the wire anode. The open end is sealed with a 'window' of mica, glass or aluminium, the thickness of which determines

Table 54. Paralysis Time Correction Table for the Conversion of Counts per Minute (cpm) Observed at 400 μsec Paralysis

Reduce the observed count to cpm and read off the corresponding corrected counting rate. Corrections of high counting rates are of doubtful validity, since the pulses are not uniformly spaced in time, and counting rates should not exceed 10 000 cpm if high accuracy is required.

Part A: 0–9950 cpm

Observed cpm	0	1000	2000	3000	4000	5000	6000	7000	8000	9000
0	0	1007	2027	3061	4109	5172	6249	7342	8450	9574
50	50	1057	2078	3113	4162	5225	6304	7397	8506	9631
100	100	1108	2129	3165	4215	5279	6358	7452	8562	9687
150	150	1158	2181	3217	4268	5333	6412	7507	8618	9744
200	200	1209	2232	3269	4320	5386	6467	7563	8674	9801
250	250	1260	2284	3321	4373	5440	6521	7618	8730	9857
300	301	1311	2335	3374	4426	5494	6576	7673	8786	9914
350	351	1362	2387	3426	4479	5547	6630	7728	8842	9971
400	401	1413	2439	3478	4532	5601	6685	7784	8898	10 028
450	451	1464	2490	3531	4586	5655	6739	7839	8954	10 085
500	502	1515	2542	3583	4639	5709	6794	7894	9010	10 142
550	552	1566	2594	3636	4692	5763	6849	7950	9066	10 199
600	602	1617	2645	3688	4745	5817	6903	8005	9123	10 256
650	653	1668	2687	3741	4798	5871	6958	8061	9179	10 313
700	703	1719	2749	3793	4852	5925	7013	8116	9235	10 370
750	754	1770	2801	3846	4905	5979	7068	8172	9292	10 427
800	804	1821	2853	3898	4958	6033	7122	8227	9348	10 485
850	855	1873	2905	3951	5012	6087	7177	8283	9404	10 542
900	905	1924	2957	4004	5065	6141	7232	8339	9461	10 599
950	956	1975	3009	4056	5118	6195	7287	8394	9517	10 656

Table 54 *continued*

Part B: 10 000–19 950 cpm

Observed cpm	10 000	11 000	12 000	13 000	14 000	15 000	16 000	17 000	18 000	19 000
0	10 714	11 870	13 043	14 233	15 441	16 666	17 910	19 172	20 454	21 755
50	10 771	11 928	13 102	14 293	15 502	16 728	17 973	19 236	20 519	21 821
100	10 829	11 987	13 161	14 353	15 562	16 790	18 035	19 300	20 583	21 886
150	10 886	12 045	13 220	14 413	15 623	16 852	18 098	19 363	20 648	21 952
200	10 944	12 103	13 280	14 473	15 684	16 913	18 161	19 427	20 713	22 018
250	11 001	12 162	13 339	14 533	15 745	16 975	18 224	19 491	20 777	22 084
300	11 059	12 220	13 398	14 594	15 806	17 037	18 287	19 555	20 842	22 149
350	11 117	12 279	13 458	14 654	15 868	17 099	18 350	19 619	20 907	22 215
400	11 174	12 337	13 517	14 714	15 929	17 161	18 413	19 683	20 972	22 281
450	11 232	12 396	13 576	14 774	15 990	17 224	18 476	19 747	21 037	22 347
500	11 290	12 454	13 636	14 835	16 051	17 286	18 539	19 811	21 102	22 413
550	11 348	12 513	13 695	14 895	16 112	17 348	18 602	19 875	21 167	22 479
600	11 406	12 572	13 755	14 956	16 174	17 410	18 665	19 939	21 232	22 546
650	11 463	12 631	13 815	15 016	16 235	17 473	18 728	20 003	21 298	22 612
700	11 521	12 689	13 874	15 077	16 297	17 535	18 792	20 068	21 363	22 678
750	11 579	12 748	13 934	15 137	16 358	17 597	18 855	20 132	21 428	22 744
800	11 637	12 807	13 994	15 198	16 420	17 660	18 918	20 196	21 493	22 811
850	11 696	12 866	14 053	15 258	16 481	17 722	18 982	20 261	21 559	22 877
900	11 754	12 925	14 113	15 319	16 543	17 785	19 045	20 325	21 624	22 943
950	11 812	12 984	14 173	15 380	16 604	17 847	19 109	20 390	21 690	23 010

Part C: 20 000–29 950 cpm

Observed cpm	20 000	21 000	22 000	23 000	24 000	25 000	26 000	27 000	28 000	29 000
0	23 076	24 418	25 781	27 165	28 571	29 999	31 451	32 926	34 426	35 950
50	23 143	24 486	25 849	27 235	28 642	30 072	31 524	33 001	34 501	36 027
100	23 210	24 553	25 918	27 304	28 713	30 144	31 598	33 075	34 577	36 104
150	23 276	24 621	25 987	27 374	28 784	30 216	31 671	33 150	34 653	36 181
200	23 343	24 689	26 056	27 444	28 855	30 288	31 744	33 224	34 729	36 335
250	23 410	24 757	26 125	27 514	28 926	30 360	31 818	33 299	34 804	36 412
300	23 477	24 825	26 194	27 584	28 997	30 433	31 891	33 374	34 880	36 489
350	23 544	24 893	26 263	27 654	29 068	30 505	31 965	33 448	34 956	36 567
400	23 611	24 961	26 332	27 725	29 140	30 577	32 038	33 523	35 032	36 644
450	23 678	25 029	26 401	27 795	29 211	30 650	32 112	33 598	35 109	36 721
500	23 745	25 097	26 470	27 865	29 282	30 722	32 196	33 673	35 185	36 799
550	23 812	25 165	26 539	27 935	29 354	30 795	32 260	33 748	35 261	36 877
600	23 879	25 233	26 609	28 006	29 425	30 868	32 333	33 823	35 337	36 954
650	23 946	25 301	26 678	28 076	29 497	30 940	32 407	33 898	35 414	37 032
700	24 013	25 370	26 747	28 147	29 569	31 013	32 481	33 973	35 490	37 110
750	24 081	25 438	26 817	28 217	29 640	31 086	32 555	34 049	35 567	37 188
800	24 148	25 507	26 886	28 288	29 712	31 159	32 629	34 124	35 643	37 265
850	24 216	25 575	26 956	28 359	29 784	31 232	32 704	34 199	35 720	37 343
900	24 283	25 644	27 025	28 429	29 856	31 305	32 778	34 275	35 796	37 421
950	24 351	25·712	27 095	28 500	29 928	31 378	32 852	34 350	35 873	37 499

Table 55. Paralysis Time Correction Table for the Conversion of Counts per Second (cps) Observed at 400 μsec Paralysis

Reduce the observed count to cps and read off the corresponding corrected counting rate. For accurate work, counting rates should not exceed 200 cps.

Observed cps	0	100	200	300	400	500
0	—	105.1	217	341	476	625
5	5	109.6	224	347	483	
10	10	115.6	229	354	490	
15	15.1	120.5	235	360	498	
20	20.2	126.1	241	367	505	
25	25.3	131.6	247	374	512	
30	30.4	137.1	253	380	519	
35	35.5	142.7	259	386	527	
40	40.7	148.3	265	394	534	
45	45.8	153.9	272	400	541	
50	51.0	159.6	278	407	548	
55	56.2	165.2	284	414	556	
60	61.5	170.9	290	421	563	
65	67.6	176.7	296	427	571	
70	72.0	182.4	303	434	579	
75	77.3	188.2	309	441	586	
80	82.6	194.0	315	448	594	
85	88.0	199.8	322	455	601	
90	93.4	205.6	328	462	609	
95	98.8	211.5	334	469	617	

the minimum energy of the β-particles which can be counted. Very weak β-emitters (^{3}H, ^{63}Ni) and α-emitters cannot be counted in this way, and weak β-emitters (^{14}C, ^{35}S) and γ-emitters can only be counted at very low efficiency.

Liquid counters (Fig. 265B). In these the annular space between the thin tube envelope and the outer wall is designed to hold a definite volume of liquid sample (usually 10 ml). The liquid counter makes a useful sub-standard instrument for calibrating sources, since its efficiency can be determined (about 7–10% for energetic β-particles) and the source and counter are in constant geometric relationship.

Scintillation Counters

Scintillation counters, the constituent parts of which are shown in Fig. 295, depend upon the fact that when certain materials (phosphors) are exposed to radiation, they emit flashes of light. Since the light yield is extremely small, the method is practicable only if the phosphor is associated with a photomultiplier tube, which is a highly photosensitive detector (Fig. 146). A pulse discriminator is essential to enable pulses due to radiation to be distinguished from those due to interference (noise) in the amplifier and photomultiplier: it also permits energy measurement. Scintillation counters are valuable since they can be used

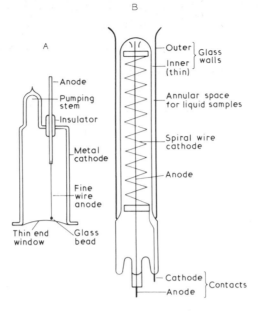

Fig. 265. Types of Geiger-Müller counter. (A) end-window; (B) liquid sample

with high efficiency for γ-rays and low energy β-particles. Further, their use avoids many difficult techniques of sample preparation.

Phosphors. For satisfactory results the light emission from the phosphor should be maximal at the peak sensitivity of the photomultiplier tube (*ca.* 4200 Å), its optical properties should be good and the decay time of the luminescence should be short to permit counting of separate pulses at high rates.

Inorganic phosphors have a decay time of about 10^{-7} sec. The most popular is thallium-activated sodium iodide (NaI/Tl) in the form of a large crystal. This is the phosphor of choice for γ-detection since it has a high absorption coefficient for X- and γ-rays. Further, the light output is proportional to the energy of the radiation, and the system can be used to measure γ-energies: this technique is used in pharmacopoeial identity tests for preparations of ^{131}I and [51]Cr. Since sodium iodide is hygroscopic, crystals are sealed in an aluminium capsule with one transparent glass side which faces the photo-multiplier.

Organic phosphors have a decay time of about 10^{-8} sec, and are more efficient for β-particles. The most common organic phosphors are anthracene, in the form of large crystals, and *p*-terphenyl. The latter is more versatile since it can be used in solution (8 % in toluene or dioxan) or as a constituent of a solid polystyrene plastic. Low energy β-particles are unable to penetrate the wall of a sample container or the light shield (Fig. 295) but may be counted with high efficiency if the sample can be dissolved in a phosphor solution so that all disintegrations take place within the phosphor.

Spark Chambers

These detectors have been developed for the rapid location of activity on radiochromatograms, especially thin-layer plates, but they can be used also for sections of tissue and for plant leaves.

The areas of activity may be located by scanning the chromatogram with a Geiger-Müller or scintillation counter with the windows masked to a slit to give good resolution. This is not too tedious with a single dimensional chromatogram up to 50 mm wide. However, if large thin-layer plates (200×200 mm) or two dimensional paper chromatograms are used, the scanning procedure becomes very time consuming, even though automatic scanning apparatus is available, e.g. the Panax RTLS-1A, which scans successively each one of a series of adjacent strips. It is thus highly desirable to have a detector capable of monitoring the whole area of a TLC plate at one time, and the spark chamber was developed for this purpose.

One successful form is used in the Panax 'Beta-Graph' and consists of a thin windowless counter ($200 \times 200 \times 7$ mm) having the same area as a TLC plate, with an electrode system consisting of 50 parallel cathode wires each surrounded by a helical, stainless steel, thin wire anode. The counter is polarised with an EHT in the range $0.2–2$ kV and an argon/methane (90/10) mixture is passed through the counter at a rate of about 200 cm^3/min. Ionisations due to disintegrations in the active areas of the plate cause sparks to jump between the electrodes, the flashes being recorded on Polaroid film. Depending on the activity in the sample, exposures of $1–60$ min may be required, compared with up to 4 hr by scanning. Typical prints from a 'Beta-Graph' are shown in Fig. 266.

Pre-amplifiers

Pre-amplifier units are not always used, but are essential if the main amplifier has a low sensitivity or if the cable joining the detector to the amplifier is so long that the pulse is unduly attenuated. They are frequently employed with proportional and scintillation counters.

Pre-amplifiers may be of two types. *Emitter followers* have a very low gain (amplification factor) but permit the use of long connecting cables. The polarity of the input pulse is preserved, i.e. a negative input pulse results in a negative output signal. Higher gain amplifiers usually give phase reversal, i.e. a negative input results in a positive output signal, though it is sometimes possible to choose the polarity of the output. A form of preamplifier which is important with G-M tubes is the *quenching probe unit* referred to above. This gives a gain of 20 to 30 times and also provides a quenching pulse of known duration for G-M tubes, so that proper paralysis time corrections can be made. Some scalers incorporate a quenching probe unit.

Scalers

Scalers are pulse counters and are the commonest form of device used with proportional, G-M and scintillation counters, since the quantity to be measured is usually the number of pulses emitted by the detector in unit time. Most scalers comprise a number of sub-units in one box, but sometimes each sub-unit is housed separately.

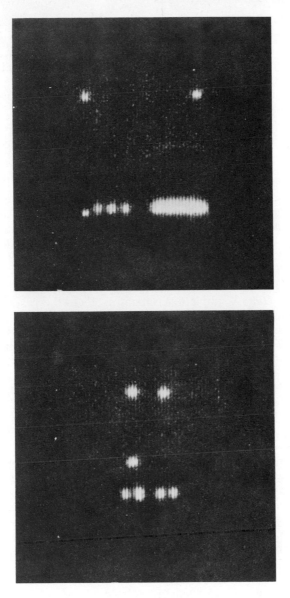

Fig. 266. Two thin layer chromatogram plates showing the areas of activity, as recorded by a spark chamber

High voltage supply. A suitable high voltage supply, usually referred to as 'HV' (High Voltage) or 'EHT' (Extra High Tension) to distinguish it from normal mains voltage, is always required in order to enable the response of a detector to be observed, and this must be stabilised against fluctuations in the mains supply.

For G-M counting, a moderately stabilised ($\pm 0.5\%$ for 10% mains fluctuation) EHT supply up to 2000 V is adequate, and this may be suitable also for some work with scintillation counters. However, for high accuracy, highly stable supplies are required, regulated to $\pm 0.02\%$ or better, and this is usually essential with scintillation counters.

Most detectors require a positive EHT supply, so scalers usually provide this only, but some EHT units can deliver either a negative or a positive output. Settings may be observed on a meter or set with potentiometers, and range switches or a '$\times 2$' multiplier switch are often incorporated. *It is essential that both the polarity and the magnitude of the EHT supply are set correctly, otherwise extensive damage may result.*

Amplifier. Simple scalers use non-linear amplifiers of fixed gain, but linear amplifiers of variable gain which amplify all pulses to the same desired extent and keep the relative pulse sizes constant, are essential for energy determinations. The gain of amplifiers may be stated as a simple amplification factor, the range $25 \times$ to $1000 \times$ being normal, or in decibels (dB, $20\,dB = 10 \times$; $40\,dB = 100 \times$; $80\,dB = 10\,000 \times$). Sometimes the degree of amplification is set by varying the attenuation (1/gain), i.e. the factor by which the maximum output of the amplifier is reduced.

Discriminator Unit. A simple discriminator blocks the passage of all pulses below the set amplitude (referred to as discriminator threshold or discriminator bias voltage), and is used to prevent unwanted pulses entering the scaling circuit. Such pulses are often random 'noise' signals which arise in the amplifiers and not in the detector, and are usually smaller than those which it is desired to count. Thus it is possible to exclude 'noise', or the weak pulses from low energy β-particles, in order to observe separately the larger pulses from higher energy β-particles in a mixture of radionuclides, e.g. to count ^{14}C in the presence of ^3H.

An *analyser* comprises two discriminators set at different levels and arranged to reject all pulses below the level set on the lower discriminator and above that set on the higher discriminator. Thus only pulses with a magnitude between the two levels are passed to the scaling circuit. This arrangement constitutes a *pulse height analyser*, (PHA, 'kicksorter') and permits determination of the distribution of pulse heights in the input signal, and hence the energy spectrum of the radiation entering the detector. The difference between the two discriminator levels is the gate width (channel width) and may be a fixed amount to be added to the threshold voltage or a fixed proportion of that voltage.

Discrimination is unnecessary with G-M counters, since all pulses have the same size, and discriminator settings should merely be arranged so as not to interfere, well below the pulse height, and a pulse height analyser should not be used.

Scaling Unit. This records the pulses transmitted from the amplifier and discriminator. Most scalers employ number-indicating tubes, which have electrodes capable of forming the numerals 1–0, though several different arrangements are used. These tubes are connected in series: in the first tube the glow steps on one position as each pulse enters the scaling circuit, the tenth pulse returns the glow to the zero position and a signal is sent out to the following tube, which steps on one in turn. Thus, successive tubes in the series count units, tens, hundreds etc. Sometimes a cheaper arrangement is used consisting of two number tubes followed by a mechanical register.

For proportional or scintillation counters, which have very short paralysis times, an input resolution time, the minimum time interval required between pulses for every pulse to be counted, of about 1 μsec is required, even at moderate counting rates, since radioactive decay is a random process and the resultant pulses are not spaced evenly but come in bursts with a high probability of short intervals between successive pulses. Thus short input resolution times avoid 'lost' counts with proportional and scintillation counters and the necessity for paralysis corrections. Suitable resolving times are achieved using transistors, the condition of which is displayed on the number indicating tubes.

An autoscaler incorporates a timing unit and auto-stop facilities to stop the counting process after a pre-set time has elapsed or when a pre-set number of counts has been accumulated. Apart from eliminating switching and timing errors due to the operator, pre-set times are useful to avoid unduly long counts with samples of low activity and the preset count facility permits a pre-determined nominal statistical error (see Experiment 7) to be attained.

Fig. 267 shows a typical autoscaler (this model also incorporates a ratemeter). The instrument has a single register used in conjunction with 'Preset Time' and 'Present Count' controls, so that the counter will stop automatically either after a preset count has been accumulated, when the register displays the elapsed time, or after a preset time, when the register displays the accumulated count. The meter which displays the counting rate also has a dual function and can be used to display the high voltage output (EHT or HV). Although the EHT is set accurately with the appropriate potentiometer, the meter gives an immediate visual indication of the approximate EHT level and so both warns that the EHT supply is on and helps to avoid applying the wrong voltage to a detector and so damaging it.

Ratemeters

The scaler is a pulse counter, which enables the number of pulses emitted by the detector in a suitable time, and so the average counting rate in counts per minute (cpm) or counts per second (cps) to be observed. Alternatively, a counting rate meter may be used in which the average voltage developed in an integrating circuit (a capacitor shunted with a resistor) is proportional to the average counting rate. The output from the integrating circuit is displayed on a meter or is used to operate a recorder. Ratemeters are often arranged to give an audible indication of the counting rate and so can be used as *monitors* to give a continuous indication of the level of activity in a laboratory.

The controls available replicate in part those of a scaler,* e.g. mains switch, EHT, input selector, discriminator. Count-stop-reset facilities are not required, since the counting rate is displayed continuously, and input resolution time is a characteristic of the circuit design. An important control is the *time constant* (integrating time) of the circuit. Low settings of this control permit a rapid indication of the changes in counting rate, so that the meter indication changes rapidly with minor fluctuations of counting rate, and high settings give a slow rise to the average counting rate with slow response to change. This latter

---

* See *General notes on the operation of equipment*, p. 471.

condition smooths out the fluctuations and enables readings to be taken more easily, but this control must be used with discretion since high settings may give false impressions of accuracy at low counting rates due to lack of responsiveness. At high counting rates, the random fluctuations are inherently small so that short time constants can be used, with the benefits of increased accuracy and responsiveness.

BODY SCANNING

This has already been mentioned in connection with radiochromatogram scanning. Exactly the same principle is applied in many diagnostic medical uses of radionuclides, in that the detector is moved across the area of interest repeatedly, each traverse detecting the activity from one of a series of contiguous strips. In medical radio-diagnosis, a suitable chemical form of a low energy γ-emitter (150–200 keV is preferred) is injected and, after uptake, the organ is scanned. Using suitable nuclides (e.g. ^{99m}Tc, ^{113m}In, ^{87m}Sr, ^{75}Se, ^{32}P, ^{125}I, ^{59}Fe, ^{58}Co, ^{51}Cr, ^{197}Hg, ^{133}Xe), the location and size of tumours and internal organs such as the spleen and liver can be determined, or the functional ability of organs (kidney, lungs) can be checked, by detecting the γ-radiation which emerges from the body from its location and concentration in the target organ.

Two methods are used for detection. In scanning, a collimated scintillation counter is used, i.e. a heavily shielded NaI/Tl counter with the lead shield towards the source drilled precisely with one or more holes to give a high resolution image of the organ of interest without undue interference from any background radiation from nuclide in the surrounding tissues. Improved resolution is sometimes obtained by using two directly opposed counters scanned simultaneously on opposite sides of the body. Figure 268 shows a lung scan obtained using Macroaggregated Iodinated (^{131}I) Human Serum Albumin which shows clearly the presence of a pulmonary embolism in the left lung. The figure illustrates also the form of dot diagram in which scanning information is usually presented, though multicolour outputs are common.

As with radiochromatogram scanning, the procedure is slow and the *gamma-camera* (scintillation camera) has been developed for organ imaging. This uses one very large NaI/Tl crystal, or a number of smaller crystals, backed by an array of photomultiplier tubes and is capable of examining an organ such as the brain or kidneys in a single stationary exposure. The gamma camera has not replaced scanning completely, the two techniques being complementary.

Fig. 267. A typical autoscaler (by courtesy of Nuclear Enterprises Ltd). A, switch to select meter function; B, meter which displays counting rate (cps) for the ratemeter or high voltage output; C, number tube register which displays total counts or elapsed time as selected by the switch D; E, high voltage on–off switch and setting potentiometer; F, pulse height analyser channel width selector; G, disc bias setting potentiometer; H, switches for ratemeter range and time constant; J, preset count switch to select the count at which counting stops automatically; K, switch to start counting; L, manual stop switch; M, switch to reset the number tubes to zero after a count or to select the test pulse; N, switch to select automatic operation or data print out; O, preset time switch, to select the time after which the counting stops automatically

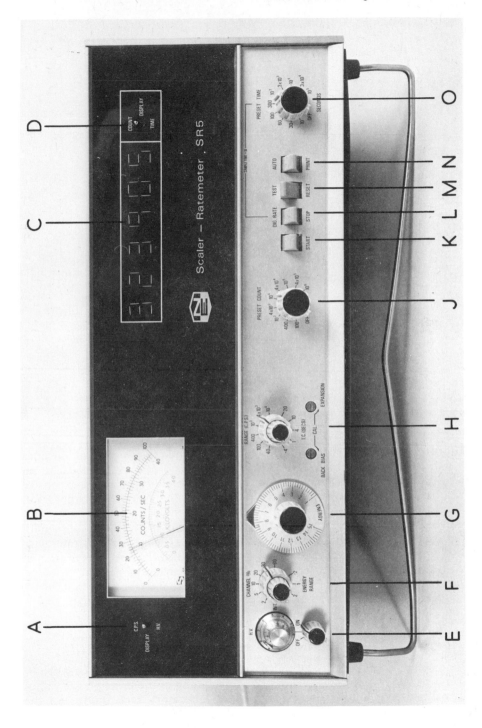

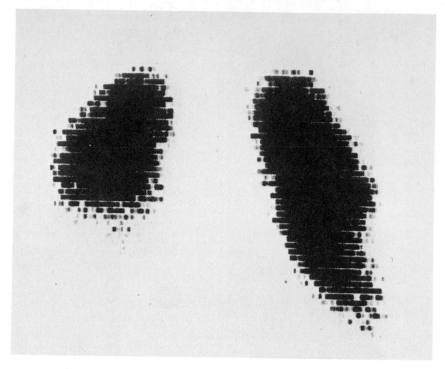

Fig. 268. A lung scan using Macro-Aggregated Iodinated (^{131}I) Human Serum Albumin, showing an embolism of the left lung

Whole Body Counting

The determination of the total human body burden of one or more radionuclides is undertaken only rarely, either for research purposes or as a sequel to an accident. Such estimations can be done only with highly specialised, expensive facilities maintained by a few institutions, e.g. the National Radiological Protection Board.

In small animal experiments, the determination of the body burden of γ-emitters may be determined simply using a NaI/Tl scintillation counter with a suitably designed shielded sample chamber in which the animal is restrained to avoid any variation in counts due to the animal's movement, giving a change in source-detector geometry. The Panax 'Gamma One-Sixty' automatic γ-spectrometer can count the γ-activity in up to 20 mice fully automatically.

SOURCE PREPARATION

Radioactive substances are all potentially hazardous, but safe working conditions are achieved through a proper understanding of the principles of radiological protection, by strict observance of laboratory regulations and the application of careful techniques.

Laboratory Organisation

Three separate areas are normally provided: the radiochemical laboratory, in which all manipulations of radioactive materials are carried out, the counting room, in which all measurements of activity are made, and the decontamination area. The latter two are *clean areas* where there should be no detectable radioactive contamination. This is essential to avoid health hazards and interference with measurements.

The Radiochemical Laboratory

All preparative work should be confined to designated individual working areas, which may be defined conveniently by a stainless steel tray, lined with absorbent paper to catch accidental spillage. It is usual also to protect all the bench surfaces in the laboratory from contamination by covering with waterproof and absorbent paper. These precautions limit the extent of contamination in case of a spill and facilitate disposal of the spilled material.

Solid and liquid active wastes must be placed in special containers. Active liquid waste must not be poured into the sinks, nor may contaminated apparatus be washed in the sinks unless the activity, as detected by the radiation monitor, is not sensibly above the normal background counting rate in the laboratory. It is common to segregate liquid waste according to half-life, i.e. up to 1 month, 1 to 3 months, greater than 3 months, in order to facilitate disposal. A ratemeter type monitor, giving a continuous audible indication of the level of background radiation, should always be in use.

It is good practice always to carry unsealed source containers inside a second container which has sufficient packing to absorb the contents in case of a spill. Non-rigid plastics are very suitable for this purpose, since they afford good mechanical protection against accidental breakage.

Personal Protection

Protective gloves should always be worn when handling unsealed sources and must be decontaminated thoroughly by washing before removal, decontamination being checked with the monitor. The techniques for putting on the gloves, and removing them, should prevent the transfer of contamination from the outsides of gloves to the insides, or to the skin of the hands. Heavily contaminated gloves must not be washed into the sinks and should be treated as solid active waste. Disposable plastic gloves are very suitable and are cheap enough to be discarded after a single wearing. Also, special laboratory coats are always provided and canvas overshoes may need to be worn if hazardous or highly active materials are being handled.

Before leaving the laboratory for any reason, gloves and overshoes must be monitored and removed and the hands washed and monitored. Gloves are not normally worn in the counting room, but if this is thought to be necessary separate gloves must be used and all apparatus then handled via disposable paper tissues. At the conclusion of a working period the hands, clothing and feet should be monitored using a *laboratory monitor*. This is a ratemeter with an audible indication of the level of activity and fitted with a detector (probe) suitable for the type of radiation known to be present; thin glass walled G-M tubes (e.g. B12H) are commonly used for β-γ monitoring, but if low energy β-particles (^{14}C, ^{35}S) are to be detected, a thin end-window counter must be used, e.g. MX123, or 2B2. A monitor capable of operating scintillation counter probes is highly desirable, since alternative probes are available to enable the detection of α and β particles and γ-rays with high efficiency. Suitable monitors are the Panax 5067 and the Nuclear Enterprises RM5.

The *maximum permissible dose* of radiation for occupationally exposed persons is 5 rem per year, corresponding to an average weekly dose of 100 millirem, assuming 50 weeks exposure. Procedures should be arranged so that this latter level is not exceeded and that the dose received is minimal. All workers must always wear a *film badge dosemeter* when in the laboratory. The film is worn in a special holder (Fig. 269), with areas of different thickness and type, so that the dose received at the badge can be estimated accurately from the pattern of blackening. Film badges are convenient and provide a permanent record, but the results are not known until some time after the period of exposure (normally 2 weeks). In the U.K., film badges are provided by the National Radiological Protection Board, who maintain records of radiation exposure for all radiation workers. Film badges are to be replaced with lithium chloride luminescent dosemeters.

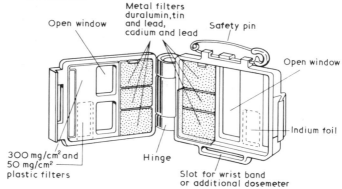

Fig. 269. A film badge dosemeter holder (the Radiological Protection Service)

An immediate indication of the dose of X- or γ-radiation received can be obtained using a *pocket electroscope* (Fig. 270), but the sensitivity is such that

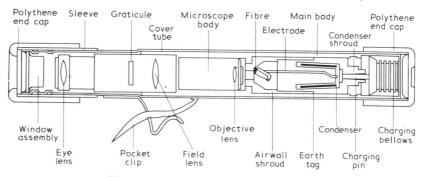

Fig. 270. A pocket electroscope dosemeter

these are of value only when the dose approaches 50% of the maximum permissible level. They are insensitive to α- and β-particles. Electronic pocket monitors are also available but are expensive.

Before commencing any new experiment, calculations should be made of the probable radiation dose and, when the apparatus, etc., is set up, the actual dose rate should be measured with an ionisation chamber type monitor (e.g. Eberline RO2, Fig. 271, or the Nuclear Enterprises 0500), to ensure that it is below the permitted level. The experimental design should ensure the minimum possible dose.

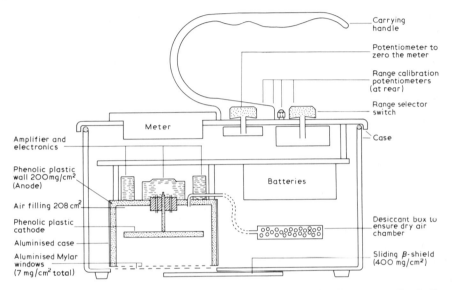

Fig. 271. An ionisation chamber monitor. (Courtesy of Eberline Instrument Co. Ltd)

Rules for the Radiochemical Laboratory

(1) Do not eat, drink, smoke or use cosmetics in the laboratory.

(2) Wear the laboratory coats provided at all times and remove before leaving.

(3) Wear a film badge at all times in the laboratory.

(4) Wash your hands and monitor hands and clothing when leaving. Report any detectable contamination.

(5) Report immediately any injuries, especially to the hands, whether received in the laboratory or before you enter it. Do not use cracked or chipped apparatus.

(6) All active materials must be retained in the areas allocated to them.

(7) Gloves must be worn when handling unsealed sources and they must be decontaminated after use (plastic gloves are rejected).

(8) Do not pipette or carry out any operation by mouth.

(9) Deposit all radioactive waste in the receptacles provided. Do not discard solutions into sinks.

(10) Spills are to be avoided by using careful techniques. In the event of accident take immediate action.

    (a) Blot up wet spills; cover dry spills with damp tissues. Use gloves and forceps. Place contaminated materials in the active waste containers.

    (b) Mark the area of the spill and the type of contamination clearly.

    (c) Report the nature of the accident and the action taken.

(11) Report immediately major spills, personal contamination, wounds or other emergencies.

Additional Rules for the Counting Room

(1) No source other than those prepared ready for counting may be taken into the counting room.

(2) All sources for counting must be transported in the containers provided.

(3) Gloves and overshoes worn in the laboratory must not be worn in the counting room.

(4) Report immediately any spillage or the occurrence of an abnormally high background.

The Toxicity of Radionuclides

Radioactive substances are graded in toxicity according to three main criteria.

Effective half-life, i.e. the time taken for the amount absorbed in the body to be reduced to one half of its initial value. This depends both on physical half-life and the rate at which the element is excreted. The rate of excretion of most ingested radionuclides cannot be increased.

The type and energy of radiation. If the material is ingested then α-emitters are very toxic, since the whole of the energy is absorbed in the immediate vicinity of the source and intense local damage results, but β-emitters are less toxic, since the energy is absorbed over a larger volume. If γ-emitters are ingested, only a part of the energy is absorbed within the body. When considering hazards from sources external to the body, the three classes of emitter are graded in the reverse order, since γ-rays are most penetrating and are most likely to reach the tissues and cause damage.

The degree of selective localisation. Where an element is absorbed quantitatively by particular tissues and its rate of turnover is low, this leads to high toxicity. Thus, radioiodides selectively radiate thyroid tissue and ^{90}Sr and ^{226}Ra are 'bone-seekers' and irradiate erythropoietic tissue from their absorption sites in bone: these are in the 'High' or 'Very High' toxicity classes.

Classification of some Radionuclides according to Toxicity

Very high	^{90}Sr + ^{90}Y, ^{210}Pb + ^{210}Bi*(Ra D + E), ^{226}Ra + daughter products*.
High	^{45}Ca, ^{56}Co*, ^{59}Fe*, ^{89}Sr, ^{125}I*, ^{131}I*, ^{140}Ba + ^{140}La, ^{170}Tm*, ^{234}Th* + ^{234}Pa*, natural thorium*, natural uranium.
Moderate	^{22}Na*, ^{24}Na*, ^{32}P, ^{35}S, ^{36}Cl, ^{42}K*, ^{54}Mn*, ^{55}Fe, ^{55}Co*, ^{59}Te*, ^{60}Co*, ^{64}Cu*, ^{65}Zn*, ^{74}As*, ^{75}Se*, ^{76}As*, ^{82}Br*, ^{132}I*, ^{137}Cs* + ^{137}Ba*, ^{182}Ta*, ^{097}Hg*, ^{198}Au*.
Slight	^{3}H, ^{14}C, ^{51}Cr*, ^{87m}Sr*, ^{99m}Te*, ^{113m}In*, ^{133}Xe*.

* γ-emitters.

The sign '+' is used to indicate a nuclide in equilibrium with its daughter products.

Sample Preparation

Mounts. Before a sample can be counted it must be prepared in a suitable form. The simplest technique is to place the sample on a *planchet* for counting under an end-window G-M counter.

Planchets may be made of aluminium or stainless steel and are of two types, plain (Fig. 272A) and dimpled (Fig. 272B). The dimpled type has the advantage of restricting small quantities of fluid to the centre, thus giving greater reproducibility. Small watch glasses may also be used satisfactorily. Where counts are to be compared, planchets of the same material must be used to obtain constant backscattering conditions.

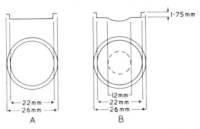

Fig. 272. Counting planchets (A) plain; (B) dimpled

Solid materials may be placed or weighed directly into a flat planchet. Solutions are measured into a planchet and then dried, and it may be advantageous to add a small amount of wetting agent (e.g. 0.1% Teepol) to the planchet in order to obtain reproducible spreading of the solution. If the dry material tends to detach from the planchet, it may be fixed by the addition of collodion (or Durofix) diluted in acetone, though this must be used with care since it may introduce undesired absorption effects.

Drying of samples is done most conveniently using an infra-red lamp clamped to a retort stand above the sample. This gives even drying without boiling and it is important not to try to hasten the process excessively, so producing splashing and contamination. Boiling of solutions is to be avoided at all times and disposable tissues should be used under a planchet to catch accidental contamination. A hot air blast from a hair dryer may be useful, but should be used cautiously. Volumes greater than 0.1 ml are best handled by drying successive small portions on the same planchet, rather than attempting to deal with the whole volume at once.

Volatile materials must be handled with especial care, e.g. radioiodine samples should be dried at a low temperature in the presence of a fixative solution (NaOH, 0.8 g; $Na_2S_2O_3$, 0.5 g; KI, 0.03 g; water to 1 litre).

Sealing. Whenever possible, planchets should be sealed to avoid accidental spillage. Cellophane adhesive tape is convenient for this purpose, but a more permanent seal may be made with aluminium foils. With low energy β-particles, such seals may absorb too much of the radiation, and thin mica or plastic film may provide an acceptable seal.

Pipetting. Most pipetting of radioactive solutions is done with micro-pipettes. Although these can be obtained in sizes down to 1 microlitre (μl) it is difficult to manipulate accurately sizes less than 50 μl without considerable practice. Micropipettes may be calibrated to 'deliver' or to 'contain'. The latter type must be rinsed out and the rinse liquid added to the sample. To avoid contaminating large volumes of rinse liquid, one or two drops may be placed on a small watch glass, a piece of cellophane or a dimpled planchet, which is discarded into the active waste container after use. Safety pipettes are used for larger volumes.

Suitable pipettes are sold under the trade names 'E-mil' (Auto-zero micro-pipettes of the G8116 series) and 'Exelo' (types P5/S, P11/S). Also normal pip-ettes can be used with a 'Griffin' pipette filler or a 'Fisons' micropipetter, and satisfactory control can also be obtained with a syringe attached to a pipette with tubing, provided the pipette is clamped rigidly. Ordinary rubber teats are unsuitable, since they give inadequate control and the inadvertent expulsion of liquid from the tip of a pipette causes spray formation and undesirable contamination of surprisingly large areas. Hamilton microsyringes are invalu-able for small volumes and Eppendorf and similar pipettes are especially useful, since the plastic tips are disposable, though cost is a problem.

Remote pipetting devices require skill in handling and their use should not be necessary in normal work. The radiation dose received is best reduced by sensible experimental design and the use of simple, speedy techniques rather than by using elaborate equipment with which inexpert handling may lead to accidental spillage.

Decontamination. Careful techniques are necessary to ensure not only that the minimum of apparatus is contaminated, but also that apparatus contamin-ated unavoidably should be decontaminated with the minimum of delay. Soil which is allowed to dry on may be extremely difficult to remove. Decontamina-tion is assisted by using 'carriers', i.e. solutions containing an inactive isotope of the same element and in the same chemical form, usually at a concentration of 1 g/litre. When carriers cannot be used, e.g. with uranium, cleaning mixtures based on nitric, nitro-sulphuric or chromic acid and powerful detergents (e.g. RBS 50) are also useful.

The bulk of a carrier or other decontamination solution must not be con-taminated by dipping into it pipettes or other contaminated items. The smallest amount appropriate to the size of the article to be decontaminated should be used from a small watch glass or beaker and washing carried on until no activity can be detected on the monitor. It is important to clean the outsides of pipettes etc. using small plastic wash bottles and tissues soaked in carrier. A convenient technique is to support a funnel, small enough to hold the pipette firmly, in a filter flask so that washings from the pipette are caught in the flask. Contamin-ated washings and tissues must be discarded into the active waste containers. If the radiation emitted has a low energy, e.g. ^{14}C, ^{35}S, it cannot be detected through the glass walls of apparatus and contamination by such materials is best tested for by wiping with a tissue, or rinsing onto a tissue, and monitoring the tissue with a suitably sensitive detector. Very low energy emitters such as ^{3}H cannot be detected readily and so their handling requires the greatest possible care, since contamination with them may not be suspected until the results of experiments are seen to be anomalous. All apparatus used in the radiochemical laboratories should be set aside for use exclusively within each laboratory and none should be handled outside the unit.

Filtration. It is often convenient to collect the active material in the form of a precipitate, and a filter apparatus suitable for this purpose is illustrated in Fig. 273. This consists of a demountable sintered glass filter capable of holding filter paper discs or membrane filters 22 mm in diameter. The precipitate is collected on the filter, washed, if possible, with acetone or ethanol to assist drying and the filter paper plus precipitate placed in a plain planchet and dried. A small amount of dilute Durofix fixative may be necessary to prevent flaking and detachment of the precipitate.

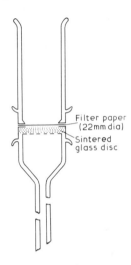

Fig. 273. Glass filter apparatus

Shielding. Generally, experiments should be arranged so that heavy shielding and the use of remote devices is not necessary. Where shielding is essential for γ-emitters a 5 cm lead wall with handling tongs should be erected inside a fume cupboard, or under a fume hood, with a mirror fixed to the back wall to permit observation of objects behind the wall. It is emphasised that, where such shielding is necessary, careful monitoring in front of, above, beside and below the level of the shield is essential to ensure that the protection is adequate. Satisfactory shielding against β-emitters can be obtained using a thick Perspex or glass safety screen (e.g. the Fisons Safeguard) or a dry box.

PRACTICAL EXPERIMENTS

Many of the experiments described below can be done with ready-prepared sources, thus reducing substantially the amount of time required, and in such cases the work given under 'source preparation' may be omitted without detriment to the principles involved, although the student will thus acquire only a very limited experience of handling radioactive materials.

There is a considerable range of equipment available which could be used for the experiments described below and it is anticipated that established laboratories will have a variety of types in use. The experiments are designed to illustrate principles, rather than the characteristics of individual items of equipment, and it is assumed that the simplest suitable equipment will be used. New equipment will inevitably be introduced within the life of this text and complex instruments can be used without in any way affecting the basic principles involved.

Few departments of Pharmacy will find the time for the inclusion of all the experiments described and a sufficient range is described from which a selection can be made.

Source mounts. All the Geiger-Müller end window counting described below is based on the use of the Labgear castle (D. 4019/E, Fig. 274). This is no longer in general manufacture. However, its use is retained here since the design provides exceptional simplicity and convenience. It can be easily and economically constructed in local workshops or may be available to order from Electrotech of Cambridge. A similar experimental arrangement can be based on the Panax PH 1 Planchet Holder.

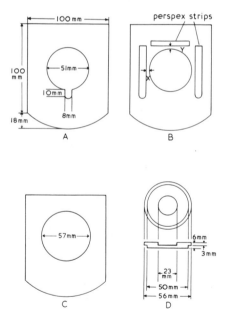

Fig. 274. Perspex shelves and a source disc for the Labgear castle. For standard aluminium absorbers, X = 10 mm, Y = 14 mm; for lead absorbers, X = 6 mm, Y = 5 mm

The castle is very suitable for student use since it provides easy access and manipulation. It is convenient to modify the castle by cutting additional shelf slots, making seven in all, and to provide Perspex shelves and a source disc of the form shown in Fig. 274. Shelf numbers are given from the top down.

General notes on the operation of equipment.

(*a*) Electronic equipment is usually left switched on permanently, since damage can result from the surge currents which occur with switching on and from the effects of expansion and contraction as equipment heats and cools. If an instrument is to be left unused for 48 hours or more, it is usually switched off, but after switching on again a period of 1 to 2 hours should be allowed for complete stabilisation. Thus equipment is always switched on and ready to operate at the beginning of a work period and should be left in the same condition when the experiment is finished.

The following controls are usually present on the scaler (Fig. 267).

(i) Mains switch and pilot lamp.

(ii) EHT coarse and fine adjustments, the output being displayed on a meter or set with a graduated control knob. ON–OFF and range switches and a polarity control may also be included and *it is essential that these be set correctly to avoid damage to associated equipment.*

(iii) COUNT, to commence counting.

(iv) STOP, to stop counting after a suitable period.

(v) RESET, to reset the registers to zero ready for another count.

(vi) INPUT SELECTOR, to select the circuits appropriate to the type of input pulse, e.g.

Positive or *Direct*, for positive going pulses, *Negative*, for negative going pulses,

GM, where a GM tube is connected directly into the scaler without an intermediate amplifier or probe unit,

Probe, for pulses carried in the 6-core lead from a probe unit (sometimes called 'External' or 'Direct'),

Test, to provide pulses, usually of mains frequency (50 or 100 cps), to test the scaling circuits.

(vii) DISCRIMINATOR (*Disc* or *Bias* or *Disc Bias*) to set the threshold for the minimum size of pulse which will be counted.

(viii) DELAY TIME (or *Paralysis Time*), to set the minimum time interval for observing successive pulses in the scaler.

Additional controls may be present but will not be mentioned here since they are best dealt with in the counting room using the particular instrument concerned.

The normal condition of a scaler when starting work or finishing, is as follows: power, ON; count key, ON; counting on TEST; EHT, discriminator and delay time, OFF or lowest settings.

(*b*) It is important to check systematically the setting of each control before starting an experiment and to record the settings used. Most apparent defects are due to incorrect control settings. The equipment details should also be noted, e.g. manufacturer, type and serial number of the counter, scaler, pre-amplifier and castle, along with details of the quenching of Geiger-Müller counters, window weights and the composition of scintillation phosphors. This enables a check to be kept on the day-to-day performance of equipment and helps to avoid apparently anomalous results arising from a change in instrumentation.

(c) The correct functioning of equipment must always be verified at the beginning of each work period, and some or all of the following checks should be done, as appropriate for the instrument being used.

(i) Mains pilot lamp glows.
(ii) Count, stop and reset functions operate.
(iii) The tubes count correctly on each of the test pulses available. The glow steps on in an orderly fashion from one number tube to the next.
(iv) If appropriate, the longer paralysis time settings give progressive loss of counts on 'fast test'.
(v) With autoscalers, that both pre-set time and pre-set count facilities operate.
(vi) Where meter displays are used, that the meter responds to a small movement of the appropriate control.

When correct functioning has been verified, the instrument should be set up for counting and the following further checks carried out.

(vii) Determination of the background count, i.e. the count in the absence of a source.
(viii) Determination of the counting rate from a standard source in a fixed position relative to the detector.

The equipment records should be consulted to ensure that these results are in accordance with the known performance, since counter characteristics change with age, temperature and other factors and counters may fail suddenly. At the end of a work period, checks (vii) and (viii) should be repeated and the instrument left in the condition indicated at the end of section (a) above. It is useful also to repeat checks (vii) and (viii) at intervals during the work, e.g. backgrounds may be counted conveniently during meal breaks.

Counting rate corrections. All counts must be corrected for paralysis, if large enough, and for background as follows:

(i) Correct all observed counts to cpm or cps.
(ii) Note the appropriate paralysis time correction (Tables 54 and 55).
(iii) Subtract the corrected background counting rate from the corrected sample counting rate.

Handling the results. When counting, it is desirable to correct a result for paralysis and to plot it on the graph as each result is obtained, and there should be ample time to do this if at least 2 min or 100 sec counts are used and the exact form and extent of the required data is decided in advance. Anomalous results are then immediately apparent and can be repeated under identical conditions. Also, it is easy to decide whether additional observations are required to define the slope of a curve more closely and these observations can be made before any settings are changed.

Counting time. For accuracy total counts of at least 3000, and preferably of 10 000, should be obtained (see Experiment 7) and counting times should be increased if necessary to allow for this. With manually operated scalers, counting times shorter than 1 min should not be used, and 2 min counts are preferable to minimise switching errors. With such scalers it is convenient to have a stopclock which runs continuously, the scaler being switched on or off when the second hand reaches a desired point. Initially it is useful to do a series of counts of a slow test pulse (3000 cpm) until consistency in switching is obtained.

Experiment 1. *The operating characteristics of Geiger-Müller counters*

The Geiger-Müller counter detects radiation by responding to the ionisations produced, mainly in the filling gas. The number of ions collected under working conditions is almost independent of the number liberated in the initial ionising event (Fig. 261).

As the voltage applied to the counter is increased a point is reached at which counts are just registered: this is the *starting voltage* (V_s). The counting rate then increases rapidly with applied EHT until a plateau is reached at the threshold voltage (V_t), after which the counting rate increases only slowly with increase in EHT. At about 100 to 200 V above V_t the counting rate increases rapidly and the counter goes into continuous discharge. This characteristic curve is shown in Fig. 275.

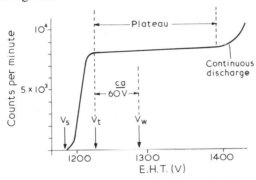

Fig. 275. The characteristic curve of a Geiger-Müller counter

The counter is operated approximately in the centre of the plateau region at the *working voltage* (V_w), normally $V_t + 60$ V.

In this experiment a uranium oxide (U_3O_8, half-life *ca* 4.5×10^9 years) standard reference source will be prepared and used to determine the characteristics of an end-window Geiger-Müller counter (Fig. 265A).

Method. (*a*) *Source preparation.* Weigh approximately 0.5 g of black uranium oxide into a plain planchet. Wet the sample with a strong solution of Durofix in acetone (50 mg/ml), ensuring that the powder is distributed evenly on the planchet, and dry under the heat lamp. Moisten and dry again a few times to ensure that the powder is firmly bonded. Seal a piece of aluminium foil (0.2 mm, 54 mg/cm^2) to the planchet with Durofix to filter out low energy particles and trim with scissors.

(*b*) Check the operation of the scaler and the settings of the controls (Note 1).

(*c*) Place the reference source on shelf 3 of the castle (Note 2) and, with the key in the 'count' position, increase the EHT slowly until the scaler just operates. Stop counting and reset the scaler.

(*d*) Record the voltage and count for 2 min.

(*e*) Increase the voltage by three steps of 10 V and then by steps of 25 V, taking 2 min counts at each setting (Note 3).

(*f*) Return the EHT to zero and place the source in the container provided.

(*g*) Plot the observed, uncorrected counting rate (Note 4) against EHT and determine V_s, V_t, V_w and the plateau slope (Note 5). Record the counting rate at V_w.

Note 1. Ensure the setting of the disc bias is the lowest available.

Note 2. Use Plate A, Fig. 274. Sources must not be handled. Always use forceps.

Note 3. Do not allow the counter to go into continuous discharge. Do not exceed $V_t + 150$ V. If the counting rate increases sharply at any time after the plateau has been reached, reduce the EHT to zero immediately and report.

Note 4. In this experiment it is not necessary to correct the observed counts for paralysis and background. Explain why this is so.

Note 5. The plateau slope is calculated as follows:

$$\text{slope} = \frac{C_v - C_t}{V - V_t} \times \frac{100}{C_t} \text{ per cent per volt,}$$

where C_v is the counting rate on the plateau at V volts and C_t is the counting rate at V_t. The plateau should be at least 100 V long and its slope should be less than 0.1 per cent per volt. The slope is usually lower for organic quenched tubes than for halogen quenched ones. The latter often have slopes of about 0.1 % per volt and usually start very sharply, so the steeply rising part of the characteristic curve, from V_s to V_t, is not observed.

Experiment 2. *Background radiation and the effect of shielding*

There is a natural background radiation derived from cosmic rays and from naturally occurring radioactive materials in soil, building materials, air and the human body. This must be corrected for when counting, especially where it is desired to examine materials of low activity.

Method. (*a*) Count the background for 10 min with the counter shielded and with the shield removed (CARE, see note).

(*b*) Record the counting rates and your conclusions.

Note. Be careful when removing and replacing the lead shield. Remember that it is heavy and that the counter is delicate and electrically live.

Experiment 3. *The absorption of β-particles*

The absorption of β-particles in matter results primarily from collisions with orbital electrons and, if electron density is allowed for (Note 1), is almost independent of atomic number.

The absorption process is approximately exponential, but the exact shape of the absorption curve (Fig. 276) depends on the relative importance of the energy dissipation processes (ionisation and excitation), the fact the β-particles have a continuous energy spectrum (Fig. 277), so that there are particles of all energies up to a characteristic maximum (E_{max}, Note 2), and the geometrical relationships between source, absorber and counter (Note 3). If β-particles of more than one energy are present, there will be an inflexion in the curve at about the absorber thickness corresponding to total absorption of the low energy β-particles. Although β-particles have a finite range, there is a tail to the curve due to Bremsstrahlung (Note 4) plus any γ-activity which may be present.

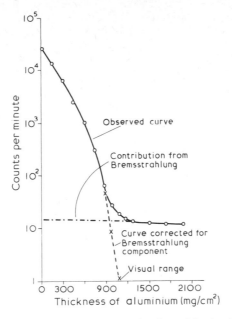

Fig. 276. A typical absorption curve for β-particles in aluminium

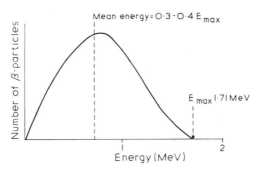

Fig. 277. The energy spectrum of β particles from ^{32}P

Method (*a*) *Source preparation*. Mark two dimpled planchets for subsequent identification and add to them 0.1 μCi and 10 μCi respectively of ^{32}P. Dry under the heat lamp and seal with cellulose tape. Decontaminate (Note 5).

(*b*) Check the operation of the scaler and place a metal shield in shelf 1 of the castle (Note 6).

(*c*) Place the weaker source in shelf 4 and an absorber holder (Plate B, Fig. 274) in shelf 3. Remove the metal shield, count for 2 min (Note 7), and replace the shield.

(*d*) Take counts with suitable thicknesses of aluminium absorber (Notes 8 and 10).

(*e*) When the counting rate drops to about 100 cpm, change to the strong source (CARE, Note 6) and repeat the counts with the last three absorbers used with the weaker source (Note 9).

(*f*) Continue counting with increasing thicknesses of absorber until the counting rate is approximately constant or declines slowly and linearly (Notes 4 and 10).

(g) Place the sources in the container provided, correct all counting rates for paralysis, background and change in source strength (Note 9) and plot the counting rate against absorber thickness (Note 11) on semi-logarithmic graph paper.

(h) Extrapolate the Bremsstrahlung background to zero absorber thickness, subtract the Bremsstrahlung contribution from the corrected counting rate at each absorber thickness and plot the corrected curve (Fig. 276). Estimate the visual range of the β-particles in aluminium, expressed in terms of superficial density (mg/cm^2) and linear thickness (mm).

(i) Repeat steps (b) to (h) using the Perspex (Note 11) and lead absorbers provided (Note 8). Compare the results obtained with the three types of absorber.

(j) Check the operation of the scaler.

Note 1. The electron density of a material is reflected in its relative density and so is allowed for when thicknesses are expressed in terms of 'superficial density', i.e. the weight of absorber behind 1 cm^2 of surface. Thus,

$$\text{Absorber thickness} = \rho x \text{ g/cm}^2$$

where ρ is the relative density of the absorber and x is its thickness in cm.

Note 2. The value of E_{max} is characteristic of a given nuclear change, so determination of E_{max} can be used as one factor in the identification of radio-nuclides (see Experiments 10 and 11).

Note 3. The absorbers should be as close as possible to the counter to avoid scattering effects.

Note 4. Bremsstrahlung is a form of low energy electromagnetic radiation produced by deceleration of the β-particles in the atomic field and is analogous to the X-radiation produced by the interaction of the cathode rays with the target of an X-ray tube. The absorption of X- and γ-radiation and Bremsstrahlung follows an exponential law (see Experiment 6), so the tail to the β-absorption curve is linear and slopes towards the abcissa. However, aluminium and Perspex absorb the radiation only feebly, so that the lines are almost parallel to the abcissa, but lead absorbs more strongly so that the line has an appreciable slope.

Note 5. The full precautions regarding handling and protection which have already been discussed must be adopted. Decontamination with phosphate carrier must be done immediately, before proceeding to counting, to avoid active solutions drying on the apparatus.

Note 6. The metal shield is a 3 mm thick sheet of brass, the function of which is to protect the counter from unnecessary exposure to very high activities. All the time the EHT is on and the source is in position, the counter is responding, whether or not the counts are recorded on the scaler, and there may be after effects giving false results if exposure to high activities precedes a low count. Also, the life of the counter is shortened unnecessarily. The shield must be in position under the counter at all times except when actually counting and especially before an absorber or a source is changed.

Note 7. Counting rates should not exceed 30 000 cpm and should be less than 10 000 cpm for accuracy. If counting rates are too high, the source should be placed on a lower shelf or reduced in activity.

Note 8. Suitable thicknesses for preliminary investigation are, approximately, 50, 100, 200, 300, 400, 500, 550, 600, 650, 700, 725, 750, 775, 800, 825, 850, 875, 900, 1000, 1300, and 1700 mg/cm^2.

Note 9. The purpose of using the stronger source is to avoid unduly long counting times when the count from the weaker source becomes very low. The ratio of activities of the two sources can be calculated at each of the three absorber thicknesses and a mean ratio obtained, by which all subsequent counts from the stronger source are divided. All counts can then be plotted on one graph as if a single (weaker) source had been used throughout.

Remember that, for accuracy, total counts of at least 3000, and preferably 10 000, should be accumulated if time permits.

Note 10. Calculate corrections and plot the points as you go so that you can decide what additional points are required and if any results are anomalous.

Note 11. The nominal absorber thickness must be corrected to allow for the absorption in the source seal, air gap and counter window. The thicknesses of these must be added to the nominal thicknesses before plotting.

Counter windows: consult the data sheet for the counter used.

Air: 1.3 mg/cm²/cm depth.

Source seal: cellulose tape, 10 mg/cm².

Experiment 4. *The determination of the maximum energy of β-particles*

Determination of the maximum energy (E_{max}) of the β-particles emitted in a nuclear transformation gives some indication of the identity of the radioactive material, although this information cannot alone give a positive identification. Precise energy determinations are done using a β-spectrometer but good approximations are obtained using absorption data. The visual range cannot be used for energy determinations since it is usually appreciably less than the true range.

The most rapid method is to use the *half-thickness*, i.e. the thickness of absorber required to reduce the intensity to one half of its original value, but the accuracy is relatively poor, about 5–10% with care.

A more accurate method (2–5%) is to determine the maximum range (R_{max}) of the β-particles using the technique of *Feather analysis* on the absorption curve data. The value of E_{max} can then be determined graphically using the range/energy curve (Fig. 278) or it can be calculated using the equations of Glendenin:

$$R_{max} = 542E_{max} - 133 \text{ mg/cm}^2 \tag{1}$$

for $R_{max} > 300$ mg/cm² or $E_{max} > 0.8$ McV, and

$$R_{max} = 407E_{max}^{1.38} \text{ mg/cm}^2 \tag{2}$$

for lower energies.

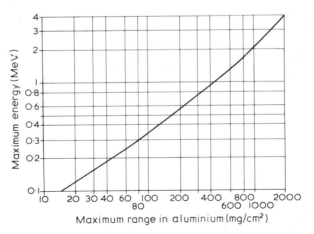

Fig. 278. The relationship between the range of β-particles in aluminium and their energy

Feather analysis involves comparison of the observed absorption curve with that of a known standard and it is preferable to check an unknown using an authentic curve of the radionuclide which it is believed to be, since the two curves should then have the same shape. However, it is common to use the absorption curve of ^{210}Bi (RaE, E_{max} 1.17 MeV), which gives a good enough approximation and is a useful standard, since it decays only slowly with an apparent half-life of 22.2 years. The ^{210}Bi is used as a secular equilibrium mixture of ^{210}Pb, ^{210}Bi and ^{210}Po (Ra D, and E and F respectively, often referred to as a RaDEF source).

$$^{210}\text{Pb} \xrightarrow[\text{22·2 years}]{\text{Weak } \beta} {}^{210}\text{Bi} \xrightarrow[\text{5·0 days}]{\text{1·17 MeV } \beta} {}^{210}\text{Po} \xrightarrow[\text{138·4 days}]{\text{5·3 MeV } \alpha} {}^{206}\text{Pb}$$

(Stable)

Method 1. Half-thickness technique. (*a*) Place the weaker ^{32}P source prepared for Experiment 3 in the castle so that the counting rate without added absorber is about 10 000 cpm. Count for 5 min.

(*b*) Count for 5 min with an aluminium absorber of approximately 80 mg/cm² in shelf 1. Repeat with an absorber of approximately 150 mg/cm².

(*c*) Correct all counts for paralysis and background and plot against the total absorber thickness (Note 1) on semi-logarithmic graph paper. Determine the thickness required to reduce the unshielded count by one half, and determine the E_{max}, using the half-thickness/energy curve (Fig. 279).

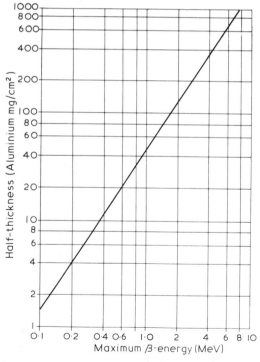

Fig. 279. The relationship between the energy of β-particles and their half-thickness in aluminium

Method 2. Feather analysis. (*a*) With the standard ^{210}Pb sources provided (Note 2) plot the absorption curve in aluminium, using the identical arrangement to that used in Experiment 3 (Note 3).

(*b*) Plot the corrected counting rate against absorber thickness on graph paper identical with that used for Experiment 3, and using the same scales. Subtract the Bremsstrahlung contribution and draw the corrected curve, extrapolating to 501 mg/cm^2 (Note 4) and to zero total absorber thickness.

(*c*) Divide the abscissa into fractions of one tenth of the range, i.e. at intervals of 50.1 mg/cm^2, and project lines from each of these intervals to the curve and then to the ordinate. This gives the ordinate values corresponding to each decimal fraction of the range (Fig. 280A).

(*d*) Lay a piece of thin card alongside the ordinate and mark on it the intercepts corresponding to zero total absorber thickness and each decimal fraction of the range.

(*e*) Place the Feather analyser so produced alongside the ordinate of the ^{32}P absorption curve determined in Experiment 3, aligning the two points corresponding to zero total absorber thickness, and mark off the fractions of range on the ordinate (Fig. 280B).

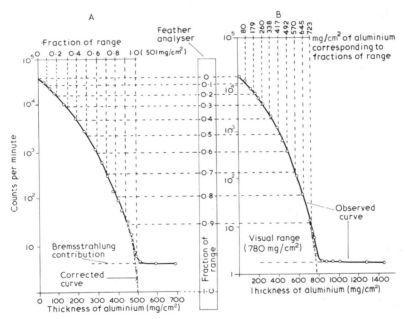

Fig. 280. The determination of β-particle range by Feather analysis. A, standard absorption curve for ^{210}Pb; B, Feather analyser and the absorption curve of a radio-nuclide of unknown β-energy

(*f*) Project lines from each of these points to the ^{32}P absorption curve and from there to the abcissa, giving the absorber thickness corresponding to each decimal fraction of range (Table 56, column 2).

(*g*) Calculate the apparent ranges (Table 56, column 3) by dividing each thickness by the corresponding fraction of range and plot the apparent ranges against the fractions of range (Fig. 281). Extrapolate the curve to unit range to obtain the *Feather range* (Note 5).

(*h*) Determine the maximum energy of the ^{32}P β-particles by calculation from equation (1) and from the range energy curve (Fig. 278).

Table 56. Calculation of Ranges for Feather Analysis

Fraction of range (A)	Absorber thickness (mg/cm²) (B)	Apparent range (mg/cm²) (B/A)	Feather range* (mg/cm²)
0.1	80	800	
0.2	179	895	
0.3	260	867	
0.4	338	845	802
0.5	417	834	
0.6	492	820	
0.7	570	813	
0.8	646	808	
0.9	723	803	

* Determined by extrapolation of the curve in Fig. 281.

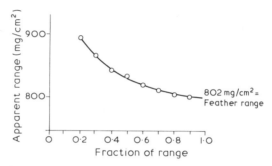

Fig. 281. Feather plot for the data from Fig. 280 and Table 56. The relationship of the apparent range to the fraction of range

Note 1. Allow for absorption due to the source seal, the air gap and the counter window.

Note 2. The RaDEF sources should be sealed with 0.005 in aluminium foil (34 mg/cm²) to absorb the low energy β-particles from the ^{210}Pb and the α-particles from the ^{210}Po, so leaving only the 1.17 MeV β-particles from the ^{210}Bi. If this is not done the shape of the curve will be complicated owing to the other components.

Note 3. All the details of the experimental set-up, e.g. source mounts, relative positions of source, absorbers and counter, must be identical for this method to be valid. Also the same graph paper and scales must be used. If any variations are made, the method as given above cannot be used, although allowance could be made for such variations of technique by suitable calibration and calculations.

Note 4. This is the known R_{max} of the ^{210}Bi β-particles.

Note 5. The curve obtained tends to become parallel to the abscissa and should be extrapolated accordingly.

Experiment 5. *The backscattering of β-particles by aluminium and other metals*

When β-particles pass through matter they are scattered by collision with orbital electrons. The extent of scattering, including backscattering, increases with atomic number, and this relationship is shown in Fig. 282.

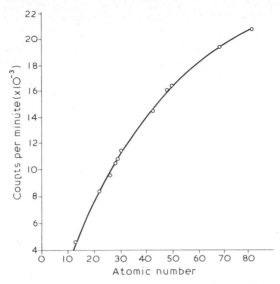

Fig. 282. The dependence of the extent of backscattering of β-particles on atomic number (see Experiment 5)

If a backing material is placed behind a source, the observed counting rate increases with the thickness of the backing and reaches a maximum at the *saturation thickness*, which is about one fifth of the maximum range of the particles in the backing material (Fig. 283). The proportion of particles backscattered is almost independent of energy above 0.6 MeV.

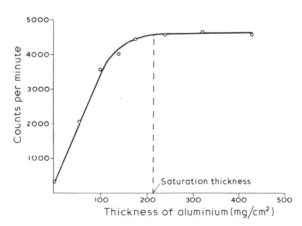

Fig. 283. The relationship between the extent of backscattering of β-particles and the thickness of the backscattering material

Backscattering effects are important in protection problems and in measurement. Changing the material of a source mount may change a count appreciably.

The method used in this experiment is similar to one of the techniques used industrially to measure film thicknesses, e.g. the thickness of a metallic foil on a plastic, of a plastic coating on a metal or of a corroded layer. The arrangement of the apparatus is shown in Fig. 284.

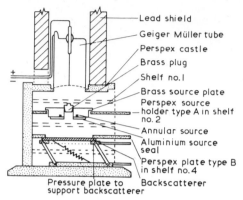

Fig. 284. The arrangement of the apparatus for backscattering experiments

Method (*a*) *Source preparation.* Pipette 0.1 ml of 0.1 % Teepol into the annular channel (Note 1) of the special source disc (Fig. 285). Pipette carefully and evenly into the channel about 2 μCi of ^{32}P in 0.1 to 0.2 ml (Note 2). Dry under the heat lamp and allow to cool. Seal with cellulose tape, leaving the central hole clear.

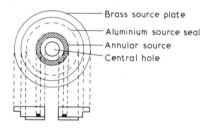

Fig. 285. Source disc for backscattering experiments

(*b*) Place the source, seal downwards (Note 3) and with a metal plug in the central hole, on shelf 2 and a Perspex retaining plate (Fig. 274c) on shelf 4. Count the background for 10 min (Note 4).

(*c*) Remove the central plug and count again in the absence of added backscatterer.

(*d*) Take counts with about eight thicknesses of aluminium in the range 50 to 450 mg/cm^2, placing each foil so that it is held firmly against the bottom of the Perspex retaining plate (Note 5).

(*e*) Count also with other metal backscatters available (Note 6) and with a 6 mm thick Perspex sheet.

(*f*) Place the source in the container provided. Correct all counting rates for paralysis and background and plot on linear graph paper:

  (i) cpm against thickness of aluminium, and
  (ii) cpm against atomic number.

Determine the saturation thickness of aluminium and comment on the results.

Note 1. This is to ensure wetting of the channel and so assist even distribution of the source.

Note 2. Care is required to avoid contaminating the surface of the disc. Even distribution can be checked by autoradiography as follows. Cover the source with cellophane film and place, annular source downwards, on a piece of Ilford G radiography film. Expose for 1 hr, develop for 8 min in ID19 developer, wash for 1 min and fix for 10 min. Wash and dry. All operations must be done under a suitable safelight, e.g. No. 6B.

Note 3. The activity of the source is such that the operation of the counter may be impaired if it is exposed directly to the source, owing to hysteresis effects. It is essential that the source is directed away from the counter.

Note 4. The background in this case consists of the natural background plus the Bremsstrahlung generated by the absorption of the β-particles in the source disc.

Note 5. For satisfactory results there should be a constant distance between the source and the surface of the backscatterers.

Experiment 6. *The absorption of γ-radiation and the determination of its energy*

The energy of γ-radiation is lost by three main processes, photo-electric absorption, Compton scattering and pair production. However, the net effect is that there is an exponential loss of intensity (Fig. 286) and this relationship is

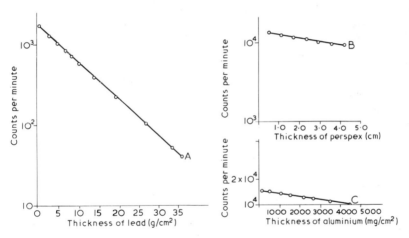

Fig. 286. Absorption curves for the γ-radiation from ^{137}Cs in lead, aluminium and Perspex

expressed by the equation,

$$\psi_x = \psi_0 \, e^{-\mu x} \tag{3}$$

where ψ_0 is the incident radiation flux, ψ_x is the radiation flux after traversing a thickness of absorber x, and μ is the *linear absorption coefficient*. The form of the relationship shows that, unlike β-particles, γ-radiation does not have a finite range, and so can only be attenuated, but not completely stopped, by matter.

The magnitude of the linear absorption coefficient decreases with increasing energy of the radiation and increases as the atomic number of the absorbing material increases. From eq. (3),

$$\mu = \frac{\log_{10} \psi_0 - \log_{10} \psi_x}{0.4343x} \tag{4}$$

The *half-thickness* (half value layer, HVL, $x_{1/2}$) is that thickness of material required to reduce the intensity of the incident radiation flux by one half. A knowledge of the half-thickness in lead can be used to determine the energy of the radiation and to calculate the thickness of shielding required for protection from a source (see Fig. 287).

$$x_{1/2} = \frac{0.693}{\mu} \tag{5}$$

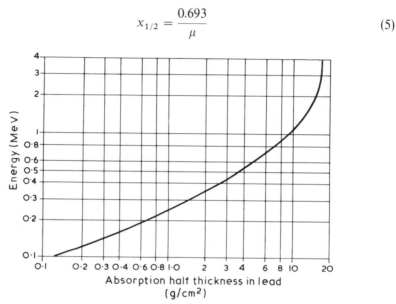

Fig. 287. The relationship between the energy of γ-radiation and its absorption half-thickness in lead

This method is satisfactory for determining γ-energy when only a single γ-ray is emitted or when the energy of one γ-ray predominates over a number of low energy ones. If several γ-rays are emitted only the average energy will be determined. The energies of mixed γ-rays can be determined precisely using a scintillation γ-spectrometer (see Experiment 13).

The half thickness can also be used to determine the thickness of shielding required for protection from a source, e.g. assuming we have γ-radiation of 0.7 MeV and a dose rate of 10 mrem/hr at the working area, we need to reduce the dose rate to less than 0.1 mrem to be well within the maximum permissible dose level for whole body exposure. From Fig. 287 we find that the half-thickness of lead is 7 g/cm². The dose reduction required is 100 (10/0.1) and this is produced by about 7 half-thicknesses ($2^7 = 128$), which will err on the side of safety. Thus, the thickness of shielding required is 49 g/cm² of lead. Lead 1 cm thick gives 11.3 g/cm², so the thickness of lead required is 49/11.3 = 4.34 cm. Therefore, a 5 cm (2 in) lead wall should be adequate. Note that having built the wall and set up the experiment, it is essential to monitor with a dose rate meter (e.g. Eberline RO2, Nuclear Enterprises 0500) to ensure that the actual dose rate is safe and is as calculated.

Method. (*a*) *Source preparation.* Pipette 10 μCi of the γ-emitting radionuclide provided (Note 1) into a dimpled planchet, dry under the heat lamp and seal with cellulose tape. Decontaminate the apparatus.

(*b*) Place the source in shelf 7 of the castle (Note 2) and a suitable thickness of aluminium absorber in shelf 6 (Note 3).

(*c*) Take counts with increasing thicknesses of lead absorber on top of the aluminium (Note 4).

(*d*) Correct all counts for paralysis and background and plot the corrected counts against the thickness of lead on semi-logarithmic graph paper.

(*e*) Repeat steps (*b*) to (*d*) using first the thick aluminium absorbers and then the thick Perspex absorbers (Note 5).

(*f*) Calculate the linear absorption coefficients for lead, aluminium and Perspex and the corresponding half-thicknesses, using equations (ii) and (iii). From the half-thickness in lead determine the γ-energy using the half-thickness/energy curve (Fig. 287). Compare the results with those obtained in Experiment 3.

Note 1. Suitable radionuclides with mono-energetic γ-radiation are ^{51}Cr, ^{54}Mn, ^{89}Sr, ^{137}Cs and ^{198}Au. ^{42}K (half-life 12.5 hr) can also be used, but if the experiment extends over an appreciable time, allowance may have to be made for radioactive decay. ^{198}Au can also be used in the form of irradiated gold foil.

The dose from the stock solution used may be appreciable, so the dose-rate should be monitored and appropriate shielding and handling techniques used. It is assumed that an end-window Geiger-Müller counter will be used. If a more sensitive detector is available, e.g. a sodium iodide scintillation counter, the source activity may be reduced to 0.1 to 0.5 μCi with corresponding advantages in economy and safety.

Note 2. Good 'geometry' requires the absorbers to be close to the counter and the source some distance from it, to minimise interference from scattered photons. These conditions can only rarely be achieved and this is one of the principal sources of error in the experiment.

Note 3. It is required to count only the γ-radiation from the source so it is essential to absorb any associated β-radiation to avoid errors due to the simultaneous absorption of β-particles. Consult the radioisotope tables to find the E_{max} of the β-particles from the source provided and select a suitable thickness of aluminium using the range/energy curve (Fig. 278).

Note 4. Appreciable thicknesses of lead are required. Use absorbers thicker than 5 g/cm^2 and increase to the maximum thickness that can be fitted into the available space. Thicknesses are merely additive, e.g. a thickness of 10 g/cm^2 may be given by a single absorber or by 2 or 3 absorbers whose thicknesses add together to make 10 g/cm^2.

Note 5. Aluminium and Perspex sheets approximately 6 mm thick are most suitable and the thickness of each sheet should be measured with a micrometer before use. Note that, although superficial densities *or* linear thicknesses can be used for plotting γ-absorption curves, linear thicknesses must be used for calculating linear absorption coefficients. The use of superficial densities gives the *mass absorption coefficient* (μ/ρ).

Experiment 7. *The statistics of counting*

The statistics of pulse counters (*scalers*). Radioactive decay is a random process and, even when the rate of decay is negligible, there is a continual fluctuation in the number of nuclei disintegrating in unit time. If a number of observations of the counting rate are made, there is a variation about the mean value and, if the sample is large enough, the frequencies with which the different values occur can be shown to have a *Poisson distribution* about the mean. This distribution is markedly asymmetric when the mean is small, but when the value of the mean is 20, the Poisson distribution closely resembles the *normal distribution* (Fig. 288), and this agreement increases as the mean increases.

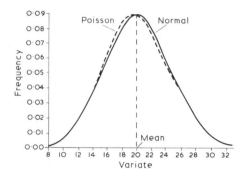

Fig. 288. Normal and Poisson distributions with a mean value of 20

Since the magnitude of counts in work with radioisotopes is rarely less than 100, the normal distribution is a good representation of the actual situation. This is advantageous, since the normal distribution is the simpler to handle mathematically, because it is symmetrical. The most probable value is the mean ($\bar{x}$) and positive and negative deviations from $\bar{x}$ of a given magnitude are equally likely, with a greater probability of small deviations than large ones. The mean is the best estimate that can be obtained of the true value, which cannot be determined directly.

The scatter of results about the mean is expressed in terms of the *standard deviation* (σ), the square root of the *variance* (V):

$$\sigma = \sqrt{V} = \left\{\frac{\Sigma (x - \bar{x})^2}{N - 1}\right\}^{\frac{1}{2}} \tag{6}$$

where the values of x are those observed and N is the total number of observations. For a large series of observations 68.3% lie within the range $\bar{x} \pm \sigma$ and 95.5% within the range $\bar{x} \pm 2\sigma$ (Fig. 289). For large values of x, a good approximation is given by the expression:

$$\sigma = \sqrt{x} \tag{7}$$

or, for a series of observations,

$$\sigma_{\bar{x}} = \sqrt{(\bar{x})} \tag{8}$$

The magnitude of σ varies with that of $\bar{x}$. If the variability of several sets of data with different means are to be compared, the effect of the magnitude of $\bar{x}$ can be eliminated by calculating the *coefficient of variation* (cv),

$$cv = \frac{100\sigma}{\bar{x}} \%, \tag{9}$$

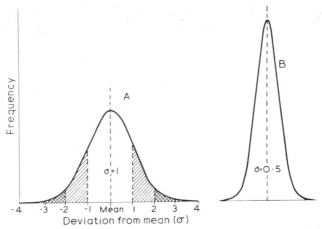

Fig. 289. Normal distributions with the same mean but different standard deviations (σ). (A) larger σ; the unshaded area is 68.3% of the total area under the curve, the unshaded area plus the single-shaded area is 95.5% of the total, the cross-hatched area is 4.5% of the total. (B) smaller σ

and the various coefficients of variation can be compared directly. The magnitude of σ (or cv) defines the precision of the determination: the greater the accuracy the lower the value of σ (Fig. 289). The relationship between the coefficient of variation and the observed count is shown in Fig. 290, from which it is clear that the precision of a count increases substantially with its magnitude.

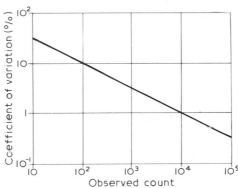

Fig. 290. The relationship between an observed count and its coefficient of variation

The *mean deviation* (*md*, sometimes called the average error), is the sum of all deviations, regardless of sign, divided by the number of observations, i.e.

$$md = \frac{\Sigma |x - \bar{x}|}{N} = 0.798\sigma \qquad (10)$$

The *probable error* (*pe*) is that error which has a 50% probability of occurrence, i.e. 50% of (a large number of) observations lie within the range $\bar{x} \pm pe$, and

$$pe = 0.6754\sigma \qquad (11)$$

The parameters described above (σ, md, cv, pe) may be used to determine whether a counting set is behaving properly and if a change has improved its performance or has impaired it. They are also useful in comparing replicate observations and in defining the precision of determinations.

The counting rate for a sample always includes the background counting rate and the net counting rate due to the sample (R_{s-b}) is the difference between the total observed counting rate (R_s) and the background counting rate (R_b), i.e.

$$R_{s-b} = R_s - R_b \qquad (12)$$

In these circumstances the precision of the determination of R_{s-b} is influenced by that of R_b, the variances being additive according to the equation,

$$\sigma_{R_{s-b}} = (\sigma_{R_s}^2 + \sigma_{R_b}^2)^{\frac{1}{2}} \qquad (13)$$

or substituting from equation (7),

$$\sigma_{R_{s-b}} = (R_s + R_b)^{\frac{1}{2}} \qquad (14)$$

It is clear that this has little importance when R_b is small relative to R_{s-b}, but when the two counts are similar the precision of a determination is affected markedly, and prolonged counting times are required for accuracy. The most efficient distribution of counting time (to minimise the error) between background and sample is given by the relationship

$$\frac{t_s}{t_b} = \left(\frac{R_s}{R_b}\right)^{\frac{1}{2}} \qquad (15)$$

where t_s and t_b are the counting times for sample and background respectively. If the counting times are apportioned in any other way, the precision of the estimate of R_{s-b} is reduced.

A more rigorous test of the performance of a counting set is given by the *index of dispersion*, which is computed as the statistic 'chi-squared' (χ^2).

$$\chi^2 = \Sigma \frac{(x - \bar{x})^2}{\bar{x}} \qquad (16)$$

Each value of χ^2 is associated with a number of *degrees of freedom* (F). For a Poisson distribution, with which we are concerned here,

$$F = N - 1$$

where N is the total number of observations, since one degree of freedom is lost in specifying the mean, which defines the distribution. Table 57 gives the distribution of χ^2 and shows the probabilities (P) with which the values of χ^2 could occur by chance. Acceptable operation is indicated by values of P in the range 0.05 to 0.95, i.e. the observed value of χ^2 could occur by chance on between 5 and 95% of occasions in which it is determined. Values of P less than 0.05 may be taken to indicate faulty equipment. Values greater than 0.95 may indicate that some factor is operating to suppress the expected

Table 57. Values of χ^2 according to Probability and Degrees of Freedom

Degrees of Freedom	Probability*						
	0.99	0.975	0.95	0.10	0.05	0.025	0.01
1	0.0002	0.001	0.0039	2.71	3.84	5.02	6.64
2	0.02	0.05	0.10	4.61	5.99	7.38	9.21
3	0.12	0.22	0.35	6.25	7.82	9.35	11.35
4	0.30	0.48	0.71	7.78	9.49	11.14	13.28
5	0.55	0.83	1.15	9.24	11.07	12.83	15.09
6	0.87	1.24	1.64	10.65	12.59	14.45	16.81
7	1.24	1.69	2.17	12.02	14.07	16.01	18.48
8	1.65	2.18	2.73	13.36	15.51	17.53	20.09
9	2.09	2.70	3.33	14.68	16.92	19.02	21.67
10	2.56	3.25	3.94	15.99	18.31	20.48	23.21
12	3.57	4.40	5.23	18.55	21.03	23.34	26.22
14	4.66	5.63	6.57	21.06	23.69	26.12	29.14
16	5.81	6.91	7.96	23.54	26.30	28.85	32.00
18	7.02	8.23	9.39	25.99	28.87	31.53	34.81
20	8.26	9.59	10.85	28.41	31.41	34.17	37.57
30	14.95	16.79	18.49	40.26	43.77	46.98	50.89
40	22.16	24.43	26.51	51.81	55.76	59.34	63.69
50	29.71	32.36	34.76	63.17	67.50	71.42	76.15
60	37.48	40.48	43.19	74.40	79.08	83.30	88.38
70	45.44	48.76	51.74	85.53	90.53	95.02	100.4
80	53.54	57.15	60.39	96.58	101.9	106.6	112.3
90	61.75	65.66	69.13	107.6	113.1	118.1	124.1
100	70.06	74.22	77.93	118.5	124.3	129.6	135.8

* The probability (P) with which the observed value of χ^2 could occur by chance: total probability, 1.0. E.g. if $\chi^2 = 18.31$ for 10 degrees of freedom, $P = 0.05$, i.e. this result could occur by chance on 5 occasions in 100 (5%). By convention, an occurrence with a probability of < 0.05 is unlikely to occur by chance.

variability of the data, possibly that the results are being 'cooked'. However, if a large number of determinations of χ^2 are made, it must be expected that values with a very low or a very high probability will occur occasionally, even when the apparatus is functioning normally.

The statistics of integrating counters (ratemeters). The uncertainty in the counting rate measured by a ratemeter derives not only from the random decay process, but also from the dependence of an observed rate on that immediately preceding it and the difficulty of reading the meter accurately. A counting rate may be determined as a single instantaneous reading, as a mean of several successive readings or, more satisfactorily, as a mean based on a continuous recording obtained over a period.

Readings are taken when the steady state output has been reached. The time required for this depends upon the magnitude of the change in rate, the new average counting rate and the *integrating time* (I) of the integrating circuit. If the initial rate is zero, the time (t) required to reach the steady state counting rate (R) for a constant source S given by the expression

$$t = I[0.394 + 0.5 \ln 2RI] \qquad (17)$$

The steady state is approached exponentially and this value of t corresponds to approaching the steady state to within less than one probable error.

The standard deviation of a single observation is related inversely to the square root of the integrating time, since if the latter is large more pulses are integrated over a longer time, resulting in greater precision. However, the larger the integrating time, the longer is the time required to reach the steady state and the time required for a reading to decline. Because of this 'memory' effect, replicate observations of the counting rate are not truly independent unless they are separated by intervals of time at least five times the value of the integrating time.

The benefits both of the quick response given by a short integrating time and of the greater accuracy afforded by a longer integrating time can be obtained by the following procedure. Switch to the shortest integrating time available and observe the maximum and minimum deflections of the pointer and so its median value. When the pointer is at its median value, switch to the next higher integrating time, allow to stabilise and repeat the procedure. Continue in this manner until the longest integrating time is reached.

The integrating time is also known as the *time constant*.

Experiment 7(a). *An examination of some aspects of the statistics of pulse counters*

Method. (*a*) Using the reference source prepared in Experiment 1 and an autoscaler (Note 1), take multiple replicate counts without disturbing the apparatus in any way (Note 2).

(*b*) Repeat the above procedure using a much shorter counting time (Note 3).

(*c*) Repeat section (*a*) using the same counting time but a source giving about 100 counts in that time.

(*d*) If time permits, repeat sections (*a*) and (*b*) using manual operation of the scaler in place of the autoscaler facility.

(*e*) In each case determine the following:
- (i) the mean count ($\bar{x}$)
- (ii) the standard deviation (σ)
- (iii) the coefficient of variation (cv)
- (iv) the mean deviation (md)
- (v) md/σ
- (vi) $\sigma/\sqrt{(\bar{x})}$
- (vii) the percentages of observations falling within the ranges $\bar{x}\pm\sigma$ and $\bar{x}\pm2\sigma$ (Note 4)
- (viii) χ^2 and the probability of its occurrence.

(*f*) Compare the results from each section and comment on your findings and on the operation of the equipment.

(*g*) Count the background and calculate
- (i) the standard deviation of the background counting rate,
- (ii) the standard deviation of the counting rate from the weaker source used in section (*c*), corrected for background,
- (iii) the best distribution of counting time between the background and the weaker source.

Experiment 7(b). *Statistical aspects of counting ratemeters*

Method. (*a*) Zero the meter of the instrument (Note 5).

(*b*) Using the same counter as in Experiment 7(a), determine the background counting rate and the time to reach the background steady state at each setting of the integrating time.

(*c*) Place the reference source in the same position as in Experiment 7(a) (section *a*) and a 1500 mg/cm² absorber between the source and the counter (Note 6). Select the appropriate range setting (Note 7) and the shortest integrating time.

(*d*) Remove the absorber and determine the counting rate and the time required to reach the steady state. Replace the absorber and determine the time required for the meter reading to reach zero.

(*e*) Repeat section (*d*) at each of the integrating time settings available. If a recorder is available make a record of each of the situations.

(*f*) Make ten replicate observations of the steady state count at the highest integrating time setting, calculate the standard deviation and compare with that obtained for the same source in Experiment 7(a), using ten consecutive counts taken at random from those available.

(*g*) Comment on your results.

Note 1. An autoscaler is advisable for this experiment to eliminate systematic and progressive errors in switching time due to the operator and to permit the accurate use of relatively short counting times. The apparatus must be allowed adequate time to stabilise at the operating conditions before starting to count.

Note 2. Each count should be at least 3000 and preferably about 10 000. A minimum of 50 counts should be done and 100 counts is preferable. To avoid errors arising from long term drift in the apparatus, or from changes in geometry, all the counts should be done in one block without interruption. It is not necessary to correct for paralysis or background.

Note 3. If possible the new time should be less than one tenth of the original time. If a time interval $\frac{1}{100}$th of the original one can be used and the original counting time gives counts of about 10 000, a good comparison with the results from the weaker source will be obtained directly. If the time available is limited, this and subsequent sections of the experiment can be done using 10 replicate counts for each and comparing the results with those from a block of 10 chosen at random from those done under section (*a*).

Note 4. Omit this if less than 25 observations have been made.

Note 5. Ensure that the ratemeter has been allowed to stabilise and carry out any checks recommended by the manufacturer. It is preferable to check the zero before making any observation.

Note 6. This absorbs all the β-particles from the source and permits the meter zero to be obtained. The starting time for an observation is when the absorber is removed (or replaced if decline is to be observed). The absorber is not required if a meter switch is fitted, when starting times are those when the meter is switched in and out of circuit.

Note 7. Decide on the correct setting from the results in Experiment 7(a). If a counting rate is unknown, always find the appropriate setting by starting at the highest counting rate, integrating time settings available, and working downwards, so that the meter is protected from overloading.

Experiment 8. *The determination of the paralysis time of a Geiger-Müller counting set-up*

The phenomenon of paralysis in a Geiger-Müller counter has been described above and the relationship by which corrections for paralysis are made is,

$$N = \frac{M}{1 - MT} \tag{18}$$

where N is the true counting rate, M is the observed counting rate and T is the paralysis time. However, paralysis times vary markedly between different types of counter and also between individual counters of one type and with the characteristics of the associated circuit. The use of a quenching probe unit overcomes the variations by introducing a known preset time. Tables 54 and 55 give the corrections calculated according to eq. (18) for a paralysis time of 400 μsec, which is the standard time used for setting up most probe units initially.

It is important to realise that eq. (18) assumes that pulses are evenly spaced, whereas radioactive disintegrations, and so the pulses derived from them, are randomly distributed. Thus the corrections given by the equation are only approximate. The errors introduced at low counting rates are small, but at high counting rates there is an appreciable discrepancy. Geiger-Müller counters should not be used at counting rates greater than 10 000 cpm if accurate results are required.

Methods of determining paralysis times

The method of paired sources uses two sources which are held in a jig so that the position of each relative to the counter and each other is fixed accurately, and each gives a counting rate of 5000 to 10 000 cpm. Counts are made with each source in turn and with both together. The paralysis time can then be calculated using the approximate equation,

$$T = \frac{R_x + R_y - R_{xy} - R_b}{R_{xy}^2 - R_x^2 - R_y^2} \tag{19}$$

where R_x and R_y are the counting rates from each of the sources, R_{xy} is the counting rate of the two sources combined and R_b is the background counting rate. In practice, this method tends to give unsatisfactory results unless prolonged counting times are used, since the method relies on measuring small differences between large numbers and the variance of such an estimate is very large.

The method of incremental sources utilises the fact that, if a series of radioactive sources of increasing concentration are made, the relationship of the observed counting rate to the concentration is of the form shown by curve ABC in Fig. 291. If the counting rates from the weakest of the sources are low enough, there will be a negligible loss due to paralysis and the theoretical relationship between the true counting rate and concentration is given by the line AD, a straight line drawn through the initial points on the graph and extrapolated. For each concentration, a theoretical true counting rate can be determined using the line AD, the observed counting rate is known, and the paralysis time can be calculated using the equation

$$T = \frac{N - M}{NM} \tag{20}$$

which is derived readily from equation (18).

Although end-window counting can be used for this purpose it is difficult to prepare the sources with sufficient accuracy and the method is better adapted to liquid counters, which have an accurately reproducible counting geometry.

Measurement of scaler delay times. For this a scaler is required which can provide a wide range of delay times in the range 1 to 10 000 μsec, and although scalers providing this facility are not generally available they may readily be

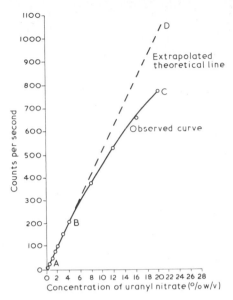

Fig. 291. The relationship of the observed counting rate to source concentration in a Geiger-Müller liquid counter. (ABC) observed curve; (AD) theoretical (extrapolated) line, assuming no paralysis

built locally using the principles of for example AERE Type 1009E. The delay time controls the minimum interval within which successive pulses from the amplifier are admitted to the scaling circuits. While the delay time setting is less than the paralysis time there is no effect on the counting rate, but if delay time settings longer than the paralysis time are used, the counting rate falls progressively as the delay time increases. A plot of counting rate against delay time yields a straight line parallel to the abcissa at low delay time settings and a straight line of negative slope as the delay time is increased (Fig. 292), the point of intersection being at the paralysis time of the counting apparatus.

Experiment 8(a). *The determination of the paralysis time of an end-window Geiger-Müller counter using paired sources*

Method. (a) Source preparation. Pipette 0.01 ml of the radioactive solution into the centres of each of two of the special mounts provided (Fig. 293), to give a counting rate of about 100 cps for each. Dry under the heat lamp. Decontaminate the pipette.

(b) Place the jig (Fig. 293) in the castle with two blank mounts in position (Note 1) and determine the background counting rate (Note 2).

(c) Remove one blank mount and place the first source in the jig and count for at least 10 min.

(d) *Without disturbing the source in any way,* remove the blank mount, place the second source in position and determine the combined counting rate.

(e) *Without disturbing the second source in any way,* remove the first source, replace with a blank mount and determine the counting rate from the second source.

(f) Calculate the paralysis time using eq. (19).

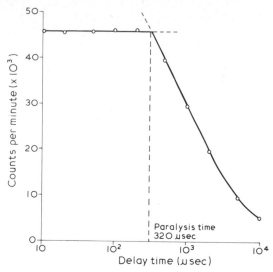

Fig. 292. The relationship between the observed counting rate and scaler delay time for an end-window Geiger-Müller counter

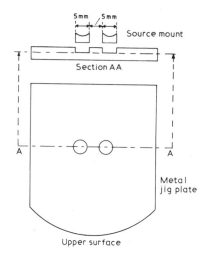

Fig. 293. Source mounts and jig for the determination of the paralysis time of an end-window Geiger-Müller counter by the method of paired sources

Experiment 8(b). *The determination of the paralysis time of a liquid counter by the method of incremental sources*

Method. (*a*) *Source preparation.* Using a solution of suitable activity (Note 3), transfer to a series of 10 ml graduated flasks the following volumes of solution: 0.10, 0.15, 0.3, 0.6, 0.9, 1.2, 1.5, 3.0, 4.5, 6.0 and 7.5 ml. Dilute each solution to 10 ml with the appropriate carrier solution, stopper and mix thoroughly (Note 4).

(*b*) Determine the background count (see Experiment 9), transfer the weakest solution to the counter and determine the counting rate, using a counting time sufficient to accumulate at least 3000 counts.

(*c*) Return the solution to its original flask, wash the counter three times with carrier solution and pour the washings into the liquid active waste container. Dry the counter, fill with the next stronger solution and determine the counting rate as before (Note 5).

(*d*) Repeat procedure (*c*), counting each solution in turn *in order of ascending concentration* and finishing with the undiluted solution. When the strongest solution has been counted, return it to its original flask, decontaminate the counter and determine the background counting rate.

(*e*) Plot the observed counting rate for each solution against its concentration and draw the curve. Assuming that paralysis time corrections are negligible below c. 150 cps, draw the best straight line representing the relationship between the counting rate and concentration below that level.

(*f*) Extrapolate this straight line, which gives the theoretical relationship in the absence of losses due to paralysis, and determine the predicted counting rate appropriate to each concentration. Calculate the paralysis time for each concentration using eq. (20) and determine the mean paralysis time.

Experiment 8(c). *The determination of paralysis times by a delay time technique*

Method. (*a*) Using a Geiger-Müller counter set-up incorporating a quenching probe unit and a scaler providing a suitable range of delay times, count a source giving a counting rate of approximately 8000 cps, using each available delay time setting. Plot the observed counting rate against delay time and determine the paralysis time. Count at least 3000 counts.

(*b*) Repeat the experiment omitting the quenching probe unit and determine the natural paralysis time of the apparatus.

Note 1. The blank mounts are used to ensure that scattering conditions are identical throughout the experiment.

Note 2. It is essential to accumulate as high a count as possible in each case, and at least 3000, as far as time allows. This results in long background counting times. It is convenient to start this count before commencing the preparation of the sources, or to determine the background during a meal break.

Note 3. A suitable solution, counted undiluted in the liquid counter gives a counting rate of about 750 cps. A convenient series of permanent standards can be provided using the following concentrations of uranyl nitrate: 0.25, 0.5, 0.75, 1.0, 2.0, 3.0, 4.0, 8.0, 12.0, 16.0 and 20.0% w/v.

Note 4. Gloves must be worn. When shaking the flasks, the stoppers should be wrapped in a tissue to trap any leakage, especially if glass stoppers are used. The tissue is monitored after use and placed in the active waste container if contaminated. The carrier must be the same element in the same chemical form as the source.

Note 5. Rigorous decontamination of the counter to background level between each solution is time consuming and unnecessary, if successive counts are made from solutions of increasing concentration and the procedure given is used. The residual contamination from the weaker solution is then negligible relative to the count from the stronger one.

Experiment 9. *Liquid Geiger-Müller counters and their use as sub-standard assay instruments*

The construction of the liquid Geiger-Müller counter is shown in Fig. 265B. The characteristics of such counters are generally similar to those of end-window counters. Halogen quenched counters have spiral wire cathodes of corrosion resistant wire, and those organically quenched have a carbon or metal film cathode. The anode contact may be made via a spring or a mercury cup and the cathode contact usually rests on an earthed metal support plate (Fig. 259). It is important not to drop the counter onto the support plate as even small shocks may damage the cathode seal and make the counter useless.

The annular space between the inner and outer walls of the counter is designed to hold (usually) 10 ml of a liquid source. Since there is a constant geometrical relationship between the source and the sensitive volume, these counters can be used as sub-standard assay instruments.

The use of liquid Geiger-Müller Counters. The counters are fragile and must be handled with care. They should always be carried in a suitable container, e.g. a plastic cylinder. They are placed in the castles or removed without switching off the EHT. This does not involve any electrical hazard and avoids errors which may result from constant switching or re-setting of the EHT. Counters containing active solutions should be handled with gloves and, when wearing the gloves, apparatus, instrument controls etc., must be handled via the medium of a tissue to avoid spreading contamination. Gloves used in the radiochemical laboratory must not be taken into the counting room. If it is desired to wear gloves in the counting room, these must be kept separately in the counting room for that purpose. A tissue should be folded and wrapped round the counter about 1 cm from the top when pouring into the counter or out of it, in order to trap any drips or creepage of liquid which could contaminate the outside of the counter and so the interior of the castle, thus causing unduly high backgrounds. The counter must be held firmly through the tissue and *not supported by the tissue alone.* The rim of the counter should be wiped with the tissue when the manipulations are complete and the tissue should then be monitored and, if contaminated, discarded into the active waste container. Tissues must be discarded immediately if they are seen to become contaminated at any stage in the manipulations.

Counter contamination is a major problem since the active material is adsorbed from samples onto the counter walls. Thus, counters must be decontaminated down to background level, i.e. 20 to 30 cpm or less, before any count is done, and it should be realised that the background count is subject to continual small changes and must be determined before each sample is counted.

The decontamination procedure involves repeated washing with an appropriate carrier solution, sequestering agents (sodium citrate, oxine, EDTA), acids, including nitrosulphuric or chromic acid in stubborn cases, or powerful detergents, e.g. RBS50. The decontamination solution is directed in a fine jet from a plastic wash bottle around the walls of the counter, using about 5 ml of solution and washing three times. The outside of the counter should also be wiped in a tissue soaked in the decontaminant. The counter is then washed with water and dried internally using filter or blotting paper strips, taking care not to break the strip and leave paper in the counter, and externally with a tissue. This helps to remove adherent material by mechanical action. The counting rate of the empty counter is then checked briefly and the procedure continued until the count is seen to be acceptably low. Finally, the counter is washed with water and dried, and the background counting rate is determined accurately. Counters should always be dried before counting the background to avoid any contamination of the subsequent solution with residual liquid and any change in background which might occur on drying. Obstinate contamination may require the counter to be filled with a powerful decontaminant solution and left to soak.

Method. (a) *Source preparation*
   (i) Place exactly 1 ml of the solution of ^{131}I provided in a standard vial (Note 1). Seal.
   (ii) Place the β-shield and the appropriate jig of the type 1383 ionisation chamber (Fig. 262) in position. If a standard source is available, check the calibration of the chamber.
   (iii) Determine the activity of the sample from the electrometer reading and the calibration chart.
   (iv) Dilute the sample to a specific activity of about 0.02 μCi/ml.

   (b) Set the correct working voltage for the counter (Note 2) and allow 5 min to stabilise. Count the background.

   (c) Place 10 ml of the standard solution of ^{131}I provided in the counter and make three replicate counts from this one filling of the counter (Note 3). Note the details of the solution from the label.

   (d) Pour the standard solution into the special container provided, decontaminate the counter with iodide carrier, dry and count the background.

   (e) Place 10 ml of the dilute solution of the sample of ^{131}I prepared in section (a) in the counter and repeat the operations given in sections (c) and (d) above.

   (f) Calculate the activity of the standard at the time of assay (Note 4) and the activity of the sample. Compare the results obtained by the two assay methods and comment on the applicability of the two instruments.

   Note 1. Proper health precautions must be taken, e.g. monitoring, lead protection, handling tools and remote control pipettes, as appropriate. The standard Radiochemical Centre vial is the normal vaccine bottle sealed with a rubber closure and an aluminium seal, and the security of this seal is preferable to using a rubber bung. Take care not to contaminate the lip or mouth of the vial when filling or the closure when counting and opening. Wipe the outside of the vial with a tissue soaked in iodide carrier and monitor the tissue to ensure the absence of external contamination.

   Note 2. Each counter is labelled with the relevant details.

   Note 3. For accuracy each count should be at least 10 000. Check for statistical accuracy by ensuring that the three counts lie within the range $\bar{x} \pm 2\sqrt{(\bar{x})}$ (see Experiment 7).

   Note 4. Obtain the requisite information from the tables of radio-isotope data.

Experiment 10. *The determination of the radiochemical purity of sodium iodide* (^{131}I) *solution*

Paper chromatography may be combined with autoradiography in a test for radiochemical purity, since impurities have R_F values which differ from that of the main component. The impurity may be present as a different chemical form of the main nuclide or as a different nuclide.

In this experiment the absence of Na ^{131}IO$_3$ from a solution of Na ^{131}I is checked by comparison of a chromatogram of the sample with those of known iodide and iodate solutions. It is important to ensure the absence of iodate since in tests of thyroid function, or in the therapy of thyroid conditions, carrier-free ^{131}I is used and the total mass of iodide taken up by the thyroid gland is very small. Any oxidation of iodide to iodate would cause a serious loss in the activity absorbed by thyroid tissue.

Method. (*a*) Cut a sheet of Whatman No. 1 chromatography paper 10 cm wide and of a suitable length to fit the apparatus available. Mark a line 2.5 cm from one end and on it mark three points, the first 2.5 cm from the edge of the paper and the others at 2.5 cm intervals (Note 1).

(*b*) Label the points 'sample', 'iodide' and 'iodate'. Rule lines equidistant between the points as guides for cutting the paper subsequently into three strips.

(*c*) On the sample point, place 0.01 ml of a solution containing 0.1 % w/v of potassium iodide, 0.2 % w/v of potassium iodate and 1 % w/v of sodium bicarbonate, keeping the spot as small as possible (Note 2). Dry thoroughly under the heat lamp.

(*d*) To the other points, apply similarly 0.01 ml of solutions containing 1 % w/v potassium iodide or 0.5 % w/v potassium iodate, as appropriate.

(*e*) Apply to the 'sample' spot 0.01 ml of the sample of radio-active sodium iodide, diluted to a specific activity of about 2 μCi/ml. Dry all the spots thoroughly.

(*f*) Place some 75 % v/v methanol in the chromatography jar, place the paper in the jar so that it hangs *above the level of the solvent*, and allow the paper to equilibrate with the solvent vapour for at least 15 min or until it hangs vertically. Immerse the edge of the paper in the 75 % v/v methanol to a depth of about 5 mm and develop the chromatogram by ascending chromatography for 4 to 5 hr (Note 3).

(*g*) Remove the paper from the jar, mark the position of the solvent front and dry rapidly. Cut into three strips (Note 4).

(*h*) Fold the sample strip between cellophane, place in a film holder with a strip of X-ray film, taking care to align the bottom edge of the chromatogram with the edge of the film (Note 5), and leave to expose for a suitable time (Note 6). Develop, fix and wash suitably for the film used, dry the film and mark on it the point of application and the solvent front.

(*i*) If suitable apparatus is available, count the activity in the sample chromatogram using a collimated thin end-window Geiger-Müller counter (Note 7).

(*j*) Spray the 'iodide' strip with acid iodate solution and the 'iodate' strip with acid iodide solution (Note 8). Mark the outlines of the spots as soon as they appear and dry the strips.

(*k*) Calculate the R_F values for the iodide, iodate and the sample. The radioactivity should appear in one spot only with an R_F value corresponding to that of iodide. Comment on your results.

Note 1. These points are those to which solutions are to be applied. Mark with a lead pencil, *not a ball point or ink*, and avoid undue handling of the paper.

Note 2. This is a carrier solution and is used to avoid failure of the active material to migrate on the paper owing to adsorption since the solution is, chemically speaking, extremely dilute.

The solution should be applied in three or four portions, drying between each application, to keep the spot small so that the spots resulting from development of the chromatogram will be as discrete as possible. The paper should be suspended in air to avoid loss of solution by contact with surfaces. This is especially important with the active solution.

Note 3. Ensure that the paper is immersed evenly in the solvent, that the jar is not disturbed during development and that the temperature remains reasonably constant; attention to these points will result in good flow characteristics and a straight solvent front.

Note 4. Gloves must be worn when handling the chromatogram to avoid the possibility of absorption of ^{131}I which has a high toxicity due to its highly selective localisation in the thyroid gland.

Note 5. This is to ensure that the positions of application and solvent front relative to the film are known.

Note 6. Suitable exposure times for fast X-ray film may be calculated by determining the time required to collect 2×10^7 counts/cm^2 of film.

Note 7. If radio-chromatogram scanning apparatus is not available, the chromatogram can be cut into 1 cm strips and the strips counted individually under a thin end-window counter. In this case the paper should be marked out at numbered 1 cm intervals from the origin before application of the solutions, to assist subsequent cutting. The developed chromatogram must be handled with gloves and strips collected on cellophane.

Note 8. The spraying solutions consist of 1 % w/v solutions of the substances, acidified just before use with some acetic acid. These strips are the reference chromatograms prepared using inactive iodide and iodate. *Do not spray the radioactive strip.*

Experiment 11. *The determination of the half-life of* ^{234m}Pa(UX$_2$)

The activity of a radionuclide decays exponentially with time, the situation being represented by the equation

$$A_t = A_0 \, e^{-\lambda t} \tag{21}$$

where A_0 is the original activity, A_t is the activity after a time t, e is the base of natural logarithms and λ is the *decay constant*. The value of λ expresses the probability of decay of a nucleus in unit time, is constant for a given radionuclide and is unaffected by external conditions. The decay curves for some common radionuclides are given in Fig. 294.

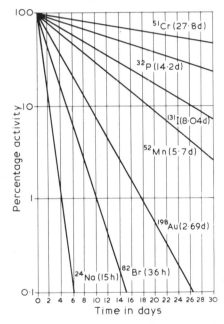

Fig. 294. Decay curves for some common radionuclides, with their half-lives

It is convenient to express the rate of decay in terms of the *half-life* ($T_{\frac{1}{2}}$), the time taken for the activity to decay to one half its original value. From eq. (21) we can derive that

$$T_{\frac{1}{2}} = \frac{0.693}{\lambda} \tag{22}$$

The half-life may be determined by counting at intervals and plotting the corrected counting rates against time on semi-logarithmic graph paper. The best straight line which fits the data is drawn and the half-life is calculated from the expression

$$T_{\frac{1}{2}} = \frac{0.301t}{\log A_0 - \log A_t} \tag{23}$$

This technique assumes that there is no appreciable contamination with other radionuclides, that $T_{\frac{1}{2}}$ is large compared with the time required to make an observation, that observations extend over several half-lives, and that the sample is chemically stable.

A simplified decay scheme for ^{238}U is given below.

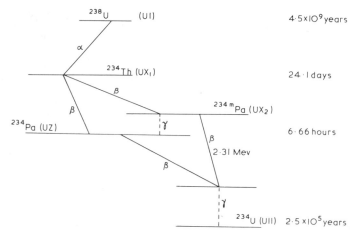

The object of this experiment is to separate the ^{234m}Pa (from ^{238}U and its other decay products) and to determine its half-life. Note that there are two forms of ^{234}Pa with identical mass numbers but different half-lives: these are *nuclear isomers*.

Method. (a) Using the automatic burettes, transfer 12 ml each of 40% uranyl nitrate in hydrochloric acid and isobutyl methyl ketone to a small separating funnel. Shake vigorously for about 5 min (Note 1).

(b) Separate the solvent layers as quickly as possible (Note 2), running the (upper) organic solvent layer into a liquid counting tube (Note 3).

(c) Place the counter in the castle, start a stopclock, and take as many as 10 sec counts as possible (Note 4) over a period of at least 5 min until the counting rate falls to about 20 cps.

(d) Wait a further 10 min and count the background (Note 2).

(e) Plot the corrected counting rate (Note 5) against time on semi-logarithmic graph paper, draw the best straight line which fits the points and determine the half-life of the material. Comment on the result.

Note 1. Wrap tissues round the stopper and tap of the separating funnel to catch any leaks. Gloves must be worn and solutions placed in the appropriate waste bottles.

Note 2. The correct volume for counting is 10 ml, which fills the counter to about 2 mm above the top of its central dome. 12 ml of each liquid is used to avoid the necessity of waiting for complete separation and the possibility of transferring some of the aqueous layer with the solvent. Should such contamination occur accidentally, it is allowed for by taking a background count when the ^{234m}Pa activity has decayed to a negligible level.

Note 3. The whole operation of separation and transfer should be completed within about 45 sec.

Note 4. An autoscaler is essential for successful counting. Note the actual time of the start of each 10 sec count using the stopclock.

It is convenient to adopt a definite routine of commencing each count exactly 15 sec after the commencement of the previous count. This gives 5 sec for noting the count, resetting the scaler and preparing to count. Although this is adequate, it is advisable to practice the technique of recording and switching before starting the experiment, using a test pulse.

Note 5. Paralysis time corrections may be determined using Table 55.

Experiment 12. *The characteristics of scintillation counters*

The basis of the scintillation counter is the conversion of the kinetic energy of an α- or β-particle, or of a secondary electron produced by the interaction of γ-radiation with matter, into a pulse of light which is converted into an amplified electrical pulse by a photomultiplier (PM) tube.

Scintillation counters are important detection instruments because they have a very short resolving time, and so can count at high rates. They are stable, and have a high efficiency of detection for X- and γ-radiation, or for low energy β-particles, depending on the system used. They may be used to determine particle or radiation energy, and simple techniques of sample preparation are usually adequate.

The general arrangement of the equipment is shown in Fig. 295.

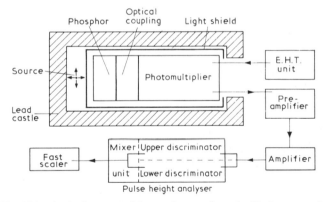

Fig. 295. Block diagram of the equipment for scintillation counting

The scintillation process. When ionising radiations traverse an absorbing material, electrons are excited to a higher energy state. This excitation energy is usually lost as heat, but in certain substances return of the electrons to their ground energy state is accompanied by the emission of light. Such substances are known as *phosphors*, those useful for scintillation counting being the ones in which the duration of the excited state is less than 10^{-6} sec.

Phosphors. Many types of phosphor are used according to the desired application (see also p. 510). They may be inorganic or organic, the delay in light emission being about 10^{-7} and 10^{-8} sec respectively. The size of the light pulse depends greatly on the purity and transparency of the phosphor, which should be of high optical quality.

The commonest inorganic phosphor is thallium-activated sodium iodide (Na I/Tl), which contains 0.1 % of thallium iodide and is used in the form of single, large crystals. Because sodium iodide is hygroscopic, crystals are sealed in an aluminium capsule with a glass or, occasionally, quartz face on the side in contact with the PM tube, the other sides being coated with magnesium oxide reflector. NaI/Tl is the phosphor of choice for the detection of X- and γ-radiations, since its high density results in a high absorption coefficient for such radiations. However, the encapsulation makes it unable to detect α-particles and low energy β-particles, and high energy β-particles are detected with low efficiency. It has a high light output and excellent optical properties.

Zinc sulphide activated with silver (ZnS/Ag) was the original scintillator but is of limited use, since large crystals are extremely difficult to produce, so this phosphor is used as a multi-crystalline material with poor optical properties. A very thin layer of ZnS/Ag on a plastic base is often used to detect α-particles, since, having a low efficiency for fast β-particles and γ-radiation, it will detect α-particles in the presence of high fluxes of these other radiations. Although the long delay in light emission (*ca.* 10^{-5} sec) precludes fast counting, such detectors are very suitable for use in monitoring instruments.

Organic phosphors are more efficient for the detection of β-particles. The standard organic phosphor is anthracene, which is used as the standard for light emission with the arbitrary value of 1.0. Large crystals of anthracene of good optical quality are difficult to grow so this material is often used in the form of thin slices mounted on a plastic base.

Other popular organic phosphors are *p*-terphenyl and 2,5-diphenyloxazole which retain their ability to act as phosphors in solution, e.g. in toluene or dioxan, and so can be used to count low energy β-particles with high efficiency if the source can be dissolved in the phosphor solution (see Experiment 14). Solid solutions of these phosphors in polystyrene or polyvinyltoluene plastics can be used in 'windowless' counters to detect weakly penetrating radiations with good efficiency. A combined α-β detector can be made using a thin layer of ZnS/Ag on a plastic phosphor base.

Light collection. To obtain good energy resolution, or to detect the very weak scintillations produced by low energy β-particles, it is essential to collect the maximum amount of light onto the photocathode of the PM tube. Reflectors are commonly used and light losses due to total reflection at the phosphor/ PM interface are minimised by using a thin film of silicone oil as a coupling fluid. A plastic light pipe coupling may also be used to conduct the light when it is necessary to separate the phosphor from the PM tube, but this gives a much reduced efficiency. Because the scintillations are extremely small and the PM is highly photosensitive, the phosphor and PM must be enclosed in a totally light-proof shield.

Photomultiplication. The photomultiplier acts as a high gain amplifier, the gain varying markedly with applied voltage (EHT). However, as the EHT is increased, the 'noise' in the PM also increases. Noise consists of non-specific pulses which do not result from light photons falling on the photocathode and arises from thermal emission of electrons from the photocathode and from stray electrons on insulators etc. Noise can be reduced by careful selection of PM tubes, by cooling the photocathode and by using two PM tubes 'in coincidence'. In coincidence counting two PM tubes view the same phosphor, and only pulses which occur simultaneously in both tubes are passed to the scaler, the two coincident pulses being recorded as a single count. Generally simultaneous pulses will arise only from scintillations in the phosphor, since noise pulses occur randomly in both PM tubes and will be coincident only rarely. PM tubes should be exposed only to low light levels and *must never be exposed to room lighting with EHT on*, since this results in immediate destruction of the photocathode. Tubes should be allowed 24 hr for dark adaptation after exposure to light.

Amplification. A pre-amplifier is normally mounted immediately adjacent to the PM tube, to keep the input capacity low and so obtain maximum pulse sizes. The pre-amplifier may have a gain of up to 1000 (40 dB), with phase reversal of the pulse, or may be an emitter follower, with a gain of about unity and no phase reversal. The main amplifier should have a gain of about 10^4 (80 dB) and good non-overloading and linear characteristics over a wide frequency range. Noise arises in the amplifiers as well as in the PM, but only that part arising in the pre-amplifiers and photo-multipliers is eliminated by coincidence counting techniques.

Pulse analysis and recording. A pulse height analyser is essential to make energy measurement possible and to minimise unwanted counts due to noise or radionuclides other than that of interest. The scaler must have a short input resultion time of about 1 μsec.

Counter characteristics. When a scintillation counter is set up, it is necessary to determine the correct working conditions. This is more complex than for Geiger-Müller counters, since the optimum settings vary with different radionuclides, with source strength and with the type of phosphor. Further, the background counting rate varies with the settings chosen. The general situation for a NaI/Tl counter is shown in Fig. 296, which shows the presence of the well-marked plateau which occurs with sources of moderate or high activity and makes it possible to select suitable operating conditions by inspection. However, the plateau slope increases, and its length decreases, with decreasing source strength and statistical criteria are essential to select optimum conditions with weak sources.

There are three criteria of merit which can be used to select the optimum instrumental settings: R_s^2/R_b, $\sqrt{R_s} - \sqrt{R_b}$ and R_{s-b}/R_b, where R_s is the counting rate due to the source, R_b is the background counting rate and R_{s-b} is thus the nett counting rate to be measured due to the source activity. Plots of the values of these three functions for the data shown in Fig. 296 are presented in Fig. 297, and the operating conditions should be adjusted to give maximum values of the desired function.

The criterion R_s^2/R_b is best suited to weak sources. Although reasonably independent of source strength it depends upon the radiation spectrum of the radionuclide. It gives the same results as the commonly used R_{s-b}^2/R_b, or

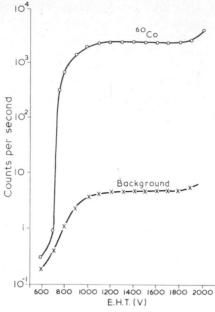

Fig. 296. The relationship between the counting rate and the applied potential in a scintillation counter. Source, ^{60}Co; phosphor, NaI/Tl, $1\frac{3}{4}'' \times 2''$; discriminator bias, 10 mV; channel width, ∞

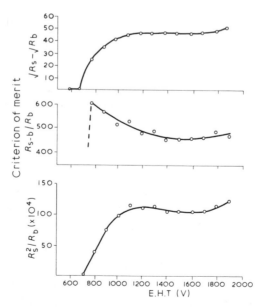

Fig. 297. Criterion of merit plots for the data from Fig. 295

(Efficiency)$^2/R_b$, and is simpler to calculate. The conditions indicated by the function $\sqrt{R_s} - \sqrt{R_b}$ are dependent on source strength, but both functions give similar results when $R_s \geqslant 100\,R_b$. Use of the function R_{s-b}/R_b leads to lower EHT settings and so to longer counting times for a given statistical error, and this is most pronounced for weak sources.

These criteria can be used to determine the best operating conditions for any proportional counter when energy differentiation is not required and must be assessed afresh for each instrument or after any modification to an instrument. They are of no value in setting up Geiger-Müller counters, with which counting rates vary only slightly throughout the possible working region. The relatively high background counting rate observed with simple scintillation counters may make these less suitable than Geiger-Müller counters for weak β-sources of an activity approaching natural background, and the type of counter should be chosen for which the value of function $\sqrt{R_s} - \sqrt{R_b}$ is maximal.

Arrangements for scintillation counting. It is implicit in the above discussions that the scintillation counting technique is very flexible and Fig. 298 illustrates some of the wide variety of arrangements which can be used for different purposes.

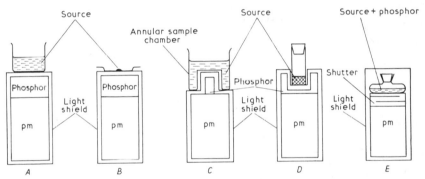

Fig. 298. Some arrangements for scintillation counting, (A) end-on counting of solutions; (B) end-on counting of solid sources; (C) annular counting (for liquids only); (D) well counting (solid or liquid sources); (E) liquid counting with phosphors in solution

Method. (*a*) Place a ^{60}Co source (Note 1) in the castle of a NaI/Tl detector, with a suitable aluminium absorber to prevent all β-particles from reaching the crystal.

(*b*) Set the discriminator bias voltage to an appropriate low setting (Note 2) and increase the EHT until the scaler starts to register counts. Take 1 minute counts at 50 V intervals up to the maximum recommended voltage of the photomultiplier and plot the observed counts on semi-logarithmic graph paper.

(*c*) Repeat this procedure at a suitably higher setting of the discriminator bias.

(*d*) Remove the source from the castle and count the background similarly to the ^{60}Co source.

(*e*) Determine the optimum working voltages using the criteria of merit and by inspection of the curves.

(*f*) Repeat the measurements using a weak source of ^{60}Co (Note 3), a source of ^{51}Cr, and a pure β-emitter (Note 4).

(g) Count the four sources using a thin end-window Geiger-Müller counter, with an absorber with ^{60}Co to absorb the β-radiation, and no absorbers with ^{51}Cr and the β-emitter (Note 5). Calculate the net source counting rate in each case and calculate the relative efficiencies of the two counters. Comment on your results.

(h) If time permits, repeat the procedures of sections (b) to (f) using a counter with a plastic phosphor (Note 6).

Note 1. The source should give a nett counting rate of about 40 000 cpm due to γ-emission, when placed as close to the crystal as possible.

Note 2. If the scaler is fitted with a pulse height analyser, this should be switched out of action, or, if this is not possible, the gate width should be set to the maximum available.

Note 3. This source should have an activity about 1 % of that of the first.

Note 4. The ^{51}Cr source should be similar to the stronger ^{60}Co source in activity. The β-emitter should give a counting rate of 10 000 to 20 000 cpm in a thin end-window Geiger-Müller counter. Suitable pure β-emitters are ^{36}Cl, ^{204}Tl and ^{32}P. The absorber should not be used with ^{51}Cr, since ^{51}Cr is a pure γ-emitter, or with the β-emitter, since it is desired to detect the β-particles from the latter.

Note 5. The counter should have a similar diameter to the phosphor and the sources should be placed as near to the counter as possible to obtain similar optimum geometry in both cases.

Note 6. The plastic phosphor should have the same dimensions as the NaI/Tl crystal, and be used with the same PM if direct comparisons are to be made. The following precautions should be used when changing the phosphor.

Switch off the EHT, disconnect the crystal/PM assembly from the apparatus and remove the assembly to a dark room under red light. Remove the NaI/Tl crystal from the PM tube, place a few drops of silicone oil on the face of the plastic phosphor and slide it onto the PM tube, rotating the phosphor to remove any air bubbles. Re-assemble with an aluminium light cap of the same thickness as the capsule of the NaI/Tl crystal. Reconnect to the apparatus, allow 10 min for dark adaptation, set the EHT to 1500 V and count the background for 2 min. Repeat the background count twice at 5 min intervals. If the three counts are within statistical limits ($\bar{x} \pm 2\sigma$) or show only a small decline with time, the PM and phosphors are sufficiently dark-adapted to proceed with counting.

If PM tubes or phosphors are exposed to bright daylight or to fluorescent lighting, several hours may be needed for dark adaptation.

Experiment 13. *The determination of the energy of gamma radiation using a sodium iodide scintillation spectrometer*

The process of gamma-ray absorption. When γ-radiation is absorbed in a NaI/Tl crystal the total light output is proportional to the energy of the incident radiation.

Gamma-radiation loses energy in matter by three principal mechanisms. In *photoelectric absorption*, which is most important at energies below 0.5 MeV, the whole of the γ-energy is absorbed by an orbital electron. If all the secondary photo-electrons so produced are absorbed within the crystal, the light yield is proportional to the γ-energy, giving a sharp absorption peak. *Compton scattering* is most important at moderate energies (0.3 to 2 MeV) and involves a series of elastic collisions with orbital electrons, producing electrons of all energies up to a certain maximum value (the Compton continuum, Fig. 299). If the secondary electrons are absorbed within the crystal, the total light output is again proportional to the γ-energy, and when the γ-photon has lost sufficient energy by Compton scattering, the whole of the residual energy is lost by photo-electric absorption within the crystal. This leads to an addition to the

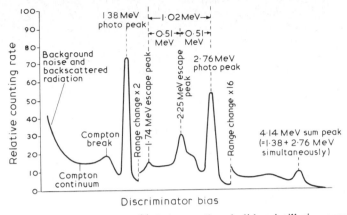

Fig. 299. The energy spectrum from ^{24}Na in a sodium iodide scintillation spectrometer

photo-electric absorption peak with a consequent decrease in the Compton scattered electrons, which occurs just before the photo-electric peak. This is the *Compton break* (Fig. 299). *Pair production* involves the creation of a positron-electron pair. The positrons have a very short life and annihilate almost immediately with the electron to yield two γ-rays, each of 0.51 MeV, in opposite directions. This annihilation radiation results from the conversion of the rest masses of the positron and the electron into energy (the mass of each = 0.00055 atomic mass units $\equiv$ 0.51 MeV). The total energy loss is thus 1.02 MeV plus the kinetic energy of the pair, so pair production is possible only at energies greater than *ca* 1.5 MeV. Whenever such high energy radiation is involved a 0.51 MeV peak is to be expected and the spectrum is further complicated due to partial escape of the annihilation radiation from the crystal, resulting in subsidiary peaks at points corresponding to energies 0.51 and 1.02 MeV less than that of the high energy γ-peak (Fig. 299). Where γ-radiation of more than one energy is emitted there will occasionally be a simultaneous emission of the various energies. This leads to a *sum peak* reflecting an energy equal to the sums of the energies of the coincident radiations.

General procedure. The form of apparatus is shown in Fig. 300 and all components should be of high quality to give stable performance. For

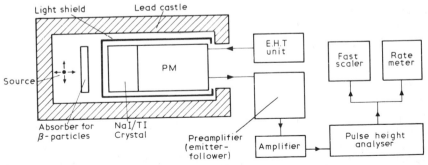

Fig. 300. Block diagram of the apparatus required for scintillation γ-spectrometry

satisfactory results there are a number of criteria. The diameter of the crystal should be approximately equal to its height, and the crystal should be of high optical quality and in good optical contact with the photomultiplier. Crystals for γ-spectrometry are specially selected for good resolution. The amount of thermionic noise in the photomultiplier may be reduced by operating it at a low EHT and with efficient collection of the electrons from the photocathode ensured by having the voltage across the first dynode double that used across subsequent dynodes. The output from the photomultiplier is fed via an emitter follower pre-amplifier into the main amplifier, and the latter must be truly linear so that the relationship between pulse-heights in the crystal is accurately maintained. A single channel-pulse-height analyser (see p. 528) is used to scan the pulse spectrum normally using the minimum channel width (gate width).

The apparatus is calibrated with radionuclides of known energy (Fig. 301A) and, once this has been done, the γ-energies of an unknown can be determined from the calibration curve (Fig. 301B) and the bias voltages at which the photo-peaks occur, e.g. the unknown giving the spectrum shown in Fig. 301C may

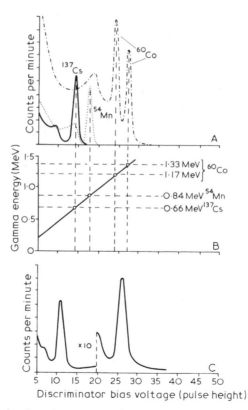

Fig. 301. The determination of the energy of γ-radiation with a sodium iodide scintillation spectrometer. (A) spectra of standard radio-nuclides; (B) calibration curve derived from A; (C) unknown spectrum giving peaks at 0.51 and 1.28 MeV (= ^{22}Na)

be identified as ^{22}Na from the peaks corresponding to 1.28 MeV and, from positron annihilation, 0.51 MeV.

Resolution. The overall resolution of the equipment defines its ability to discriminate between closely adjacent peaks. Two criteria are available.

The resolution for the ^{137}Cs peak is the width of the ^{137}Cs peak at half the maximum peak height expressed as a percentage of the bias voltage at which the peak maximum occurs. Resolutions of 10% are acceptable but modern instruments should be able to yield values of 8% or slightly better.

A more sensitive criterion is the peak/valley ratio for ^{60}Co, which is the ratio of the counting rate at the maximum of the 1.33 MeV photo-peak to the counting rate at the minimum of the trough between the 1.17 and 1.33 MeV peaks. Values of 3/1 are acceptable but ratios of 6/1 are obtainable and give much superior results.

If large numbers of spectra are to be determined routinely, semi-conductor detectors with multi-channel analysers give very high resolutions and rapid results.

Method. (*a*) Check the interconnections between the units of the instrument and set the controls as directed (Note 1), with the channel width (gate width) at the minimum available (Note 2).

(*b*) Place a 1000 mg/cm^2 aluminium absorber in the castle immediately adjacent to the crystal (Note 3).

(*c*) Place a ^{137}Cs source in the castle and scan the whole range of discriminator bias using the ratemeter, noting the approximate position of each peak (Note 4).

(*d*) Take counts at increasing values of the discriminator bias voltage using suitable steps (in the immediate vicinity of a peak use smaller intervals). Cease taking counts when the counting rate drops to a very low level after the highest energy peak noted with the ratemeter.

(*e*) Repeat this procedure with sources of ^{60}Co and ^{54}Mn (Note 5) and the unknown.

(*f*) Plot the γ-energies of the standard sources (Note 6) against the bias voltages at which the photo-peaks occur (Note 7) using linear graph paper, and draw the calibration curve.

(*g*) Determine the γ-energies of the unknown radionuclide from the calibration curve. Consult the tables of radio-isotope data and suggest a possible identity for the unknown. Comment on your results.

(*h*) Express the resolution of the spectrometer in terms of the ^{137}Cs photo-peak and the peak/valley ratio for ^{60}Co.

(*i*) If automatic recording facilities are available, record the spectrum for each source.

Note 1. It is impossible to give exact details since the types of equipment available vary considerably and are undergoing rapid development. The amplifier should have a rise time of 0.1 μsec and a fall time of 0.3 μsec. A high speed scaler with an input resolution time of about 1 μsec is essential and a ratemeter is highly desirable.

The settings of EHT and amplifier gain are directly related: any increase in amplification or EHT will spread the spectrum over a wider range of discriminator bias voltages, and vice versa. Settings which produce the ^{137}Cs peak at about 12 to 16 mV are convenient, and these must not be changed during the experiment.

Note 2. Within reason, the resolution improves as the channel width is reduced. If the channel width is changed during the experiment the counting rates will change accordingly, although the peaks will still occur in the same positions, with reduced resolution.

It is not possible to indicate appropriate discriminator and window width settings, owing to the variations between instruments.

Note 3. The absorber is used to prevent interference from any β-particles which may be emitted by the sources. The thickness recommended is effective up to a β-energy of 2 MeV.

Note 4. It is essential to scan *slowly* and to cover the whole range, otherwise peaks may be missed. This technique is used to speed up the subsequent accurate location of the photo-peaks and to indicate the number of peaks to be expected. If a ratemeter is not available, the same result can be achieved by judging the counting rate on the scaler.

Note 5. The choice of suitable standards depends on the energies to be determined and the expected energies of the unknown should fall within the range of these standards. The standards recommended are suitable for the range of 0.5 to 1.5 MeV and the calibration could be extended down to 0.1 MeV by using ^{170}Tm, provided the equipment has adequate stability and low noise characteristics. Possible unknowns are ^{22}Na, ^{58}Co, ^{46}Sc, ^{95}Nb, ^{59}Fe, ^{65}Zn, and ^{134}Cs.

Note 6. Consult the tables of radio-isotope data if necessary.

Note 7. The peaks should be quite symmetrical with the sides almost rectilinear. The position of a peak centre (though not its height) can be located precisely at the intersection of the extrapolated straight lines drawn through the points on the two sides of the peak.

Experiment 14. *Liquid scintillation counting*

Scintillation counting with phosphors in solution is particularly useful for counting β-emitters of low energy, such as ^{14}C (0.155 MeV), ^{3}H (0.018 MeV) and ^{35}S (0.167 MeV), since the sample is dissolved in the phosphor solution and the whole of the radiation is absorbed within the phosphor. Liquid phosphors are very cheap. They can be made any desired size and they have very short decay times (about 3×10^{-9} sec), thus permitting extremely fast counting. The sample cell may be coupled to the PM tube with a layer of silicone oil, the whole being enclosed in the lead castle which gives support, some shielding from background radiation and also acts as a light shield.

Phosphors. These usually contain three components: solvent, primary solute and secondary solute. The energy of the radiation is imparted primarily to the solvent molecules, from which the excitation energy is transferred to the primary solute and emitted as light. However, the peak emission wavelength of the fluorescence is normally in the range 350 to 380 nm, while the peak sensitivity of PM tubes occurs at a wavelength of about 420 nm, and the secondary solute acts as a wavelength shifter so that the final light emission has spectral characteristics which match approximately those of the PM photocathode.

The most efficient solvents are alkylbenzenes, of which toluene is the most widely used and xylene less so, but these are not miscible with water so special solvents must be used for aqueous materials.

Typical primary solutes are 2,5-diphenyloxazole (I, PPO) and *p*-terphenyl (II) at a concentration of about 4 g/l. Although the latter is one of the most efficient in toluene, it has a limited solubility at low temperatures and in the presence of aqueous samples. The most satisfactory secondary solute is 2-*p*-phenylenebis[4-methyl,5-phenyloxazole] (III, dimethyl-POPOP), about 0.05 to 1.0 g/l being used, and this has a peak emission wavelength of 430 nm. There is no fixed rule, however, and the phosphor for a particular source is formulated to give the most satisfactory results under the conditions to be used. The secondary solute may not give any advantage in some circumstances and formulations should be the simplest compatible with the required efficiency. Ready prepared liquid phosphor formulations are available commercially.

Homogeneous systems. The method of choice is to use a homogeneous solution of the active material in the phosphor. Some samples may be dissolved directly

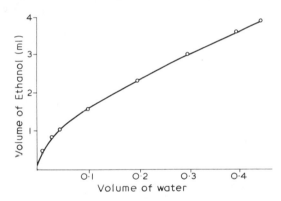

(I)
(II)
(III)

in a toluene-based phosphor, but with aqueous samples, solvents may be toluene–ethanol mixtures (see Fig. 302), dioxane containing 70 to 120 g/l of naphthalene, or dioxane–ethanol–xylene mixtures in the ratio 5:3:5.

Fig. 302. The mutual solubility of water and ethanol in toluene. The curve gives the minimum volume of ethanol required to retain a desired volume of water in solution in 10 ml of toluene–ethanol–water mixture

Acidic substances must be neutralised and Primene (a mixture of tertiary alkylamines) or Hyamine hydroxide [*p*-(di-isobutylcresoxyethoxyethyl)dimethylbenzylammonium hydroxide] can be used to solubilise carbon dioxide and many organic acids and proteinaceous materials.

Heterogeneous systems. Insoluble materials containing ^{14}C and ^{35}S can be counted with reasonable efficiency in the form of fine, uniform suspensions in the phosphor, provided the sample is not coloured. Stable suspensions in toluene-based phosphors can be prepared with aluminium stearate, Thixcin (a castor oil derivative) and Cab-O-Sil M-5 (a finely divided silica). Powders and minced biological tissues have been counted in this way. However, it is possible to add a solid, e.g. alumina from a TLC plate, to a liquid phosphor and count directly. The efficiency is acceptable even though the whole sample is on the base of the counting vial.

Quenching. The light output may be quenched by the presence of coloured substances, by a variety of solutes and by the inclusion of ethanol or aqueous samples in the phosphor. The most reliable method of correcting for quenching effects is to count the sample twice, the second time after the addition of a known amount of standard internally.

When sample activity is reasonably high and quenching is not too severe the effect of a change in efficiency, i.e. quenching, can be measured by the change in the ratio of the net counts in two channels of pulse height analysis adjusted to span certain portions of the spectrum, since the channel ratios may be correlated directly with efficiency (Fig. 303). Corrections may also be made

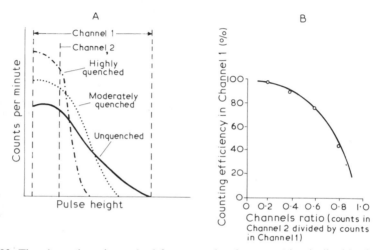

Fig. 303. The channels ratio method for correcting for quenching in liquid scintillation counting. (A) the effect of quenching on the pulse height spectrum; (B) channels ratio correction curve

by measuring changes in the count due to an external standard, and this is done automatically in modern instruments. Once again the sample is counted twice, the second time after a ^{137}Cs or other suitable source has been placed automatically in a fixed position relative to the vial. The instrument determines the degree of quenching by comparing the count with that obtained in phosphor in the absence of sample.

Double labelled samples. It is often desirable to count samples which contain two different radionuclides, e.g. ^{3}H and ^{14}C, and to determine the amounts of each. The pulse height spectra for these are shown in Fig. 304, from which it is clear that a major part of their activities can be counted in two distinct channels of pulse height analysis. Usually only the relative activities are required, but if absolute activities must be determined, the efficiency of counting in each channel may be calculated using suitable standards.

The labelling of different groups in a molecule with ^{3}H and ^{14}C is often undertaken to elucidate the course of synthetic reactions or metabolic pathways, and to study the distribution of drugs in the body and their metabolism and

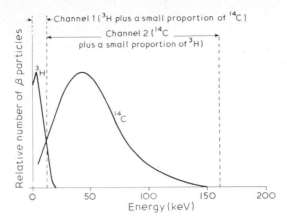

Fig. 304. The pulse height spectra of ^{3}H and ^{14}C and their separation into two counting channels

excretion. Many other combinations of isotopic label are possible, the criterion being the ability to differentiate the radionuclides on the basis of the energies of their radiations. Typical examples are ^{3}H and ^{35}S, ^{14}C and ^{32}P, ^{55}Fe and ^{59}Fe, and triple labelling is also possible using ^{3}H, ^{14}C and ^{32}P.

Background and noise. Pulses due to background and noise are small relative to those from ^{14}C and ^{35}S but are important in relation to those from ^{3}H, especially with samples of low activity. In the latter case it is essential to reduce PM and amplifier noise by cooling (see Fig. 305) or by using a coincidence counting system (see '*Photomultiplication*', p. 502).

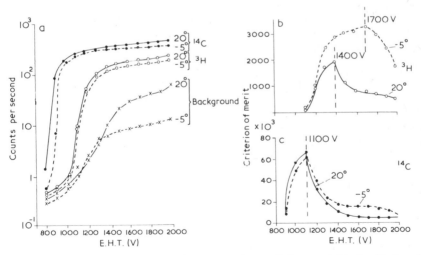

Fig. 305. The relationship of the counting rate to the applied potential in liquid phosphor counting. (A) the effect of cooling (affects the background but not the sources); (B) the related criterion of merit plots showing optimum EHT settings: ^{3}H, 1400 V at 20°C, 1700 V at −5°C; ^{14}C, 1100 V at both 20° and −5°C

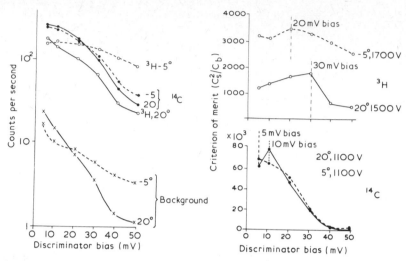

Fig. 306. The selection of discriminator bias settings in liquid phosphor counting. (A) the relationship of the counting rate to the bias voltage; (B) the related criterion of merit plots showing optimum bias settings: ^{3}H, 30 mV at 20°C, 20 mV at −5°C; ^{14}C, 10 mV at 20°C, 5 mV at −5°C

Experiment 14(a). *The characteristics of a liquid scintillation counter and the effect of temperature.*

Method. (a) With the counter and sources at laboratory temperature, set the discriminator bias to a low value and check that there is sufficient silicone oil in the counter.

(b) Place a background standard source (Note 2) in the counter without trapping air underneath it and allow 5 min for dark adaptation (Note 3).

(c) Increase the EHT slowly until counts are registered and count for 1 min. Determine the change of counting rate with EHT over the whole range up to the maximum voltage recommended for the PM tube (see Fig. 305). Repeat this procedure using a higher discriminator bias (about 50% of full range). Plot the results on semi-logarithmic graph paper.

(d) Repeat the procedure of sections (b) and (c) with the standard sources of ^{3}H and ^{14}C.

(e) Switch off the EHT, or reduce it to zero, and circulate refrigerant to reduce the temperature of the apparatus to about 5°, cooling the samples at the same time (Note 4). Allow 30 min to stabilise at the low temperature.

(f) Repeat the procedures of sections (b), (c) and (d) at the lower temperature.

(g) Decide the optimum EHT settings for counting ^{3}H and ^{14}C, using the criteria of merit (see Experiment 12).

(h) Determine the change of counting rate with discriminator bias voltage, using small intervals of about 5–10% of full range for the ^{3}H and ^{14}C sources at their appropriate EHT settings and for the background at *both* of these EHT settings. Plot the curves as before.

(i) Decide the optimum discriminator settings for each radionuclide, using the criteria of merit.

(j) Compare the counting rates obtained under the various conditions and comment on your results.

Experiment 14(b). *The assay of a sample of ^3H by liquid scintillation counting and the quenching effect*

Method. (a) Prepare, in the sample vials provided, sources for liquid scintillation counting using the toluene-based phosphor, the toluene solution of the standard and the aqueous solution of the sample (Note 5), as shown in the following table.

Volumes of materials to be used (ml)

Source	Phosphor	Ethanol	^3H standard	Toluene	Water	Sample
A	5	—	0.02	1.02	—	—
B	5	1	0.02	0.02	—	—
C	5	1	0.02	—	0.02	—
D	5	1	—	0.02	0.02	—
E	5	1	—	0.02	—	0.02
F	5	1	0.02	—	—	0.02

(b) Allow the sources to cool in the dark to counter temperature and count each source in turn under the optimum conditions indicated for ^3H in Experiment 14(a), accumulating at least 5000 counts for each. Calculate the counting rates in cpm.

(c) *Interpretation of the results.* Let $a, b \ldots f$ represent the counting rates for samples A, B ... F. Then,

(i) $\dfrac{\text{activity of sample}}{\text{activity of standard}} = \dfrac{e-d}{f-e}$,

(ii) the counts lost due to quenching by the alcohol $= a - b$,

(iii) the counts lost due to quenching by the alcohol plus the water $= a - c$,

(iv) if the sample causes quenching other than that due to its water content, $f - e < c - d$.

(d) Calculate the activity of the sample and comment on your results.

Note 1. For this experiment a simple discriminator should be used and not a pulse height analyser. The oil level should be such that the base of the sample cell is completely immersed.

Note 2. The background source consists of phosphor without added sample, and the background counting rate observed consists of the scintillation background in the phosphor plus 'noise' in the photomultiplier and amplifiers.

The trapping of air may be avoided by inserting the cell initially at an angle.

Note 3. The sources should be kept in the dark and exposed to the minimum amount of light. Fluorescent lighting is undesirable in the counting room since it induces considerable luminescence in the glass of the sample cell and in the phosphor.

Note 4. This experiment is designed for liquid scintillation counters in which refrigerant is circulated through the castle wall around the photomultiplier, e.g. Panax Type SC-LP. It is convenient to have a jacketted, light-proof sample cooling chamber through which the refrigerant passes before entering the castle. The cooling process is hastened considerably if the refrigerant is initially at about 10°C below the desired working temperature: the refrigeration unit is then reset to the working temperature after the temperature of the refrigerant has risen due to absorption of heat from the apparatus.

Note 5. The phosphor must be formulated suitably for use at the desired working temperature. A suitable source is urine containing a suitable tritiated compound, and with this a definite quenching effect is obtained due to both water and colour. This particular quenching problem is often encountered in medical diagnostic applications and in drug metabolism experiments.

Experiment 15. *The extraction of carrier-free thorium nitrate from uranyl nitrate and the determination of its partition coefficient between chloroform and water*

The decay scheme for ^{238}U is given in the introduction to Experiment 11. The compounds present, provided that the uranyl compound has not been chemically processed, are in *secular equilibrium* and the proportions of each are constant and depend upon their decay constants. If the ^{234}Th is separated from the equilibrium mixture, it decays to ^{234m}Pa and equilibrium between these is reached rapidly (within 10 min) because of the short half-life of ^{234}Pa. The ^{234}Th has only low energy β-particles (0.19 MeV) but can be detected conveniently by counting the 2.31 MeV β-particles of ^{234m}Pa.

The technique used here illustrates the principles of solvent extraction with complex formation, which is frequently used in radiochemical separations. The ^{238}U is retained in solution as ammonium uranyl carbonate, while the ^{234}Th is complexed with Cupferron reagent (the ammonium salt of nitrosophenyl-hydroxylamine) and extracted into chloroform.

Method. (a) Place 10 ml of 10% uranyl nitrate solution (Note 1) in a 250 ml beaker, add 3 ml N NH$_4$OH, then saturated ammonium carbonate solution dropwise until the precipitate dissolves.

(b) Adjust the pH to 8.0 to 8.5 with 0.880 NH$_4$OH or nitric acid as necessary.

(c) Transfer to a 50 ml separating funnel, add 2 ml of 6% Cupferron reagent and 10 ml of chloroform. Shake well and allow to separate.

(d) Run the chloroform layer into the beaker and collect the aqueous layer in a 25 ml measuring cylinder as 'Uranium residues'.

(e) Add to the chloroform solution 20 ml of water and 1 ml of Cupferron reagent, adjust to pH 8.0 to 8.5 and warm on a hot plate.

(f) Shake well and separate as before into a chloroform solution and 'Uranium washings'.

(g) Add to the chloroform extract 10 ml of $3N$ HNO$_3$ and 4 ml of saturated bromine water (Note 2). Shake well, separate as before into 'aqueous thorium solution' and 'chloroform solution of thorium nitrate'. Measure the volume of each and making up to 10 ml if necessary.

(h) Using a liquid Geiger-Müller counter (Note 3) determine the activities of 10 ml portions of the following:

   (i) uranium residues,
   (ii) uranium washings,
  (iii) aqueous thorium nitrate,
  (iv) chloroformic thorium nitrate,
   (v) 10% uranyl nitrate.

(i) Determine the total activities in each fraction and compare with the activity of the original solution. Comment on your results.

(j) Determine the partition coefficient of thorium nitrate between chloroform and water (Note 4).

Note 1. Uranium solutions are chemical poisons and precautions should be taken to avoid internal absorption. Gloves must be worn and tissues wrapped around the stoppers, taps etc. of the separating funnels to trap any leakage which may occur. Organic solvents are very difficult in this respect. Tissues should be discarded frequently and care taken that gloves do not rupture as a result of solvent action.

Note 2. This breaks down the Cupferron complex.

Note 3. The solutions can cause troublesome contamination. Decontaminate the counter thoroughly between each count, using nitric acid or RBS50 if necessary, and count backgrounds.

Note 4. The partition coefficient is defined as the ratio of the concentrations in the two solvents, concentrations in this context being measured in cpm/ml.

Experiment 16. *Radioactivation analysis. The determination of copper in aluminium*

When a substance is irradiated in a nuclear reactor each element present undergoes nuclear transformation and radioactivity is induced. The activity induced from each component depends upon a number of factors whose relationship is expressed by eq. (24).

$$mS = \frac{0.6\psi\sigma}{3.7 \times 10^{10}\,A}(1 - e^{-0.69t/t_{\frac{1}{2}}}) \tag{24}$$

where m is the mass of element present in the sample in grams, S is the specific activity in curies/g, ψ is the neutron flux in neutrons/cm^2/sec, σ is the *excitation cross section* in barns, A is the atomic weight of the target element, t is the irradiation time and $t_{\frac{1}{2}}$ is the half-life of the product radionuclide. After removal from the reactor the activity decays as usual. Thus it is theoretically possible to determine the mass of each element in a substance by measuring the activity produced from it under known conditions. However, it is usual to carry out the determination by comparison with a standard containing a known amount of the element of interest, *both standard and sample being irradiated simultaneously under identical conditions.*

Since the element of interest is normally present in a mixture of radionuclides a chemical separation is usually essential, both standard and sample being treated identically. However, it is sometimes possible to identify and measure components by γ-spectrometry (Experiment 13) or other physical techniques.

Method. (a) Weigh a piece of irradiated aluminium foil (Note 1) and a similar mass of unirradiated foil from the same source (Note 2).

(b) Place each foil in a 100 ml beaker and add to each 3 ml of 5N hydrochloric acid and a small crystal of potassium chlorate (Note 3).

(c) When dissolved, add to each exactly 1.0 ml of 0.25M copper sulphate and 10 ml of water (Note 4).

(d) Add exactly 1.0 ml of standard copper sulphate solution (Note 5) to the *inactive* foil to make the standard.

(e) Add to each 1 ml of 10% sodium metabisulphite solution (Note 6), dilute to 50 ml, heat to boiling on a hot plate (CARE) and add slowly, with stirring, 1 ml of 10% ammonium thiocyanate. Cover with a watch glass, boil gently for 5 min (CARE) to coagulate the precipitate and allow to cool.

(f) Dry and weigh two filter papers which fit the demountable filter apparatus and filter the solutions. Wash the precipitates, first with water containing a few drops of ammonium thiocyanate and sodium metabisulphite solutions and finally with a few ml of water.

(g) Dry the precipitates slowly under the heat lamp, add 2 drops of dilute Durofix in acetone and weigh (Note 1). Dry to constant weight.

(h) Mount the samples (Note 7) and count accurately using a thin end-window Geiger-Müller counter. Correct the counts as usual.

(*i*) *Calculations.* Determine:

(i) the fractional chemical yields of the precipitation process (Note 8) for both standard and unknown,

(ii) the counting rate equivalent to 1 μg of copper standard,

(iii) the counting rate equivalent to the *total* copper content of the sample of aluminium foil,

(iv) the number of micrograms of copper in the sample, and so

(v) the amount of copper impurity in the sample, expressed in parts per million and as a percentage.

(*j*) If time permits, establish the radiochemical purity of the sources by determining their half-thicknesses in aluminium. These should be identical and correspond to the half-thickness for ^{64}Cu.

Note 1. Weigh on a piece of tarred polythene sheet to avoid contamination of the balance. Monitor the polythene and dispose of it accordingly.

Note 2. This is to allow for any effects of the material on the chemical separation used.

Note 3. This encourages solution of the foil.

Note 4. The copper added here acts as a carrier to give a weighable mass of precipitate and so ease handling, and to prevent adsorption losses.

Note 5. The standard solution is prepared as follows.

Weigh exactly about 10 mg of activated copper (*W* mg), irradiated identically to the sample of foil, and dissolve in 1 ml of conc. HNO_3 in a 50 ml beaker covered with a watch glass. When solution is complete, dilute and transfer quantitatively to a 1 litre graduated flask and dilute to the mark. Mix thoroughly. This solution then contains *W* μg of copper in 1 ml.

Care. The radiation dose involved in the preparation of this solution is appreciable. Wear protective gloves, manipulate behind a lead wall and use handling tools until the final stages in dilution are reached.

Note 6. This ensures that the copper is in the cuprous state. The solution should smell strongly of sulphur dioxide.

Note 7. It there is a tendency for the filter papers to curl, weigh them down with brass retaining rings. Count on a suitable shelf, accumulating at least 10 000 counts.

Note 8. The mass of copper present as impurity in the foil, or added in the standard *active* solution, can be ignored.

Let

a = weight of precipitate from the standard
b = weight of precipitate from the sample
d = observed counting rate from the standard
e = observed counting rate from the sample
f = weight in μg of Cu in 1 ml of standard solution
g = weight in g of the sample of aluminium foil
h = weight in μg of Cu in the sample
r = total counting rate (100% yield) from the standard
s = total counting rate (100% yield) from the sample
t = counting rate from 1 μg of Cu
x = Cu content of the sample of aluminium foil

Atomic weight of Cu = 63.54

Molecular weight of CuCNS = 121.63

Now 1 ml of $0.25M$ $CuSO_4$ contains

$$\frac{63.54 \times 0.25}{1000} \text{ g Cu} \equiv \frac{121.63}{4000} \text{ g CuCNS}$$

$\therefore$ Fractional yield of CuCNS from standard $= \dfrac{4000a}{121.63}$

and

fractional yield of CuCNS from sample $= \dfrac{4000b}{121.63}$

$$\therefore \quad r = \frac{121.63d}{4000a}$$

and

$$s = \frac{121.63e}{4000b}$$

and

$$t = r/f = \frac{121.63d}{4000af}$$

$$\therefore \quad h = s/t = \left(\frac{121.63e}{4000b}\right)\left(\frac{4000af}{121.63d}\right) \mu g \text{ Cu}$$

$$= aef/bd \; \mu g \text{ Cu}$$

$$\therefore \quad x = h/g$$

$$= aef/bdg \text{ ppm}$$

$$= aef/10^4 \, bdg \text{ per cent}$$

Experiment 17. *The separation of mixed radionuclides*
When NaCl is irradiated in a reactor to produce ^{24}Na a number of side reactions occur in addition to that desired: e.g.

Target isotope	Type of reaction*	Product radionuclide and half-life
^{23}Na	n,γ	^{24}Na (15.0 hr)
	n,α	^{20}F (11.2 sec)
	n,p	^{23}Ne (40 sec)
^{35}Cl	n,γ	^{36}Cl (3.0×10^5 yr)
	n,α	^{32}P (14.2 d)
	n, p	^{35}S (87.2 d)
^{37}Cl	n,γ	^{38}Cl (37.3 min)
	n,α	^{34}P (12.4 sec)
	n,p	^{37}S (5.1 min)

* The absorption of a neutron by the nucleus, followed by the emission of γ-radiation, α-particles or protons, as indicated.

After about 14 days' decay, the residual activity is due primarily to ^{32}P and ^{35}S, with a small contribution from ^{36}Cl, and these may be detected by precipitation as magnesium ammonium phosphate, barium sulphate and silver chloride respectively.

Method. (a) Place 10 ml of a solution of irradiated sodium chloride (10 mg/ml) in a 50 ml centrifuge tube (Note 1), add 10 ml each of phosphate and sulphate carriers (Note 2), and two drops of hydrogen peroxide (Note 3). Transfer 3 ml of the mixture to a 15 ml centrifuge tube.

(b) *Separation of* ^{32}P

  (i) To the 3 ml portion add 0.5 ml of magnesia mixture and 1 ml of *5N* NH$_4$OH. Stir and scratch the tube gently and heat in a water bath for 15 min.

  (ii) Centrifuge and transfer the supernatant to a 50 ml centrifuge tube.

  (iii) Add to the precipitate 0.5 ml *N* HCl and 1 ml of sulphate carrier, and reprecipitate the ^{32}P (Note 4) as in section b(i).
    Centrifuge and add the supernatant to that reserved under section b(ii).

  (iv) Wash the precipitate with 1 ml of sulphate carrier and add the washings to the combined supernatants.

  (v) Transfer the precipitate to a counting tray by making a slurry with water and using a capillary pipette. Dry and seal with cellulose tape.

(c) *Separation of* ^{35}S

  (i) Add to the combined supernatants and washings, prepared under sections b(ii), (iii) and (iv), 5 ml of *5N* HCl, 5 ml of phosphate carrier, and 1 ml of *0.25M* BaCl$_2$. Stir well and heat in a water bath for 5 min.

  (ii) Centrifuge and discard the supernatant (Note 5).

  (iii) Wash the precipitate with 5 ml each of *0.25M* BaCl$_2$ and phosphate carrier and discard the supernatant.

  (iv) Prepare a source from the precipitate by slurrying and drying (Note 6).

(d) *Separation of* ^{36}Cl

  (i) To the larger portion reserved under section (a) add 1 ml of concentrated HNO$_3$ and 5 ml of *0.05M* AgNO$_3$. Heat in a water bath for 5 min.

  (ii) Centrifuge and discard the supernatant liquid (Note 5).

  (iii) Dissolve the precipitate in 1 ml of *5N* NH$_4$OH, add 5 ml each of sulphate and phosphate carriers and reprecipitate as in sections d(i) and (ii) (Note 7).

  (iv) Repeat the procedure given in section d(iii), wash the precipitate twice with 10 ml of water plus 1 ml of concentrated HNO$_3$ and 1 ml of *0.05M* AgNO$_3$. Wash finally with 10 ml of water.

  (v) Prepare a source from the precipitate as before.

(e) *Counting.* Confirm the identities of the separated radionuclides by determining the energies of the radiations emitted (Note 8) from their absorption in aluminium.

Note 1. Take normal precautions for handling β-γ emitters. Look up the energies of the radiations involved. Check the dose-rate from the original solution (^{24}Na has a high K-factor) and monitor carefully for ^{35}S, which is difficult to detect owing to the low energy of its β-particles.

Note 2. The carriers are *0.025M* Na$_2$SO$_4$ and *0.017M* KH$_2$PO$_4$ and are used here in two distinct ways. When they are present with the same element which is being isolated, they bulk out the minute amount of active material, make it easier to handle and they prevent losses by adsorption. If they are present when a different element is being isolated, they act as *hold-back carriers* which prevent contamination of precipitates by minimising co-precipitation and the adsorption of active material onto the precipitate.

Note 3. This ensures that the ^{32}P and ^{35}S are present as phosphate and sulphate respectively.

Note 4. This is a purification step, the ^{32}P being dissolved and reprecipitated in the presence of sulphate hold back carrier to prevent contamination of the precipitate with ^{35}S from the solution.

Note 5. Discard into the active waste container provided.

Note 6. Do not seal this source with cellulose tape. Check the activity of the precipitate during slurrying and, if the activity is unduly high, use only a part of it to prepare the source.

Note 7. This and subsequent steps are careful purification procedures to remove contaminants, since the ^{36}Cl is present in very small amounts and so is difficult to detect.

Note 8. Half-thicknesses may be used but, if time permits, the full absorption curves for the three sources should be plotted, since contamination due to other radionuclides will show as inflexions in the absorption curves or by the occurrence of counts at absorber thicknesses greater than the R_{max} expected.

Experiment 18. *The preparation of amino acids labelled with* ^{131}I

Amino acids labelled with ^{131}I can be prepared either by exchange or by direct iodination.

Labelling by exchange. The iodine in many organic iodides can be exchanged with iodide ions, under suitably defined conditions, the reaction being of the general form illustrated by the equation.

$$RI + {}^{131}I' \rightleftharpoons R^{131}I + I'$$

In this experiment, use is made of this process to prepare a sample of labelled thyroxine of reasonably high specific activity.

It should be noted that, since such exchange is possible, labelled compounds should be used with discretion and the possibility considered that exchange reactions may occur in the system of interest, resulting in difficulties of interpretation.

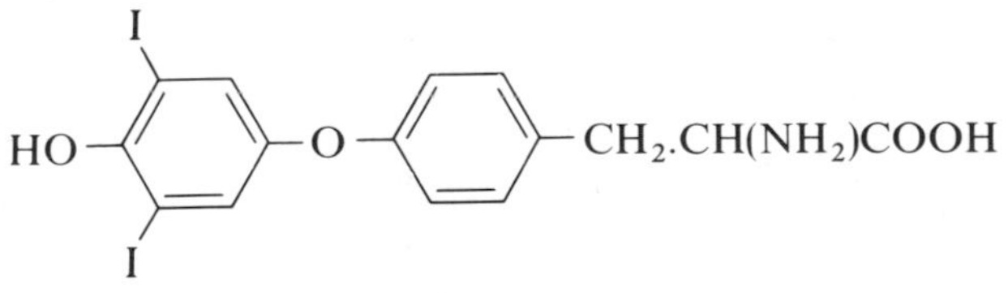

Thyroxine

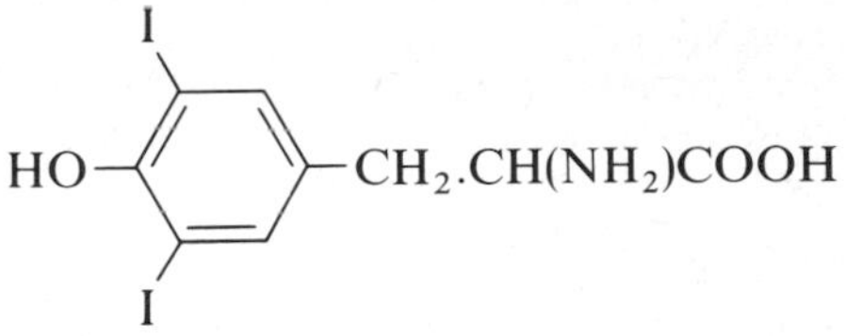

3,5-Di-iodotyrosine

Labelling by direct iodination. Tyrosine can be labelled by iodination in ammonia solution, using an ethereal solution of iodine, to produce 3,5-diiodotyrosine. Since rather large amounts of carrier iodide are required the product is of low specific activity.

Experiment 18(a). *The preparation of thyroxine labelled with* ^{131}I *by an exchange reaction*

Method. (a) Mix 45 ml of butanol with 5 ml of water and adjust to pH 5 with glacial acetic acid.

(b) Place the mixture in a 100 ml flask and add 10 mg of thyroxine, 1 ml of a solution containing 40 μc of ^{131}I and 0.1 ml of 0.1 % w/v potassium iodide.

(c) Reflux at boiling point for at least 8 hr or overnight (Note 1).

(d) Cool and filter off any undissolved material (Note 2).

(e) Place the solution in a separating funnel, add 50 ml of water (Note 3), shake well and run off the lower aqueous layer.

(f) Wash the butanol layer with 10 ml of water, transfer the butanol to a distillation flask and distil to dryness under reduced pressure.

(g) Mix the residue with the precipitate obtained in section (d) and purify by crystallisation from 0.1N sodium carbonate.

(h) Weigh the product, calculate the chemical yield, and determine its specific activity in cpm/mg by counting a suitable aliquot with a thin end-window Geiger-Müller counter (Note 4).

(i) Pack in a suitable specimen tube and deliver to the lecturer in charge.

Experiment 18(b). *The preparation of* 3,5-di-iodotyrosine *labelled with* ^{131}I *by direct iodination*

Method. (a) Add to a 50 ml separating funnel, in order, 5.5 ml of N potassium iodate, 8 ml of 10% w/v potassium iodide, 0.2 ml of a solution of K^{131}I containing 20 μCi, 5 ml of ether and 1 ml of 5N hydrochloric acid.

(b) Shake well and separate, retaining the aqueous layer (CARE, Note 5).

(c) In a stoppered test tube prepare a solution of 0.25 g of tyrosine in 5 ml of 0.880 ammonium hydroxide and add the ethereal solution of ^{131}I dropwise with thorough shaking.

(d) Pour into the separating funnel, shake again, separate and run the aqueous layer containing the labelled material into a second separating funnel.

(e) Wash the solution with 2 ml of ether to remove unreacted iodine.

(f) Transfer the solution to an evaporating basin and heat to dryness on a water bath.

(g) Dissolve the residue in 0.05N sodium hydroxide, shake with activated charcoal to decolourise and filter off the charcoal.

(h) Add N hydrochloric acid dropwise until the solution is neutral to universal indicator and then 5 drops of 2N acetic acid to complete the precipitation of the 3,5-di-iodotyrosine.

(i) Recrystallise by repeating procedures (g) and (h). Dry and weigh the product, determine its specific activity (Note 4) and pack as before.

(j) Measure the activity of ^{131}I in the aqueous layer from section (b) and 0.01 ml of the solution of K^{131}I. Draw up a radioactivity balance and calculate the percentage loss of activity.

Note 1. It is important to carry out all manipulations involving free iodine or heat in a fume cupboard with an adequate draught. Inhalation of iodine vapour can lead to an undesirable thyroid dose of ^{131}I.

Note 2. Retain the solid for subsequent use.

Note 3. This results in separation into aqueous and butanol layers.

Note 4. Make a brief check of activity, transfer a suitable weight to a dimpled planchet, slurry with a small amount of water in a capillary pipette and dry to give a thin source.

Note 5. Gloves must be worn and tissues used to trap any leakage from stoppers, taps etc. Ether will attack most gloves and care must be taken that the ethereal solution of ^{131}I does not penetrate the gloves and reach the skin. If this should occur, treat the site *immediately* with sodium thiosulphate solution, monitor and report to the lecturer in charge.

Experiment 19. *The distribution of iron in the tissues of the rat*

It is frequently desired to determine the distribution of a substance in the body tissues. This can be done readily using radionuclides, which may be used as cations of interest or organic compounds labelled suitably with ^{14}C, ^{3}H, ^{35}S etc. Following parenteral or oral administration the selected tissues are dissected out, extracted or treated suitably, and the activity is counted. If the radionuclide is a γ-emitter it may be counted readily with a NaI/Tl well counter, but if it emits β-particles of low energy then more elaborate techniques of sample preparation may be required to permit liquid scintillation counting. Alternatively, tissue extracts may be subjected to chromatographic, ion-exchange or electrophoretic separations to isolate and identify metabolites. The technique is of the greatest value in metabolic studies of all kinds.

The experiment described below is best carried out as a long-term group experiment to obtain data at different times after administration of the radionuclide and to provide sufficient replication to eliminate gross variations in response between animals.

Method. (*a*) Weigh a rat which has been injected previously with 1 μCi of ^{59}Fe in citrate buffer and determine the total blood volume from the tables provided (Note 1).

(*b*) Anaesthetise the animal with ether and remove 1 ml of blood by cardiac puncture. Place the blood in a labelled, stoppered 16 mm specimen tube and dilute to 5 ml with anticoagulant solution.

(*c*) Sacrifice the rat by over-anaesthesia and dissect out, as free from other tissues as possible, the bones of the hind limbs, the spine, pelvis, spleen and liver. Place the tissues in separate, small labelled beakers, add a few ml of concentrated nitric acid to each, cover with a watch glass and digest gently until completely dissolved (Note 2).

(*d*) Transfer each solution quantitatively to separate, labelled specimen tubes and dilute each to 5 ml. Stopper the tubes.

(*e*) In another specimen tube place 0.1 μCi of the original solution of ^{59}Fe, dilute to 5 ml and stopper (Note 3).

(*f*) Count each sample in a NaI/Tl well counter, using the instrumental settings appropriate to the instrument and ^{59}Fe.

(*g*) Express the activity of each tissue and of the total blood volume as a percentage of that injected originally. Plot the percentage activity in each organ against the elapsed time since administration and comment on your results.

Note 1. As previously indicated, this experiment is best conducted by a group, different individuals using rats injected with ^{59}Fe at 5, 10, 15, 25, 50 and 100 hr previously, if possible. The carrier iron content of the injection should be 1 μg. A portion of the original ^{59}Fe solution should be retained as a standard.

Note 2. Carry out the digestion carefully in a fume cupboard, avoiding undue foaming or boiling. If time permits other organs may be treated similarly but, in this case, the rat should be exsanguinated as completely as possible by injecting saline into the aorta whilst withdrawing venous blood. This avoids counting the activity of the blood remaining in the tissues.

Note 3. This acts as a standard which, counted at the same time as the samples, automatically allows for any radioactive decay (half-life 45 d.) since the injection was made. This activity represents 10% of that injected originally.

Experiment 20. *The study of exchange kinetics with radioactive tracers*

Sodium iodide will exchange its iodine with *n*-butyl iodide and this reaction can be followed using Na ^{131}I, the situation being represented by the following equation.

$$C_4H_9I + Na\ ^{131}I \rightleftharpoons C_4H_9\ ^{131}I + NaI$$

Let the concentrations of the reactants be represented by symbols as follows.

$$[C_4H_9I] + [C_4H_9\ ^{131}I] = a$$

$$[NaI] + [Na\ ^{131}I] = b$$

$$[C_4H_9\ ^{131}I] = x\ \text{cpm}$$

$$[Na\ ^{131}I] = y\ \text{cpm}$$

Also when the reaction time t is infinite (t_∞) the reaction is in equilibrium so $x = x_\infty$, and $y = y_\infty$.

The rate of appearance of labelled butyl iodide is a function of a and b, i.e.

$$\frac{dx}{dt} = \frac{V}{ab}(ay - bx) \tag{25}$$

where V is the reaction velocity constant. Now

$$x + y = x_\infty + y_\infty \tag{26}$$

Therefore

$$y = x_\infty + y_\infty - x \tag{27}$$

Also $ay_\infty = bx_\infty$ and thus $y_\infty = x_\infty(b/a)$. Substituting for y_∞ in eq. (27)

$$y = x_\infty + x_\infty(b/a) - x \tag{28}$$

and substituting for this value of y in equation (25)

$$\frac{dx}{dt} = \frac{V}{ab}[a\{x_\infty + x_\infty(b/a) - x\} - bx]$$

or

$$\frac{dx}{(x_\infty - x)} = \{V(a+b)/ab\}\ dt \tag{29}$$

Integrating eq. (29) gives

$$-\ln(x_\infty - x) = \{Vt(a+b)/ab\} - C \tag{30}$$

where C is the integration constant. Since at $t_0, x = 0; C = -\ln x_\infty$ and

$$-\ln(x_\infty - x) = \{Vt(a+b)/ab\} - \ln x_\infty$$

or

$$-\ln(1 - x/x_\infty) = Vt(a+b)/ab \tag{31}$$

Using common logarithms eq. (31) becomes

$$\log\left(1 - \frac{x}{x_\infty}\right) = \frac{Vt}{2.303} \cdot \left(\frac{a+b}{ab}\right) \tag{32}$$

This is the equation of a straight line with slope V.

From this the half-time of the reaction, i.e. the time required for the butyl iodide to receive one-half of its activity at equilibrium, can be determined.

$$\text{Half-time of the reaction} = \frac{0.693ab}{V(a+b)} \tag{33}$$

Also, if the experiment is carried out at two temperatures, the experimental activation energy (ΔE) may be calculated.

$$\Delta E = 2.303R\left(\frac{T_2 - T_1}{T_2 T_1}\right) \log\frac{V_2}{V_1} \tag{34}$$

where R is the gas constant (1.98 cal/K°/mole), V_1 is the reaction velocity constant at the lower temperature T_1 °K, and V_2 is the reaction velocity constant at T_2 °K.

Method. (a) Place some *0.08M* sodium iodide in 90% v/v acetone and *0.08M* n-butyl iodide in 90% v/v acetone in stoppered tubes in a thermostat with another empty stoppered tube (Note 1) and allow to equilibrate.

(b) Place five stoppered centrifuge tubes in a beaker of ice and add to each 1 ml of *0.08M* aqueous sodium iodide and 1 ml of benzene.

(c) To the prepared empty tube, add 1 ml of the acetone solution of sodium iodide, 0.2 μCi of ^{131}I in 0.1 ml (Note 2) and 1.1 ml of the butyl iodide–acetone, starting a stop-clock at the time of addition of the latter reagent (zero time). Shake gently for about 1 min.

(d) At appropriate intervals (Note 3) remove 0.1 ml of the reaction mixture and add to one of the centrifuge tubes containing the sodium iodide solution and benzene. Shake vigorously for 2 min and centrifuge to separate the layers.

(e) Remove the benzene layer quantitatively (Note 4) to a stoppered, labelled specimen tube using a capillary pipette and treat the aqueous layer similarly. Count each sample using a NaI/Tl well-type counter.

(f) At the end of the experiment reflux the residual reaction mixture gently for 30 min, cool in the thermostat and treat a 0.1 ml sample identically to the other samples. This gives the values of x_∞ and y_∞.

(g) *Analysis of the data*

   (i) Plot values of $(1 - x/x_\infty)$ against time on semi-logarithmic graph paper.
   (ii) Calculate the value of V from the data for each sample using equation (32) and determine its mean value.
   (iii) Confirm the validity of eq. (26) and determine and explain the relationship between x_∞ and y_∞.
   (iv) Calculate the half-time of the reaction.
   (v) If the experiment has been conducted using two or more different temperatures, calculate the activation energy (ΔE) using eq. (34).

Note 1. Suitable temperatures are 25°, 30° and 35° and group working to cover this range is convenient. It is advantageous if the empty tube can be fitted directly with a condenser.

Note 2. The ^{131}I should be in the form of aqueous sodium iodide. The activity given assumes a counting efficiency of 25% and if the efficiency differs from this appreciably, the activity may be adjusted accordingly, still added in a volume of 0.1 ml.

Note 3. Suitable intervals are 25, 20 and 15 min at temperatures of 25°, 30° and 35° respectively.

Note 4. The benzene layer contains the butyl iodide and the aqueous layer the sodium iodide. It is important to transfer the samples as accurately as possible and, if desired, the residual aqueous layer may be washed with a further 1 ml of benzene, adding the washings to the benzene sample. The aqueous layer is then transferred to its specimen tube and the centrifuge tube washed out with 1 ml of water, combining washings and sample. Volumes must be measured accurately and all samples for counting must have the same volume, or variations in counting efficiency will result.

If time permits, it is instructive to observe the effect of sample volume on counting efficiency, as follows. Place 0.1 ml of a solution containing 0.02 μCi of ^{131}I in a 16 mm specimen tube and count. Add 0.4 ml of water, mix and count again. Add a further 0.5 ml, then further 1 ml volumes, mixing and counting after each addition, until the tube contains at least 5 ml or is three quarters full. Express all counts as a percentage of the initial count with 0.1 ml of sample and plot the percentage counts against sample volume. Explain the results.

Experiment 21. *The production of* ^{99m}Tc *using the* ^{99m}Tc *Sterile Generator and its use for the preparation of a* ^{99m}Tc *colloid suitable for liver scanning*

Radionuclide generators are a means of storing short-lived radionuclides in a form from which they can be obtained readily when required, without having to re-order constantly or incurring the losses due to decay during delivery. The nuclides currently so used are ^{132}I ($t_{1/2}$ 2.3 hr), ^{99m}Tc ($t_{1/2}$ 6 hr), ^{87m}Sr ($t_{1/2}$ 2.8 hr) and ^{113m}In ($t_{1/2}$ 100 min). In all cases chromatographic columns are used.

The generators make use of the fact that when a radioactive parent nuclide decays to give a radioactive daughter, and the half-life of the parent is much longer than that of the daughter, an equilibrium is reached and the concentration of daughter present depends solely on the relative half-lives. Further, the daughter activity decays with the relatively long half-life of the parent. If the daughter is extracted from the mixture, its activity is regenerated rapidly so that approximately the original activity is soon restored, e.g. 12 hr with the ^{132}I generator and 24 hr with the ^{99m}Tc generator. However, useful amounts of activity can be obtained after much shorter times and the ^{132}I generator can be eluted twice daily, e.g. 9 a.m. and 1 p.m.

The most widely used generator is the ^{99m}Tc Sterile Generator which utilises the system

$$^{99}\text{Mo} \xrightarrow[t_{1/2}\ 67\ \text{hr}]{\beta^-,\,\gamma} {}^{99m}\text{Tc} \xrightarrow[t_{1/2}\ 6\ \text{hr}]{\gamma,\,0.14\ \text{MeV}} {}^{99}\text{Tc} \xrightarrow[t_{1/2}\ 2.1 \times 10^5\ \text{yr}]{\beta^-} {}^{99}\text{Ru}$$

The form of the generator is shown in Fig. 307. Since the ^{99m}Tc is always administered parenterally, the column is sterilised by autoclaving after assembly. It is closed with standard multi-dose vial closures at the top, through which sterile eluant can be added, and at the bottom, through which sterile eluate containing ^{99m}Tc is discharged into a sterile vaccine bottle. In most

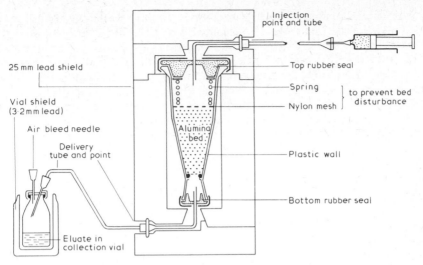

Fig. 307. The Radiochemical Centre ^{99m}Tc Sterile Generator

generators, a biocide is included to guard against the effects of chance microbial contamination, but none are compatible with the ^{99m}Tc generator. The column is transported and maintained inside a 25 mm lead shield at all times. The features of construction illustrated in Fig. 307 make for uniformly reliable, rapid elution without breakthrough of the alumina bed or the ^{99}Mo parent. Alumina particles in the eluate interfere with many of the medical diagnostic applications of ^{99m}Tc.

Preparations of ^{99m}Tc are used principally to delineate the liver, heart, spleen, placenta, brain, thyroid, lung and bone marrow. The features of ideal energy for scanning, low patient dose and chemical versatility have made ^{99m}Tc exceptionally useful and new applications appear regularly. The procedure described in Experiment 22b is typical of those used in hospitals to produce colloids labelled with ^{99m}Tc. This method was devised by R. J. French at the Royal Marsden Hospital, London.

It is stressed that the techniques used in this experiment require a thorough appreciation of the principles both of aseptic and radiochemical handling.

Radiological Hazards

The ^{99m}Tc generators may be obtained with a range of activities from 25 to 500 mCi. Although the generator and eluate vial are retained within lead shields, *the radiation dose to the operator may be appreciable if handling techniques are inadequate* and the experiment should not be attempted unless the operator has the appropriate handling skills. Long tongs must be used whenever possible to increase the distance from the source and a small lead wall is useful to prevent accidental exposure. If the operator can be rehearsed using a spent generator, this is valuable training.

The dose rate at the surface of the 25 mm lead shielding of a 25 mCi generator will vary with position, but will be of the order of 125 mrad/hr. The activity of the eluate from a new 25 mCi generator at equilibrium will be about 20 mCi, and the dose rate due to this (unshielded) will be about 2 mrad/hr at 1 m or 20 mrad/hr at 30 cm. If shielded with 3.2 mm lead the dose rate from the eluate at the surface of the lead pot will be about 0.2 mrad/hr. These dose rates are only a very rough guide and measurements should be made around the working area using an ión chamber dosemeter, e.g. Eberline RO2 or Nuclear Enterprises 0500, to determine the actual dose rates at various points in the working situation. Pocket dosemeters should also be worn.

The maximum permissible doses for occupational exposure are as follows:

Organ	In one week	Dose (rem) In any period of 13 weeks	In one year
Whole body	0.1	3	5
Hands, forearms, feet	1.5	20	75
Skin	0.6	8	30

Thus, assuming a 40 hr week, the whole body permissible dose is 2.5 mrem/hr and, clearly, doses at the working position must be less than this.

Experiment 21(a). *The elution characteristics of the* ^{99m}Tc *Sterile generator*

Method. (*a*) Set up the generator as directed in the instruction leaflet (Note 1). All setting up and experimental procedures must be carried out aseptically.

(*b*) *Elution profile.* Swab the closure of a sterile eluate collection vial (Note 2), place in the lead pot and insert an air bleed needle (Note 3) and the delivery needle. Swab the rubber closure of the injection tube insert the needle of a syringe containing 20 cm^3 of eluent (Injection of Sodium Chloride) and inject 4 cm^3 steadily (Note 4). When the flow into the collection vial ceases, transfer the delivery and air bleed needles to a second collection vial and again collect 4 cm^3. Proceed in this way until you have five 4 cm^3 eluate fractions in vials.

Finally, place the delivery needle up to its hub into a fresh sterile eluate collection vial, to protect the needle against contamination, and discard the syringe.

Measure the activity of each fraction (Note 5) and report on which fractions you would use to obtain an eluate of high specific activity.

(*c*) *The effect of elution time.* Allow the generator to stand for at least 20 hr. Swab a series of collection vials, insert an air bleed needle into one of them and transfer the delivery needle to the same vial.

Swab the rubber closure of the injection tube and injection 15 cm^3 of eluent steadily. Discard the syringe and note the time since the last elution. When the flow stops withdraw the delivery needle from the collection vial and place in a second collection vial together with the bleed needle. Measure the activity of the eluate.

Repeat the procedure at suitable times after the first elution (Note 6) and plot a graph showing the effect of elapsed time on the activity of the eluate. Comment on your results.

Experiment 21(b). *The preparation of a* ^{99m}Tc *colloid for liver scanning*

Method. (*a*) To a multidose injection vial containing 5 mg of gelatin and 10 mg sodium thiosulphate in 1 cm^3 (Note 7) add aseptically 5 cm^3 of ^{99m}Tc eluate containing 5 mCi.

(b) Add to the vial 1 cm³ of *1.4M* HCl (Note 8), shake and heat in a boiling water bath for 5 min, with occasional shaking.

(c) Cool. Inject 1 cm³ of phosphate buffer (Note 9).

(d) Measure the activity in the vial and label as follows (Note 10).

> WARNING—RADIOACTIVE
> The name of the product
> Volume in container
> Activity, reference date and time
> PREPARED ASEPTICALLY
> NO PRESERVATIVE
> Your name

State the quality control procedures appropriate to this product.

Note 1. Proceed as follows for the Radiochemical Centre generator.

(a) Place the generator upright and remove the bottom tear-off tape strip labelled *Delivery Point.*

(b) Remove the plug from the connection point thus exposed by rotating a quarter turn anti-clockwise with the key provided.

(c) Swab the connection point with a sterile swab, avoiding excess liquid, and connect the sterile delivery tube to the generator, rotating clockwise until finger tight.

(d) Fit a sterile delivery needle to the delivery tube, remove the protective sleeve from the needle and insert up to its hub in the swabbed collection vial.

(e) Remove the tear-off tape strip labelled *Injection Point* to expose the appropriate connection point. Connect the *Injection Tube* as in (c) and (d) above. The generator is now ready for elution.

(f) *WARNING. No tapes other than the two referred to above should be removed* or severe radiation hazard may result.

Note 2. Use the sterile swabs provided with the generator or their equivalent.

Note 3. The air bleed needle consists of a sterile hypodermic needle, the mount being plugged with cotton wool. The needles are inserted in the caps of injection vials so that large volumes can be injected into the vial or extracted from it without creating excessive pressure or vacuum in the vial, so preventing liquid flow.

Note 4. Excessive flow rates must be avoided as these may damage the column and cause breakthrough of ^{99}Mo or alumina.

This method of elution, by injecting eluent under pressure, is probably preferable to the vacuum method, in which a bag of sterile eluent is permanently attached to the top of the column and elution occurs by pulling a vacuum on the collection vial. A vacuum system could result in sucking in contaminants, and having a bag of infusion fluid in a warm room with the seal broken for up to 14 days is undesirable. However, the vacuum system avoids the necessity for the repeated swabbings and injections of the pressure method.

Note 5. Use a suitable calibrated ionisation chamber, e.g. the Type 1383A re-entrant chamber manufacturered by GEC-Elliott Process Instruments Ltd. London SE13.

Note 6. It is desirable to elute at 1, 3, 6, 10 and 16 hr after the first elution. These times are clearly impracticable, but it may be possible to obtain all the data by using two columns eluted at different times.

Note 7. Gelatin stabilises the colloid.

All solutions must be sterile, pyrogen free and suitable for parenteral use. Use aseptic procedures throughout.

Note 8. This gives *0.2M* HCl in the total volume. A sulphur colloid forms under the acid conditions, resulting in a milky suspension.

Note 9. In 1 cm³; Na_2HPO_4, 60 mg; NaH_2PO_4, 180 mg. This raises the pH to an acceptable level for parenteral use.

Experiment 22. *The radioimmunoassay of Insulin*

The immunoassay procedure is now widely used for the assay of minute quantities of a variety of substances, e.g. Digoxin, Insulin and L-thyroxine, especially when they are present in serum or other tissue fluids which contain many compounds likely to interfere with the assay procedure. The method utilises the high specificity of an antigen-antibody reaction to precipitate the compound of interest selectively from a heterogeneous solution, the high sensitivity and ease of detection obtained with radioactive labelling. These two features confer an exceptional sensitivity, e.g. in the insulin immunoassay it is possible to detect about 10^{-3} microunits (0.4×10^{-4} ng).

The first steps are to prepare a precipitating immunoglobulin (antibody) against the substance of interest, to act as a specific binding agent, and a suitably labelled form of the substance. Various labels may be used, but for a protein such as insulin, ^{125}I is simple and convenient. If the substance is not itself antigenic, it may be coupled to a protein carrier. Injection of this protein complex will elicit the production of an immunoglobulin which will react with the pure substance, acting as a hapten. To determine the concentration, or presence, of unlabelled substance in a sample, suitable volumes are mixed with standard solutions of labelled compound and the binding agent. The unlabelled substance in the sample and the labelled standard compete for the combining sites on the binding agent, giving a precipitate which is filtered off or centrifuged, and counted. The greater the concentration of substance in the sample, the less the labelled standard is bound and *vice versa*, so the concentration of substance in the sample is related inversely to the amount of radioactivity precipitated and measured.

To speed the procedure and increase the very small bulk of precipitate, thus making it easier to handle, the binding agent is often in the form of a complex with its specific immunoglobulin. The principles of the two forms of binding agent are illustrated in Fig. 308. The drug (substance of interest) is first injected into a guinea pig to obtain the specific immunoglobulin in the guinea pig serum. Then guinea pig serum is injected into a rabbit to obtain rabbit–anti-guinea pig serum. The immune guinea pig and rabbit sera are

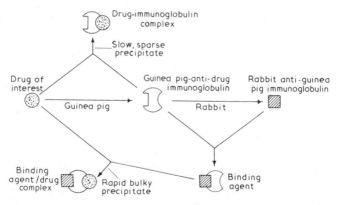

Fig. 308. The preparation of specific immune binding agent for radioimmune assays

then reacted to obtain a precipitate of guinea pig–rabbit complex, which will still bind the drug since the combining sites for the drug on the guinea pig immunoglobulin are not filled. The complex is then freeze dried and may be stored almost indefinitely as a specific binding agent for the drug.

The radioactivity of precipitates may be determined using planchet-type counters or well-type or liquid phosphor scintillation counters, according to the type of radioactive label.

The experiment described below is based on the use of the Radiochemical Centre Insulin Immunoassay Kit (Code IM39), using their modified rapid method.

Method. (a) Reconstitute a vial of insulin binding reagent with 8 cm³ of distilled water (Notes 1 and 2).

(b) Prepare the phosphate-albumin buffer and the phosphate-albumin isotonic buffer (Note 4).

(c) *Standard insulin.* Dissolve the contents of the vial of insulin in 2–3 cm³ of isotonic buffer and dilute to a sufficient volume to give a solution containing 160 microunits/cm³. Then prepare further dilutions in isotonic buffer to give solutions containing 80, 40, 20 and 10 microunits/cm³.

(d) *Iodinated insulin-*[125]*I.* Dilute 1.0 cm³ of the solution provided (0.1 µg in 5 cm³ of phosphate-albumin buffer) with 7.0 cm³ of phosphate albumin buffer (Note 5).

(e) Set up tubes as shown in the following table, using polythene or polypropylene (NOT glass) tubes of 8.5–9.0 mm i.d. and suitable micro-syringes and/or repetition micro-pipettes (Note 6).

	For Total Counts	Blank	Standard Curve for (microunits/cm³) 0	10	20	40	80	160	Unknown Sample	
Tube number	1	2	3	4	5	6	7	8	9	10
Phosphate-albumin buffer	—	100	—	—	—	—	—	—	—	—
Isotonic buffer	—	100	100	—	—	—	—	—	—	50
Appropriate standard dilution of insulin	—	—	—	100	100	100	100	100	—	—
Unknown sample	—	—	—	—	—	—	—	—	100	50
Binding reagent	—	—	100	100	100	100	100	100	100	100
Iodinated insulin-[125]I (2.5 ng/cm³)	100	100	100	100	100	100	100	100	100	100
Procedure	—	Mix, refrigerate for 3 hr, mix again, add the following								
Phosphate-albumin buffer	—	500	500	500	500	500	500	500	500	500

Scheme for Insulin Assay (all volumes in microlitres)

(f) After addition of the iodinated insulin-^{125}I, mix the tubes thoroughly (Note 7) and refrigerate at 2–4° for 3 hr. Mix the tubes again, add 500 µl of phosphate-albumin buffer to each, mix, centrifuge at 1500–2000 g for 30 min and decant the supernatant liquor carefully (Note 8), discarding it into the active waste container.

(g) Invert the tubes onto tissues and allow to drain for 15 min, remove any liquid from the mouths of the tubes with a tissue and discard the tissues to active waste.

(h) Count each tube in a well-type scintillation counter set up for [125]I (Note 9).

(j) *Calculations.* Determine the count per unit time for each tube and plot the counting rates (proportional to the amount of bound iodinated insulin) against the standard concentrations. The amount of immunoreactive insulin in the unknown is then determined directly from the graph.

(k) Decontaminate and monitor thoroughly.

Note 1. All glass apparatus should be cleaned thoroughly by overnight soaking in 4% 'Pyroneg', followed by thorough washing in distilled, then deionised, water.

Note 2. When reconstituted, the product contains 1/16 000 guinea pig serum precipitate and 0.03M EDTA in phosphate albumin buffer (Note 3).

Note 3. Phosphate-albumin buffer contains: NaH_2PO_4, $2H_2O$, 0.62%; thiomersal, 0.025%; bovine serum albumin, 1.7% of a 30% solution; aqueous NaOH to give pH 7.4.

Note 4. The isotonic buffer is a 0.9% solution of NaCl in phosphate-albumin buffer.

Note 5. The diluted solution contains 250 pg of iodinated insulin in 0.1 cm^3.

Note 6. (a) Glass tubes absorb the low energy radiation of ^{125}I excessively.

(b) Add the components in the order given in the table.

(c) It is desirable to set up the tubes in triplicate for accuracy.

Note 7. The 'total counts' tube should not be processed but counted directly. Use a 'Whirlimixer' or 'Rotamixer' for all mixing operations.

Note 8. Any multi-tube centrifuge will do, and refrigeration is not necessary provided temperature rises are modest. Do not use any braking or the pellet may be disturbed and counts lost. Note that the pellet is often invisible.

Note 9. It is desirable to avoid any possibility of contaminating the counter by placing each tube inside another plastic (not glass) tube.

Note 10. The most sensitive range is 20–80 microunits/cm^3. If the result indicates a concentration > 160 microunits/cm^3, the sample should be diluted and the assay repeated. Serum samples are preferred and plasma is likely to give aberrant results.

References

Analytical Control of Radiopharmaceuticals, Proceedings of a Panel, Vienna, 7–11 July 1969. International Atomic Energy Agency, Vienna. HMSO, London, 1970.

Belcher, E. H. and Vetter, H. *Radioisotopes in Medical Diagnosis*, Butterworths, London, 1971.

Code of Practice for the Protection of Persons Exposed to Ionising Radiations in Research and Teaching, HMSO, London, 1968.

Diagnostic Uses of Radioisotopes in Medicine, Hospital Medical Publications, London, 1969.

Hendee, W. R. *Radioactive Isotopes in Biological Research*, Wiley, New York, 1973.

New Developments in Radiopharmaceuticals and Labelled Compounds, Symposium, Copenhagen, 1973. International Atomic Energy Agency/World Health Organisation. HMSO, London, 1974.

Oliver, R. *Principles of the use of Radio-Isotope Tracers in Clinical Research Investigations*, Pergamon, Oxford, 1971.

Overman, R. T. and Clark, H. M. *Radioisotope Techniques*, McGraw-Hill, New York, 1960.

Radioactivity and Radioisotopes. *Bentley's Textbook of Pharmaceutics*, 8th edn., Ch. 28, Baillière Tindall, London. To be published in 1975.

Radiopharmaceuticals from Generator-produced Radionuclides, Proceedings of a Panel, Vienna, 11–15 May 1970. International Atomic Energy Agency, Vienna. HMSO, London, 1971.

Radioimmune Assay and Saturation Analysis, *Brit. Med. Bull.* (1974) **30**, Number 1.

Radioimmunoassay and Related Procedures in Medicine Vol. 1 (1974). Proceedings of a symposium, Vienna, International Atomic Energy Agency. HMSO, London.

Radiochemicals, 1974/5. The Radiochemical Centre, Amersham, Bucks, HP7 9LL.

Radiopharmaceuticals and Clinical Radiation Sources, 1974/5. The Radiochemical Centre, Amersham, Bucks, HP7 9LL.

Radiochemical Centre Reviews. The Radiochemical Centre, Amersham, Bucks, HP7 9LL.

Radiochemical Centre Medical Monographs, The Radiochemical Centre, Amersham, Bucks. HP7 9LL.

The Radiochemical Manual, 2nd edn., 1966. The Radiochemical Centre, Amersham, Bucks, HP7 9LL.

Index

Abbé refractometer, 17
Absorbance, 236, 237
Absorptiometry, 235
Absorption coefficient, 236
Absorption
  frequencies, 331
  fundamental, 331
  infrared, 331
  overtones, 331, 356
  ultraviolet, 223, 253
Absorption spectra, 221
  analytical applications of, 235
  liquids, determination, 344
  solids, determination, 345, 351
Acetaldehyde
  mass spectrum, 436
  NMR, 371
Acetic Acid
  titration, 166, 169, 179
  ultraviolet absorption, 253
Acetone
  NMR estimation, 407
  ultraviolet absorption, 253
Acetonitrile, infrared spectrum, 347
Acetylacetone
  mass spectrum, 437
  quantitative NMR, 402
Acetylcholine, conformation, 401
Acetylene, NMR, 377
Acetylsalicylic Acid
  determination by infrared, 352
  infrared spectrum, 339
Acidity, 150
Aconite, Belladonna, and Chloroform Liniment
  camphor in, 129
  chloroform in, 129
Aconite Liniment, camphor in, 129
Activation analysis, 517
Active sites, 111
Adipic acid, solubility in water, 7
Adrenaline
  fluorescence, 320
  Injection, determination, 131, 238
  ultraviolet absorption, 253

Adrenaline Injection, GLC determination, 131
Adsorbents, 79
  activity of, 81
Adsorption, 77
  chromatography, 75
  isotherms, 78
Ajowan, TLC, 106
Alcohol (95%), determination of other alcohols in, by GLC, 124
Alcohols, mass spectra, 435
Alkalinity, 150
Alkylbenzenes, fragmentation, 434
Alkylhalides, mass spectrum, 436
3-Alkylpyridines, fragmentation, 434
Allopurinol, 103, 124
Allyl alcohol, infrared spectrum, 337
Allyl cation, 436
Allylmorphine, fluorescence, 320
Alumina, 81, 97
Aluminium, determination by radioactivation analysis, 517
  fluorescence, 319
Aluminium chelate, fluorescence, 325
Aminacrine, fluorescence, 319
Aminco-Bowman spectrophotofluorimeter, 314
Amines, determination by GLC, 130
  mass spectra, 435
Amino acids
  estimation, 440
  paper chromatography, 93
  sequence analysis, 440
p-Aminobenzoic acid, titration, 121
  ultraviolet absorption, 258, 259, 273
2-Aminoethanol, mass spectrum, 435
3-Aminophthalimide, fluorescence, 325
p-Aminosalicylic Acid
  fluorescence, 320
  ultraviolet absorption, 258
Amitriptyline Hydrochloride, TLC, 100
Ammonium hydroxide, titration, 179
Ammonium sulphate, titration, 179
Ammonium vanadate, titration, 170
Amperometric titrations, 211

Amphetamine
  in urine, 130
  ultraviolet absorption, 257
Ampicillin Capsules, determination by HPLC, 144
Ampicillin Sodium, Dichloromethane in, by GLC, 124
Amplifiers, nucleonic, 503
Amytal, fluorescence, 320
Andreasen pipette, 53
Anethole, 106
Aneurine Hydrochloride, determination, 319
Aniline, ultraviolet absorption, 273
Aniline Hydrochloride, titration, 169
Anise, TLC, 106
Anisotropy, 372
C-18-Annulene, NMR, 378
Anthracene, ultraviolet absorption, 260
Antimony electrode, 152
Antimony Sodium Tartrate, titration, 170
A.P.C., NMR assay, 408
Apiezon L, 112
Apiole, 106
Aquametry, 164
Arachis Oil
  fatty acid composition (GLC), 130
  fatty acids in, 130
Areas, measurement of, 92, 122
Argon
  carrier gas, 117
  ionisation detector, 117
Aromatic coupling constants, 389, 391
Aromatic systems, NMR, 377
Ashton and Tootill, 252
Aspirin and Caffeine Tablets
  Aspirin determination by HPLC, 139
  Caffeine determination by HPLC, 139
Asymmetry potential, 160, 163
Atomic absorption, 301
  apparatus, 301
  exercises in, 306
  interferences in, 304
  procedure in, 305
Atropine Sulphate
  determination in urine, 274
  Eye Ointment, TLC, 100
  Injection, TLC, 100
  ultraviolet absorption, 272
Aureomycin, fluorescence, 320
Autoradiography, 91
Auxochrome, 254
Azatropylium ion, 436

Bacitracin Zinc, amino acids, 102
Background radiation, 474
Back pressure, 188
Baljet's reagent, 242
Band spectrum, 223
Barbiturates
  in body fluids, 275
  TLC, 107
Barrier-layer cell, 230
  output, 231
  sensitivity, 231
Base-line technique, 341
Base peak, 431
Bases in urine, 130
Bathochromic shift, 254
Beckmann thermometer, 12
Beclomethasone Dipropionate
  by absorptiometry, 239
  foreign steroids in, 108
Beer's Law, 236
Bemigride, 277
Bendrofluazide Tablets, determination, 240
Bentonite, particle size, 42
Bent tip electrode, 194
Benzene, NMR, 377
  ultraviolet absorption, 257, 260, 261, 273
Benzhexol Tablets, determination, 240
Benzocaine, by HPLC, 138
  infrared spectrum, 339
Benzoic Acid, by HPLC, 140
  infrared spectrum, 338, 351
  molecular association, 13
  ultraviolet absorption, 261, 273
Benzoin, in boron determination, 319
6,7-Benzomorphans, NMR, 378
Benzyl Alcohol, 337
Benzylic protons, non-equivalence, 400
Bephenium Hydroxynaphthoate, particle size, 42
Betamethasone Valerate, 85
Beta-particles
  absorption of, 474
  backscattering of, 480
  energy, determination of, 477
  energy, spectrum of, 475
  half-thickness/energy curve, 478
  range/energy relationships of, 477
Bisacodyl
  Suppositories, TLC, 100
  Tablets, TLC, 100
Blake titrimeter, 181
Body fluids, determination of drugs in, 252, 274

Body scanning, 460
Boiling point elevation, 9
Bolometer, 233
Boltzmann distribution, 364
Bond angles, NMR determination, 391
Bond lengths, 331
Boric acid, titration, 169, 179
Boron, by fluorescence, 319
Bremsstrahlung, 476
Brilliant Green, detection in a paint, 96
Bromocresol Green reagent, 92
Brucine in Nux Vomica Mixture, 108
Buffer solution
  McIlvaine, 162
  potassium hydrogen phthalate, 160
Buoyancy correction, 1
Butadiene, ultraviolet absorption, 254
Butanols, determination in admixture by GLC, 127
N-sec-Butylaniline, NMR spectra, 368

Cadalene, 79
Caffeine, ultraviolet absorption, 262
Calamine by atomic absorption, 307
Calciferol
  solution, determination, 240
  Tablets, determination, 240
  ultraviolet absorption, 254
Calcium
  determination in Magnesium Chloride, 300
  specific ion electrode titration, 171
Calcium Carbonate, 81
  lead in, 306
  particle size, 52
Calcium Hydroxide, 81
Calcium Phosphate, 81
Calomel electrode, 156, 167
Camphor, determination in liniments, 129
Camphorated Opium Tincture
  Benzoic acid by HPLC, 140
  morphine determination, 241, 263
Capacity factor, 132
Capillary characteristic, 187, 210
Capillary constant, 188
Capreomycin Sulphate, paper chromatography, 95
Caraway, TLC, 106
Carbachol Injection, determination, 241
Carbazole, 84
Carbenoxolone Sodium, TLC for related compounds, 100

Carbimazole, methimazole in, 109
Carbon disulphide, 81, 348
Carbon-13 NMR Spectroscopy, 412
Carbon number, 127
Carbon tetrachloride, 81, 348
Carbonyl compounds, mass spectra, 436
Carbowax, 112
Carrier gas, 111, 116
  argon, 117
  carbon dioxide, 118
  helium, 116
  hydrogen, 116
  nitrogen, 116, 118
Carvone, 106
Cascara Tablets, test for Frangula, 100
Castles for radionuclide counting, 447
Catalytic hydrogen wave, 191
Catechol, 351
Cathode ray polarograph, 203
Cells
  for infrared, 345, 358, 359
  for ultraviolet, 266
  path length, determination, 345
  polarographic, 206
  polishing, 359
Cellulose powder, 97
Cephaëline in Ipecacuanha Mixture, 108
Cephaloridine, residual solvent by GLC, 124
Cetrimide, 127
Charcoal, 81
Chemical shift, 371
Chlorcyclizine Hydrochloride
  determination in skin cream, 307, 308
  Tablets, TLC, 100
Chlordiazepoxide, by fluorescence, 319
Chlormerodrin (^{197}Hg) Injection, radio-chemical impurity, 95
p-Chloroacetanilide
  in Paracetamol, 107
  in Phenacetin, 107
Chlorocresol, ultraviolet determination in injections, 248
2-Chloroethanol, mass spectrum, 435
Chloroform, by GLC, 128
  by infrared, 355
  hydrogen-bonding, 348
  in preparations, 128
  Liniment, determination, 355
  Spirit, determination, 355
Chloroquin, fluorescence, 320
Chlorpromazine fluorescence, 320

Chlorpromazine metabolism, mass spectrum, 442
Chlorpromazine sulphoxide, fluorescence, 320
Chlorpropamide Tablets, impurities in, 100
Chlorproquanil Tablets, ultraviolet determination, 241
Chromatography
 adsorption, 75
 column, 75
 gas, 109
 high pressure, 131
 partition, 77, 82
 thin-layer, 96
Chromophore, 253
Cinchonidine, fluorescence, 320
Cinchonine, fluorescence, 320
Cinnamic acid, ultraviolet absorption, 273
Circular dichroism, 23
Clioquinol, GLC assay, 124
Clofibrate, volatile substances in, 124
Clomiphene Citrate, determination of isomer ratio, 357
CMR Spectroscopy, 412
Codeine, fluorescence of, 327, 328
Codeine Phosphate, determination in a linctus by HPLC, 141
Colchicine, TLC, 102
 ethyl acetate or chloroform in, by GLC, 124
Colistin Sulphate, 93
Colour wheel, 265
Column chromatography, 75
 apparatus, 80
 preparation, 80
Compensation circuits, 199
Complex formation reactions, 175
Compound Chloroform and Morphine Tincture, 355
Computer of average transients (C.A.T.), 369
Concentration polarisation, 186
Concentrations, interconversion, 2
Condenser current, 185
Conductance, 175
Conductimetric titrations, 172
Conformational isomerism, NMR, 398
Copper, determination by radioactivation analysis, 517
Copper Sulphate, light absorption, 263
Coriander, TLC, 106
Correlation charts, 373–6
Cortisone Acetate

crystal forms, 333
 determination, 239
 particle size, 42
 ultraviolet absorption, 239
Coulometric analysis, 217
Coulometric cell, 218
Coulometric titrations
 hydrochloric acid, 218
 use of integrating motor, 219
Coulter counter, 49, 58
 calibration, 58
 particle size measurement, 61
Counting room, 466
Coupling constants, 384, 390, 392
Cracking pattern, 417
Craig tube, 104
Criteria of merit for counting of radionuclides, 503, 513
Critical micelle concentration, 20, 38
Crotonaldehyde, ultraviolet absorption, 254, 256
Crystal Violet, detection in a paint, 96
Cumin, TLC, 106
Cuminaldehyde, 106
Cyanide, by fluorescence, 319
Cyanocobalamin, fluorescence, 320
Cyanogen chloride, use in fluorimetry, 319
Cyclohexane, 81
 isomerism, 398
Cyclohexanol, 344, 346
Cyclophosphamide Tablets, 358

Danthron, test for Hg in, 310
Dead-stop end point, 164
Debye and Falkenhagen theory, 180
Decay, radioactive, 499
Decomposition potential, 185
Decontamination of radioactive apparatus, 494
Decoupling, spin-spin, 392
Deformation modes, 332
Density, 1
 of liquids, 3
 of solids, 5
Deoxycortone
 implants, 108
 ultraviolet determination, 239
Deoxycortone Acetate, ultraviolet determination, 239
Derivative polarography, 201
Derivatives for GLC, 123
Desipramine Hydrochloride, TLC for iminodibenzyl, 100

Detection of compounds in chromatography, 80, 92, 99
Detectors
  argon ionisation, 117
  barrier-layer, 230
  bolometer, 233
  electron capture, 117
  flame, 116
  flame ionisation, 116
  for HPLC, 134
  Golay, 233
  in gas chromatography, 114
    photoconductive, 232
    photo-emissive, 232
    sensitivity, 118
    thermistor, 233
    thermocouple, 233
  radiation, 447
Detectors, discriminators in, 458
  Geiger-Müller, 450
  ionisation chambers, 449
  ionisation type, mechanisms, 447
  proportional counters, 449
  ratemeters, 459, 490
  scintillation, 454, 514
Dexamethasone
  determination, 239
  related foreign steroids in, 108
Dexamethasone Acetate, 239
  determination, 239
  related foreign steroids in, 108
Dextrose, effect of in flame photometry, 300
De Zeeuw, 104
Diamagnetic anisotropy, 372
Diamagnetic shielding, 372
1,2-Diaminopropane, mass spectrum, 435
Diamorphine Hydrochloride Injection, determination, 242
Diastereoisomers, NMR, 403
Diazepam metabolism, mass spectrum, 441
Dichlorodimethylsilane, 82
Dichlorophen, TLC for 4-chlorophenol, 101
Dielectric constant, 180
Dienoestrol Tablets, determination, 242
Diethylcarbamazine Citrate
  TLC for N-methylpiperazine, 101
Differential curve, 163
Differential polarography, 201
Differential scanning calorimetry, 64
  of Carnauba Wax, 67
  of polymers, 67
  instrumentation, 69
  instrument calibration, 68

  practical experiments, 72
  reference materials, 70
  sample form, 66
Differential thermal analysis, 64
  factors affecting, 65
  instrument calibration, 68
  quantitative, 67
  sample form, 66
Diffusion current, 185, 209, 210
Diffusion current constant, 204
Diffusion potential, 157
2,2-Difluorotetrachloroethane, NMR, 399
Digitalis, glycosides in, 84, 94
Digitoxin, determination, 242
  Tablets, determination, 84
Digoxin
  Injection, determination, 242
  Tablets, determination, 242
Digoxin Tablets, dissolution of, 328, 329
Dihydroxydiphenyl, 351
Di-iodothyronine, determination in Liothyronine Sodium, 95
Dill, TLC, 106
Dill-apiole, 106
Dimethylaminobenzaldehyde, ultraviolet absorption, 273
Dimethylformamide, 349
  restricted rotation, 398
4-Dimethyl-4'-nitrostilbene, fluorescence, 325
2,2-Dimethyl-2-silapentane-5-sulphonate sodium salt, 383
Dimethylsulphoxide, 349
Dinonyl phthalate, 112
Dinitrobenzoic acid reagent, 92
Dioxan, 349
Diphenhydramine, determination in body fluids, 278
Diphenylamine, 82
Diphenylhydantoin, determination in body fluids, 277
  mass spectrum, 439
Dipole moment, 180
Direct comparison method, 204
Discs, preparation, 351
Dispersion
  by grating, 229
  by prism, 227
Displacement analysis, 76
Dissolution rate of Digoxin Tablets, 328, 329
Dodecyl Gallate, TLC identity test, 101
Dosemeter, film badge, 464

Double resonance, 364, 394
Doxycycline Hydrochloride
  GLC for ethanol, 125
  TLC for related compounds, 101
Dragendorff's reagent, 92, 184
Dropping mercury electrode, 184
  care of, 193
DSS, 383

EEL atomic absorption spectrophotometer,
  301
EEL Flame Photometer, 298
Efficiency of GLC columns, 126
Electrocapillary maximum, 189
Electrocapillary zero, 189
Electrodes
  antimony, 152
  calomel, 156, 186
  cracked tube, 194
  dropping mercury, 184, 193, 205
  glass, 152, 160, 162
  horizontal mercury, 194
  hydrogen, 151
  indicator, 151
  inverted mercury, 195
  mercurous sulphate, 157
  mercury, 184
  mercury pool, 184
  multiple tip, 195
  platinum, 177, 196
  reference, 151, 190, 206
  silver-silver chloride, 157
  Smoler, 194
  specific ion, 153, 171
  streaming mercury, 196
Electromagnetic spectrum, 221
Electron capture detector, 47
Electrons
  α, 253
  n, 253
  π, 253
Electroscope dosemeter, 464
Emetine
  fluorescence, 328
  in Ipecacuanha Mixture, 108
Emetine Hydrochloride, TLC for other
  alkaloids, 101
Emf, 150, 158
Emission spectra, 221, 312, 324
  determination, 324
  fluorescence, 221, 313, 324
  line, 297
Empirical calibration curves, 204

Enantiomer ratio, NMR determination, 403
Energy, absorption of, 222
Energy levels
  infrared, 331 .
  ultraviolet, 223
Ephedrine, ultraviolet absorption, 275, 272
Epinephrine, fluorescence, 320
Equipment, radiochemical, operation of,
  471
Ergometrine Maleate
  fluorescence, 319
  Injection, determination, 242
  ultraviolet absorption, 247
Ergometrine Tablets, determination, 242
Ergot, 92
Ergotamine Aerosol Inhalation, particle
  size control, 42
Ergotamine Tartrate
  Injection, determination, 95, 243
  Tablets, determination, 243
Erythromycin Estolate, identity, 95
Esters, mass spectra, 435
Ethambutol Hydrochloride Tablets, TLC
  for identity and impurity, 101
Ethanol
  GLC determination in Stramonium Tinc-
    ture, 128
  effect in flame photometry, 300
  NMR spectra, 384, 387, 395
Ether, 81
Ethers, mass spectra, 436
Ethinyloestradiol Tablets, determination,
  327
Ethisterone Tablets, determination by
  HPLC, 143
Ethylbenzene, NMR, 370
Ethylene, NMR, 377
2-Ethylpyridine in γ-picoline, 342
Eu(FOD)$_3$, 381
Exchange effects, NMR, 387
Exchange kinetics, 524
Excitation effects, NMR, 387
Excitation spectrum, 313, 324
Extinction coefficient, 236, 270
Extinction scale, calibration, 266

Falkenhagen, 180
Faradaic current, 185
Fatty acids by GLC, 130
Feather analyser for β-particle energy, 479
Fenfluramine Hydrochloride Tablets, for-
  eign substances in (GLC), 125
Fennel, TLC, 106

Ferrous ammonium sulphate titration, 170
Filters
  choice of, 263
  gelatin, 226
  glass, 225
  interferometric, 226
First order coupling, NMR, 387
Flame
  for atomic absorption, 303
  for photometry, 297
Flame detector, 116
Flame ionisation detector, 116
  effect of water on, 117
Flame photometry, 297
  interference effects, 300
  potassium determination, 299
Flow rate, 126
Flucloxacillin Sodium Capsules, determination by HPLC, 144
Fludrocortisone, determination, 239
Fludrocortisone Acetate, foreign steroids
  in, 108
Fluocinolone Acetonide
  foreign steroids in, 108
  in formulated products, 85
Fluocortolone Hexanoate, foreign steroids
  in, 109
Fluocortolone Pivalate, foreign steroids in,
  109
Fluorescence, 311
  applications, 318
  effect of concentration, 316
  effect of oxygen, 318
  effect of pH, 318
  effect of temperature, 318
  effect of viscosity, 318
  methods of illumination, 317
Fluorescence efficiency, 312
Fluorescent compounds, table of, 320
Fluorimetry, 311
  experiments in, 321
Folic Acid
  determination, 243
  fluorescence, 320
  Tablets, determination, 243
Force constant, 342
Free indication decay, 369
Freezing point depression, 8, 12
Frequencies, infrared
  fundamental, 331
  group, 331
  overtone, 331
Fresnel, 222

Frontal analysis, 76
Fullers earth, 60
Fusidic Acid Mixture, particle size control,
  42

Gamma radiation, 460
  absorption of, 483
  Compton scattering of, 506
  half-thickness/energy curve, 477
  pair production by, 507
  photo-electric absorption of, 506
  spectrometry of, 506
Gas cell, infrared, 355
Gas chromatography, 109
  apparatus, 110
  carrier gas, 111, 116, 117, 118
  column efficiency, 126
  column performance, 121
  columns, 111
  detectors, 114
  experiments, 126
  injection systems, 114
  internal standard, 121
  optimum flow rate, 126
  preparation of support, 112
  stationary phases, 112, 124
  supports, 110
  temperature programming, 120
Gas Chromatography—Mass Spectrometry, 126
Geiger-Müller counters, 450, 473
  characteristics, determination of, 451, 491
  end-window, 452
  liquid, 454, 495
Gelatin, specific rotation, 28
Geminal coupling constants, 391
Gentamycin Sulphate, identity, 95
Glass electrodes, 152, 160, 162
  linearity, 162
Glutaric acid, paper chromatography, 94
Glycerin of Thymol, Compound
  Sodium Benzoate in by HPLC, 141
  Sodium Salicylate in by HPLC, 141
Glyceryl Trinitrate Tablets, determination,
  243
Glycine titration, 169
Golay columns, 133
Golay detector, 233
Gold (^{198}Au) Injection, particle size control,
  42
Gradient elution, 133
Grating dispersion, 177
Grating, 333

Griseofulvin Tablets, particle size control, 42

Haemodialysis solutions, determination of Na^+ and K^+, 300
Half-cell, 151
Half-life
  of protoactinium-234m, 499
  of radionuclides, 500
Half-wave potential, 184, 193
Halothane, volatile related compounds (GLC), 124, 125
Hamilton syringe, 114
Helmholtz, 185
  coils, 366
Henderson equation, 168
Heteronuclear decoupling, 395
Hexane, infrared spectrum, 350
Heyrovsky, 183
High frequency titration, 178
  applications, 181
High pressure liquid chromatography
  apparatus, 133
  column packing, 137
  columns, 135
  detectors, 134
  experiments, 137–146
  injection, 135
  packing materials, 135, 136
  pumps, 134
  theory, 131
High voltage supplies for radiochemical equipment, 457
Histamine Acid Phosphate
  histidine impurity, 103
Homologous series, chromatography of, 127
Homonuclear decoupling, 395
Horizontal mercury electrode, 194
Hyaluronidase, tyrosine in, 246
Hydrallazine metabolism, mass spectrum, 442
Hydrocarbons, mass spectra, 432
Hydrochloric acid titration, 165
Hydrochloric acid, vibration-rotation spectrum, 342
Hydrocortisone, determination, 239
Hydrocortisone acetate, determination, 239
Hydrocortisone Hydrogen Succinate, foreign steroids in, 109
Hydrocortisone Sodium Succinate, foreign steroids in, 109
Hydrocortisone preparations, particle size control, 42

Hydroflumethiazide Tablets, determination, 240
Hydrogen analysis by NMR, 410
Hydrogen bonding, infrared, 347
  NMR, 379, 401
Hydrogen chloride
  bond length, 344
  force constant, 342
  infrared absorption, 342
  moments of inertia, 342
Hydrogen overvoltage, 191
Hydroxocobalamin, coloured impurities in, 95
Hydroxonium ion, 150
2-Hydroxy-2-methylpentan-4-one, NMR spectrum, 367
Hydroxyprogesterone Hexanoate, foreign steroids in, 109
2-Hydroxypropionitrile, NMR spectrum, 393
Hydroxysteroids, determination in plasma, 329
Hyperchromic effect, 254
Hypochromic effect, 254
Hypsochromic shift, 254

Ilkovic equation, 187
  factors affecting, 187
Immunoassay, 530
Impurity detection, mass spectrometry, 437
Index of refraction, 14
  factors affecting, 15, 16
  measurement, 24
  of water, 22
Indicator electrode, 152
INDOR spectroscopy, 396
Infrared radiation, 224
Infrared spectrophotometry, 331
  comparison of spectra, 336
  experiments, 342–358
  instrumentation, 332
  qualitative uses, 333
  quantitative analysis, 340
  quantitative uses, 340
  structure correlation charts, 334, 335
Injection systems, 114
Insulin, immunoassay, 530
Interference fringes, 345
Integration, NMR, 366
Internal reflectance, 351
Internal standard, GLC, 122
  method, 294

Intraperitoneal Dialysis Solution, determination of Na$^+$, 300
Inulin, 81
Inverted mercury electrode, 194
Iodide, polarographic determination, 216
  effect on quinine fluorescence, 322
Iodine values, NMR determination, 411
Iodine vapour, 92
Ionic conductivity, 173
Ionic mobility, 172
Ionic product of water, 150
Ionisation chambers, 449
α-Ionone, ultraviolet absorption, 256
Ion re-arrangement, 437
Ipecacuanha and Opium Powder, morphine determination, 187
Ipecacuanha Mixture
  cephaëline in, 108
  emetine in, 108
Ipecacuanha Powder, 328
Ipecacuanha Tincture, 328
Irrelevant absorption, 250, 274
Isobestic point, 273
Isocarboxazid Tablets, 244
Isocratic elution, 133
Isoelectric point, 189
Isoniazid, ultraviolet absorption, 261
Isoprenaline, detection in aerosol, 96
Isoprenaline Tablets, determination, 239
Isopropenyl acetate, NMR estimation, 407
Isopropyl alcohol
  NMR estimation, 407
  NMR spectrum, 407
Isotope abundances, 431
Isotope dilution technique, 439

J quantum number, 343
Jensen and Parrack titrimeter, 180

Kaolin, particle size control, 42
Karl Fischer, 164
Karplus relationship, 391
Katharometer, 115
Keller-Kiliani reaction, 242
Keto-enol tautomerism, NMR, 401
Ketones
  isotopic exchange, 392
  mass spectra, 436
Kieselguhr, 97
Kováts, 119

Labelled compounds, radioactive preparation of, 521

Laboratories
  radiochemical, 463
  rules for, 465
Lambert's Law, 235
Lanthanide shift reagents, 380
Larmor equation, frequency, 363
Law of Mass Action, 150
Lead, in Calcium Carbonate, 306
Lead nitrate titration, 212
Lederer and Lederer, 75
Leucine ethyl ester, mass spectrum, 439
Levodopa, optical activity, 28
Levodopa Capsules
  identity in Capsules, 125
Tablets, 125
Light
  circularly-polarised, 21
  dispersion, 227
  filters, 225
  plane-polarised, 20
  sources, 224
Light sources
  atomic absorption, 302, 303
  fluorimetry, 324
  spectrophotometry, 224
Linalol, 106
Lincomycin Hydrochloride Capsules, GLC assay, 125
Lincomycin Hydrochloride Tablets, GLC assay, 125
Liothyronine Sodium, di-iodothyronine in, 95
Liquid junction potential, 158
Liquorice-Liquid Extract, TLC identity test, 101
Llewellyn-Littlejohn separator, 425
Loading effect, 180
Longitudinal relaxation, 365
2,6-Lutidine, in γ-picoline, 342
Lymecycline and Procaine Injection, 244
Lysergic acid diethylamide, fluorescence, 320

Macrisalb (^{131}I) Injection, particle size control, 42
Macroaggregated Iodinated (^{131}I) Human Serum Albumin, 460
Magnesia, 81
Magnesium carbonate, 81
Magnesium Chloride, Calcium in, 300
Magnesium oxide, 81
Magnesium silicate, 81
Magnetic equivalence, 381

Magnetogyric ratio, 362
Mandrax Tablets, 278
Martin, 86, 88, 114
Martin and Synge, 77
Mass fragmentography, 426
Mass peak, 429
Mass spectrometer
  AEI MS9, 420
  AEI MS12, 418
  double-focusing, 420
  in combination with GLC, 424
  quadrupole, 421
  single-focusing, 418
  time-of-flight, 418
Mass spectrometry
  amino acid sequence analysis, 440
  chemical ionisation, 426
  clinical application, 444
  drug metabolism, 440
  forensic application, 444
  fragmentation pathway, 432–7
  impurity detection, 437
  quantitative analysis, 438
  resolution, 422
  sample introduction, 422
  sample purity, 423
  sample size, 422
  structure elucidation, 429
  temperature effects, 423
Maxima, polarographic, 189
Mean size of particulate system, 47
Measurement of emf, 150
Measurement of pH, 150
Megesterol Acetate, foreign steroids in, 109
Menadione, fluorescence, 320
Menthol in Peppermint Oil, determination
  by GLC, 129
Mercaptans, mass spectra, 435
Mercurous sulphate electrode, 157
Mercury, determination by atomic absorp-
  tion, 308
Mercury pool anode, 184
Metastable ions, 431
Meter bridge, 165
Metformin Hydrochloride, dicyandiamide
  in, 103
Metformin Tablets, 241
Methadone, ultraviolet absorption, 257, 272
Methandienone Tablets, determination, 244
Methane, ultraviolet absorption, 253
Methanol, determination by infrared, 355
Methaqualone, determination in body
  fluids, 278

Methimazole
  determination in plasma, 145
  limit of Carbimazole, 109
Methotrexate, 93
Methotrexate Injection, 95
Methotrexate Tablets, 95
Methoxyflurane, volatile related com-
  pounds in, GLC, 125
Methylacetylene, NMR, 381
Methylamphetamine
  in urine, 130
  ultraviolet absorption, 257, 272
Methyldopa Tablets, 239
Methylene chloride, infrared spectrum,
  350
Methylergometrine Injection, 243
Methylergometrine Maleate, 243
N-Methylformamide, NMR, 379
Methyl orange, 79
Methylpentanes, fragmentation, 432
Methylprednisolone, 239
  determination, 239
  foreign steroids in, 109
Methyl Salicylate
  determination by GLC, 129
  in liniments and ointments, 129
Methyltestosterone Tablets
  determination by HPLC, 143
  ultraviolet determination, 244
Microcells for NMR, 369
Micrometer, 222
Micron, 222
Migration current, 186
Millimicron, 222
Mixtures, ultraviolet analysis, 248
Moisture analysis, NMR, 413
Molar conductance, 172
Molar refraction, 16
Molecular asymmetry, 399
Molecular ellipticity, 22
Molecular rotation, 22
Molecular weight, 8
Moment of inertia, 342
Monitors, radiation
  electroscope, pocket, 464
  film badges, 464
  ionisation chamber, 465
Monochromatic radiation, 225
Monochromator
  atomic absorption, 304
  ultraviolet, 227
Morphine, fluorescence, 328
  in Camphorated Opium Tincture, 241

Morphine Sulphate Injection, determination, 241
Morton, 161
Morton-Stubbs correction, 250
Mull technique, 345
Multiple tip electrode, 195
McLafferty re-arrangement, 437

Nandrolone Decanoate, see Part 1
Nanometer, 222
Naphthalene, ultraviolet absorption, 260
α-Naphthol, fluorescence, 325
β-Naphthol, fluorescence, 325
Natural products, spectrophotometry of, 250
Near infrared, 357
Nernst equation, 151
Neutralisation reactions, 163
Nialamide Tablets, 244
Nickel, polarographic determination, 216
Nicotinamide
  fluorescence, 319
  ultraviolet absorption, 261
Nicotinic Acid, ultraviolet absorption, 261
Nikethamide, ultraviolet absorption, 261
Ninhydrin reagent, 92
Nitrazepam Tablets, TLC for impurities, 101
Nitriles, isotopic exchange, 392
Nitrogen rule, 431
m-Nitromethylaniline, fluorescence, 325
Nitrosophenols, tautomerism, 262
NMR
  acidic solvents, 380
  double-coil instruments, 366
  drug-macromolecular interactions, 406
  dynamic properties of molecules, 398
  fourier transform, 369
  instrumentation, 365
  integrator, 366
  magnets, 365
  micelle formation, 404
  microcells, 369
  optical purity determination, 403
  quantitative analysis, 406
  resolution, 369
  sample spinning, 367
  scales of measurement, 382
  signal, 364
  single-coil instruments, 366
  solvent effects, 379, 403
  solvents, 367
  spectrum calibration, 383
  spin-spin decoupling, 383, 394
  structure elucidation, 397
  theory, 361
  time-averaging computer, 369
  variable temperature, 369
Noradrenaline, fluorescence, 320
Norepinephrine, fluorescence, 320
Novobiocin Calcium, solvents in, 125
Novobiocin Mixture, particle size control, 43
Novobiocin Ointment, particle size control, 43
Novobiocin Salts, residual solvent (GLC), 125
Novobiocin Sodium, solvents in, 125
Nuclear double resonance, 394
Nuclear electric quadrupole moment, 362
Nuclear induction, 366
Nuclear magnetic double resonance, 394
Nuclear Overhauser effect, 396
Nujol mull, 345
Nux Vomica Mixture, strychnine in, 108
Nystatin Ointment, particle size control, 43

Oct-3-ene, ultraviolet absorption, 253
Oestradiol Monobenzoate Injection, ultraviolet determination, 244
Ohm's Law, 172
Oils, iodine values by NMR, 411
Olefins
  coupling constants, 390
  mass spectra, 434
Optical activity, 20
  effects of concentration, 21
  effects of temperature, 21
  effects of solvent, 21
Optical density, 326
Optical purity, NMR determination, 403
Optical rotation, 21
  factors affecting, 22
Optical rotatory dispersion, 23
Optimum flow rate, 126
Orciprenaline, detection, 96
Orciprenaline Sulphate
  methanol in by GLC, 125, 355
Organic polarography, 192
Osmotic pressure, 10
Overtones, 331
Overvoltage, 191
Oxalic acid, paper chromatography, 94
  titration, 169
Oxidant, 164, 189
Oxychloroquin, fluorescence, 320

Oxygen maximum, 208
Oxygen wave, 192, 208

Pamquin, fluorescence, 320
Paper chromatography, 86
  apparatus, 88
  detection of compounds, 91
  methods, 91
  two-dimensional, 90
Paper for chromatography, 87
Paracetamol, p-chloroacetanilide in, 107
Paralysis time
  correction, 452
  correction tables, counts per minute, 452
  correction tables, counts per second, 454
  determination, 491, 492
Parent peak, 431
Parsley, TLC, 106
Particle
  shape, 45
  size, 44
  size distribution, 45
Particle size
  concepts, 44
  distribution, 45
  mean, 47
  Pharmacopoeial standards, 41
Particle size analysis methods, 48
  conductivity, 49, 58, 61
  microscope, 48, 54
  permeability to gas flow, 50, 57
  sampling procedures, 51, 57
  scope of methods, 51
  sedimentation, 49, 52
  sieve analysis, 48
Partition chromatography, 77, 82
Partition coefficient, 77
  determination of, with radionuclides, 516
Pascal triangle, 386
Path length, determination, 345
P-branch, 342
Peak area by triangulation, 122
Peak height, 122
Peak matching, 430
Penicillamine, mercury in, 310
Penicillins, by HPLC, 145
Pentaerythritol Tablets, 244
Pentagastrin, TLC identity, 101
Pentobarbitone, fluorescence, 320
Pentothal, fluorescence, 320
Peppermint Oil, menthol in (GLC), 129
Peptides, amino acid sequence, 440
Pethidine, ultraviolet absorption, 248, 257

Pethidine Hydrochloride, ultraviolet absorption, 257, 272
Pethidine Injection
  chlorocresol determination, 248
  pethidine determination, 248
Petroleum Ether, 81
pH, 150, 160
pH meter, 159, 167, 169
Phenacetin
  infrared determination, 352
  infrared spectrum, 339
  TLC for p-chloroacetanilide, 107
Phenanthrene, ultraviolet absorption, 260
Phenazone, by HPLC, 138
Phenformin Hydrochloride
  related biguanides in, 95
Phenformin Tablets, 241
Phenobarbitone
  determination by infrared, 352
  fluorescence, 320
  in body fluids, 275
  mass spectrum, 439
  polymorphism, by DTA, 66
  ultraviolet absorption, 261, 262
Phenobarbitone Tablets, determination, 352
Phenol, solubility in water, 7
  ultraviolet absorption, 271
Phenols, ultraviolet absorption, 257
Phenothiazine, determination, 84
α-Phenylethylamine, NMR solvents, 403
β-Phenylethylamine, in urine, 130
Phenylmercuric Nitrate, ultraviolet absorption, 257, 272
Phenytoin, determination in body fluids, 277
Phosphoric Acid, titration, 169
Phosphor, radiation detection, 502, 510
Phosphors, 454, 502, 510
Photoconductive detectors, 232
Photodecomposition, in fluorescence, 318
Photoemissive cells, 232
Photomultipliers, in scintillation counting, 502
Phthalylsulphathiazole, TLC for related substances, 101
Physostigmine
  determination in Eye-Drops, 96
  fluorescence, 320
Phytomenadione
  chromatographic determination, 85
Phytomenadione Injection, 86
Phytomenadione Tablets, 86

β-Picoline, in γ-picoline, 342
γ-Picoline, impurities, 342
Pilocarpine, detection, 96
Pilot ion, 294
Piperidine, titration, 169, 179
Piperidine,2,6-dimethyl-1-benzyl-,NMR, 400
Piperidine, tetradeutero-, 394
Plancket, 467
Planck's constant, 361
pOH, 150
Poiseuille's Law, 30
Polarimeter, 25
Polarimetry, 25
Polarisation, 186
Polarised light, 20
Polar molecules, 180
Polarogram, 183, 207
Polarograph, 183
  cathode ray, 203
  potentiometric manual, 198
  recording, 199
  simple manual, 196
  square, curve, 203
Polarographic analysis
  direct comparison method, 204
  empirical calibration curves, 204
  internal standards, 204
  pilot ion method, 204
  quasi-absolute method, 205
  standard addition method, 205
Polarographic cells, 206
Polarographic experiments, 205
Polarographic indicator, 214
Polarographic wave, 184, 210
  analysis, 210
Polarography, 183
  derivative, 201
  differential, 201
Polishing of sodium chloride blanks, 359
Polyamide powders, 97
Polyamino alcohols, mass spectra, 440
Polyethyleneglycol adipate, 112
Polymyxin B Sulphate, hydrolysis, 93
Polynomials, 252
Porapak polymers, 111
Potassium, determination by flame photometry, 299
Potassium bromide, titration, 170
Potassium chloride
  disc, 351
  for infrared, preparation, 360
Potassium dichromate, spectrophotometric

standards, 266
Potassium iodate, paper chromatography, 93
Potassium iodide, 93
  paper chromatography, 93
  titration, 170
Potassium permanganate, titration, 170
Potassium permanganate reagent, 92
Potassium thiocyanate, titration, 170
Potential energy diagram, 223
Potentiometer, 158
Potentiometric titrations, 162
Powders, 44
  mean size, 47
  moisture analysis, 411
  surface of, 47
Pre-amplifiers, in radiochemical equipment, 456
Pre-calciferol, ultraviolet absorption, 255
Precipitation reactions, 164, 170
Prednisolone
  particle size control, 42
  ultraviolet absorption, 256
  ultraviolet determination, 239
Prednisolone Acetate, ultraviolet determination, 239
Prednisolone Pivalate
  foreign steroids in, 109
  particle size control, 43
Prednisone
  ultraviolet absorption, 256
  ultraviolet determination, 239
  particle size control, 43
Prednisone Acetate
  ultraviolet absorption, 256
  ultraviolet determination, 239
Primidone Tablets, impurity in (GLC), 125
Prisms
  dispersion, 229
  for infrared, 333
Procainamide
  fluorescence, 320
  ultraviolet absorption, 258
Procaine, fluorescence, 320
Procaine Hydrochloride, ultraviolet determination, 248
Proflavine, fluorescence, 319
Proflavine Cream, determination, 323
Proflavine Hemisulphate, determination, 323
Progesterone, ultraviolet absorption, 256
Projected Area, 45

Promethazine Hydrochloride, NMR, 407
Promethazine Hydrochloride Injection, TLC for impurities, 101
Proportional counters, 449
Protection
 radiological, 463
 shielding in, 469
Pseudoephedrine Hydrochloride, determination in linctus, by HPLC, 141
Pulse analysis, 503
Purity, determination by DSC, 70
Purity, radiochemical, 497
Pyridine, 81
 titration, 169
 ultraviolet absorption, 261, 272

Quadrupole moment, 362, 365
Quanta, number of, 325
Quasi-absolute method, 294
Q-branch, 342
Quenching, 316, 322, 512
Quinalbarbitone, mass spectrum, 439
Quinidine Sulphate, other Cinchona alkaloids, 102
Quinine
 fluorescence, 319, 320, 324
 in Ferrous Phosphate Syrup with Quinine and Strychnine, 323
Quinine Dihydrochloride, titration, 179
Quinine salts, other Cinchona alkaloids, 102

Radiation, maximum permissible doses of, 464
 β-particles, 474
 personal protection, 463
 shielding, 474
Radiationless transitions, 365
Radioactivation analysis, 517
Radioactive sources
 mounts for, 470
 preparation of, drying in, 467
 preparation of, by filtration, 469
 preparation of, by pipetting, 466
 preparation of, general, 462, 467
 sealing of, 467
Radioactive tracers, in exchange kinetics, 524
Radiochemical
 experiments, 469
 laboratory, 463
 purity, 497
 rules, 465

techniques, 447
Radiochemical counter
 characteristics, 503
 contamination, 496
 decontamination, 496
Radiochemical techniques, 447
Radio-labelling, 521
Radiological hazards, 527
Radionuclide generators, 526, 528
Radionuclides,
 Arsenic-74, 466
 Arsenic-76, 466
 Barium-137, 466
 Barium-140, 466
 Bismuth-210, 466, 478
 Bromine-82, 466
 Caesium-134, 510
 Caesium-137, 466
 Calcium-45, 466
 Carbon-14, 466
 Chlorine-36, 466
 Chlorine-38, 519
 Chromium-51, 460, 466
 Cobalt-55, 466
 Cobalt-56, 466
 Cobalt-58, 460
 Cobalt-60, 466
 Copper-64, 466
 Fluorine-20, 519
 Gold-198, 466
 Indium-113m, 460, 466
 Iodine-125, 460, 466
 Iodine-131, 466, 497
 Iodine-132, 466
 Iron-55, 466
 Iron-59, 460, 466, 510
 Lanthanum-140, 466
 Lead-210, 466, 478
 Manganese-52, 510
 Manganese-54, 466
 Mercury-197, 460, 466
 Neon-23, 519
 Nickel-63, 466
 Niobium-95, 510
 Phosphorus-32, 460
 Phosphorus-34, 519
 Polonium-210, 478
 Potassium-42, 466
 Protoactinium-234m, 499
 Radium-226, 466
 Scandium-46, 510
 Selenium-75, 460, 466
 Sodium-22, 466, 510

Radionuclides—(continued)
  Sodium-24, 466
  Strontium-87m, 460, 466
  Strontium-89, 466
  Strontium-90, 466
  Sulphur-35, 466
  Sulphur-37, 519
  Tantalum-182, 466
  Technecium-99m, 460, 466
  Thallium-204, 506
  Thorium, natural, 466
  Tritium, 466
  Uranium, natural, 466
  Xenon-133, 460, 466
  Yttrium-90, 466
  Zinc-65, 466, 510
Radionuclides
  classification, 466
  distribution of, in animal tissues, 523
  mixed, separation of, 519
  sample preparation, 467
  toxicity, grading of, 466
Raman emission, 317
Rast method, 14
Rast molecular weight, determination, 14
Ratemeters, for radiochemistry, 459, 490
Rayleigh, criterion for resolution, 228
R-branch, 342
Reagents, for chromatography, 92
Recovery of compounds, 103
Redox titrations, 164
Reduced mass, 343
Reductant, 164, 189
Reduction potential, apparent, 193
References
  atomic absorption, 310
  chromatography, 146
  electrode, 156
  flame photometry, 310
  fluorimetry, 330
  general physical methods, 40
  infrared, 360
  mass spectrometry, 445
  NMR, 414
  particle size analysis, 63
  radio-chemical techniques, 532
  thermal analysis, 74
  ultraviolet, 296
Reflectance, total internal, 347
Refractive index, 14
  of water, 17
  specific increment, 16
Refractometry, 14

Relaxation processes, 364
Relaxation time, 178
Rescinnamine, fluorescence, 320
Reserpine
  determination, 245
  fluorescence, 320
Reserpine Tablets, determination, 245
Residual current, 185
Resolution
  GLC, 121
  HPLC, 133
  of scintillation γ-spectrometers, 509
  Rayleigh's criterion, 228
Restricted rotation, NMR, 398
Retention index (Kováts), 119
Retention time, 118
Retention volume, 118
Retro-Diels Alder fission, 434
Riboflavine
  fluorescence, 319
  ultraviolet absorption, 247
Rigden apparatus, 57
Ringing, 366
Rotational quantum number, 343
R value, 77
R_F value, 86
R_G value, 88
R_X value, 88
R_M value, 88
ΔR_M value, 88
Ryhage separator, 424

Saccharin, impurities in, 102
Salbutamol, detection, 96
Salbutamol Aerosol Inhalation, particle
  size control, 43
Salicylic acid, fluorescence, 320
Salt bridge, 157
Saturation, NMR, 364
Scalers, for radiochemistry, 456
Scintillation counters, 454, 514
  characteristics of, 501
  liquid phosphor, 510
  possible arrangements, 505
SCOT columns, 120
Screening constant, 372
Selenium, by fluorescence, 318
Selenomethionine (^{75}Se) Injection
  radiochemical purity, 95
Separation factor, 133
Sephadex, 97
Shielding constant, 371
Shift reagents, NMR, 380

Shoolery's rules, 381
Sieves, 48
  diameter, 44
Silica gel, 81, 97
Silver nitrate, titrations, 179
Silver-silver chloride, electrode, 157
Silylation, 123, 131
Skeletal vibrations, 336
Slit width, 267
Soap bubble meter, 110
Sodium acetate, titration, 179
Sodium p-aminosalicylate, ultraviolet absorption, 258, 260
Sodium arsenite, titration, 170
Sodium benzoate, determination by HPLC, 141
Sodium borate, titration, 169
Sodium carbonate, 81
  titration, 169
Sodium Chromoglycate Cartridges, particle size control, 43
Sodium iodide (^{125}I), radiochemical purity, 95
Sodium Iodide (^{131}I) Injection, radiochemical purity, 94, 95
Sodium Iodide (^{131}I) Solution, 94
Sodium Iodohippurate (^{131}I) Injection, radiochemical purity, 95
Sodium nitrite, titration, 169
Sodium oxalate, titration, 179
Sodium pertechnetate, radiochemical purity, 95
Sodium Phosphate (^{32}P) Injection, 94
Sodium salicylate, determination, by HPLC, 141
Sodium stibogluconate, polarographic determination, 211
Sodium sulphate, titration, 179
Solubility, 6
  factors affecting, 6
  of liquids in liquids, 7
  of solids in liquids, 7
  products, 164
Solvent effects, NMR, 372, 379
  spectrophotometry, 270, 271
Solvents
  chromatography, 81
  infrared absorption, 347
  removal of oxygen, 367
  residues in chemicals, 356
Solvents for NMR, 367, 380
Spark chambers, 456
Specific conductance, 175

Specific ellipticity, 22
Specific gravity, 3
Specific rotation, 22
  concentrated dependence, 27
Spectra
  absorption, 221
  emission, 297
  infrared, 331
  infrared, interpretation, 336–40
  near infrared, 355
  orders of, 230
  ultraviolet, 253
  use in analysis, 235
  use in determination of structure, 253
  visible, 238
Spectrofluorimeter, 314
Spectrofluorimetry, 312
  analytical factors, 315
  experiments, 321
  instrumentation, 313
Spectrophotometer
  infrared, 331, 332
  ultraviolet, 233
Spectrophotometry
  cell matching, 266
  determination of mixtures, 248
  effect of pH, 271
  effect of slit width, 267
  effect of solvent, 270
  solutions for, 266
  stray radiation, 268
  wavelength scales, 267
Spectropolarimeter, 26
Spectrum
  electromagnetic, 221
  formation of, 228
Spekker photofluorimeter, 321
Spin-cooling, 365
Spin-lattice relaxation, 365
Spinning side-bands, 367
Spin quantum number, 361
Spin-spin coupling, 383, 394
  notation, 388
Spin-spin decoupling, 392
Spin-spin relaxation, 365
Spin tickling, 396
Standard addition, 205
Standard potential, 151
Standards, fluorescent, 324
Starch, 81
Stationary phases,
  Apiezon L, 112
  carbowax, 112

Stationary phases—(*continued*)
  dinonyl phthalate, 112
  polyethylene glycol adipate, 112
  SE 30, 112
  Versamide, 112
Statistics
  chi-squared, table of, 489
  normal and Poisson distribution, 486
  of radionuclide counting, 485
Stilboestrol
  ultraviolet irradiation, 245
Stilboestrol Tablets, determination, 245
Stokes' diameters, 44, 54
Strain, 79
Strammonium Tincture, ethanol in, by
    GLC, 128
Stray radiation, 268
Streaming mercury electrode, 196
Structure of compounds
  by infrared, 333
  by ultraviolet, 253
Strychnine, determination
  chromatographic, 83
  in Nux Vomica Mixture, 108
Strychnine Hydrochloride, titration, 179
Succinic acid, 94
Sucrose, 81
Sulphadiazine, TLC for related substance,
    102
Sulphanilamide, ultraviolet absorption,
    254, 259, 271
Sulphonamides, separation of, 107
Sulphur, Precipitated, particle size, 43
Supporting electrolyte, 209
Surface tension, 35
  factors affecting, 35
  by drop weight, 38
  by Wilhelmy plots, 37
Surfactant chain-lengths, NMR determina-
    tion, 410
Symmetry factor, 121
Synge, 77

Talc, 81
Talc, Purified particle size, 43
Temperature
  compensation, 160
  programming, 120
Terramycin, fluorescence, 320
Tetracosactrin, acetic acid in, by GLC, 125
Tetramethylsilane, 382
Theophylline, ultraviolet absorption, 262
Theoretical plate, 77, 120

Thermal analysis, 63
  differential, 64
Thermistor, 233
Thermocouple, 233
Thiambutosine Tablets, determination, 246
Thin-layer chromatography, 96
  detection of compounds, 99
  materials, 97
  preparation of chromatogram, 99
  preparation of plate, 98
  quantitative, 105
  recovery of components, 103
  sensitivity, 102
  vapour-programmed, 104
Thiochrome, 319
2-Thioethanol, mass spectrum, 435
Thiomersal, determination by atomic ab-
    sorption, 308
Thiotepa Injection, 357
Thymol
  fluorescence, 320
  in Halothane, 246
  in Tetrachlorethylene, 246
  in Trichlorethylene, 246
  Thin-layer chromatography, 106
Time-averaging computer, 369
Tiselius, 76
Titration curves, calculations, 168
Titrations
  acetic acid, 166, 169, 179
  p-aminobenzoic acid, 169
  ammonium hydroxide, 179
  ammonium sulphate, 179
  ammonium vanadate, 170
  amperometric, 211
  aniline hydrochloride, 169
  Antimonyl Sodium Tartrate, 170
  barium chloride, 179
  boric acid, 169, 179
  calculations, 168
  chelating compounds, 215
  conductimetric, 170
  copper ions, 215
  ferrous ammonium sulphate, 170
  glycine, 169
  high frequency, 178
  hydrochloric acid, 165, 169, 179
  iodide, 216
  lead acetate, 179
  lead nitrite, 213, 216
  nickel, 216
  oxalic acid, 169
  phosphoric acid, 169

Titrations—(*continued*)
 piperidine, 169, 179
 potassium bromide, 170
 potassium chloride, 170, 179
 potassium iodide, 170
 potassium permanganate, 170
 potassium thiocyanate, 170
 potentiometric, 162
 pyridine, 169
 Quinine Dihydrochloride, 179
 redox, 164, 169
 silver nitrate, 179
 sodium acetate, 179
 sodium arsenite, 170
 sodium borate, 169
 sodium carbonate, 169
 sodium hydroxide, 179
 sodium nitrite, 169
 sodium oxalate, 179
 sodium sulphate, 179
 Strychnine Hydrochloride, 179
 triethanolamine, 169
 triethylamine, 169
Toluene, 81
Transitions
 electronic, 224, 253
 fluorimetric, 311
 infrared, 331
 ultraviolet absorption, 224
Transverse, relaxation, 364
Tranylcypromine Sulphate
 cis-isomer, by GLC, 125, 126
 TLC for cinnamylamine, 102
Triethanolamine titration, 169
Triethylamine titration, 169
Triglycerides, NMR, 411
Triphenyltetrazolium chloride, 239
Triprolidine Hydrochloride, determination
 by HPLC, 141
Tritium, scintillation counting, 515
Troplyium ion, 437
Tung oil, iodine values, 411
Twsett, 75
Tyrosine, in Hyaluronidase, 246

Ultraviolet absorption of drugs in common
 use
 by name, 278
 by wavelength, 283
Ultraviolet radiation, 225

Umbelliferous fruit, oil components, 105
Units, spectrophotometric, 221
Unsaturation level, calculation, 305
Urine volatile bases in, by GLC, 130

Vancomycin Hydrochloride, identity, 95
Van Deemter equation, 126
Versamide, 112
Vibrational energy, 331
Vibration levels
 infrared, 331
 ultraviolet, 221
Vibration modes, 331
Vicinal coupling constants, 391, 401
Viscometer
 capillary, 30, 33
 Couette, 32
 falling ball, 32, 34
Viscosity, 29
 factors affecting, 31
 units, 31
Visible radiation, 225
Vitali reaction, 274
Vitamin A
 correction formulae, 250, 251
 determination, 250, 251
 fluorescence, 320
 ultraviolet absorption, 254, 255
Vitamin A Ester Concentrate, retinol in, 96
Vitamin B_1, determination, 274
Iso-Vitamin D_2, ultraviolet absorption, 255
Volume
 molar, 1
 partial molar, 1, 5

Warfarin Sodium (Clathrate), isopropanol
 in, by GLC, 125, 356
Water, hardness determination, 171
Watson-Biemann separator, 424
Wavelength, 221, 266, 267
Wave-number, 222
Wheatstone bridge, 175
Wijs method, 411
Woodward, 256
Wool Alcohols, antioxidants in, by GLC,
 125
Wool Fat, antioxidants in, by GLC, 125

Xylene, fragmentation, 437